Information Processing by Neuronal Populations

Models and concepts of brain function have always been guided and limited by the available techniques and data. This book brings together a multitude of data from different backgrounds. It addresses questions such as: How do different brain areas interact in the process of channeling information? How do neuronal populations encode the information? How are networks formed and separated or associated with other networks? The authors present data at the single-cell level both *in vitro* and *in vivo* and at the neuronal population level *in vivo* comparing field potentials (EEGs) in different brain areas, and also present data from spike recordings from identified neuronal populations during the performance of different tasks. Written for academic researchers and graduate students, the book strives to cover the range of single-cell activity analysis to the observation of network activity, and finally to brain area activity and cognitive processes of the brain.

CHRISTIAN HÖLSCHER is an Assistant Professor at the University of Ulster in Northern Ireland. He has published widely in international journals on topics of memory formation, synaptic plasticity, neurodegeneration, and information processing in neuronal populations. He was the editor of *Neuronal Mechanism of Memory Formation* in 2001 which investigated processes of synaptic plasticity that might underlie memory formation.

MATTHIAS MUNK is a Research Scientist at the Max Planck Institute for Biological Cybernetics in Tübingen, Germany and an Assistant Professor (Privatdozent) at the University of Darmstadt, Germany. He has conducted extensive research in the area of information processing in the visual cortex of primates, using electrophysiological, pharmacological, and imaging techniques. He has published his research widely in a range of top scientific journals.

Information Processing by Neuronal Populations

Edited by

CHRISTIAN HÖLSCHER
University of Ulster

MATTHIAS MUNK
Max Planck Institute for Biological Cybernetics, Tübingen

CAMBRIDGE UNIVERSITY PRESS
Cambridge, New York, Melbourne, Madrid, Cape Town,
Singapore, São Paulo, Delhi, Mexico City

Cambridge University Press
The Edinburgh Building, Cambridge CB2 8RU, UK

Published in the United States of America by Cambridge University Press, New York

www.cambridge.org
Information on this title: www.cambridge.org/9781107411296

© Cambridge University Press 2009

This publication is in copyright. Subject to statutory exception
and to the provisions of relevant collective licensing agreements,
no reproduction of any part may take place without the written
permission of Cambridge University Press.

First published 2009
First paperback edition 2012

A catalogue record for this publication is available from the British Library

Library of Congress Cataloguing in Publication Data
Information processing by neuronal populations / edited by Christian Hölscher, Matthias Munk.
   p. ; cm.
   Includes bibliographical references and index.
   ISBN 978-0-521-87303-1 (hardback)
   1. Neural networks (Neurobiology)   2. Neural circuitry.   I. Hölscher, Christian.
   II. Munk, Matthias.   III. Title.
   [DNLM: 1. Neurons–physiology.   2. Brain–physiology.   WL 102.5 I43 2009]
   QP363.3.I43 2009
   612.8′2–dc22
                                                                       2008022721

ISBN 978-0-521-87303-1 Hardback
ISBN 978-1-107-41129-6 Paperback

Cambridge University Press has no responsibility for the persistence or
accuracy of URLs for external or third-party internet websites referred to in
this publication, and does not guarantee that any content on such websites is,
or will remain, accurate or appropriate.

This book is dedicated to the memory of Professor Werner Jürgen Schmidt who tragically and unexpectedly died on 16 April 2007. He will not be forgotten.

Contents

List of contributors x

Part I Introduction 1

1 How could populations of neurons encode information? 3
CHRISTIAN HÖLSCHER

Part II Organization of neuronal activity in neuronal populations 19

2 Cellular mechanisms underlying network synchrony in the medial temporal lobe 21
EDWARD O. MANN AND OLE PAULSEN

3 Cell assemblies and serial computation in neural circuits 49
KENNETH D. HARRIS

4 Neural population recording in behaving animals: constituents of a neural code for behavioral decisions 74
ROBERT E. HAMPSON AND SAM A. DEADWYLER

5 Measuring distributed properties of neural representations beyond the decoding of local variables: implications for cognition 95
ADAM JOHNSON, JADIN C. JACKSON, AND A. DAVID REDISH

6 Single-neuron and ensemble contributions to decoding simultaneously recorded spike trains 120
Mark Laubach, Nandakumar S. Narayanan, and Eyal Y. Kimchi

Part III Neuronal population information coding and plasticity in specific brain areas 149

7 Functional roles of theta and gamma oscillations in the association and dissociation of neuronal networks in primates and rodents 151
Christian Hölscher

8 Theta rhythm and bidirectional plasticity in the hippocampus 174
James M. Hyman and Michael E. Hasselmo

9 Distributed population codes in sensory and memory representations of the neocortex 192
Matthias Munk

10 The role of neuronal populations in auditory cortex for category learning 224
Frank W. Ohl and Henning Scheich

11 The construction of olfactory representations 247
Thomas A. Cleland

Part IV Functional integration of different brain areas in information processing and plasticity 281

12 Anatomical, physiological, and pharmacological properties underlying hippocampal sensorimotor integration 283
Brian H. Bland

13 A face in the crowd: which groups of neurons process face stimuli, and how do they interact? 326
Kari L. Hoffmann

14 Using spikes and local field potentials to reveal computational networks in monkey cortex 350
KRISTINA J. NIELSEN AND GREGOR RAINER

15 Cortical gamma-band activity during auditory processing: evidence from human magnetoencephalography studies 363
JOCHEN KAISER AND WERNER LUTZENBERGER

Part V Disturbances of population activity as the basis of schizophrenia 385

16 Neural coordination and psychotic disorganization 387
ANDRÉ A. FENTON

17 The role of synchronous gamma-band activity in schizophrenia 409
CORINNA HAENSCHEL

Part VI Summary, conclusion, and future targets 431

18 Summary of chapters, conclusion, and future targets 433
CHRISTIAN HÖLSCHER AND MATTHIAS MUNK

Index 470

Contributors

Brian H. Bland
Department of Psychology, University of Calgary, Canada

Thomas A. Cleland
Department of Psychology, Cornell University, USA

Sam A. Deadwyler
Department of Physiology and Pharmacology, Wake Forest University School of Medicine, Winston-Salem, USA

André A. Fenton
Department of Physiology and Pharmacology, State University of New York, Downstate Medical Center, Brooklyn, USA

Corinna Haenschel
Max Planck Institute for Brain Research, Frankfurt/Main, Germany

Robert E. Hampson
Department of Physiology and Pharmacology, Wake Forest University School of Medicine, Winston-Salem, USA

Kenneth D. Harris
Rutgers University, Newark, USA

Michael E. Hasselmo
Department of Psychology Center for Memory and Brain, Boston University, USA

Kari L. Hoffman
Max Planck Institute of Biological Cybernetics, Tübingen, Germany

Christian Hölscher
School of Biomedical Sciences, University of Ulster, UK

James M. Hyman
Department of Psychiatry, University of British Columbia, Vancouver, Canada

Jadin C. Jackson
Department of Neuroscience, University of Minnesota, USA

Adam Johnson
Department of Neuroscience, University of Minnesota, USA

Jochen Kaiser
Institute of Medical Psychology, University of Frankfurt, Germany

Eyal Y. Kimchi
The John B. Pierce Laboratory, Interdepartmental Neuroscience Program, Yale University School of Medicine, New Haven, USA

Mark Laubach
The John B. Pierce Laboratory, Department of Neurobiology, Yale University School of Medicine, New Haven, USA

Werner Lutzenberger
MEG-Centre, Institute of Medical Psychology and Behavioural Neurobiology, University of Tübingen, Germany

Edward O. Mann
Department of Physiology, Oxford University, UK

Matthias Munk
Max Planck Institute for Biological Cybernetics, Tübingen, Germany

Nandakumar S. Narayanan
The John B. Pierce Laboratory, Interdepartmental Neuroscience Program, Yale University School of Medicine, New Haven, USA

Kristina J. Nielsen
Salk Institute for Biological Studies, La Jolla, USA

Frank W. Ohl
Leibniz Institute of Neurobiology, Magdeburg, Germany

Ole Paulsen
Department of Physiology, Oxford University, UK

Gregor Rainer
Max Planck Institute for Biological Cybernetics, Tübingen, Germany

A. David Redish
Department of Neuroscience, University of Minnesota, USA

Henning Scheich
Leibniz Institute of Neurobiology, Magdeburg, Germany

Part I INTRODUCTION

1

How could populations of neurons encode information?

CHRISTIAN HÖLSCHER

Information representation in neuronal populations: what is the "machine language" of the brain?

Research in the area of neuroscience and brain functions has made extraordinary progress in the last 50 years, in particular with the advent of novel methods that enables us to look at the properties of neuroanatomy and neurophysiology in much finer detail, and even at the activity of living brains during the performance of tasks. However, the question of how information is actually represented and encoded by neurons is still one of the "final frontiers" of neuroscience, and surprisingly little progress has been made here. How information is encoded in the brain has captivated medics, scientists, and philosophers for centuries. Scholars such as Leonardo da Vinci or René Descartes had already an astonishingly detailed knowledge of the anatomy of the brain, and had made suggestions that it is the brain that processes information and even harbors the seat of the personality or of the soul. However, whenever suggestions are brought forward how information might be processed and represented in the brain, these often turn out to be simplistic and idealistic. These rarely add up to more than a kind of "homunculus" that somehow receives information that is received via the eyes or the ears. This model only transfers the problem of information representation from the brain to the homunculus.

One problem with the research of information encoding is that it is completely counter-intuitive. Often it is very helpful to explain complex anatomical and functional processes with mental images or sketches that compare

Information Processing by Neuronal Populations, ed. Christian Hölscher and Matthias Munk.
Published by Cambridge University Press. © Cambridge University Press 2009.

a difficult, unknown process with objects and machines of everyday use that people can easily picture. Unfortunately, it is impossible to do so with the topic of information encoding without ending up with comparisons that convey a completely wrong message. In the past, the brain had been compared to machines that were in use at the time, e.g. Descartes compared the nervous system to water pipeline networks that convey information via liquid-filled tubes to the brain ventricles, where the information is gathered and presented to the pineal gland, the proposed seat of the soul. Later, the brain had been compared to a telephone exchange, with information arriving from the outside at the "central switchboard" where it is processed, and leaving the brain through outgoing lines. More recently, the brain has been compared to a computer.

However, none of these images actually really explains how the brain processes information, since the architecture and the actual algorithms that govern information processing in neuronal populations are completely different from a computer that runs a piece of software on a silicon chip. The reason for this is based in the very counter-intuitive process of translating information about the real world (e.g. the color of a plant) into abstract symbols or codes that have nothing in common with the original information. It is difficult to explain to the lay person that a digital camera translates an image into rows of zeroes and ones. These rows encode the information "somehow" and can be translated again by a monitor or a printer into a two-dimensional picture that we can "understand." It is impossible to grasp intuitively where the information of the picture is located in the string of zeroes and ones. In a similar fashion, it will not be possible to provide an intuitive model of how information about the real world is encoded in neuronal activity. Neuronal activity consists of discrete "digital" states such as action potentials, but also encompasses analogue states of membrane potentials on dendrites and cell bodies. Any model that wants to explain neuronal encoding of information will have to work within these parameters.

In the 1930s, techniques became available that permitted to record the activity of single neurons, and of large brain areas (electroencephalograms, EEGs). This very much formed and influenced the models and concepts developed in those times. The main concepts that were emphasized then were the models of information coding in neurons by rate coding. Even though several other mathematical concepts and network models did exist back then, the available techniques did not provide any data that could be used to underpin such models.

With the development of single-cell recording in freely moving animals it became possible to actually observe neurons during the process of information processing and storage. The data of single-cell recording had been mainly interpreted in the traditional view that neurons code information by modulating firing rates. In parallel, large-scale activity of brain areas became observable with the development of PET scan technology. However, the resolution of these techniques did not allow for any analysis of neuronal population activities and network properties, and the available data very much biased the views and models greatly towards a single-cell information encoding theory, even though it was understood at the time that information was most likely represented in a distributed form in networks.

With the advent of modern technology that allows the recording of groups of single-cell activity of large populations of neurons, as well as the recording of local field potentials with electrode arrays in several brain areas simultaneously, it has become possible to observe how large networks behave during the performance of memory or recognition tasks. We now can ask specific questions of how exactly networks are coordinated and how neurons are associated or dissociated in their activity during information processing. We can test network models and adapt them or discard them. The gap between single neuron activity and large-scale cortical activity can now be filled, and the experimental results allow us to make specific statements on how information is processed and stored in the brain.

We are now at a threshold where the novel observations and experimental data need to be integrated into new concepts and models of how the brain processes information. While the technology has made extraordinary advances, the concepts in people's minds still lag behind. We can still find books, reviews, papers that reiterate single neuron encoding concepts and do not mention network properties that have been observed and described in neuronal networks in the living brain. The time has come to get people from different backgrounds together to synthesize the available results and information that have accumulated in the last 10 years.

This book will try to do exactly that. The different authors that have contributed chapters in this book present their research findings and formulate a set of theories and concepts that will encompass the latest findings. The authors present data at single-cell level in vitro and in vivo, and at neuronal population level in vivo comparing field potentials (EEGs) in different brain areas, and also data from spike recordings from identified neuronal populations during the performance of different tasks.

The results obtained over the last decade cast new light on how the brain acts as a system, bringing together separate areas of research that could not have been seen in functional context before. For example, neuronal firing is affected by activating projections from the basal brain, e.g. the nuclei that use acetylcholine as their main neurotransmitter. The neurophysiology and pharmacology of acetylcholine receptors have been known for many decades. However, the functional context of their activity could only have been guessed at. We now know how the state of cortical activation is modulated by these projections that basically activate inhibitory feedback oscillation loops. These oscillations impose temporal patterns on neuronal activity that are crucial for information processing and memory formation. This controls how neuronal functional assemblies are brought together or are kept separate and how synaptic weights are changed through synchronous excitatory input, and they dynamically switch associations of different cortical areas and of different information. This concept then can be applied to states of cognitive dysfunctions or disorders, such as schizophrenia. Here, the disturbance of the dynamic network association control could well be responsible for the inability to separate disconnected events in time and space or to bring together segments of events that belong together.

This book strives to cover the range of single-cell activity analysis through to observation of network activity and finally to brain area activity and cognitive processes of the brain, discussing and proposing mechanisms that are of importance for the system to function.

Questions that will be addressed in the book are:

> *How do different brain areas interact when processing information?* As recent research has shown, brain areas such as the hippocampus are not isolated units that can be studied independently of other brain areas that relay already highly processed information to the hippocampus where it is associated with higher-level representations.
>
> *How do neuronal populations encode information?* More and more neurophysiologists employ multi-electrode recording techniques to study how information is encoded in networks. It has become increasingly clear that single-cell recording and analysis is limited and is insufficient in the assessment of what information is encoded in the particular neuronal populations of interest. We know information is distributed in networks that are best analyzed *in toto* and in behaving animals. Modern techniques make it possible to use electrode arrays, and the computing power and software available today makes it feasible to analyze large sets of data.
>
> It is now possible to investigate how information is distributed over networks, and that this distribution does follow "holistic" network and parallel processing rules rather than linear data relay and analysis. Convincing evidence for this are the error correction properties, pattern completion properties, and distributed information properties of specific brain areas.
>
> *How are networks formed and separated from or associated with other networks?* The observation that neurons fire in temporal patterns, and that

EEG field potential oscillations have powerful influences on the firing probability of neurons, seems to have been accepted in the last 10 years, with some initial reluctance. While there are still schools of thought that completely ignore these findings and do not incorporate them into their models, the large increase of available data from different laboratories and from different species, brain areas, and task performances in this area clearly emphasizes the need to do so. We will incorporate these findings into our models, and they will allow us to develop specific models of information processing and representation in the brain.

A brief overview of current ideas how information might be encoded in the brain

To help interested readers who have only a limited background in this topic to understand the very specific reports in the following chapters, we provide a brief overview of theories and concepts of how information might be encoded in neuronal activity.

Frequency coding

The first and main theory of how information is translated from input systems such as temperature receptors in the skin to neuronal activity is that of frequency coding. Recordings of neurons in the skin have shown that temperature or pressure is represented in the firing rate of the neuron that gets the input from the sensors in the skin. A low-temperature stimulus causes the relay neuron to fire slowly, and increasing temperature will increase the firing rate of that neuron. There is a direct relation between sensory quality (temperature) and neuronal activity (see Fig. 1.1).

Another good example for frequency coding is found in the auditory system. Sound waves activate the eardrum which in turn transmits these waves to the cochlear system. The basilar membrane is a resonant structure and is deflected in response to these sound waves. Each location along the basilar membrane responds best to a small range of sound frequencies and acts as a filter. When sound waves activate the basilar membrane of the inner ear, these vibrations are transmitted to the hair cells. This triggers the release of neurotransmitter at the base of the hair cell and the excitation of primary afferent neurons. The neurons encode the information for transmission to the CNS. The firing frequency of these neurons is a direct related to the sound frequency up to 1 kHz. This means a 500-Hz sound will activate a 500-Hz neuronal firing, a direct frequency encoding of the sound waves into neuronal firing activity. However,

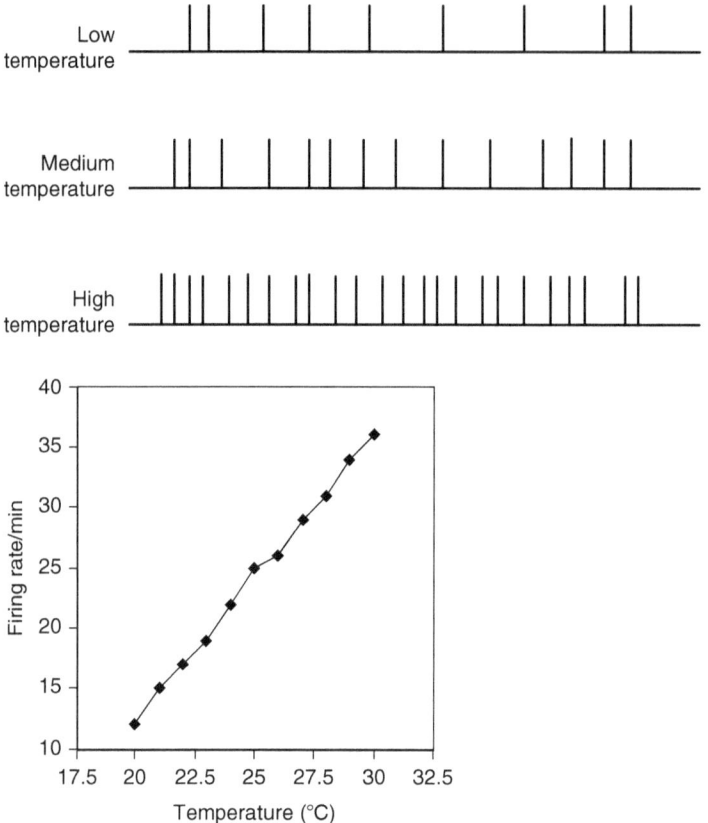

Figure 1.1 (Top) Schematic representation of frequency coding of a temperature-sensitive peripheral neuron in the skin. Recordings of neuronal activity are made while heating the skin to higher temperatures. The change in temperature is directly correlated with the firing rate of the neuron (tonic type of response). (Bottom) Linear correlation between stimulation of skin with a temperature probe and the firing rate of a neuron in the skin.

the maximal firing frequency is limited to about 1 kHz. Above this threshold, different encoding strategies must be chosen.

Frequency coding is seen by a majority of researchers as the main (if not only) type of information encoding.

Topological coding (labeled line)

It is important to note that the firing activity of a neuron that conveys temperature is in no way different to the firing activity of a neuron that relays touch or pain information. The rate coding and the action potential is similar. The reason why a sensory sensation is identified as touch or pain is that it is defined from which type of sensory input system the information comes.

Another example is the projection of a touch receptor in the thumb that will relay information to the primary somatosensory cortex that represents the thumb. No special encoding of location in the firing activity of neurons is required here, since the anatomical projections are fixed. This is sometimes called the "labeled line" type of coding. Proof for this concept is that these projections can be wrong in some instances, and phantom pain can be observed. An example is the typical pain in the arm that patients with angina pectoris (blood supply problems to the heart) experience. The pain receptors that originate in the heart also appear to project to somatosensory areas that represent the arm, and cause pain sensation in the wrong body part.

Another good example of topological coding comes from the auditory system. As described above, each location along the basilar membrane responds best to a small range of sound frequencies and acts as a filter. Therefore, it is sufficient for a neuron that innervates an area of a defined frequency to fire when it is activated in order to transmit the information that this defined frequency had arrived. No further information is required, since these sensory neurons can only be activated if the particular frequency has been received. The firing frequency of the neuron is of secondary importance, since due to the physical properties of the cochlea the neuron can only be activated by this defined frequency. Instead, the firing frequency of neurons can be used to indicate the intensity of sound. A low-frequency activity of a 5-kHz neuron will indicate that the received sound intensity is low, while high-frequency firing activity will indicate a loud 5-kHz sound. Here, topological and frequency coding are combined.

Phase coding

Phase coding is a different type of information coding that is independent from firing rate. Here, the timing of the neuronal action potential is of importance. A simple example is given in Fig. 1.2, where pressure changes in

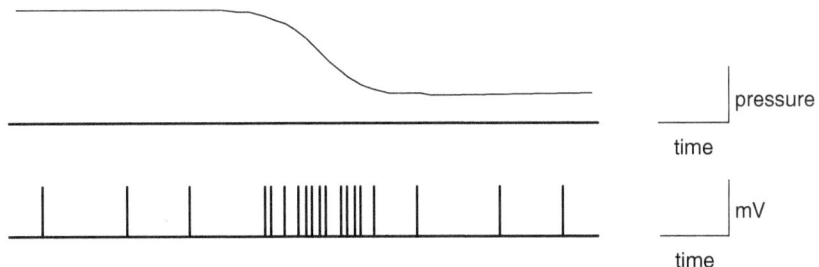

Figure 1.2 Schematic representation of a neuronal response to touch. The top trace shows the change of pressure applied to skin. Only the changes of pressure on the skin are coded by this type of neuron, not the absolute values of pressure (phasic type of response).

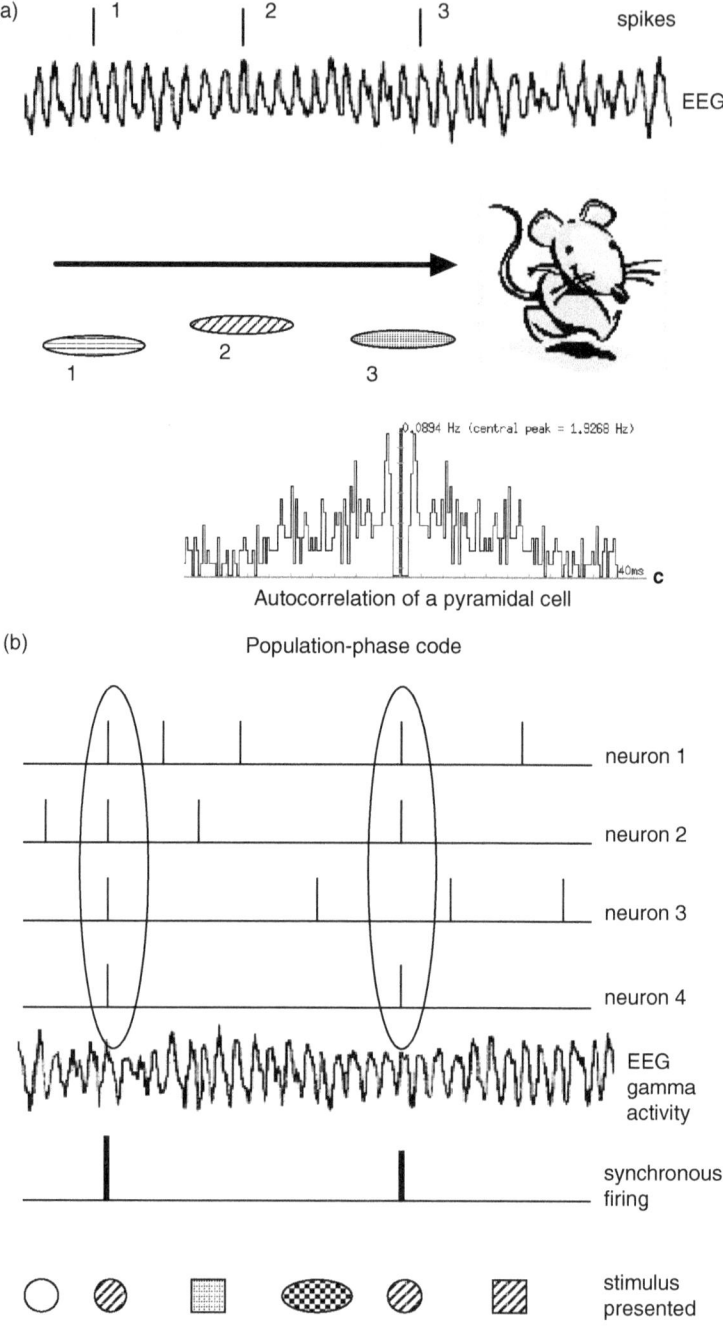

Figure 1.3 (a) Phase coding in the hippocampus. Pyramidal neurons in the hippocampus often fire in a burst mode, with one to five spikes followed by inactivity. The firing activity is tightly coordinated in time in relation to theta activity in the EEG. A neuron recorded from a rat that travels through space will fire in a particular

skin receptors are coded by neurons that only respond to changes in pressure (phasic response), and not to the absolute amount of pressure. A single spike would be sufficient to relay the information "pressure change," since the spike will only fire under those conditions. Other examples are given in Chapters 7 and 16, where the timing of neurons signify a particular information content, e.g. the location of an animal in space. Figure. 1.3 (below) shows an example of how hippocampal neurons appear to encode space. The firing activity of these neurons rarely exceeds 10 Hz, suggesting that frequency encoding is not of importance in this system. In fact, most hippocampal neurons only fire bursts of about five spikes and then remain silent for some time until the next burst appears. Such a firing mode would not be suitable for frequency coding. It is ideal for phase coding, and the highly controlled temporal relationship of firing activity of these neurons with the theta EEG rhythm strongly points into the direction of phase coding. The advantage of this coding principle that it is very fast and can relay information with only one spike. The disadvantage is that it is a digital type of information coding and cannot differentiate between various states of pressure, which would require an analogue type of encoding as seen in the frequency modulation technique. An overlay of frequency with phase coding therefore allows a combination of digital and analogue information encoding.

Population coding

Single neurons can only convey a very limited amount of information. In addition, single neurons are noisy and unreliable and therefore cannot be depended on in absolute terms. Therefore, information is often encoded in populations that code similar information. Instead of relying on one noisy neuron,

Caption for Figure 1.3 (cont.)
area in space (the place field), and the spikes will be highly coordinated in relation to the phase of the theta wave. Spikes of three neurons are shown that fire at different place fields while the rat is moving through space. This is reproducible, and the neurons will (almost) always fire in this confined way. An autocorrelation of the spike activity will also show a temporal order. Shown is the time between spikes in the firing activity of this neuron. The firing in time is not random, but follows a temporal ordered in the theta frequency (see also Chapter 7 for details). (b) Phase coding of neuronal activity. Neurons of a population are brought together by firing synchronously. The synchronous activity is controlled and induced by EEG field potentials (in this case in the gamma frequency). The EEG field potential oscillations control the firing probability of excitatory neurons and confine neuronal spikes to defined time windows. Populations of neurons can be assembled or separated by this method.

12   Christian Hölscher

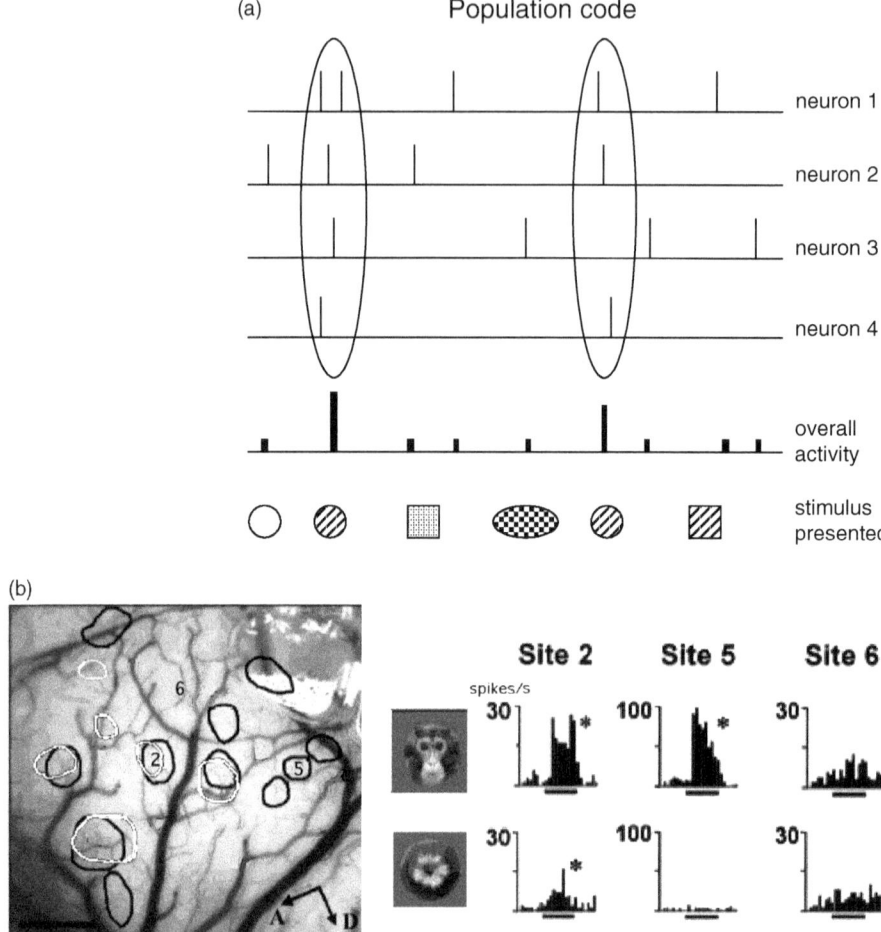

Figure 1.4 Population coding. (a) In order to represent information in a more reliable manner, a group of neurons can represent the information as a population. As shown in this figure, a visual image can be represented by a group of neurons. The neurons hardly fire when an image is shown that the neurons are not tuned to. The schematic representation shows two important principles of population coding: spontaneous activity of neurons that do not contain information can be easily filtered out, and the failure of one neuron to respond can be compensated for by the other neurons in the population. (b) Example of population coding of information in the cortex. Images are shown to a primate, and the neuronal activity in the visual cortex is made visible using voltage-sensitive dyes. When the image of a face is shown, several areas in the inferotemporal cortex are activated (shown as black circles in the photograph). Showing an image of an apple activates a different set of areas (shown as white circles). Note that some of the areas overlap, indicating that some neuronal populations are shared between different information encoding networks.

When recording from single neurons within the activated areas, neurons can be found that specifically respond to the image. Site 2 (marked in the photograph of the cortex), only the sum of all neurons within this population can induce a response. This way, the noise can be filtered out, and also the lack of response of one neuron when it should fire can be compensated for (see Fig. 1.3a). Optical recording makes use of voltage-sensitive dye can make such populations visible. Figure 1.3b gives an example where images shown to a primate activate neuronal population in the visual cortex. Single neurons can be recorded from these active spots that respond to the images in similar manner, demonstrating that such populations of neurons do indeed encode similar information (Fig. 1.3b).

Populations can also represent a range of information. In order to code the whole spectrum of temperature that a person might encounter, neurons with different temperature tuning are found in the skin. They have a range of ideal response that differs slightly from other neurons. The overall range of temperature can be represented by the overall population of neurons. As shown in Fig. 1.5, a few simple subtractions of different temperatures can produce further refined information. The complete information of temperatures in the body is encoded in the overall network system, not in individual neurons nor in one population of neurons.

A similar situation is found in the auditory system. The accuracy with which a listener can locate sounds is much greater than the accuracy of single neuron. When recording from neurons in different areas in the brain, the accuracy increases as the information ascends through the auditory system. The computing of such higher accuracy is performed by subtraction of several broadly tuned neurons to obtain a much more precise value, similar to the process explained in Fig. 1.5.

Encoding of sequences and of time

Another important aspect is the encoding of time sequences and association over time, or of the development of movement sequences (as in a motor program that has been learned by heart). Here, individual neuronal representations of information have to be associated in time. The previously discussed encoding processes all happen instantly, and cannot provide a mechanism for sequence learning. For this, a new mechanism has to be postulated that can store temporal associations. Recordings from neurons in the motor system show us that during the execution of a motor task, highly repetitive firing patterns

Caption for Figure 1.4 (cont.)
which is activated by both images, contains neurons that respond to both images also. Site 5 only is selectively activated by the image of the face, and most neurons within this area respond to the face only. When recording from site 6 that is not activated by either image, no selective neurons are found. (Adapted from Tsunoda et al., 2001.)

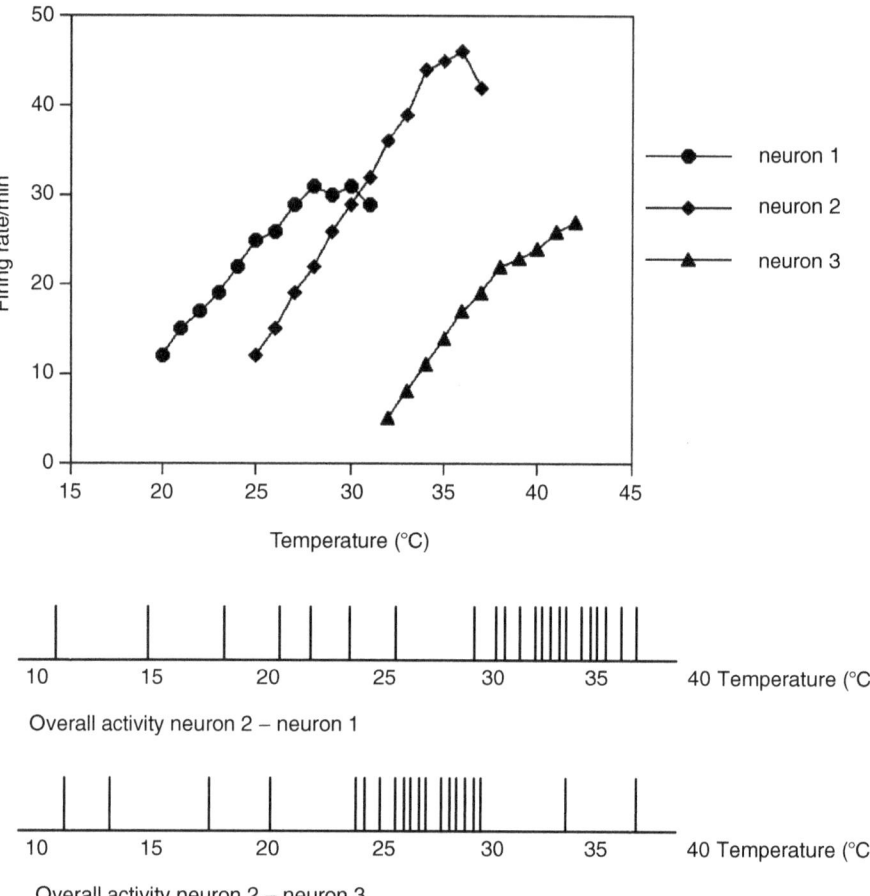

Figure 1.5 Distributed encoding of information in populations. (Top) Schematic representation of different "tuning" in frequency coding of temperature-sensitive neurons in the skin. Not all neurons respond equally well to all temperatures. Most neurons have an ideal temperature range and are "tuned" to this temperature. (Bottom) The overall activity of all neurons contains the information of all temperatures across the biological range. For example, by subtracting the activity of neuron 1 from that of neuron 2, the range between 32 and 37 degrees Celsius will be identified. When subtracting the activity of neuron 3 from that of neuron 2, the range between 24 and 32 degrees Celsius will be encoded.

of groups of neurons exist that drive different muscles that are involved in the activity. Other brain areas show very similar patterns during the learning of a spatial memory task. Here, neurons or groups of neurons that represent a particular part of the maze appear to drive following neuronal groups. These associations are specific and can be repeated with little alterations after rehearsal. This shows that high-order population encoding exists that includes

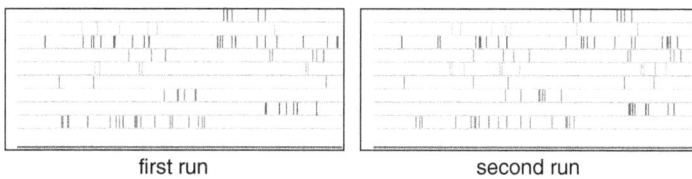

first run   second run

Figure 1.6 Encoding of sequences and time coding in the hippocampus: each line represents the spike trains of a different neuron. (Left) Nine neurons were recorded simultaneously during a spatial task in which the animal had to turn left in a square maze. (Right) When the animal has to turn left again at the same place in the second run, a very similar sequence of neuronal network activity is observed from the same nine neurons. Each neuron fires in a specific reproducible pattern, such as tonic of burst activity, and the relation of the firing activities of all neurons to each other is also very similar. More impressively, these specific firing patterns in time are reproduced during sleep of the animal. It has been suggested that this could be part of a rehearsal process of previously experienced information required to encode it in long-term memory. This indicates that networks encode time sequences and network firing patterns that are stored and can be reproduced. For details see Hölscher (2003), and Chapter 7.

several populations and furthermore can "stretch" across time to include not only space but time as a dimension. Figure 1.6 shows an example of neuronal activity that can be "stored" and replayed when required.

Reassembling the image: how are the individual elements brought together again?

Different properties of objects are encoded separately. In the visual system, separate channels exist for direction, color, depth, and edge/surface properties. These "packets" of information remain separate even in the visual cortex area V1. Here, areas can be identified where neurons respond to brightness and color but not to direction of movement (the so-called "blobs"), while other areas encode the direction of movements, but not the color (the "inter-blobs"). The separation of labor goes even further than this. Higher visual areas are specialized for either movement analysis, hand–eye coordination, color perception, or depth perception. Color processing is accomplished within the so-called "ventral pathway": from the retina, projections reach the lateral geniculate nucleus (LGN) parvo layers, project further to the V1 layer 4Cβ, then to the V1 inter-blobs, → to the V2 interstripes → V4 and inferior temporal cortex (IT). For motion processing, a similar information pipeline has been described. It is processed in the "dorsal pathway": LGN magno layers → V1 layer 4Cα → V1 layer 4B → V2 thick stripes → area MT. Color processing is also processed within the "ventral pathway": LGN parvo layers → V1 layer 4Cβ → V1 blobs → V2 thin stripes → area V4 (see Zeki (1993) for a detailed review).

This opens up the question of how these individual bits of information are brought together again. If we see a white duck on a blue lake, we are aware that the color white is associated with the duck, not with the lake. If the lake was frozen and covered in snow, we now would correctly associate the color white with the lake. The brain has to have a mechanism by which information can be assembled quickly and in a reversible manner. The color white, which is processed independently of shapes and depth, has to be connected somehow to the correct object. This has been called the "binding problem" (Malsburg, 1981). How is this binding achieved? What mechanisms are in place, and what rules control the association of dissociation of the individual parts of relevant information?

In order to perform this information processing correctly, new mechanisms and rules have to be introduced. The set of rules described so far are not sufficient for this. We are at a threshold to a new level, and modern neuroscience has embraced the challenges that we have to meet in order to start to understand how the brain works as a system.

As one would expect, there are several competing theories in existence that try to address these questions. Some try to solve the problem by the discrete influence of attention on neuronal activity. In this proposal, attentional mechanisms support the activity of some sets of population while suppressing others. This way, the false association of a color to an object is suppressed (see Reynolds and Desimone (1999) and Maunsell and Cook (2002) for details on this concept). Others make use of dynamic mechanisms in which neuronal networks can be temporarily assembled by "binding" them together in time. EEG oscillations can shift the firing probability of neurons, and can separate or "connect" neurons by enabling them to fire simultaneously. This mechanism could be used to reversibly associate feature information when required (for a review see Singer, 1999; Engel et al., 2001). Other people argue that such a complex system is not required for enabling the correct binding of features, and that conventional rate coding and hierarchical systems could be sufficient (Ghose and Maunsell, 1999; Shadlen and Movshon, 1999). Yet others do not really see the need for such systems at all, and propose that the available mechanisms of information encoding in the brain should be sufficient (Riesenhuber and Poggio, 1999; Serre et al., 2007).

The authors in this book will present data and concepts to address these and other central issues that govern neuroscience research at present, and that will have to be solved in order to crack the code of how large populations of neurons function together and process information. Suggestions will also be made on how to tackle these problems in the future.

We hope that the reader will find this wide selection of ideas and results of interest, and will gain a deeper understanding in the current scientific questions, and in the challenging plans for the future.

References

Engel, A. K., Fries, P., and Singer, W. (2001). Dynamic predictions: oscillations and synchrony in top-down processing. *Nat Rev Neurosci* **2**:704–716.

Ghose, G. M. and Maunsell, J. (1999). Specialized representations in visual cortex: a role for binding? *Neuron* **24**:79–85.

Hölscher, C. (2003). Time, space, and hippocampal functions. *Rev Neurosci* **14**:253–284.

Malsburg, C. von der (1981). *The Correlation Theory of Brain Function*, Technical Report 81-2. Frankfurt, Germany: Biophysical Chemistry, Max Planck Institute.

Maunsell, J. H. and Cook, E. P. (2002). The role of attention in visual processing. *Philos Trans R Soc Lond B* **357**:1063–1072.

Reynolds, J. H. and Desimone, R. (1999). The role of neural mechanisms of attention in solving the binding problem. *Neuron* **24**:19–29.

Riesenhuber, M. and Poggio, T. (1999). Are cortical models really bound by the "binding problem"? *Neuron* **24**:87–93.

Serre, T., Wolf, L., Bileschi, S., Riesenhuber, M., and Poggio, T. (2007). Robust object recognition with cortex-like mechanisms. *IEEE Trans Pattern Anal Mach Intell* **29**:411–426.

Shadlen, M. N. and Movshon, J. A. (1999). Synchrony unbound: a critical evaluation of the temporal binding hypothesis. *Neuron* **24**:67–77.

Singer, W. (1999). Striving for coherence. *Nature* **397**:391–393.

Tsunoda, K., Yamane, Y., Nishizaki, M., and Tanifuji, M. (2001). Complex objects are represented in macaque inferotemporal cortex by the combination of feature columns. *Nat Neurosci* **4**:832–838.

Zeki, S. (1993). *A Vision of the Brain*. Oxford, UK: Blackwell Scientific Publications.

Part II ORGANIZATION OF NEURONAL
         ACTIVITY IN NEURONAL POPULATIONS

2

Cellular mechanisms underlying network synchrony in the medial temporal lobe

EDWARD O. MANN AND OLE PAULSEN

Introduction

The hippocampus lies at the apex of the hierarchical organization of cortical connectivity, receiving convergent multimodal inputs that are funneled through the adjacent entorhinal cortex (Fig. 2.1). The output of the hippocampus is relayed back through the entorhinal cortex, and thus these structures are ideally placed to both store novel associations and detect predictive errors (Lavenex and Amaral, 2000; Witter et al., 2000). Indeed, while memories are likely to be stored across distributed brain regions, the learning and consolidation of explicit memories appear to depend upon the hippocampus and surrounding parahippocampal regions (Morris et al., 2003; Squire et al., 2004). However, while the anatomical substrate of such learning is becoming increasingly well defined, it remains unclear how cells act collectively within these neuronal networks to extract and store salient input correlations.

Over 50 years ago, Donald Hebb postulated a simple cellular learning rule, whereby the strength of the synaptic connection between two neurons would be increased if activity in the presynaptic neuron persistently contributed to discharging the postsynaptic neuron (Hebb, 1949). It has since then been shown that such repeated pairings of synaptic events with postsynaptic action potentials (spikes), within a window of tens of milliseconds, can produce long-term changes in synaptic efficacy in many different neuronal systems, both in vitro

Information Processing by Neuronal Populations, ed. Christian Hölscher and Matthias Munk.
Published by Cambridge University Press. © Cambridge University Press 2009.

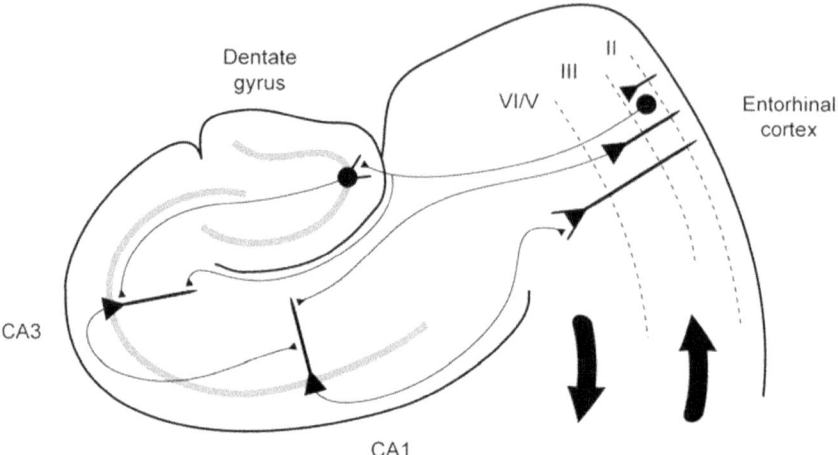

Figure 2.1 The entorhino-hippocampal circuit. Schematic diagram of the excitatory pathways into and out of the hippocampus via the entorhinal cortex. The bold arrows represent the connections with the neocortex, with inputs converging on the superficial layers of the entorhinal cortex, and outputs emanating from the deep layers. The connections with the subiculum have been removed for clarity.

and in vivo (Paulsen and Sejnowski, 2000; Bi and Poo, 2001). Moreover, the relative timing and temporal order of pre- and postsynaptic spikes not only determine the size of these plastic changes, but also whether synaptic transmission is enhanced or depressed (Markram et al., 1997; Bi and Poo, 1998). Such spike timing-dependent plasticity (STDP) is an attractive mechanism for storing the directional association between events, enabling sequence learning and predictive encoding, as well as competitive learning (Song et al., 2000) (see also Chapter 7 for more details). However, STDP requires spikes to occur in precise orders within short time windows, and, therefore, it is not intuitively obvious how this mechanism could operate successfully if the converging inputs are affected by the inherent spike jitter in polysynaptic pathways or when the behaviorally relevant temporal associations to be encoded occur over more protracted time scales.

The existence of brain oscillations could provide a mechanism to naturally organize spike times in cortical networks, and thus set the conditions for cellular learning rules based on STDP (Paulsen and Sejnowski, 2000). The entorhino-hippocampal system displays a variety of network oscillations whose frequency and spatial coherence vary with the behavioral state of the animal (Chrobak et al., 2000). These brain rhythms could act to synchronize spike times as activity propagates through different levels of the network, and/or act as a reference for temporal encoding, and thereby compress sequences at behavioral timescales into spike sequences within individual oscillatory cycles (Skaggs et al., 1996).

It has not yet been resolved whether network oscillations serve these, or any other, functions in the cortex (Sejnowski and Paulsen, 2006). However, oscillations are a clear feature of cortical network activity, and one important step in understanding their functional roles will be to elucidate the underlying cellular and network mechanisms. Here, we will review the principal mechanisms by which network oscillations can be generated, and our current knowledge of how these mechanisms are specifically involved in different oscillatory patterns in the entorhino-hippocampal system. This will provide the platform for a critical evaluation of how network oscillations could provide the temporal structure for STDP, and thus play a role in learning and memory.

Basic cellular mechanisms contributing to cortical network oscillations

Cortical microcircuits, including individual hippocampal subfields and cortical layers, appear to follow a stereotypic organizational principle, whereby ~80% of the neurons comprise a relatively homogeneous population of excitatory cells, with the remaining ~20% forming a diverse array of inhibitory GABAergic interneurons (Somogyi et al., 1998). Interneurons can be classified into subtypes based on their axonal arborization, with the simplest distinction being between those interneurons that selectively target the perisomatic region of excitatory neurons, those that target the dendritic tree of excitatory cells, and those that selectively target other interneurons (Freund and Buzsáki, 1996) (Fig. 2.2). While information storage is likely to involve the synaptic connections

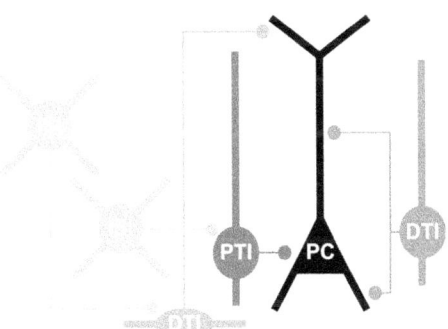

Figure 2.2 Interneuronal subtypes. GABAergic interneurons can be classified into subtypes based on their synaptic targets, expression of neurochemical markers, intrinsic properties and firing patterns in vivo. For the purpose of this review, the simplest distinction is made on the basis of local axonal arborization, distinguishing three main subclasses of interneuron: perisomatic-targeting interneurons (PTI), dendritic-targeting interneurons (DTI), and interneuron-selective interneurons (ISI).

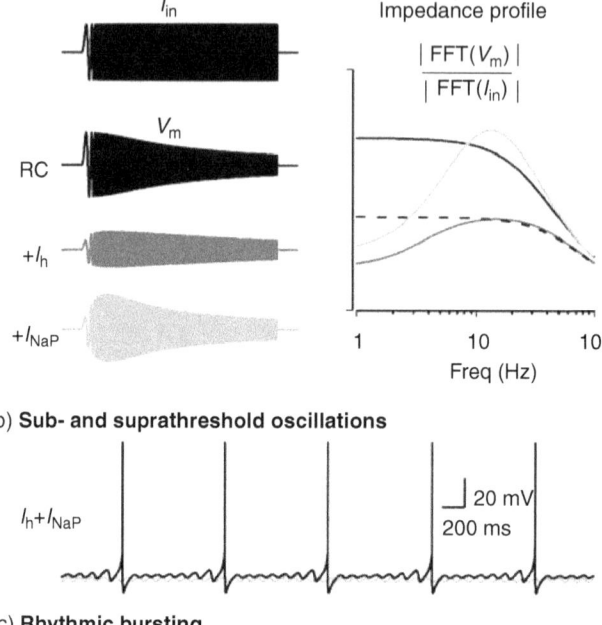

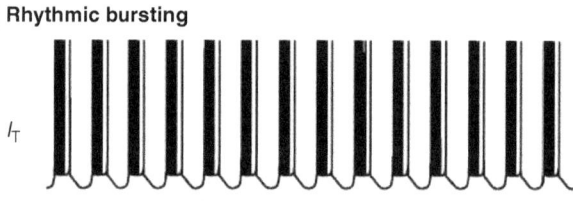

Figure 2.3 Intrinsic resonance and oscillatory properties of individual neurons. The responses elicited in neurons vary with input frequency, and this input–output relationship varies across both cell types and neuromodulatory states. The short-term dynamics of synaptic release is an important component of this frequency dependence, but the intrinsic properties of neurons also shape synaptic integration and endow neurons with frequency preferences. (a) A neuron's frequency preference is often characterized by injecting a sinusoidal input current of varying frequency (I_{in}; here, linearly increasing frequency from 0 to 100 Hz) and measuring the subthreshold changes in membrane potential (V_m). These measurements can then be used to calculate an impedance profile, which displays the impedance magnitude as a function of frequency. The impedance profile of a passive cell shows frequency selectivity, as the plasma membrane consists of a resistor and capacitor in parallel (RC circuit), which act to passively filter out high-frequency inputs. However, the neuron can act as a band-pass filter, if there is also a voltage-dependent current that actively opposes low-frequency changes in membrane potential, such as the hyperpolarization-activated non-specific cation current (I_h). This resonance peak can be amplified by currents that activate rapidly relative to the resonant currents, which in this example is fulfilled by the persistent sodium current (I_{NaP}).

between excitatory neurons, the smaller and more diverse population of cortical interneurons appears to play the predominant role in controlling the precision of spike timing within cortical network oscillations (Mann and Paulsen, 2006). This principle appears to hold true whether the rhythm itself is imposed by subcortical nuclei, propagated from other cortical regions, or generated intrinsically within the cortical microcircuit. To understand the basic mechanisms by which oscillations could emerge in neuronal networks, and why interneurons would be the preferred mediators of cortical fast synchronization, it is important to consider both the intrinsic properties of neurons and their connections via electrical and chemical synapses.

Intrinsic cellular properties

The intrinsic properties of individual neurons can endow them with a preference for certain input frequencies, and thus shape their participation within different brain oscillations (Llinas, 1988; Hutcheon and Yarom, 2000). Such frequency preference is characterized by peaks in a neuron's impedance profile, which is often estimated using an impedance amplitude profile ("ZAP") protocol which measures the subthreshold changes in membrane potential produced by sinusoidal input current of varying frequency (Puil *et al.*, 1986) (Fig. 2.3). All neurons display some frequency dependence in their impedance profile, as the plasma membrane consists of a conductance and capacitor in parallel, which act to passively filter out high-frequency inputs. To generate a resonance peak it is also necessary to have a voltage-dependent current that actively opposes low-frequency changes in membrane potential. If the activation time constant of this current is slower than the passive membrane time constant,

Caption for Figure 2.3 (cont.)

(b) If the amplifying currents are sufficiently powerful, there may be a range of membrane potentials in which the neuron displays self-sustained oscillations. This example is based on models of stellate cells in the entorhinal cortex (Dickson *et al.*, 2000; Rotstein *et al.*, 2006), in which the combination of I_h and I_{NaP} can produce rhythmic subthreshold oscillations (gray). At more depolarized membrane potentials, the oscillatory properties of the membrane can lead to precisely controlled spike timing (black), but the amplitude and phase of subthreshold oscillation is reset by the spike. (c) The inactivation and activation processes of the low-threshold Ca^{2+} current (I_T) can mediate amplified resonance, and lead to self-sustained oscillations. Intrinsic oscillations in neurons may involve a whole array of depolarizing and hyperpolarizing currents that can mediate rhythmic burst firing, and which enable the initiation and frequency of the oscillations to be controlled by a variety of neuromodulatory pathways. Such robust oscillators are prevalent in subcortical nuclei.

then there will be a window of frequencies that are relatively unattenuated, and the neuron will act as a band-pass filter (Hutcheon and Yarom, 2000). As the kinetics of resonant currents are dependent on the membrane potential, and the membrane time constant depends on the input conductance of the neuron, such subthreshold frequency tuning would not only vary between different cell classes, but also across different physiological states.

While resonant currents determine the subthreshold frequency preference of neurons, amplifying currents, which potentiate, rather than oppose, voltage changes and activate rapidly relative to the resonant currents, can enhance neuronal responses and accentuate frequency tuning (Hutcheon and Yarom, 2000). If these amplifying currents are sufficiently strong the neuronal membrane potential will become unstable and display self-sustained oscillations (Hutcheon and Yarom, 2000). Such single-cell oscillations are prevalent in subcortical nuclei, where they often depend on the activation/inactivation of T-type Ca^{2+} currents and/or the interaction between I_h and the ionic mechanisms underlying Ca^{2+} spikes (Llinas and Yarom, 1981; Bal and McCormick, 1993, 1997; Lampl and Yarom, 1997; for review see Contreras, 2006; Llinas and Steriade, 2006). Coupled networks of such single-cell oscillators are likely to act as pacemakers for various slow brain oscillations. However, the relationship between impedance profiles and oscillatory behavior may not always be as simple. In cortical neurons, stochastic channel behavior may contribute to the oscillatory properties (Dorval and White, 2005). Moreover, while several cortical neuronal subtypes have been shown to display subthreshold resonance at different frequencies (Pike et al., 2000), it remains possible that these intrinsic properties are more important for synaptic integration and spike timing precision within network oscillations rather than in generating the oscillation itself.

Electrical synaptic coupling

A simple mechanism for synchronizing a network of intrinsic oscillators is via electrical coupling between neurons. Such coupling could occur through ephaptic interactions, resulting from current flow in the extracellular space, but stronger and more reliable coupling is achieved through gap junctions forming electrical synapses between neurons. The electrical synapses that have been studied in most detail are those that occur between excitatory projection neurons in the inferior olive (Llinas et al., 1974; Long et al., 2002; De Zeeuw et al., 2003), inhibitory interneurons in the cortex (Galarreta and Hestrin, 1999; Gibson et al., 1999; Venance et al., 2000; Meyer et al., 2002), and inhibitory cells in the thalamus (Landisman et al., 2002; Long et al., 2004; for review see Connors and Long, 2004). The gap junction coupling within all these networks appears to require connexin-36 (Deans et al., 2001; Hormuzdi et al., 2001; Landisman et al.,

2002; Long et al., 2002), which is the predominant gap junction protein expressed in neurons, and forms bidirectional electrical connections preferentially at dendro-dendritic appositions (Connors and Long, 2004). Due to the capacitance of the neuronal membrane and the relatively low conductance of these channels (Srinivas et al., 1999), combined with their distant electrotonic location, these gap junctions tend to act as low-pass filters for electrical signals transferred between coupled neurons. Accordingly, low-frequency subthreshold oscillations are propagated through electrical synapses with particular efficiency, which could enable gap-junction-coupled syncytia of intrinsic oscillators to act as pacemakers for slow cortical oscillations. However, while action potentials are strongly attenuated through electrical synapses, the resulting postsynaptic potentials ("spikelets") are also able to mediate spike synchronization in the millisecond time range. Therefore, electrical synapses between neurons may be involved in both slow and fast synchronous oscillations.

The reason that gap junction coupling is of particular interest with respect to the precise control of spike timing in the cortex is that these electrical connections appear to occur almost exclusively between interneurons belonging to the same subtype, thereby offering the potential to preserve functional diversity, whilst enhancing synchrony within each interneuronal class (Beierlein et al., 2000; Blatow et al., 2003; Connors and Long, 2004). However, it has proved difficult to establish the role of interneuronal gap junctions within cortical network oscillations. Connexin-36 knockout mice, in which electrical coupling between interneurons is all but eliminated, still display the same cohort of hippocampal oscillations as their wild-type litter mates, although the power in particular frequency bands may be reduced (Hormuzdi et al., 2001; Buhl et al., 2003). While the knockout mice may be subject to compensatory changes in neuronal connectivity, these studies suggest that electrical coupling between interneurons is at least not necessary for cortical oscillogenesis. This does not per se rule out a fundamental role for electrical synapses, as cortical neurons also express lower levels of other gap junction proteins, such as connexin-45 and the pannexins (Bruzzone et al., 2003; Maxeiner et al., 2003). Indeed, one suggestion is that these alternative proteins mediate axo-axonal gap junctions between excitatory pyramidal neurons, and that these connections are the critical components for the generation of cortical synchrony (Traub et al., 2000, 2003). The role of such putative axo-axonal gap junctions remains controversial, however, as crosstalk between projecting axons might aid synchronization in computer simulations, but is likely to degrade the specificity of neuronal encoding. Moreover, it is not yet known whether connexin-45 or pannexins form electrical synapses between neurons, and while electrical coupling between pyramidal cells has been reported on rare occasions (Schmitz et al., 2001), the ultrastructural basis for this coupling has not been elucidated. It may be that gap junctions between excitatory cortical neurons are rare but important, and the generation and evaluation of further genetically modified animal models may be required to resolve these issues in the future.

Chemical synaptic coupling

Coupling via gap junctions at intersections between neuronal processes is naturally spatially restricted, and, while these connections can be modulated by pH, $[Ca^{2+}]_i$, and phosphorylation, evidence of longer-term activity-dependent plasticity of mammalian electrical synapses is essentially non-existent (Connors and Long, 2004), and they are often regarded as a hard-wired feature of the neuronal circuitry. Neuronal communication via neurotransmitter release from axonal terminals enables coupling over more distributed areas, and is both more diverse and dynamic. Coupling through excitatory chemical synapses alone can act to synchronize intrinsic oscillators, but can also enable the emergence of rhythmic bursting in populations of non-oscillatory neurons. The positive feedback that drives the initiation of these bursts requires a mechanism to curtail activity, such as synaptic depression or Ca^{2+}-dependent K^+ currents, and thus allow another cycle of activity to commence. Such oscillations within recurrent excitatory networks, which may depend on the intrinsic neuronal properties, appear to be important for pattern generation in the spinal cord and brain stem (Grillner, 2006; Kiehn, 2006), slow oscillations in the cortex (Steriade *et al.*, 1993; Sanchez-Vives and McCormick, 2000; Shu *et al.*, 2003), and epileptiform activity in disinhibited networks (Traub and Wong, 1982; Menendez de la Prida *et al.*, 2006). However, purely excitatory mechanisms are unlikely to explain the millisecond precision of spike timing in the cortex. Excitatory terminals onto cortical projection neurons tend to be dispersed over extensive dendritic arbors, which attenuate and prolong synaptic responses. Individual excitatory inputs have weak and unreliable control over spike timing, making these connections more suitable for temporal integration.

In contrast, accumulating evidence suggests an important role for synaptic inhibition in the precise control of spike timing in principal neurons during oscillations (Whittington and Traub, 2003; Mann and Paulsen, 2005). In addition to balancing excitation, inhibition carries high-frequency signals in cortical networks (Hasenstaub *et al.*, 2005), which can be conveyed to downstream networks by synchronizing spiking in principal cells. The hyperpolarizing effect of synaptic inhibition can even be proactively involved in driving spike generation in the excitatory cells, by interacting with intrinsic voltage-dependent conductances, and thus producing "anode-break" excitation (Cobb *et al.*, 1995). This mechanism can even generate bursts of spikes, which has been particularly well studied in relation to thalamic spindle oscillations (7–12 Hz), during which hyperpolarization of thalamic relay cells activates I_h and deactivates T-type Ca^{2+} channels, both contributing to rebound bursts (von Krosigk et al., 1993; Bal et al., 1995). Thus, the divergent output from inhibitory interneurons provides a powerful mechanism to control spike timing in populations of excitatory neurons. The interneurons themselves may be synchronized by several mechanisms, including gap junctions, as discussed above, other inhibitory inputs, and synaptic excitation. Many interneurons are coupled by mutual inhibition, including those targeting perisomatic sites, and such interneuronal networks are capable of self-synchronizing their output (Bacci and Huguenard, 2006; Vida et al., 2006). Moreover, a specific set of interneurons, viz. interneuron-selective interneurons, appears to be dedicated to controlling other interneurons (Freund and Buzsáki, 1996), and, at least in some oscillatory states, are strongly phase-coupled to ongoing oscillations (Hajos et al., 2004; Oren et al., 2006). Finally, most, if not all, cortical interneurons receive excitatory input from local pyramidal neurons (Freund and Buzsáki, 1996). The coupling of excitatory and inhibitory neurons in negative feedback loops was early suggested as a route by which network oscillations can emerge (Freeman, 1968). These negative feedback loops can generate fast oscillations in cortical circuits. In contrast to excitatory cortical projection neurons, many local interneurons are electrotonically compact, with rapid integration time constants, and consequently can fire action potentials in response to excitatory input with much improved temporal precision (Fricker and Miles, 2000). Furthermore, perisomatic-targeting interneurons have their inhibitory synaptic output close to the spike initiation zone of excitatory cells, thus providing exquisite control over the timing of spike generation. Feedback loops between excitatory cells and inhibitory interneurons can therefore oscillate at frequencies of at least 40 Hz, primarily limited by the time constants of the two types of chemical synapses involved, and can synchronize excitatory cell discharges within millisecond time windows (Fisahn et al., 1998; Mann et al., 2005). Even higher frequencies can be achieved for mutually interconnected inhibitory networks (Brunel and Wang, 2003). Neurons coupled by mutual inhibition can also display more complicated oscillatory properties, such as anti-phase oscillations between subpopulations of neurons (Elson et al., 2002). With the permutations offered by intrinsic properties, gap junction coupling, synaptic kinetics, and conduction delays, it is not surprising that neuronal networks display a diverse array of rhythmic behaviors, whose mechanisms are not always easy to dissect. However, the key point here is that the temporal control of cortical activity on time scales relevant for STDP is most likely mediated ultimately through inhibitory GABAergic transmission.

Specific mechanisms underlying entorhinal and hippocampal network oscillations

The rhythmic nature of cortical activity is most easily and commonly detected using extracellular recordings of the field potential, which most likely reflects the currents flowing in the principal excitatory neurons. The average power spectrum of these field recordings resembles a $1/f$ noise distribution (Buzsáki and Draguhn, 2004), but the emergence of oscillations within different frequency bands is segregated in time, and correlated with different behavioral states (Sejnowski and Paulsen, 2006). However, oscillations at particular frequencies are not exclusively associated with any given cognitive process, and our understanding of the conditions determining if and when they arise appears to be continually refined (Dickinson, 2006; Steriade, 2006). There are nevertheless some stereotypical patterns of activity that have been extensively studied in the entorhino-hippocampal system, and which have provided the basis for understanding some of the cellular mechanisms underlying network synchrony (Fig. 2.4). These patterns include UP and DOWN states, gamma oscillations, hippocampal sharp wave–ripple complexes, and the theta rhythm. Both entorhinal cortex and the hippocampus have the capacity to intrinsically generate network oscillations at several different frequencies, as shown by the induction of such network oscillations in isolated slice preparations. A main challenge is to understand how such intrinsic network oscillations can interact with other local network oscillators and participate in global network states.

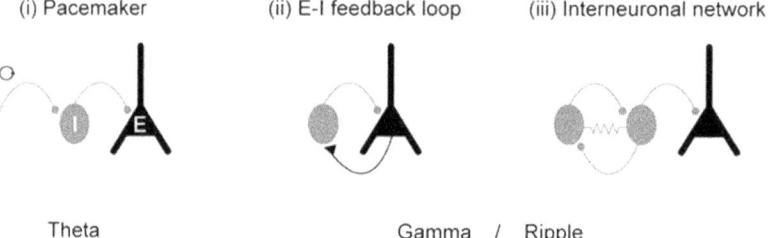

Figure 2.4 Synaptic circuits involved in the generation of inhibition-based networks oscillations. GABAergic interneurons could be synchronized within network oscillations by three principal mechanisms: (i) an external pacemaker, (ii) recurrent feedback loops between excitatory neurons and GABAergic interneurons, and (iii) synaptic and electrical coupling within an interneuronal network. Cortical oscillations within a given frequency band may involve more than a single mechanism, but here we suggest that entorhino-hippocampal theta-frequency oscillations are most likely driven by an external pacemaker, while higher-frequency oscillations may involve both recurrent feedback loops and interneuronal networks, with an increasing dependence on interneuronal networks as the frequency increases from the gamma- to ripple-frequency range.

Slow oscillations

During slow-wave sleep, neurons in the entorhinal cortex follow a similar behavior to that of neocortical neurons (Steriade et al., 1993), and engage in synchronous oscillations between periods of sustained activity (UP states) and relative quiescence (DOWN states) (Isomura et al., 2006). In the neocortex, such slow oscillations, in the ultraslow- to delta-frequency range (<4Hz), appear to be intrinsically generated through recurrent synaptic excitation (Sanchez-Vives and McCormick, 2000), with the UP state stabilized through a balanced increase in recurrent inhibition (Shu et al., 2003). Similar oscillations have been reported in isolated slices from the medial entorhinal cortex (Cunningham et al., 2006). While hippocampal pyramidal neurons do not seem to display a similar rhythmic bistability, they are not immune to the influence of this cortical activity. Slow oscillations in the superficial layers of the entorhinal cortex propagate to the distal dendritic regions of hippocampal pyramidal neurons (Wolansky et al., 2006), and are reflected in the UP and DOWN state transitions of the locally targeted interneurons (Hahn et al., 2006). Furthermore, entorhinal UP states are superposed by gamma-frequency oscillations, which then propagate to the dentate gyrus (Isomura et al., 2006), and UP states also influence the occurrence of hippocampal populations bursts, known as sharp wave–ripple complexes (Isomura et al., 2006; Molle et al., 2006). Therefore, while UP states might not be important in themselves for controlling the precision of spike timing, they both coalesce other faster rhythms and influence both the state of hippocampal neurons and those neurons that receive the hippocampal output, and could thus play an important role in setting the conditions for synaptic plasticity.

Gamma-frequency oscillations

Gamma-frequency oscillations (30–80Hz) occur in entorhino-hippocampal circuits throughout many sleeping and awake states, and are most strongly expressed superposed on theta-frequency oscillations and during cortical UP states (Bragin et al., 1995; Isomura et al., 2006). These fast oscillations are driven by two separate gamma generators in the superficial layers of the entorhinal cortex and hippocampal CA3, which entrain the dentate gyrus and CA1 respectively (Csicsvari et al., 2003). Rhythmogenesis within each of these gamma generators appears to depend on synaptic feedback loops between pyramidal neurons and perisomatic-targeting interneurons (Fisahn et al., 1998; Cunningham et al., 2003; Mann et al., 2005; Fuchs et al., 2007) with the oscillation propagated via feedforward inhibition (Csicsvari et al., 2003; Mann et al., 2005; Fuchs et al., 2007). However, these are not the only mechanisms involved, as connexin-36 knockout mice display a reduced power of gamma-frequency oscillations, both in vitro (Hormuzdi et al., 2001) and in vivo (Buhl et al., 2003).

Thus, synchronization appears to be sharpened by electrical coupling between interneurons (Traub, 2001), and may be further honed by chemical synaptic interactions between these GABAergic cells (Hajos et al., 2004; Oren et al., 2006). In fact, such interneuronal networks are capable of independently generating gamma-frequency oscillations in the absence of phasic excitatory input (Whittington et al., 1995; see review by Bartos et al., 2007), although it is not yet clear under what conditions these oscillations would emerge in vivo. Irrespective of the precise mechanism of generation, these fast network oscillations could control principal cell spike timing with a precision appropriate for STDP (Paulsen and Sejnowski, 2000; Hajos et al., 2004).

Sharp wave–ripple complexes

Sharp waves are a unique pattern of hippocampal activity that appears during slow-wave sleep and awake immobility (O'Keefe and Nadel, 1978; Buzsáki, 1986; O'Neill et al., 2006). During sleep, most ripples occur during cortical UP states (Battaglia et al., 2004; Isomura et al., 2006; Molle et al., 2006). These bursts of activity are generated by recurrent excitation in the CA3, propagate to CA1, and finally lead to the discharging of neurons in the deep layers of the entorhinal cortex (Chrobak and Buzsáki, 1996; Chrobak et al., 2000). The sharp waves in CA3 may be accompanied by variable ~100-Hz ripples (Ylinen et al., 1995; Csicsvari et al., 1999a), but the archetypal high-frequency ripple oscillations (150–250 Hz) are those that ride on sharp waves in CA1 (Buzsáki et al., 1992). These oscillations may be generated within interneuronal networks that receive a massive tonic drive, and escape from phasic excitation (Ylinen et al., 1995; Csicsvari et al., 1999a). Indeed, interneurons fire phase-locked to the ripple oscillations in vivo, and provide phasic inhibition to pyramidal neurons (Ylinen et al., 1995; Penttonen et al., 1998). It has also been shown that the frequency of ripple oscillations in vivo is sensitive to benzodiazepines, which modulate the kinetics of $GABA_A$-receptor-mediated inhibition (Ponomarenko et al., 2004). However, hippocampal networks can display very-high-frequency oscillations in the absence of GABAergic inhibition in vitro (Draguhn et al., 1998). It was suggested that pyramidal neurons could be synchronized through axo-axonal gap junction coupling (Draguhn et al., 1998) or ephaptic interactions, although it was later reported that similar spontaneous sharp wave–ripple complex-like events are sensitive to GABA antagonists (Maier et al., 2003; but see Nimmrich et al., 2005). It is possible that GABA-receptor-independent very-high-frequency oscillations represent pathological (seizure-related or inter-ictal "epileptic") events (Bragin et al., 2002; Le Van Quyen et al., 2006). While the precise mechanism of physiological ripple generation remains to be resolved, it has been shown that ripple-frequency oscillations are associated with the rapid replay of spike sequences previously observed during exploratory behavior (Skaggs and McNaughton, 1996; Nadasdy et al., 1999), and could therefore enable information stored in the hippocampus to be transferred to the neocortex.

Theta-frequency oscillations

During exploratory behavior and rapid eye movement (REM) sleep, the entorhino-hippocampal system engages in prominent theta-frequency (4–12 Hz) oscillations (Buzsáki, 2002). In contrast to faster network rhythms, it is commonly assumed that theta activity requires a subcortical pacemaker (Stewart and Fox, 1990). While many subcortical nuclei may contribute to pacing this rhythm, it is completely abolished by lesioning or inactivation of the medial septum/diagonal band of Broca (MSDB) (Green and Arduini, 1954). This nuclear complex emanates both cholinergic and GABAergic projections to hippocampus and entorhinal cortex (Lewis and Shute, 1967; Kohler et al., 1984; Jakab and Leranth, 1995). The termination pattern of the cholinergic projection is predominantly (~80%) non-synaptic (Descarries et al., 1997), relying on diffuse transmission that is likely to be most effective in slowly modulating neuronal excitability through metabotropic muscarinic acetylcholine receptors (mAChR). The smaller proportion of synaptic cholinergic terminals, which appear to selectively target GABAergic interneurons (Sudweeks and Yakel, 2000), appear more suited to transmitting synchronous activity through the activation of ionotropic nicotinic acetylcholine receptors (nAChR), but the balance of evidence suggests that the role of cholinergic transmission in theta-frequency oscillations is largely permissive or modulatory (Lee et al., 1994). The GABAergic projection from the MSDB selectively targets interneurons (Freund and Antal, 1988), and interneurons in turn provide the only feedback from the hippocampus back to the MSDB (Alonso and Kohler, 1982; Toth et al., 1993). This loop could play a more fundamental role in theta-frequency synchronization (Wang, 2002). Indeed, while the MSDB is required for theta-frequency oscillations in the entorhino-hippocampal system, and displays some inherent rhythmicity (Petsche et al., 1962), the actual rhythm generation may depend on reciprocal connections with the cortex and other subcortical nuclei, particularly the supramammilary nucleus, as well as the active properties of these target regions, with the MSDB providing a final "pacemaker" output (Fig. 2.4a) (Buzsáki, 2002).

The theta-frequency oscillations that consequently emerge within the hippocampus originate from two distinct sites, which can be pharmacologically dissociated. An atropine-sensitive, anesthetic-resistant form of the theta rhythm depends on local interneurons, which contribute to the rhythmic somatic inhibition observed in CA1 pyramidal cells (Soltesz and Deschenes, 1993). These interneurons are controlled by septal GABAergic afferents as well as local excitatory afferents from CA3 and CA1 (Buzsáki, 2002). The superficial layers of the entorhinal cortex are a second origin for hippocampal theta rhythm, whose emergence is resistant to muscarinic antagonists, but is sensitive to urethane and ketamine anesthesia, and may depend on the local activation of N-methyl-D-aspartate (NMDA) receptors (Buzsáki, 2002). This oscillation propagates, via feedforward excitation and/or inhibition, to the dentate gyrus and the distal dendrites of CA3 and CA1 pyramidal neurons (see Fig. 2.1). The complex interplay between multiple theta generators, diverse feedforward connections, and local feedback circuits makes it difficult to understand precisely how activity is routed through the entorhino-hippocampal system during theta activity, but the result is that the spiking of principal neurons throughout all hippocampal subfields is phase-coupled to the global theta rhythm. Furthermore, different interneuronal subtypes selectively couple to different phases of the theta-frequency oscillation (Klausberger *et al.*, 2003, 2004), providing a spatiotemporal pattern of inhibition that could be important in modulating dendritic integration, spike timing, and synaptic plasticity.

While global theta-frequency oscillations in vivo depend on subcortical structures, individual cortical neurons, as well as the local networks, appear tuned to participate in the theta-frequency rhythm. Subthreshold resonance in the theta-frequency range has been observed in many neuronal types, including CA1 pyramidal neurons, principal cells in the superficial layers of the entorhinal cortex, and distinct classes of hippocampal interneurons (Leung and Yu, 1998; Pike *et al.*, 2000; Hu *et al.*, 2002; Erchova *et al.*, 2004). Moreover, voltage-dependent theta-frequency oscillations have been observed in stellate cells in the entorhinal cortex (Alonso and Llinas, 1989), and in both the somata and dendrites of CA1 pyramidal neurons (Leung and Yim, 1991; Paulsen and Vida, 1996). These studies are consistent with the view that these cortical structures play an active role in generating the theta rhythm. However, although rhythmic activity in the theta-frequency range has been reported in hippocampal slices (Konopacki *et al.*, 1987, 1988; MacVicar and Tse, 1989; Gillies *et al.*, 2002), evidence suggests that at least some forms of such activity are more reminiscent of epileptiform activity (Williams and Kauer, 1997). In fact, reduced pyramidal cell subthreshold resonance in HCN1 knockout mice, is associated with an enhanced rather than reduced power of theta-frequency oscillations (Nolan *et al.*, 2004). It is possible that the ionic mechanisms revealed in these in vitro studies are more relevant to the control of spike timing within cycles of the oscillation. Indeed, ionic currents that activate/deactivate with time constants relevant for the theta rhythmicity, such as I_h and I_M, are likely to have powerful voltage-dependent effects on the coupling of synaptic input to the rate and phase of spike output, and can even enable phasic excitation to both delay and advance spike timing relative to inhibition-based oscillations (Lengyel et al., 2005). Such potential modulatory mechanisms are of particular interest, as hippocampal principal neurons do not lock their spikes slavishly to the theta-frequency oscillations, but demonstrate a phenomenon termed "phase precession" – these cells fire selectively in discrete regions of the animal's environment (their "place fields"), and as the animal passes through a neuron's "place field," it fires at progressively earlier phases of the theta-frequency oscillation (O'Keefe and Recce, 1993). Moreover, as would be expected from the tight coupling of pyramidal neurons and interneurons during oscillations (Csicsvari et al., 1999b), interneurons can also show such phase precession (Maurer et al., 2006; Ego-Stengel and Wilson, 2007), enabling a local network to escape from the global theta oscillation. It may be in the interaction between such a local network oscillator and the global theta activity that the most interesting computational phenomena arise, as this mechanism would offer the opportunity to control the local spike timing relative to that of the external afferents at the millisecond timescale, with important consequences for what type of information can be stored with an STDP rule. It was recently discovered that layer II cells of the medial entorhinal cortex, but not layer III cells, display phase precession (Fyhn et al., 2006). Since layer III cells are the origin of the direct input to the CA1, whereas layer II is the origin of the indirect input, the indirect input might contribute information encoded in phase, while the direct input maintains an oscillatory reference for phase encoding (Fig. 2.1), and the convergence of these pathways on individual CA1 pyramidal cells could lead to interesting spike timing-dependent interactions.

Functional implications

Neurons are sensitive to temporal correlations in their presynaptic inputs (Salinas and Sejnowski, 2000), and thus the synchronization of spiking within different brain oscillations is likely to have functional consequences for the propagation of activity through multilayered networks (Salinas and Sejnowski, 2001; Sejnowski and Paulsen, 2006). Network oscillations might also provide a mechanism for controlling the relative spike times both within and between neuronal assemblies. Evidence for such oscillatory control over the temporal order of activity in the entorhino-hippocampal system includes theta phase precession during exploratory behavior (O'Keefe and Recce, 1993), (increased) gamma-frequency synchronization between hippocampal and rhinal cortices during successful periods of learning (Fell et al., 2001), and the replay of spike sequences during sleep-related theta-frequency oscillations and sharp wave–ripples (Skaggs and McNaughton, 1996; Nadasdy et al., 1999; Louie and Wilson, 2001). The phenomenon of STDP provides a mechanism for storing relative spike times in cortical circuits, and thus offers a potential means by which network oscillations could play a role in learning and memory (Paulsen and Sejnowski, 2000). However, there is a limited window within which pre- and postsynaptic spiking leads to STDP, and the timescales of such spike correlations will vary with the oscillation frequency and the phase-coupling between neuronal populations. Therefore, while the empirical evidence for the functional role of brain rhythms remains tentative, it seems pertinent to ask which patterns of oscillatory activity could potentially support cellular learning within the confined windows of STDP.

STDP depends on postsynaptic action potentials back-propagating into the dendritic tree to provide the associative signal for input-specific modifications. The combined effect of pairing back-propagating action potentials with synaptic activation can lead to the relief of NMDA receptors from Mg^{2+} block and/or activate voltage-dependent Ca^{2+} channels, and thereby trigger a variety of dendritic Ca^{2+}-activated signaling cascades. The reliability of back-propagation, and its coupling to other signaling processes, are susceptible to many factors, and one would expect the precise windows of STDP to vary with cell type, the dendritic location of synaptic inputs (Sjostrom and Hausser, 2006), neuromodulatory state (Tsubokawa and Ross, 1997), and the degree of dendritic inhibition (Tsubokawa and Ross, 1996). The immediate history of pre- and postsynaptic activity is also likely to affect these interactions, and thus it may be difficult to extrapolate simple learning rules elucidated with single spike pairings to those that operate during realistic spike trains (Froemke and Dan, 2002; Dan and Poo, 2004). However, a relatively consistent result from experiments exploring STDP is that postsynaptic spiking must occur <20 ms after presynaptic activation to produce robust long-term increases in synaptic efficacy (see Fig. 2.5a(i)). This time window overlaps the period of ripple oscillations, and thus it appears that STDP mechanisms would be activated maximally during the replay of spike sequences during sharp wave–ripple complexes. However, these oscillations are only synchronized within small populations of pyramidal neurons, and it is not clear whether both pre- and postsynaptic spiking are temporally coordinated. Indeed, SDTP has been discussed most commonly in relation to plasticity within gamma- and theta-frequency oscillations.

The correspondence between the windows for STDP and the period of gamma-frequency oscillations has evoked much interest, but also introduces some problems. Within local recurrent networks, spike sequences across gamma cycles would occur at the weak extremes of the STDP window, with the magnitude of plasticity dependent on the precise frequency of the oscillation in gamma range (Fig. 2.5a(i, ii)). In contrast, spike time variability within cycles of the gamma-frequency oscillations would produce forward and reverse pairings in the

(a) Compatibility of STDP and gamma-frequency oscillations

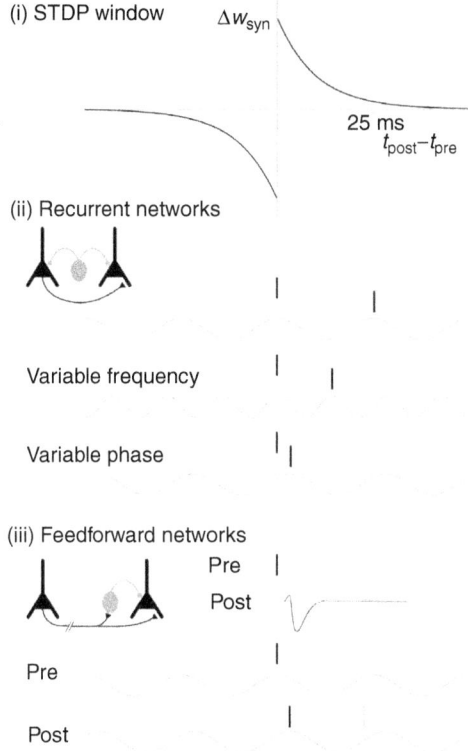

(b) Sequence compression within theta oscillations

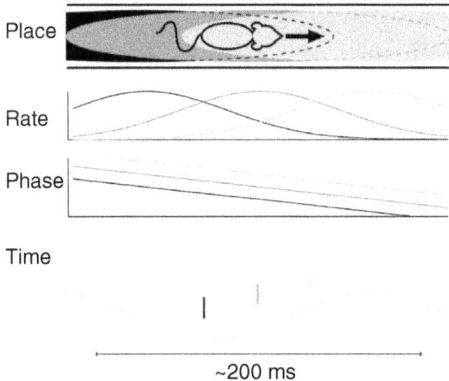

Figure 2.5 Possible links between network oscillations and spike timing-dependent plasticity. (a)(i) An example of an asymmetric spike timing-dependent plasticity window, where the timing between the pre- and postsynaptic spike ($t_{post} - t_{pre}$) determines the amplitude and direction of the change in synaptic weight (w_{syn}). It is commonly found that postsynaptic spiking must occur <20ms after presynaptic activation to produce robust long-term increases in synaptic efficacy, although the

most powerful region of the STDP window (see Fig. 2.5a(i)), and even if ultimately balanced, could interfere with encoding. This problem could be its own solution if spike time variability represented phase precession or spike sequences within gamma cycles (Axmacher *et al.*, 2006), rather than noisy jitter, but gamma phase encoding has yet to be reported. The relevance of STDP might be easier to interpret for connections between networks. For example, while there is evidence that the gamma rhythm itself might be propagated via feedforward inhibition (Csicsvari *et al.*, 2003; Mann *et al.*, 2005), causing the majority of neurons to spike only after inhibition has decayed (Csicsvari *et al.*, 2003; Mann and Paulsen, 2005), strongly activated neurons could spike in response to the initial feedforward excitation and thus fall within the window for spike timing-dependent potentiation (Fig. 2.5a(iii)). Altering the phase coupling between networks oscillating at gamma frequencies would also provide an effective means of controlling learning through STDP, but we do not yet understand the mechanisms that control phase coupling itself. Therefore, while the links between

Caption for Figure 2.5 (cont.)

window for long-term depression may be longer. (ii) It is not immediately obvious how STDP windows observed experimentally would operate during gamma-frequency oscillations. In recurrent networks, where pyramidal neurons receive synchronized inhibition, spike sequences across gamma cycles would occur at the weak extremes of the STDP window (upper trace). One would also expect that the efficacy of learning across cycles would be dependent on the frequency of the oscillation in the gamma-frequency band (middle trace), and be most powerful if information was encoded by the phase of pyramidal neuron spiking *within* gamma-frequency oscillations. (iii) If gamma-frequency oscillations propagate in feedforward networks, then the postsynaptic neuron would experience feedforward excitation rapidly curtailed by di-synaptic inhibition. In this scenario, weakly activated neurons may only fire after inhibition has decayed, and thus outside the STDP window for potentiation, while strongly activated neurons may spike on the initial feedforward excitation, and thus produce pre- and postsynaptic spike times within the window for spike timing-dependent potentiation. (b) Slower theta-frequency oscillations could enable sequences at behavioral timescales to be encoded by spike sequences at timescales relevant to STDP. Many principal neurons in the entorhinal cortex and hippocampus fire selectively in discrete regions of the animal's environment (their "place fields"). During exploration these networks also display theta-frequency oscillations, and as the animal passes through a neuron's place field, the first spike in each theta cycle occurs at a progressively earlier phase of the oscillation. Therefore, one would expect the trajectory of the animal through a region represented by overlapping place fields to be encoded as spike sequences within an individual theta cycle.

gamma-frequency oscillations and STDP windows remain tantalizing, a more detailed understanding of how both processes operate in vivo is required to determine their compatibility for organized learning.

At first glance, the synchronization of spikes within slower theta-frequency oscillations seems even less appropriate for cellular learning based on STDP. However, the reported phenomenon of phase precession suggests that cells with overlapping place fields along an animal's trajectory could fire at progressively earlier phases of the theta oscillation, and thus represent this path as a sequence of spikes within a single theta cycle (Skaggs et al., 1996) (Fig. 2.5b). The size of place fields recorded in different regions of the entorhinal cortex and hippocampus can vary from a typical 20–25 cm to over 8 m on an 18-m linear track (Jung et al., 1994; Kjelstrup et al., 2006; McNaughton et al., 2006), and so theta-frequency oscillations could enable behavioral sequences over a range of timescales to be compressed into spike orders detectable by STDP. Indeed, experience-dependent asymmetric expansion of place fields (Mehta et al., 1997, 2000), which requires NMDA receptors (Ekstrom et al., 2001), has been suggested to represent the phase-precession-driven enhancement of synaptic weights through an STDP-like process. Spiking during theta-frequency oscillations tends to be nested within gamma rhythms, which could complicate the conditions required for plasticity, but if phase encoding is robust, it provides a generally attractive mechanism for Hebbian learning.

Conclusion

Network oscillations observed during different behaviors clearly reflect which neuronal populations are active, and how they communicate with each other. It remains unclear whether this rhythmic coordination of spiking activity has an independent functional role. However, the intrinsic and synaptic properties of neurons seem tuned to embrace network rhythmicity, and it remains an important goal to understand why this is.

References

Alonso, A. and Kohler, C. (1982). Evidence for separate projections of hippocampal pyramidal and non-pyramidal neurons to different parts of the septum in the rat brain. *Neurosci Lett* **31**:209–214.

Alonso, A. and Llinas, R. R. (1989). Subthreshold Na^+-dependent theta-like rhythmicity in stellate cells of entorhinal cortex layer II. *Nature* **342**:175–177.

Axmacher, N., Mormann, F., Fernandez, G., Elger, C. E., and Fell, J. (2006). Memory formation by neuronal synchronization. *Brain Res Brain Res Rev* **52**:170–182.

Bacci, A. and Huguenard, J. R. (2006). Enhancement of spike-timing precision by autaptic transmission in neocortical inhibitory interneurons. *Neuron* **49**:119–130.

Bal, T. and McCormick, D. A. (1993). Mechanisms of oscillatory activity in guinea-pig nucleus reticularis thalami in vitro: a mammalian pacemaker. *J Physiol* **468**:669–691.

Bal, T. and McCormick, D. A. (1997). Synchronized oscillations in the inferior olive are controlled by the hyperpolarization-activated cation current I(h). *J Neurophysiol* **77**:3145–3156.

Bal, T., von Krosigk, M., and McCormick, D. A. (1995). Synaptic and membrane mechanisms underlying synchronized oscillations in the ferret lateral geniculate nucleus in vitro. *J Physiol* **483**:641–663.

Bartos, M., Vida, I., and Jonas, P. (2007). Synaptic mechanisms of synchronized gamma oscillations in inhibitory interneuron networks. *Nat Rev Neurosci* **8**:45–56.

Battaglia, F. P., Sutherland, G. R., and McNaughton, B. L. (2004). Hippocampal sharp wave bursts coincide with neocortical "up-state" transitions. *Learn Mem* **11**:697–704.

Beierlein, M., Gibson, J. R., and Connors, B. W. (2000). A network of electrically coupled interneurons drives synchronized inhibition in neocortex. *Nat Neurosci* **3**:904–910.

Bi, G. Q. and Poo, M. M. (1998). Synaptic modifications in cultured hippocampal neurons: dependence on spike timing, synaptic strength, and postsynaptic cell type. *J Neurosci* **18**:10464–10472.

Bi, G. and Poo, M. (2001). Synaptic modification by correlated activity: Hebb's postulate revisited. *Annu Rev Neurosci* **24**:139–166.

Blatow, M., Rozov, A., Katona, I., et al. (2003). A novel network of multipolar bursting interneurons generates theta frequency oscillations in neocortex. *Neuron* **38**:805–817.

Bragin, A., Mody, I., Wilson, C. L., and Engel, J., Jr. (2002). Local generation of fast ripples in epileptic brain. *J Neurosci* **22**:2012–2021.

Bragin, A., Jando, G., Nadasdy, Z., et al. (1995). Gamma (40–100 Hz) oscillation in the hippocampus of the behaving rat. *J Neurosci* **15**:47–60.

Brunel, N. and Wang, X. J. (2003). What determines the frequency of fast network oscillations with irregular neural discharges? I. Synaptic dynamics and excitation–inhibition balance. *J Neurophysiol* **90**:415–430.

Bruzzone, R., Hormuzdi, S. G., Barbe, M. T., Herb, A., and Monyer, H. (2003). Pannexins, a family of gap junction proteins expressed in brain. *Proc Natl Acad Sci USA* **100**:13644–13649.

Buhl, D. L., Harris, K. D., Hormuzdi, S. G., Monyer, H., and Buzsáki, G. (2003). Selective impairment of hippocampal gamma oscillations in connexin-36 knock-out mouse in vivo. *J Neurosci* **23**:1013–1018.

Buzsáki, G. (1986). Hippocampal sharp waves: their origin and significance. *Brain Res* **398**:242–252.

Buzsáki, G. (2002). Theta oscillations in the hippocampus. *Neuron* **33**:325–340.

Buzsáki, G. and Draguhn, A. (2004). Neuronal oscillations in cortical networks. *Science* **304**:1926–1929.

Buzsáki, G., Horvath, Z., Urioste, R., Hetke, J., and Wise, K. (1992). High-frequency network oscillation in the hippocampus. *Science* **256**:1025–1027.

Chrobak, J. J. and Buzsáki, G. (1996). High-frequency oscillations in the output networks of the hippocampal–entorhinal axis of the freely behaving rat. *J Neurosci* **16**:3056–3066.

Chrobak, J. J., Lorincz, A., and Buzsáki, G. (2000). Physiological patterns in the hippocampoentorhinal cortex system. *Hippocampus* **10**:457–465.

Cobb, S. R., Buhl, E. H., Halasy, K., Paulsen, O., and Somogyi, P. (1995). Synchronization of neuronal activity in hippocampus by individual GABAergic interneurons. *Nature* **378**:75–78.

Connors, B. W. and Long, M. A. (2004). Electrical synapses in the mammalian brain. *Annu Rev Neurosci* **27**:393–418.

Contreras, D. (2006). The role of T-channels in the generation of thalamocortical rhythms. *CNS Neurol Disord Drug Targets* **5**:571–585.

Csicsvari, J., Hirase, H., Czurko, A., Mamiya, A., and Buzsáki, G. (1999a). Fast network oscillations in the hippocampal CA1 region of the behaving rat. *J Neurosci* **19**:RC20.

Csicsvari, J., Hirase, H., Czurko, A., Mamiya, A., and Buzsáki, G. (1999b). Oscillatory coupling of hippocampal pyramidal cells and interneurons in the behaving rat. *J Neurosci* **19**:274–287.

Csicsvari, J., Jamieson, B., Wise, K. D., and Buzsáki, G. (2003). Mechanisms of gamma oscillations in the hippocampus of the behaving rat. *Neuron* **37**:311–322.

Cunningham, M. O., Davies, C. H., Buhl, E. H., Kopell, N., and Whittington, M. A. (2003). Gamma oscillations induced by kainate receptor activation in the entorhinal cortex in vitro. *J Neurosci* **23**:9761–9769.

Cunningham, M. O., Pervouchine, D. D., Racca, C., et al. (2006). Neuronal metabolism governs cortical network response state. *Proc Natl Acad Sci USA* **103**:5597–5601.

Dan, Y. and Poo, M. M. (2004). Spike timing-dependent plasticity of neural circuits. *Neuron* **44**:23–30.

De Zeeuw, C. I., Chorev, E., Devor, A., et al. (2003). Deformation of network connectivity in the inferior olive of connexin 36-deficient mice is compensated by morphological and electrophysiological changes at the single neuron level. *J Neurosci* **23**:4700–4711.

Deans, M. R., Gibson, J. R., Sellitto, C., Connors, B. W., and Paul, D. L. (2001). Synchronous activity of inhibitory networks in neocortex requires electrical synapses containing connexin-36. *Neuron* **31**:477–485.

Descarries, L., Gisiger, V., and Steriade, M. (1997). Diffuse transmission by acetylcholine in the CNS. *Prog Neurobiol* **53**:603–625.

Dickinson, P. S. (2006). Neuromodulation of central pattern generators in invertebrates and vertebrates. *Curr Opin Neurobiol* **16**:604–614.

Dickson, C. T., Magistretti, J., Shalinsky, M. H., et al. (2000). Properties and role of I(h) in the pacing of subthreshold oscillations in entorhinal cortex layer II neurons. *J Neurophysiol* **83**:2562–2579.

Dorval, A. D., Jr. and White, J. A. (2005). Channel noise is essential for perithreshold oscillations in entorhinal stellate neurons. *J Neurosci* **25**:10025–10028.

Draguhn, A., Traub, R. D., Schmitz, D., and Jefferys, J. G. (1998). Electrical coupling underlies high-frequency oscillations in the hippocampus in vitro. *Nature* **394**:189–192.

Ego-Stengel, V. and Wilson, M. A. (2007). Spatial selectivity and theta phase precession in CA1 interneurons. *Hippocampus* **17**:161–174.

Ekstrom, A. D., Meltzer, J., McNaughton, B. L., and Barnes, C. A. (2001). NMDA receptor antagonism blocks experience-dependent expansion of hippocampal "place fields." *Neuron* **31**:631–638.

Elson, R. C., Selverston, A. I., Abarbanel, H. D., and Rabinovich, M. I. (2002). Inhibitory synchronization of bursting in biological neurons: dependence on synaptic time constant. *J Neurophysiol* **88**:1166–1176.

Erchova, I., Kreck, G., Heinemann, U., and Herz, A. V. (2004). Dynamics of rat entorhinal cortex layer II and III cells: characteristics of membrane potential resonance at rest predict oscillation properties near threshold. *J Physiol* **560**:89–110.

Fell, J., Klaver, P., Lehnertz, K., *et al.* (2001). Human memory formation is accompanied by rhinal–hippocampal coupling and decoupling. *Nat Neurosci* **4**:1259–1264.

Fisahn, A., Pike, F. G., Buhl, E. H., and Paulsen, O. (1998). Cholinergic induction of network oscillations at 40 Hz in the hippocampus in vitro. *Nature* **394**:186–189.

Freeman, W. J. (1968). Relations between unit activity and evoked potentials in prepyriform cortex of cats. *J Neurophysiol* **31**:337–348.

Freund, T. F. and Antal, M. (1988). GABA-containing neurons in the septum control inhibitory interneurons in the hippocampus. *Nature* **336**:170–173.

Freund, T. F. and Buzsáki, G. (1996). Interneurons of the hippocampus. *Hippocampus* **6**:347–470.

Fricker, D. and Miles, R. (2000). EPSP amplification and the precision of spike timing in hippocampal neurons. *Neuron* **28**:559–569.

Froemke, R. C. and Dan, Y. (2002). Spike-timing-dependent synaptic modification induced by natural spike trains. *Nature* **416**:433–438.

Fuchs, E. C., Zivkovic, A. R., Cunningham, M. O., *et al.* (2007). Recruitment of parvalbumin-positive interneurons determines hippocampal function and associated behavior. *Neuron* **53**:591–604.

Fyhn, M., Hafting, T., Moser, M. B., and Moser, E. I. (2006). Theta modulation and phase precession in grid cells in the medial entorhinal cortex. *FENS Abstr* **3**:A197.130.

Galarreta, M. and Hestrin, S. (1999). A network of fast-spiking cells in the neocortex connected by electrical synapses. *Nature* **402**:72–75.

Gibson, J. R., Beierlein, M., and Connors, B. W. (1999). Two networks of electrically coupled inhibitory neurons in neocortex. *Nature* **402**:75–79.

Gillies, M. J., Traub, R. D., LeBeau, F. E., *et al.* (2002). A model of atropine-resistant theta oscillations in rat hippocampal area CA1. *J Physiol* **543**:779–793.

Green, J. D. and Arduini, A. A. (1954). Hippocampal electrical activity in arousal. *J Neurophysiol* **17**:533–557.

Grillner, S. (2006). Biological pattern generation: the cellular and computational logic of networks in motion. *Neuron* **52**:751–766.

Hahn, T. T., Sakmann, B., and Mehta, M. R. (2006). Phase-locking of hippocampal interneurons' membrane potential to neocortical up–down states. *Nat Neurosci* **9**:1359–1361.

Hajos, N., Palhalmi, J., Mann, E. O., et al. (2004). Spike timing of distinct types of GABAergic interneuron during hippocampal gamma oscillations in vitro. *J Neurosci* **24**:9127–9137.

Hasenstaub, A., Shu, Y., Haider, B., et al. (2005). Inhibitory postsynaptic potentials carry synchronized frequency information in active cortical networks. *Neuron* **47**:423–435.

Hebb, D. O. (1949). *The Organization of Behavior*. New York: John Wiley.

Hormuzdi, S. G., Pais, I., LeBeau, F. E., et al. (2001). Impaired electrical signaling disrupts gamma frequency oscillations in connexin 36-deficient mice. *Neuron* **31**:487–495.

Hu, H., Vervaeke, K., and Storm, J. F. (2002). Two forms of electrical resonance at theta frequencies, generated by M-current, h-current and persistent Na^+ current in rat hippocampal pyramidal cells. *J Physiol* **545**:783–805.

Hutcheon, B. and Yarom, Y. (2000). Resonance, oscillation and the intrinsic frequency preferences of neurons. *Trends Neurosci* **23**:216–222.

Isomura, Y., Sirota, A., Ozen, S., et al. (2006). Integration and segregation of activity in entorhinal–hippocampal subregions by neocortical slow oscillations. *Neuron* **52**:871–882.

Jakab, R. L. and Leranth, C. (1995). Septum. In: *The Rat Nervous System*, 2nd edn, ed. Paxinos, G., pp. 405–442. San Diego, CA: Academic Press.

Jung, M. W., Wiener, S. I., and McNaughton, B. L. (1994). Comparison of spatial firing characteristics of units in dorsal and ventral hippocampus of the rat. *J Neurosci* **14**:7347–7356.

Kiehn, O. (2006). Locomotor circuits in the mammalian spinal cord. *Annu Rev Neurosci* **29**:279–306.

Kjelstrup, K. B., Solstad, T., Brun, V. H., et al. (2006). Spatial scale expansion along the dorsal-to-ventral axis of hippocampal area CA3 in the rat. *FENS Abstr* **3**:A197.133.

Klausberger, T., Magill, P. J., Marton, L. F., et al. (2003). Brain-state- and cell-type-specific firing of hippocampal interneurons in vivo. *Nature* **421**:844–848.

Klausberger, T., Marton, L. F., Baude, A., et al. (2004). Spike timing of dendrite-targeting bistratified cells during hippocampal network oscillations in vivo. *Nat Neurosci* **7**:41–47.

Kohler, C., Chan-Palay, V., and Wu, J. Y. (1984). Septal neurons containing glutamic acid decarboxylase immunoreactivity project to the hippocampal region in the rat brain. *Anat Embryol (Berl)* **169**:41–44.

Konopacki, J., Maciver, M. B., Bland, B. H., and Roth, S. H. (1987). Theta in hippocampal slices: relation to synaptic responses of dentate neurons. *Brain Res Bull* **18**:25–27.

Konopacki, J., Bland, B. H., and Roth, S. H. (1988). Carbachol-induced EEG 'theta' in hippocampal formation slices: evidence for a third generator of theta in CA3c area. *Brain Res* **451**:33–42.

Lampl, I. and Yarom, Y. (1997). Subthreshold oscillations and resonant behavior: two manifestations of the same mechanism. *Neuroscience* **78**:325–341.

Landisman, C. E., Long, M. A., Beierlein, M., et al. (2002). Electrical synapses in the thalamic reticular nucleus. *J Neurosci* **22**:1002–1009.

Lavenex, P. and Amaral, D. G. (2000). Hippocampal–neocortical interaction: a hierarchy of associativity. *Hippocampus* **10**:420–430.

Le Van Quyen, M., Khalilov, I., and Ben-Ari, Y. (2006). The dark side of high-frequency oscillations in the developing brain. *Trends Neurosci* **29**:419–427.

Lee, M. G., Chrobak, J. J., Sik, A., Wiley, R. G., and Buzsáki, G. (1994). Hippocampal theta activity following selective lesion of the septal cholinergic system. *Neuroscience* **62**:1033–1047.

Lengyel, M., Kwag, J., Paulsen, O., and Dayan, P. (2005). Matching storage and recall: hippocampal spike timing-dependent plasticity and phase response curves. *Nat Neurosci* **8**:1677–1683.

Leung, L. S. and Yu, H. W. (1998). Theta-frequency resonance in hippocampal CA1 neurons in vitro demonstrated by sinusoidal current injection. *J Neurophysiol* **79**:1592–1596.

Leung, L. W. and Yim, C. Y. (1991). Intrinsic membrane potential oscillations in hippocampal neurons in vitro. *Brain Res* **553**:261–274.

Lewis, P. R. and Shute, C. C. (1967). The cholinergic limbic system: projections to hippocampal formation, medial cortex, nuclei of the ascending cholinergic reticular system, and the subfornical organ and supra-optic crest. *Brain* **90**:521–540.

Llinas, R. R. (1988). The intrinsic electrophysiological properties of mammalian neurons: insights into central nervous system function. *Science* **242**:1654–1664.

Llinas, R. R. and Steriade, M. (2006). Bursting of thalamic neurons and states of vigilance. *J Neurophysiol* **95**:3297–3308.

Llinas, R. R. and Yarom, Y. (1981). Electrophysiology of mammalian inferior olivary neurones in vitro: different types of voltage-dependent ionic conductances. *J Physiol* **315**:549–567.

Llinas, R. R., Baker, R., and Sotelo, C. (1974). Electrotonic coupling between neurons in cat inferior olive. *J Neurophysiol* **37**:560–571.

Long, M. A., Deans, M. R., Paul, D. L., and Connors, B. W. (2002). Rhythmicity without synchrony in the electrically uncoupled inferior olive. *J Neurosci* **22**:10898–10905.

Long, M. A., Landisman, C. E., and Connors, B. W. (2004). Small clusters of electrically coupled neurons generate synchronous rhythms in the thalamic reticular nucleus. *J Neurosci* **24**:341–349.

Louie, K. and Wilson, M. A. (2001). Temporally structured replay of awake hippocampal ensemble activity during rapid eye movement sleep. *Neuron* **29**:145–156.

MacVicar, B. A. and Tse, F. W. (1989). Local neuronal circuitry underlying cholinergic rhythmical slow activity in CA3 area of rat hippocampal slices. *J Physiol* **417**:197–212.

Maier, N., Nimmrich, V., and Draguhn, A. (2003). Cellular and network mechanisms underlying spontaneous sharp wave–ripple complexes in mouse hippocampal slices. *J Physiol* **550**:873–887.

Mann, E. O. and Paulsen, O. (2005). Mechanisms underlying gamma ("40 Hz") network oscillations in the hippocampus: a mini-review. *Prog Biophys Mol Biol* **87**:67–76.

Mann, E. O. and Paulsen, O. (2006). Keeping inhibition timely. *Neuron* **49**:8–9.

Mann, E.O., Suckling, J.M., Hajos, N., Greenfield, S.A., and Paulsen, O. (2005). Perisomatic feedback inhibition underlies cholinergically induced fast network oscillations in the rat hippocampus in vitro. *Neuron* **45**:105–117.

Markram, H., Lubke, J., Frotscher, M., and Sakmann, B. (1997). Regulation of synaptic efficacy by coincidence of postsynaptic APs and EPSPs. *Science* **275**:213–215.

Maurer, A.P., Cowen, S.L., Burke, S.N., Barnes, C.A. and McNaughton, B.L. (2006). Phase precession in hippocampal interneurons showing strong functional coupling to individual pyramidal cells. *J Neurosci* **26**:13485–13492.

Maxeiner, S., Kruger, O., Schilling, K., et al. (2003). Spatiotemporal transcription of connexin45 during brain development results in neuronal expression in adult mice. *Neuroscience* **119**:689–700.

McNaughton, B.L., Battaglia, F.P., Jensen, O., Moser, E.I., and Moser, M.B. (2006). Path integration and the neural basis of the "cognitive map." *Nat Rev Neurosci* **7**:663–678.

Mehta, M.R., Barnes, C.A., and McNaughton, B.L. (1997). Experience-dependent, asymmetric expansion of hippocampal place fields. *Proc Natl Acad Sci USA* **94**:8918–8921.

Mehta, M.R., Quirk, M.C. and Wilson, M.A. (2000). Experience-dependent asymmetric shape of hippocampal receptive fields. *Neuron* **25**:707–715.

Menendez de la Prida, L.M., Huberfeld, G., Cohen, I., and Miles, R. (2006). Threshold behavior in the initiation of hippocampal population bursts. *Neuron* **49**:131–142.

Meyer, A.H., Katona, I., Blatow, M., Rozov, A., and Monyer, H. (2002). In vivo labeling of parvalbumin-positive interneurons and analysis of electrical coupling in identified neurons. *J Neurosci* **22**:7055–7064.

Molle, M., Yeshenko, O., Marshall, L., Sara, S.J., and Born, J. (2006). Hippocampal sharp wave-ripples linked to slow oscillations in rat slow-wave sleep. *J Neurophysiol* **96**:62–70.

Morris, R.G., Moser, E.I., Riedel, G., et al. (2003). Elements of a neurobiological theory of the hippocampus: the role of activity-dependent synaptic plasticity in memory. *Phil Trans R Soc Lond B* **358**:773–786.

Nadasdy, Z., Hirase, H., Czurko, A., Csicsvari, J., and Buzsáki, G. (1999). Replay and time compression of recurring spike sequences in the hippocampus. *J Neurosci* **19**:9497–9507.

Nimmrich, V., Maier, N., Schmitz, D., and Draguhn, A. (2005). Induced sharp wave-ripple complexes in the absence of synaptic inhibition in mouse hippocampal slices. *J Physiol* **563**:663–670.

Nolan, M.F., Malleret, G., Dudman, J.T., et al. (2004). A behavioral role for dendritic integration: HCN1 channels constrain spatial memory and plasticity at inputs to distal dendrites of CA1 pyramidal neurons. *Cell* **119**:719–732.

O'Keefe, J. and Nadel, L. (1978). *The Hippocampus as a Cognitive Map*. Oxford, UK: Oxford University Press.

O'Keefe, J. and Recce, M.L. (1993). Phase relationship between hippocampal place units and the EEG theta rhythm. *Hippocampus* **3**:317–330.

O'Neill, J., Senior, T., and Csicsvari, J. (2006). Place-selective firing of CA1 pyramidal cells during sharp wave/ripple network patterns in exploratory behavior. *Neuron* **49**:143–155.

Oren, I., Mann, E.O., Paulsen, O., and Hajos, N. (2006). Synaptic currents in anatomically identified CA3 neurons during hippocampal gamma oscillations in vitro. *J Neurosci* **26**:9923–9934.

Paulsen, O. and Sejnowski, T.J. (2000). Natural patterns of activity and long-term synaptic plasticity. *Curr Opin Neurobiol* **10**:172–179.

Paulsen, O. and Vida, I. (1996). Sustained dendritic oscillations at theta frequencies elicited in CA1 pyramidal cells in rat hippocampal slices. *J Physiol* **495**:P50–P51.

Penttonen, M., Kamondi, A., Acsady, L., and Buzsáki, G. (1998). Gamma frequency oscillation in the hippocampus of the rat: intracellular analysis in vivo. *Eur J Neurosci* **10**:718–728.

Petsche, H., Stumpf, C., and Gogolak, G. (1962). The significance of the rabbit's septum as a relay station between the midbrain and the hippocampus. I. The control of hippocampus arousal activity by the septum cells. *Electroencephalogr Clin Neurophysiol* **14**:202–211.

Pike, F.G., Goddard, R.S., Suckling, J.M., et al. (2000). Distinct frequency preferences of different types of rat hippocampal neurones in response to oscillatory input currents. *J Physiol* **529**:205–213.

Ponomarenko, A.A., Korotkova, T.M., Sergeeva, O.A., and Haas, H.L. (2004). Multiple $GABA_A$ receptor subtypes regulate hippocampal ripple oscillations. *Eur J Neurosci* **20**:2141–2148.

Puil, E., Gimbarzevsky, B., and Miura, R.M. (1986). Quantification of membrane properties of trigeminal root ganglion neurons in guinea pigs. *J Neurophysiol* **55**:995–1016.

Rotstein, H.G., Oppermann, T., White, J.A., and Kopell, N. (2006). The dynamic structure underlying subthreshold oscillatory activity and the onset of spikes in a model of medial entorhinal cortex stellate cells. *J Comput Neurosci* **21**:271–292.

Salinas, E. and Sejnowski, T.J. (2000). Impact of correlated synaptic input on output firing rate and variability in simple neuronal models. *J Neurosci* **20**:6193–6209.

Salinas, E. and Sejnowski, T.J. (2001). Correlated neuronal activity and the flow of neural information. *Nat Rev Neurosci* **2**:539–550.

Sanchez-Vives, M.V. and McCormick, D.A. (2000). Cellular and network mechanisms of rhythmic recurrent activity in neocortex. *Nat Neurosci* **3**:1027–1034.

Schmitz, D., Schuchmann, S., Fisahn, A., et al. (2001). Axo-axonal coupling: a novel mechanism for ultrafast neuronal communication. *Neuron* **31**:831–840.

Sejnowski, T.J. and Paulsen, O. (2006). Network oscillations: emerging computational principles. *J Neurosci* **26**:1673–1676.

Shu, Y., Hasenstaub, A., and McCormick, D.A. (2003). Turning on and off recurrent balanced cortical activity. *Nature* **423**:288–293.

Sjostrom, P.J. and Hausser, M. (2006). A cooperative switch determines the sign of synaptic plasticity in distal dendrites of neocortical pyramidal neurons. *Neuron* **51**:227–238.

Skaggs, W. E. and McNaughton, B. L. (1996). Replay of neuronal firing sequences in rat hippocampus during sleep following spatial experience. *Science* **271**:1870–1873.

Skaggs, W. E., McNaughton, B. L., Wilson, M. A., and Barnes, C. A. (1996). Theta phase precession in hippocampal neuronal populations and the compression of temporal sequences. *Hippocampus* **6**:149–172.

Soltesz, I. and Deschenes, M. (1993). Low- and high-frequency membrane potential oscillations during theta activity in CA1 and CA3 pyramidal neurons of the rat hippocampus under ketamine-xylazine anesthesia. *J Neurophysiol* **70**:97–116.

Somogyi, P., Tamas, G., Lujan, R., and Buhl, E. H. (1998). Salient features of synaptic organization in the cerebral cortex. *Brain Res Brain Res Rev* **26**:113–135.

Song, S., Miller, K. D., and Abbott, L. F. (2000). Competitive Hebbian learning through spiketiming-dependent synaptic plasticity. *Nat Neurosci* **3**:919–926.

Squire, L. R., Stark, C. E., and Clark, R. E. (2004). The medial temporal lobe. *Annu Rev Neurosci* **27**:279–306.

Srinivas, M., Rozental, R., Kojima, T., *et al.* (1999). Functional properties of channels formed by the neuronal gap junction protein connexin36. *J Neurosci* **19**:9848–9855.

Steriade, M. (2006). Grouping of brain rhythms in corticothalamic systems. *Neuroscience* **137**:1087–1106.

Steriade, M., Nunez, A., and Amzica, F. (1993). A novel slow (<1 Hz) oscillation of neocortical neurons in vivo: depolarizing and hyperpolarizing components. *J Neurosci* **13**:3252–3265.

Stewart, M. and Fox, S. E. (1990). Do septal neurons pace the hippocampal theta rhythm? *Trends Neurosci* **13**:163–168.

Sudweeks, S. N. and Yakel, J. L. (2000). Functional and molecular characterization of neuronal nicotinic ACh receptors in rat CA1 hippocampal neurons. *J Physiol* **527**:515–528.

Toth, K., Borhegyi, Z., and Freund, T. F. (1993). Postsynaptic targets of GABAergic hippocampal neurons in the medial septum-diagonal band of broca complex. *J Neurosci* **13**:3712–3724.

Traub, R. D. and Wong, R. K. (1982). Cellular mechanism of neuronal synchronization in epilepsy. *Science* **216**:745–747.

Traub, R. D., Bibbig, A., Fisahn, A., *et al.* (2000). A model of gamma-frequency network oscillations induced in the rat CA3 region by carbachol in vitro. *Eur J Neurosci* **12**:4093–4106.

Traub, R. D., Pais, I., Bibbig, A., *et al.* (2003). Contrasting roles of axonal (pyramidal cell) and dendritic (interneuron) electrical coupling in the generation of neuronal network oscillations. *Proc Natl Acad Sci USA* **100**:1370–1374.

Tsubokawa, H. and Ross, W. N. (1996). IPSPs modulate spike backpropagation and associated $[Ca^{2+}]_i$ changes in the dendrites of hippocampal CA1 pyramidal neurons. *J Neurophysiol* **76**:2896–2906.

Tsubokawa, H. and Ross, W. N. (1997). Muscarinic modulation of spike backpropagation in the apical dendrites of hippocampal CA1 pyramidal neurons. *J Neurosci* **17**:5782–5791.

Venance, L., Rozov, A., Blatow, M., *et al.* (2000). Connexin expression in electrically coupled postnatal rat brain neurons. *Proc Natl Acad Sci USA* **97**:10260–10265.

Vida, I., Bartos, M., and Jonas, P. (2006). Shunting inhibition improves robustness of gamma oscillations in hippocampal interneuron networks by homogenizing firing rates. *Neuron* **49**:107–117.

von Krosigk, M., Bal, T., and McCormick, D.A. (1993). Cellular mechanisms of a synchronized oscillation in the thalamus. *Science* **261**:361–364.

Wang, X.J. (2002). Pacemaker neurons for the theta rhythm and their synchronization in the septohippocampal reciprocal loop. *J Neurophysiol* **87**:889–900.

Whittington, M.A. and Traub, R.D. (2003). Interneuron diversity series: inhibitory interneurons and network oscillations in vitro. *Trends Neurosci* **26**:676–682.

Whittington, M.A., Traub, R.D., and Jefferys, J.G. (1995). Synchronized oscillations in interneuron networks driven by metabotropic glutamate receptor activation. *Nature* **373**:612–615.

Williams, J.H. and Kauer, J.A. (1997). Properties of carbachol-induced oscillatory activity in rat hippocampus. *J Neurophysiol* **78**:2631–2640.

Witter, M.P., Wouterlood, F.G., Naber, P.A. and Van Haeften, T. (2000). Anatomical organization of the parahippocampal–hippocampal network. *Ann N Y Acad Sci* **911**:1–24.

Wolansky, T., Clement, E.A., Peters, S.R., Palczak, M.A., and Dickson, C.T. (2006). Hippocampal slow oscillation: a novel EEG state and its coordination with ongoing neocortical activity. *J Neurosci* **26**:6213–6229.

Ylinen, A., Bragin, A., Nadasdy, Z., *et al.* (1995). Sharp wave associated high-frequency oscillation (200Hz) in the intact hippocampus: network and intracellular mechanisms. *J Neurosci* **15**:30–46.

3

Cell assemblies and serial computation in neural circuits

KENNETH D. HARRIS

Introduction

Analogies between the brain and the digital computer have been out of fashion for a long time. The differences between brains and computers are numerous. Computers run pre-specified programs written by people. Computers store programs and data in specialized RAM circuits, and have one CPU (or at most a handful) which follows a coded list of instructions to the letter. A computer has a central clock, which allows all of its components to march through a program in lockstep. The brain, on the other hand, has billions of neurons operating in parallel, no central clock, no externally supplied list of instructions, and no separation of RAM and CPU. Although the inventors of the modern computer held the brain as a model, the analogy is rarely taken seriously today.

A more popular analogy for the brain in recent years has been artificial neural networks (ANNs). ANNs, although typically simulated on a digital computer, have an apparently more "brain-like" design. They consist of elements that function (at least a bit) like neurons, connected by "synapses" whose strength can be modified by the network's history. ANNs do not need an external program, but "learn" from a set of training examples. The most successful of these, the multilayer perceptron or "backprop" net, is good enough at generalizing from training examples to be used in real-world information processing tasks, by people who have no interest in how the brain works. The backprop net, like the brain, is a massively parallel device, with no separation of CPU and RAM.

Information Processing by Neuronal Populations, ed. Christian Hölscher and Matthias Munk.
Published by Cambridge University Press. © Cambridge University Press 2009.

But unlike the brain, information in this network flows strictly forward from input to output, which renders it incapable of serial computation. Each time an input is presented, it causes activity to propagate forwards through the neurons of the network, until it reaches the output neurons. The next time an input comes along, the memory of the last is forgotten. The backprop net, one could say, is a purely parallel system, incapable of serial computation.

This chapter will discuss whether, and how, the brain might perform something like serial computation. Introspection suggests that it does. We are all aware of our "train of thought," which at least feels like it proceeds sequentially from one step to the next. And everybody knows that mental problems take longer the harder they are. But do these high-level examples from humans reflect a universal principle of brain operation that applies also to simpler tasks, and other animals? In this chapter we will review classical theory of brain function, the "cell assembly" hypothesis of Donald Hebb, in which the performance of serial computations is central. We consider the theory's predictions at the level of spike trains, propose four experimental signatures of cell assembly organization, and review recent experimental results in the light of these proposed signatures.

The cell assembly hypothesis

The cell assembly hypothesis is best understood in a historical context. In the early twentieth century, the behaviorist paradigm was dominant in psychology. Behaviorist analysis had two taboos: "introspection," contemplation of ones own thoughts; and "neuralizing," the development of psychological theories based on hypothesized (and uncertain) principles of brain operation. Behaviorist methods were applied with great success to conditioned reflexes and stimulus–response learning. However, it was also widely accepted that conditioned reflexes were not a complete description of animal behavior, and that the brain held an "internal state" that could also influence behavior (Tolman, 1932; Skinner, 1938). The restrictions of the behaviorist paradigm made study of internal state difficult. Hebb's theory was radical because it broke the second taboo, and conjectured how the dynamics of cortical circuits could subserve the evolution of internal state, and consequent performance of behaviors beyond stimulus–response association. Below, we shall refer to psychological processes that are dependent on internal state as internal cognitive processes, or ICPs.

The cell assembly hypothesis proposes that the key to ICPs lies in the recurrent nature of neuronal circuits. The theory is based on the premise, now known as Hebb's rule, that synaptic connections become strengthened by synchronous activity of pre- and postsynaptic neurons. However, this was only the starting point of a much broader framework. According to this, repeated co-activation of a group of neurons during behavior will lead to the formation of a cell assembly – an anatomically dispersed set of neurons amongst which excitatory connections have been potentiated. Mutual excitation allows the assembly to later maintain its activity, without requiring continuous sensory stimulation. Consequently, the activity of assemblies can become decoupled from external events, and be initiated by internal factors such as the activity of other assemblies. A chain of assemblies, each one triggered by the last, was termed a *phase sequence* (note that "phase" in this context does not necessarily connote timing with respect to an oscillation). The phase sequence allows for complex computations, which are only partially controlled by external input, and is the proposed substrate of ICPs (see Fig. 3.1, Box 3.1). An important feature of this theory is that the same assembly might be triggered by either sensory or internal factors. Consequently, a single neuron might participate in both sensory representation and ICPs, and it is expected that neurons even in primary sensory cortices will play a role in ICPs.

It is important to distinguish the concept of a phase sequence from that of a temporal code. In a temporal coding scheme, a local neural population uses precise patterns of spike times to transmit a discrete item of information to downstream targets. The downstream targets must then decode this temporal

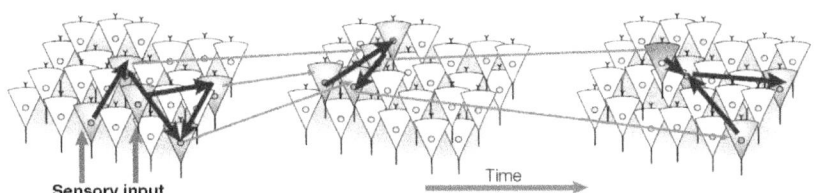

Figure 3.1 The cell assembly hypothesis. Sensory input (left) drives certain neurons to fire. Thereafter, activity evolves due to intrinsic cortical dynamics. Strengthened recurrent connections between members of a single assembly (black arrows) transiently stabilize assembly firing through mutual excitation. As the excitability of this assembly fades, inter-assembly connections (gray arrows) lead to subsequent activation of a new assembly. The resulting "phase sequence" evolves through network dynamics, and is not strictly determined by the time series of sensory inputs. The evolution of this phase sequence is the hypothesized substrate of internal cognitive processes (see also Box 3.1).

Box 3.1 Computational network models

The cell assembly hypothesis and associated Hebbian plasticity rule have inspired a number of computational network models. One of the most prominent of these is the auto-associative or "Hopfield" network (Marr, 1971; Gardner-Medwin, 1976; Hopfield, 1982; Amit, 1994). In learning mode, these networks store externally presented patterns through the modification of recurrent excitatory synapses, according to a Hebbian rule. In recall mode, the state of the network evolves from an externally supplied starting point through intrinsic dynamics, ending in a stable state ("attractor") where its activity matches one of the originally learned inputs. This network therefore exhibits the hypothesized behavior of self-sustaining assembly activity through reverberant excitation. Theoretical studies and simulations have allowed for further characterization of network properties such as the maximum number of assemblies possibly expressed in a single network (Marr, 1971; Gardner-Medwin, 1976; Hopfield, 1982; Amit, 1994).

It is also instructive to consider the limitations of the Hopfield model as an illustration of the cell assembly hypothesis. Most importantly, once the Hopfield network converges to an attractor, it stays there. A phase sequence, on the other hand, is proposed to consist of a series of semi-stable states, where each assembly is active only transiently before the next takes over. The occurrence of phase sequences would therefore have to rely on features of neural systems not captured in this simplified model. For example, in a biological network, once an assembly becomes active a variety of factors such as neuronal adaptation, inhibition, synaptic depression and channel inactivation might affect its constituent neurons, progressively dampening the excitability of the assembly until it is no longer capable of sustaining its own activity, and so facilitating the transition to a new assembly. A further consideration is that the assumption of symmetric connectivity in the Hopfield model, which implies stability of network attractor states, is not likely to hold in reality. Recently, more complex networks have been described whose activity more closely resembles phase sequence dynamics (Wallenstein and Hasselmo, 1997; Izhikevich et al., 2004).

pattern in order to extract the information. In an assembly organization, sequences of assemblies will occur; however, the fundamental currency of information processing is the firing of a single assembly, not the sequence. The progression of assemblies in the phase sequence represents successive steps in a serial computation carried out by a larger network including both the local population and its targets, and there is no need for postsynaptic neurons to "decode" the phase sequence.

Signatures of cell assembly organization

We now propose four expected experimental signatures of cell assembly organization. The relation of these signatures to recent experimental results will be discussed in the next section.

> **Signature 1.** Spike trains show temporal structure that is not present in the stimulus. If presentation of a stimulus initiates a phase sequence, the spike times of participating neurons will reflect the temporal structure of the phase sequence as well as that of the sensory stimulus. Neurons will therefore exhibit structured temporal patterns, even for temporally unstructured stimuli. Furthermore, if the particular phase sequence that is initiated depends on the stimulus, the observed temporal structure will vary across stimuli.
>
> **Signature 2.** Spiking is not strictly controlled by sensory input. The evolution of the phase sequence is only partially influenced by external stimuli. Therefore, the phase sequence following a stimulus will depend on internal factors prior to stimulus presentation, as well as the nature of the stimulus. Spike trains will therefore appear variable, even across repeated presentations of an identical stimulus. In psychological language, the nature and timing of cognitive processes will vary from trail to trial, even the presented stimulus is the same. If we could observe the underlying ICPs, we could better predict when a given cell will fire. Because we cannot, trial-to-trial variability of ICPs will be reflected in apparent variability of recorded spike trains. Note that spike trains would still appear unpredictable even if the underlying phase sequence dynamics were completely deterministic.
>
> **Signature 3.** Apparently unpredictable spikes will be coordinated to reveal an assembly organization. If apparent unpredictability of spike trains indeed reflects participation of assemblies in ICPs, trial-to-trial fluctuations in spike trains will be coordinated at the population level. It is important to distinguish coordination due to ICPs from coordination due to sensory

input. For example, in a recording of visually responsive neurons, which are stimulated by a strobe flashed every second, synchronous activity at 1Hz would be expected due to simultaneous sensory modulation. However, if neuronal populations are involved in processes beyond stimulus representation, they will be coordinated beyond the prediction of common stimulus modulation. This can be formalized as a question of conditional independence (Box 3.2). If, given the stimulus, any further fluctuations about the stimulus-driven mean response were independent between neurons, we could assume that these fluctuations were noise superimposed on a reliable stimulus representation. The cell assembly hypothesis posits the opposite: that stimulus representation is not the only role of neural populations, and conditional independence will be violated.

Box 3.2 Conditional independence and the virtual population

In probability theory, the concept of independence has more than one definition. Two events A and B are said to be *marginally independent* if the probability for them both to occur is the product of their individual probabilities:

$(P): P(A \wedge B) = P(A)P(B).$

Alternatively, dependence may exist between two variables A and B, but be explained by common dependence on a third variable C. In this case, A and B are said to be *conditionally independent*, given C. In this case, it is the conditional probabilities that multiply:

$P(A \wedge B|C) = P(A|C)P(B|C).$

It is possible for A and B to be conditionally independent given C, but not marginally independent. For example, suppose randomly selected stimuli were simultaneously presented to two animals, and a sensory responsive neuron was recorded from each animal. Because no causal influence exists from one brain to the other, the response of two neurons recorded in the two brains will be independent, for any given stimulus presentation (conditional independence). Nevertheless, because both animals experience the same stimuli, an overall correlation (that is, marginal dependence) will still be observed, due to the common effect of the stimulus.

Box 3.2 (cont.)

When only single-neuron recordings are available, one can collect the responses of sequentially recorded neurons to a stimulus, and treat them as if they were recorded simultaneously. This technique is widely used, and we shall term the data that are obtained this way a "virtual population." If the virtual population were indistinguishable from a true population, it would be fortunate, as simultaneous recordings would be unnecessary. This question is not just of technical significance: the equivalence of the real and virtual populations might be considered equivalent to the concept of independent coding (deCharms and Zador, 2000), or as a restatement of the single neuron doctrine (Barlow, 1972).

Virtual populations are often used in stimulus-reconstruction studies, and it is frequently possible to recover the presented stimulus from a virtual population, sometimes with better accuracy than from a true population (Furukawa et al., 2000; Ghazanfar et al., 2000; Oram et al., 2001). However, this does not imply that the real and virtual populations are equivalent. In particular, the responses of neurons in a virtual population will be conditionally independent given the stimulus. By contrast, in a true population conditional independence is often violated (Vaadia et al., 1995; Riehle et al., 1997; Harris et al., 2003).

While a given neuron might participate in many assemblies, not every possible combination of neurons comprises a cell assembly. The latter feature should be reflected by preferences in the probability with which a neuron joins its various peers in synchronous firing. This can only be tested with simultaneous recordings of large populations (Fig. 3.2), and will also require novel analysis methods that are capable of exploiting this data.

Signature 4. Patterns of assembly activity should correlate with ongoing ICPs. If assembly organization reflects performance of ICPs, the nature of ICPs should be reflected in the observed pattern of population coordination. Empirical verification of this prediction is challenging, because ICPs are by definition unobservable. Nevertheless, by careful control of the stimulus and behavioral task, it is possible to design experiments where one can be reasonably confident that certain cognitive events occur at specific times. We will review evidence for modulation of population activity patterns by presumed occurrence of attention, expectancy, novelty, and binding.

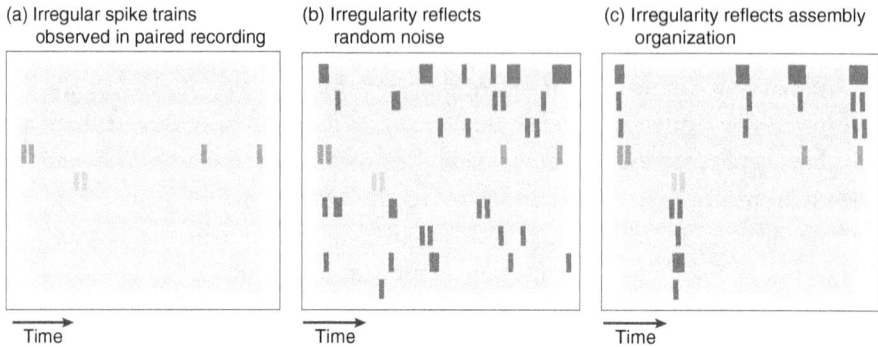

Figure 3.2 Analysis of assembly structure requires simultaneous recording of large populations. (a) Hypothetical simultaneous recording of two spike trains, which show an irregular temporal structure. The significance of this irregularity cannot be determined without the context that is provided by more simultaneously recorded cells. (b), (c) Hypothetical activity of a larger population containing the two original cells. Two possibilities are illustrated: (b) Spike train irregularity reflects random noise. (c) Apparent irregularity reflects a non-random grouping of neurons into cell assemblies.

Signature 1: spike train temporal structure

Spike times can follow the temporal structure of a stimulus in a wide range of sensory systems, often with a remarkable degree of precision (Gabbiani et al., 1996; Buracas et al., 1998). However, it has also been observed that spike timing can correlate with stimulus qualities other than temporal structure. For example, presentation of different visual patterns can lead to responses with different temporal profiles, as well as different spike counts (Optican and Richmond, 1987; Hegde and Van Essen, 2004). Similar results have been obtained in auditory (Middlebrooks et al., 1994), somatosensory (Panzeri et al., 2001), gustatory (Katz et al., 2001; Di Lorenzo and Victor, 2003), and olfactory (Laurent, 1996) modalities, in a wide range of species. In rat hippocampus, the timing of spikes relative to the ongoing theta rhythm correlates with the animal's location in space (O'Keefe and Recce, 1993; Skaggs et al., 1996; Harris et al., 2002; Mehta et al., 2002; Huxter et al., 2003).

These observations have been interpreted as evidence for temporal coding, whereby spike time patterns, in addition to firing rates, serve to communicate the stimulus to downstream neurons. Here we propose an alternative interpretation. The presentation of a stimulus initiates a phase sequence, which evolves in time through the dynamics of the cortical network. Thus, spike trains show temporal structure that is not present in the stimulus itself. Furthermore, if the specific phase sequence that is evoked depends on stimulus parameters, the temporal structure of a given neuron relative to stimulus onset will be correlated with these parameters. However, downstream neurons are not required to "decode" the phase sequence in order to determine which stimulus was presented; instead the sequence of assemblies reflects a chain of internal events that are triggered by the stimulus.

This idea is illustrated by studies of a much simpler recurrent network, the antennal lobe of the insect olfactory system. Presentation of odor stimuli evokes a complex sequence of spike patterns, which can continue evolving even after stimulus offset (Stopfer et al., 2003). Different odors evoke different sequences, and pharmacological disruption of these pattern sequences alters firing in downstream structures (MacLeod et al., 1998), and impairs fine behavioral discrimination of odors (Stopfer et al., 1997). Nevertheless, neurons that are immediately postsynaptic to this structure do not appear to detect temporal sequences of afferent spikes, but simply fire when sufficient coincident input occurs within a certain time window (Perez-Orive et al., 2004). Similar complex, stimulus-dependent, temporal responses have been observed in the mammalian cortex (Optican and Richmond, 1987; Hegde and Van Essen, 2004), and we suggest that they too result from dynamical evolution in recurrent networks.

In the hippocampus the situation is different because it is the timing of spikes relative to the ongoing theta oscillation, rather than relative to an external stimulus presentation, that is correlated with the animal's location in the environment. Does the phase of spikes with respect to the theta oscillation form a temporal code for the animal's location? If this were the case, then when location is constant, phase should either be constant, or completely random. This was investigated with spike trains that were recorded during a non-spatial behavior (wheel-running), as well as spatial behaviors (maze-running) and rapid eye movement (REM) sleep (Harris et al., 2002). In all behaviors, firing rate and phase fluctuated from moment to moment. Importantly, phase fluctuations showed the same relationship to rate fluctuations during spatial and non-spatial behaviors (Fig. 3.3), exhibiting a phase advance during periods of increased activity, even when head location was constant. The similarity of phase dynamics in spatial and non-spatial behaviors suggests that theta phase is not an explicit code for space, but is just one manifestation of a more fundamental principle that underlies the processing of both spatial and non-spatial information. We suggest that this principle is related to the evolution of assembly sequences during the theta cycle, and can only be properly characterized at the population level (see below).

A number of studies, in systems including behaving animals and brain slices, have suggested that precisely repeating temporal patterns of spikes can occur, with millisecond precision (Abeles et al., 1993; Prut et al., 1998; Abeles and Gat, 2001; Mao et al., 2001; Cossart et al., 2003; Ikegaya et al., 2004; MacLean et al.,

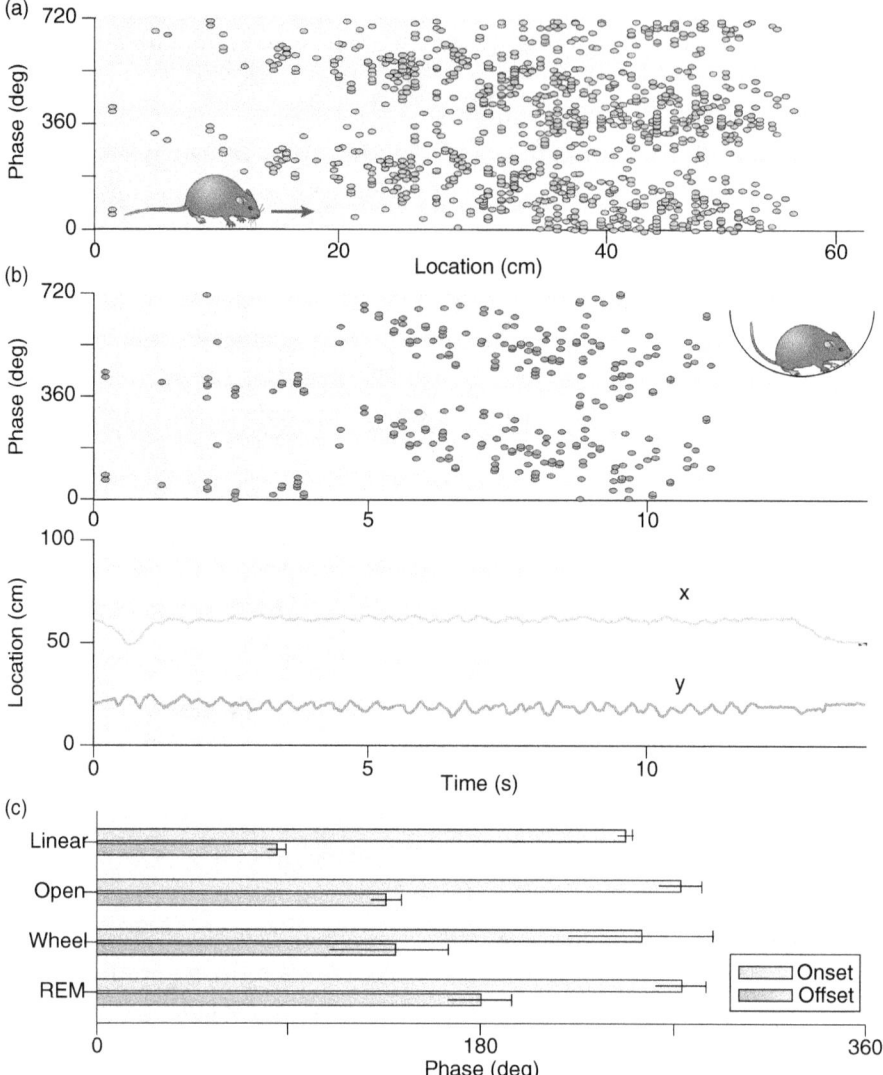

Figure 3.3 Spike timing of hippocampal neurons. (a) In a track-running task, the timing of spikes of a hippocampal pyramidal neuron is correlated with an external variable, the location of the animal's head in space. Each dot shows the phase of an action potential of a single pyramidal cell with respect to the theta rhythm, accumulated over many traversals of the place field. The phase axis is doubled to avoid introducing an arbitrary cut-off point. (b) During a wheel-running episode, a transient increase in firing rate of a CA1 pyramidal cell is accompanied by phase advance similar to that seen in spatial behaviors, even though the rat's head location does not change (lower trace). (c) The mean spike phase at onset and offset of firing periods for linear track-running (linear), open-field exploration (open), wheel-running (wheel), and REM sleep (REM). Phase advancement during periods of strong firing occurred in all behaviors that were studied. (Modified from Harris et al., 2002.)

2005). This would, at first, appear to be strong support for the occurrence of phase sequences. However, these studies have been controversial (Oram *et al.*, 1999, 2001; Luczak *et al.*, 2007; Mokeichev *et al.*, 2007). Part of the reason for this controversy might be the extraordinary nature of the finding itself: that after a delay of several seconds, a single spike can be produced by a cell timed with millisecond accuracy, even though the same neuron may have fired multiple, untimed, spikes in the intervening period. What physiological mechanism could be capable of producing such accurate timing is unclear. Furthermore, this result appears to differ from studies on behavioral timing, in which timing errors increase proportionally to the interval to be timed (*scalar timing*) (Gibbon *et al.*, 1997). A second reason for the controversy of these results has been the complex statistical methods necessary to detect such repeating sequences, and the difficulty of framing a null hypothesis in which there are no repeating sequences, but that is not overly simplistic and already known to be incorrect (such as independent Poisson process). Recently, studies have suggested that the detection of these sequences may be explained by the stereotyped latencies of individual neurons to punctuate events such as behavioral motion onsets (Oram *et al.*, 1999, 2001; Baker and Lemon, 2000), or spontaneous cortical UP states (Luczak *et al.*, 2007). Furthermore, at least in the case of UP states, the timing accuracy of the sequence does indeed appear to decay in a scalar manner, rather than maintaining millisecond precision after delays of seconds or more. Together, these observations indicate that sequential activity does indeed occur in neural circuits, and that sequences can be triggered reliably by punctuate events that may (like motion onsets) or may not (like UP states) be correlated with behavior. Furthermore, the temporal accuracy of these sequences is not greater than one might reasonably expect from neural circuits.

Signature 2: spike train variability

The variable nature of cortical spike trains has been widely discussed. In peripheral receptors, the strength of a stimulus is encoded in the firing rate of a regularly spaced spike train (Adrian and Zotterman, 1926). In cortical structures, however, spike trains are more irregular than in the periphery, and repeated presentation of identical stimuli can lead to responses that vary across presentations (Britten *et al.*, 1993; Softky and Koch, 1993; Shadlen and Newsome, 1998; Stevens and Zador, 1998; Bair *et al.*, 2001; Mazurek *et al.*, 2002). Several measures exist for quantifying the irregularity and variability of spike trains, including the interspike interval coefficient of variation (ISI CV; ratio of standard deviation of ISIs to mean ISI) (Softky and Koch, 1993), and Fano factor (ratio of spike count variance to mean spike count in a given time window) (Teich, 1989; Baddeley *et al.*, 1997). Both measures produce dimensionless numbers, which are larger for more variable spike trains. The variability of spike trains is often compared to that of a Poisson process (on which both measures take the value 1). However, nothing about neuronal biophysics suggests that neurons should emit a Poisson process; it is simply a mathematically convenient "yardstick" against which spike train variability can be measured.

Spike train variability appears to correlate with location in the processing hierarchy. For example, studies in the primary sensory cortex and the thalamus have indicated that these cells can produce spike trains that are less variable than a Poisson process, under correct stimulus conditions (Kara et al., 2000; Deweese et al., 2003; Wehr and Zador, 2003; Deweese and Zador, 2004). By contrast, spike trains in the hippocampus are typically more variable than a Poisson process (Fenton and Muller, 1998), with ISI CV typically around 3 for a wheel-running task (K. D. Harris, H. Hirase, and G. Buzsáki, unpublished observations).

Could this variability constitute "noise"? Temporally structured current injections lead to spike times reliably locked to stimulus transients in vitro, which suggests that neurons are not fundamentally stochastic devices (Mainen and Sejnowski, 1995; Fellous et al., 2004). A second possibility is that irregularity of a neuron's output spike train reflects irregularity of its inputs. Certain model neurons, when driven by balanced excitatory and inhibitory inputs of Poisson variability, can produce irregular outputs (Gerstein and Mandelbrot, 1964; Shadlen and Newsome, 1998). In vitro experiments have not so far been able to confirm this (Stevens and Zador, 1998; Chance et al., 2002); however, even if it were the case, this would only push the question back a stage: if variable outputs reflect variable inputs, what causes these? Recordings of the primary sensory receptors indicate that they are often highly reliable (Berry et al., 1997).

A third possibility is that the apparent variability of spike trains reflects incomplete control of the sensory environment. For example, trial-to-trial variability of neurons in the middle temporal (MT) area in response to moving-dot animations is reduced if a precisely repeated, rather than stochastic motion signal is presented (Bair and Koch, 1996; Bair et al., 2001). One might therefore conclude that some of the apparent variability results from trial-to-trial variability in the precise trajectories of individual dots, rather than a noisy representation of the mean motion vector. Could apparent variability always result from incompletely controlled stimuli? This seems to be unlikely: even if all sensory stimuli were controlled perfectly, one could not control the animal's ICPs. If a neuron participates in ICPs, its spike train would therefore still appear unpredictable, however well the sensory environment is controlled.

In vitro experiments have studied the characteristics of injected currents that are required to produce in-vivo-like output irregularity. Summed injection of independent Poisson inputs rarely produced supra-Poisson outputs, even when excitation and inhibition are balanced (Stevens and Zador, 1998; Chance et al., 2002). To produce irregular outputs, the input instead needed to show correlated fluctuations, which consisted of sporadic periods of synchronous input lasting for approximately 30 milliseconds (Stevens and Zador, 1998). Interestingly, this is precisely the type of input that would be expected if the presynaptic population was organized into assemblies that were synchronized at this timescale.

Signature 3: population organization

If apparently unpredictable spike timing arises from participation of assemblies in ICPs, this will be reflected by synchronous firing of neurons, beyond what is expected from common modulation by external input. We recently introduced a "peer prediction" method which aims to test this possibility (Harris et al., 2003). For each neuron, the spike train is first predicted from the external variable that it is presumed to represent. If the function of the recorded population were simply to represent this external variable, this would be the best prediction that could be made. Alternatively, if neurons are organized into assemblies and their firing is only partially determined by external factors, it should be possible to better predict when a neuron will fire given the spike times of simultaneously recorded assembly members.

This analysis was applied to populations of hippocampal CA1 pyramidal cells in rats that were performing a spatial exploration task. Figure 3.4 shows the results for a representative neuron. Prediction of firing rate from the rat's trajectory accurately captures the time-dependence of the mean firing rate, which rises and falls as the rat enters and leaves the place field (black trace, in Fig. 3.4c). Adding prediction from the simultaneously recorded population results in a more "spiky" function, which reflects a higher temporal precision of predicted firing probability. Quantitative assessment of these predictions indicates that apparently stochastic spike times are indeed better predicted by this short-scale structure, which indicates that the apparent variability might instead reflect an assembly organization that is visible only at the population level.

Could correlated assembly firing arise from simultaneous phase coding for spatial location in multiple cells? If this were the case, neuronal synchronization would arise from the dependence of the mean theta phase of each cell on position, but fluctuations in timing around this mean would be random and independent among cells. To test this possibility, the prediction of spike trains from position was refined by incorporating a position-dependent theta modulation (Harris et al., 2003). This refinement improved on prediction from place alone, as expected. However, peer prediction further improved on the refined spatial prediction, which indicates that neurons show coordinated activity

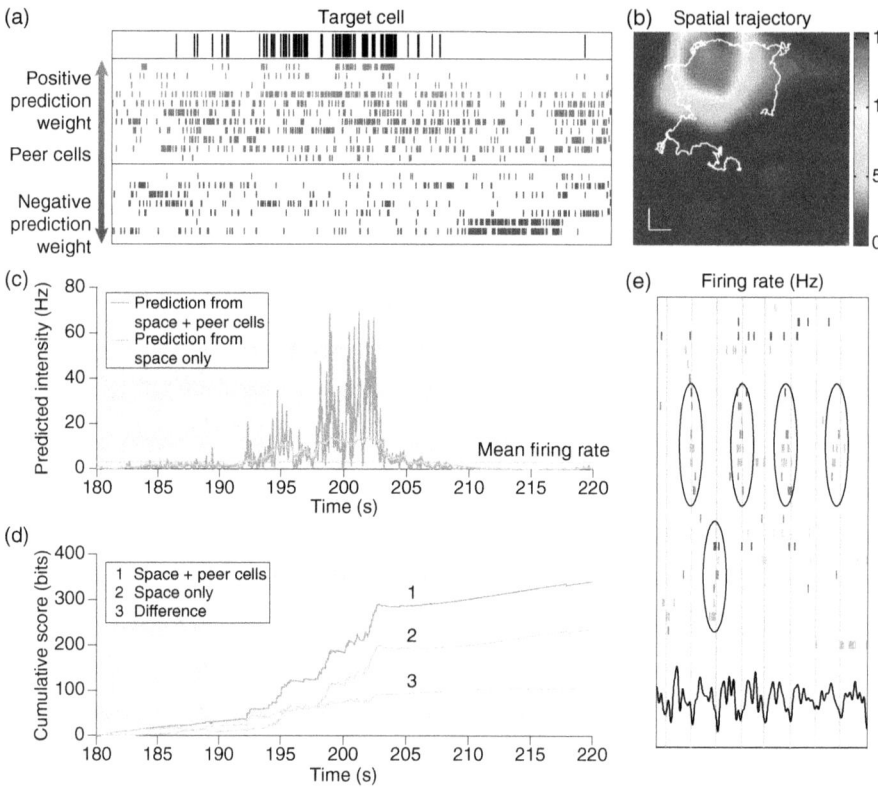

Figure 3.4 Peer prediction analysis of assembly organization in the hippocampus. (a) Activity of a "target cell" (black, top), and a population of simultaneously recorded ("peer") pyramidal cells (below). Each peer cell is assigned a prediction weight, with activity of positively or negatively weighted cells predicting increased or decreased probability of synchronous target-cell spikes. (b) The target cell's place field and animal's trajectory (white trace). Scale bar: 10 cm. (c) Target cell firing probability that is predicted from the animal's position (black), or from position and peer activity (gray). (d) Prediction quality is quantified by assessing the fit of the observed spike train against the prediction. (e) Example of one second of simultaneously recorded spike-train data. Spike rasters were arranged vertically by stochastic search to highlight putative assembly memberships (circled). (Modified from Harris et al., 2003.)

beyond what is predicted by simultaneous phase precession, and that the phase–space correlation might be only one manifestation of a more fundamental mechanism determining exact spike times. Indeed, close examination shows that spike trains are not typically characterized by a single discrete spike cluster per theta cycle, at a phase that regularly advances as the animal crosses the place field. Instead, irregular patterns are observed, with some cycles skipped altogether. However, this apparent irregularity is in fact coordinated across the population, which reflects an organization of neurons into synchronously firing groups (Fig. 3.4e; cf. Fig. 3.2c).

Could apparent assembly organization result from common modulation by an uncontrolled, non-spatial external sensory input? Firing rates of hippocampal neurons are known to correlate with external variables such as odors (Wood et al., 2000). However, these sensory variables typically change on timescales of seconds, much slower than the observed population coordination. Furthermore, to explain the relationship of assembly timing to the theta rhythm, the uncontrolled external events would themselves have to be coordinated with the theta rhythm, which seems to be extremely unlikely.

It is important to distinguish the peer prediction method from the commonly applied stimulus-reconstruction paradigm (Oram et al., 1998). The peer prediction method tests for conditional independence (Box 3.2) by predicting individual spike trains from the stimulus, and by determining whether this prediction can be further improved by predicting from peer activity. In the stimulus reconstruction paradigm, the direction of prediction is reversed, and population activity is used to predict the sensory stimulus that is presented to the animal (Fig. 3.5). Although this "decoding" paradigm can help clarify the relationship of neuronal activity to sensory input, it cannot determine the structure of assembly activity beyond what is caused by common stimulus modulation. Indeed, trial shuffling, which destroys all non-stimulus-locked correlations, does not typically impair and can even improve decoding performance (Furukawa et al., 2000; Ghazanfar et al., 2000; Oram et al., 2001). At first this might seem to imply that the real and shuffled populations are indistinguishable. However, these analyses do not imply that non-stimulus-locked correlations do not exist, or that they serve no purpose in information processing; the original and trial-shuffled populations are only equivalent if the function of the population is solely to represent the assumed sensory stimulus. Participation of neuronal assemblies in ICPs would result in correlations that would not survive shuffling, but still reflect an important aspect of information processing. However, the stimulus reconstruction method would not detect these, as they could only lower the quality of reconstruction of a sensory stimulus.

Signature 4: relation to ICPs

Our final postulated signature is that patterns of assembly activity should correlate with the performance of ICPs. Experimental assessment of this prediction is challenging, because ICPs are by definition unobservable. Nevertheless, by careful task design one can sometimes infer that certain cognitive processes are likely to occur at prescribed moments. In several sensory

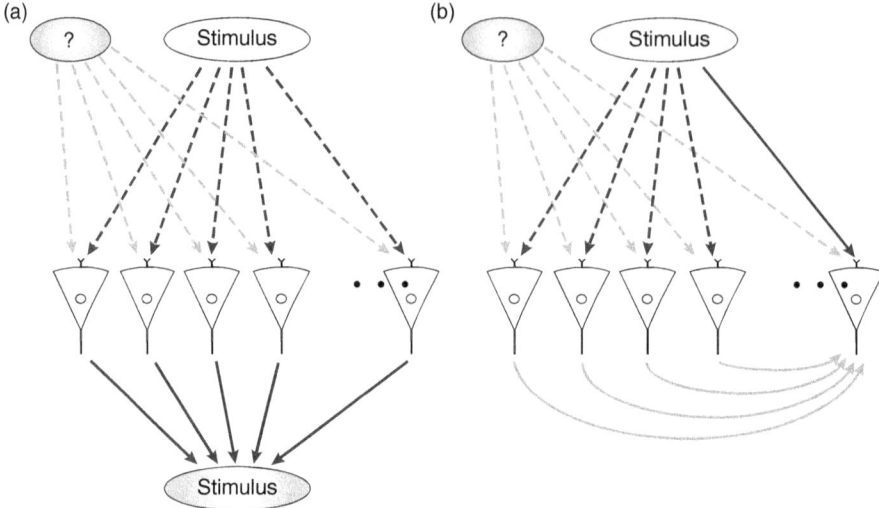

Figure 3.5 Stimulus reconstruction and peer prediction paradigms. "Stimulus" indicates predicted variable, solid lines indicate predictions, and dashed lines indicate other correlations. (a) Stimulus-reconstruction paradigm. The firing of a population of neurons (lower row) correlates with a stimulus (correlation indicated by gray dotted lines). The activity of the population is used to predict or "reconstruct" the stimulus (prediction indicated by solid lines). If these neurons also correlate with additional unobserved variables, such as internal cognitive processes (question mark), this cannot be detected by stimulus reconstruction. (b) Peer prediction paradigm. The activity of a target cell (right neuron) is predicted from the stimulus (indicated by black solid line). If the prediction of the target cell's activity can be improved using the activity of simultaneously recorded neurons (prediction indicated by gray solid lines), the population is not conditionally independent given the stimulus, which indicates that the activity of all neurons might be influenced by a further, unobserved, variable (question mark).

systems, spike patterns observed immediately after a sensory stimulus is presented correlate well with simple stimulus features, whereas later latency responses correlate better with more complex stimulus features, or required behaviors (Romo et al., 2002). Furthermore, in tasks where animals are required to hold an item in working memory, some neurons (particularly in prefrontal areas) show persistent spiking, suggesting their participation in assemblies that fire either continuously or repeatedly during the delay period.

Few studies have analyzed large numbers of cells during appropriate behavioral tasks. However, evidence from paired-unit recordings and field-potential studies have indicated that neural synchrony might indeed correlate with presumed "top–down" cognitive activity (Engel et al., 2001). Several psychological terms have been used to characterize the cognitive states in which synchrony increases, including expectancy (Riehle *et al.*, 1997), attention (Fries *et al.*, 2001), novelty (von Stein *et al.*, 2000), and binding (Gray *et al.*, 1989). Whether these multiple terms in fact refer to distinct neural processes, or reflect approximations to a single, as yet uncharacterized circuit function that is associated with increased synchrony is a matter for debate. Nevertheless, these studies all support the hypothesis that patterns of assembly coordination correlate with internal cognitive state.

During sleep, sensory input is largely blocked, so any cognition during sleep is by definition internal. A variety of studies have indicated that patterns of correlated activity are preserved between waking activity and sleep (Wilson and McNaughton, 1994; Skaggs and McNaughton, 1996; Kudrimoti *et al.*, 1999; Nadasdy *et al.*, 1999; Louie and Wilson, 2001; Hoffman and McNaughton, 2002; Lee and Wilson, 2002). Such preservation would be expected from the cell assembly hypothesis: not every subset of neurons constitutes an assembly, and reverberant activity is likely to favor synchrony between similar cell subsets during both waking and sleep. Furthermore, if assemblies are indeed formed by coincident firing during behavior (Hebb, 1949), one might expect that correlations that are observed for the first time during a given behavioral session would continue in subsequent rest or sleep, as has been claimed (Wilson and McNaughton, 1994; Kudrimoti *et al.*, 1999; Nadasdy *et al.*, 1999; Hirase *et al.*, 2001).

Timescales

The above has argued that many observations that concern spike timing could arise from an organization of neurons into cell assemblies. We now move to more quantitative questions that regard assembly structure. A primary question is the timescale at which assemblies are coordinated. Cross-correlation analysis indicates that coincident spikes can occur at two characteristic timescales: sharp correlations, which have a peak width in the order 1 millisecond (Csicsvari *et al.*, 1998; Usrey and Reid, 1999; Constantinidis *et al.*, 2001), and broader peaks, which are measured in tens of milliseconds (Vaadia *et al.*, 1995; deCharms and Merzenich, 1996; Bair *et al.*, 2001; Constantinidis *et al.*, 2001) (Fig. 3.6a). It has been hypothesized that sharp synchronization reflects monosynaptic drive between neurons (Csicsvari *et al.*, 1998; Marshall *et al.*, 2002; Bartho *et al.*, 2004) or common monosynaptic input from a third cell (Usrey and Reid, 1999). Broader synchronization, however, is likely to involve larger networks.

In the peer prediction framework, it is possible to estimate the timescale with which neurons are coordinated into assemblies by varying the temporal window within which spike times are predicted from the population. Figure 3.6b shows prediction quality as a function of timescale for an example neuron.

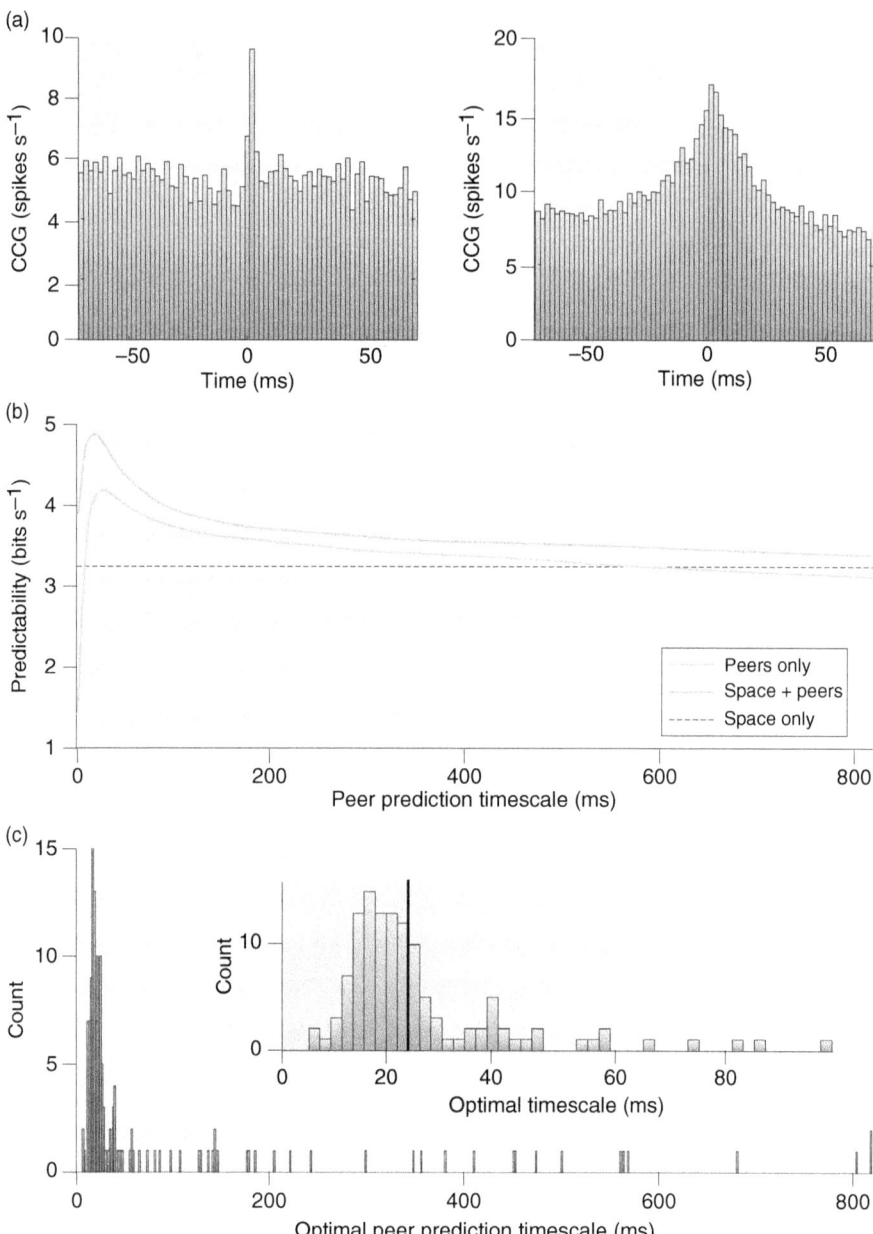

Figure 3.6 Timescales of synchronization. (a) Cross-correlograms (CCGs) of simultaneously recorded neurons in neocortex. The left CCG shows a sharp peak (width ~2ms), which reflects putative monosynaptic connection between cells. The right CCG shows a broader peak (width ~20ms), which is likely to result from more complex network processes. (b) Estimation of assembly synchronization timescale from optimal peer prediction window. Predictability is plotted against peer prediction window for an example cell. For short (~1ms) timescales, prediction is

For this neuron the optimal window was ~25 ms. Across all recorded neurons, a modal range of 10–30 ms was found (Fig. 3.6c). This timescale might be of particular physiological significance. Because it closely matches the membrane time constant and presumed excitatory postsynaptic potential (EPSP) width of pyramidal neurons in the hippocampal region (Spruston and Johnston, 1992), activity that is synchronized with this timescale might be optimal for inducing spiking in downstream neurons. Furthermore, this timescale matches the period of the gamma oscillation in hippocampal circuits (Csicsvari et al., 2003), and the effective window for synaptic plasticity (Magee and Johnston, 1997; Bi and Poo, 1998), which indicates that assembly activity at this timescale might be optimal for propagation and storage of information in local circuits.

Conclusion

We have argued that observations that are often interpreted as evidence for temporal coding might instead reflect involvement of cell assemblies in ICPs. Does this mean that a rate-coding framework is more suitable? The coding paradigm itself might be of limited use in describing the dynamics of cortical circuits; if a given circuit subserves ICPs as well as sensory representation, it is hard to define what "signal" is being coded without degenerating into mere semantics. However, as the coding paradigm is widely used, it is worth attempting to express our conclusions in these terms. In this case, a rate-coding framework seems most appropriate. Downstream cells are not expected to have a memory of spikes that occurred further in the past than the EPSP width, and the information conveyed by a population at any moment is fully determined by the assembly that is currently active, without reference to temporal patterns of previous spikes. Assemblies might reoccur periodically, for example in time with cortical oscillation patterns (Singer, 1993), but the fundamental currency of information processing is the firing of a single assembly, not the temporal sequence, and each time the assembly fires this would constitute a discrete unit of information processing.

Caption for Figure 3.6 (cont.)
poor. For long (~1 s) timescales, prediction from peers is no better than that from trajectory alone. Predictability peaks at ~25 ms, which indicates that this is the timescale of synchronization for the assemblies in which this cell participates. (c) Histogram of timescales at which peer activity best improved spike time prediction, for all cells (expanded in inset). A large mode is seen between 10 and 30 ms, with a median optimal timescale of 23 ms (solid black line). ((a): P. Bartho and K. D. Harris, unpublished data; (b, c): modified from Fries et al., 2001.)

Whether this constitutes a rate code or coincidence code is then just a matter of timescale. In this context, it is interesting to note that the estimated 10–30 ms timescale of assembly synchronization is close to the presumed EPSP width. This indicates that assembly patterns evolve at the fastest possible speed that allows integration of assembly activity by downstream neurons. The observed irregularity of spike trains might therefore reflect not temporal coding of a slowly changing sensory stimulus, but rate coding of rapidly evolving cognitive processes. The equality of these two timescales indicates that cognitive processing in neural circuits might occur at the fastest speed that is allowed by neuronal biophysics.

Acknowledgments

I thank Gyorgy Buzsáki, Sean Montgomery, and Elliot Ludvig for comments on the manuscript. KDH is supported by the National Institutes of Health (R01MH073245), and an Alfred P. Sloan Research Fellowship.

References

Abeles, M. and Gat, I. (2001). Detecting precise firing sequences in experimental data. *J Neurosci Methods* **107**:141–154.

Abeles, M., Bergman, H., Margalit, E., and Vaadia, E. (1993). Spatiotemporal firing patterns in the frontal cortex of behaving monkeys. *J Neurophysiol* **70**:1629–1638.

Adrian, E. D. and Zotterman, Y. (1926). The impulses produced by sensory nerve endings. II. The response of a single end organ. *J Physiol (Lond)* **61**:151–171.

Amit, D. (1994). The Hebbian paradigm reintegrated: local reverberations as internal representations. *Behav Brain Sci* **18**:617–626.

Baddeley, R., Abbott, L. F., Booth, M. C., *et al.* (1997). Responses of neurons in primary and inferior temporal visual cortices to natural scenes. *Proc R Soc Lond B* **264**:1775–1783.

Bair, W. and Koch, C. (1996). Temporal precision of spike trains in extrastriate cortex of the behaving macaque monkey. *Neural Comput* **8**:1185–1202.

Bair, W., Zohary, E., and Newsome, W. T. (2001). Correlated firing in macaque visual area MT: time scales and relationship to behavior. *J Neurosci* **21**:1676–1697.

Baker, S. N. and Lemon, R. N. (2000). Precise spatiotemporal repeating patterns in monkey primary and supplementary motor areas occur at chance levels. *J Neurophysiol* **84**:1770–1780.

Barlow, H. B. (1972). Single units and sensation: a neuron doctrine for perceptual psychology? *Perception* **1**:371–394.

Bartho, P., Hirase, H., Monconduit, L., *et al.* (2004). Characterization of neocortical principal cells and interneurons by network interactions and extracellular features. *J Neurophysiol* **92**:600–608.

Berry, M. J., Warland, D. K., and Meister, M. (1997). The structure and precision of retinal spike trains. *Proc Natl Acad Sci USA* **94**:5411–5416.

Bi, G. Q. and Poo, M. M. (1998). Synaptic modifications in cultured hippocampal neurons: dependence on spike timing, synaptic strength, and postsynaptic cell type. *J Neurosci* 18:10464–10472.

Britten, K. H., Shadlen, M. N., Newsome, W. T., and Movshon, J. A. (1993). Responses of neurons in macaque MT to stochastic motion signals. *Vis Neurosci* 10:1157–1169.

Buracas, G. T., Zador, A. M., DeWeese, M. R., and Albright, T. D. (1998). Efficient discrimination of temporal patterns by motion-sensitive neurons in primate visual cortex. *Neuron* 20:959–969.

Chance, F. S., Abbott, L. F., and Reyes, A. D. (2002). Gain modulation from background synaptic input. *Neuron* 35:773–782.

Constantinidis, C., Franowicz, M. N., and Goldman-Rakic, P. S. (2001). Coding specificity in cortical microcircuits: a multiple-electrode analysis of primate prefrontal cortex. *J Neurosci* 21:3646–3655.

Cossart, R., Aronov, D., and Yuste, R. (2003). Attractor dynamics of network UP states in the neocortex. *Nature* 423:283–288.

Csicsvari, J., Hirase, H., Czurko, A., and Buzsáki, G. (1998). Reliability and state dependence of pyramidal cell-interneuron synapses in the hippocampus: an ensemble approach in the behaving rat. *Neuron* 21:179–189.

Csicsvari, J., Jamieson, B., Wise, K. D., and Buzsáki, G. (2003). Mechanisms of gamma oscillations in the hippocampus of the behaving rat. *Neuron* 37:311–322.

DeWeese, M. R. and Zador, A. M. (2004). Shared and private variability in the auditory cortex. *J Neurophysiol* 92:1840–1855.

DeWeese, M. R., Wehr, M., and Zador, A. M. (2003). Binary spiking in auditory cortex. *J Neurosci* 23:7940–7949.

deCharms, R. C. and Merzenich, M. M. (1996). Primary cortical representation of sounds by the coordination of action-potential timing. *Nature* 381:610–613.

deCharms, R. C. and Zador, A. (2000). Neural representation and the cortical code. *Annu Rev Neurosci* 23:613–647.

Di Lorenzo, P. M. and Victor, J. D. (2003). Taste response variability and temporal coding in the nucleus of the solitary tract of the rat. *J Neurophysiol* 90:1418–1431.

Engel, A. K., Fries, P., and Singer, W. (2001). Dynamic predictions: oscillations and synchrony in top-down processing. *Nat Rev Neurosci* 2:704–716.

Fellous, J. M., Tiesinga, P. H., Thomas, P. J., and Sejnowski, T. J. (2004). Discovering spike patterns in neuronal responses. *J Neurosci* 24:2989–3001.

Fenton, A. A. and Muller, R. U. (1998). Place cell discharge is extremely variable during individual passes of the rat through the firing field. *Proc Natl Acad Sci USA* 95:3182–3187.

Fries, P., Reynolds, J. H., Rorie, A. E., and Desimone, R. (2001). Modulation of oscillatory neuronal synchronization by selective visual attention. *Science* 291:1560–1563.

Furukawa, S., Xu, L., and Middlebrooks, J. C. (2000). Coding of sound-source location by ensembles of cortical neurons. *J Neurosci* 20:1216–1228.

Gabbiani, F., Metzner, W., Wessel, R., and Koch, C. (1996). From stimulus encoding to feature extraction in weakly electric fish. *Nature* 384:564–567.

Gardner-Medwin, A. R. (1976). The recall of events through the learning of associations between their parts. *Proc R Soc Lond B* **194**:375–402.

Gerstein, G. L. and Mandelbrot, B. (1964). Random walk models for the spike activity of a single neuron. *Biophys J* **71**:41–68.

Ghazanfar, A. A., Stambaugh, C. R., and Nicolelis, M. A. (2000). Encoding of tactile stimulus location by somatosensory thalamocortical ensembles. *J Neurosci* **20**:3761–3775.

Gibbon, J., Malapani, C., Dale, C. L., and Gallistel, C. (1997). Toward a neurobiology of temporal cognition: advances and challenges. *Curr Opin Neurobiol* **7**:170–184.

Gray, C. M., Konig, P., Engel, A. K., and Singer, W. (1989). Oscillatory responses in cat visual cortex exhibit inter-columnar synchronization which reflects global stimulus properties. *Nature* **338**:334–337.

Harris, K. D., Henze, D. A., Hirase, H., *et al.* (2002). Spike train dynamics predicts theta-related phase precession in hippocampal pyramidal cells. *Nature* **417**:738–741.

Harris, K. D., Csicsvari, J., Hirase, H., Dragoi, G., and Buzsáki, G. (2003). Organization of cell assemblies in the hippocampus. *Nature* **424**:552–556.

Hebb, D. O. (1949). *The Organization of Behavior*. New York: John Wiley.

Hegde, J. and Van Essen, D. C. (2004). Temporal dynamics of shape analysis in macaque visual area V2. *J Neurophysiol* **92**:3030–3042.

Hirase, H., Leinekugel, X., Czurko, A., Csicsvari, J., and Buzsáki, G. (2001). Firing rates of hippocampal neurons are preserved during subsequent sleep episodes and modified by novel awake experience. *Proc Natl Acad Sci USA* **98**:9386–9390.

Hoffman, K. L. and McNaughton, B. L. (2002). Coordinated reactivation of distributed memory traces in primate neocortex. *Science* **297**:2070–2073.

Hopfield, J. J. (1982). Neural networks and physical systems with emergent collective computational abilities. *Proc Natl Acad Sci USA* **79**:2554–2558.

Huxter, J., Burgess, N., and O'Keefe, J. (2003). Independent rate and temporal coding in hippocampal pyramidal cells. *Nature* **425**:828–832.

Ikegaya, Y., Aaron, G., Cossart, R., *et al.* (2004). Synfire chains and cortical songs: temporal modules of cortical activity. *Science* **304**:559–564.

Izhikevich, E. M., Gally, J. A., and Edelman, G. M. (2004). Spike-timing dynamics of neuronal groups. *Cereb Cortex* **14**:933–944.

Kara, P., Reinagel, P., and Reid, R. C. (2000). Low response variability in simultaneously recorded retinal, thalamic, and cortical neurons. *Neuron* **27**:635–646.

Katz, D. B., Simon, S. A., and Nicolelis, M. A. (2001). Dynamic and multimodal responses of gustatory cortical neurons in awake rats. *J Neurosci* **21**:4478–4489.

Kudrimoti, H. S., Barnes, C. A., and McNaughton, B. L. (1999). Reactivation of hippocampal cell assemblies: effects of behavioral state, experience, and EEG dynamics. *J Neurosci* **19**:4090–4101.

Laurent, G. (1996). Dynamical representation of odors by oscillating and evolving neural assemblies. *Trends Neurosci* **19**:489–496.

Lee, A. K. and Wilson, M. A. (2002). Memory of sequential experience in the hippocampus during slow wave sleep. *Neuron* **36**:1183–1194.

Louie, K. and Wilson, M. A. (2001). Temporally structured replay of awake hippocampal ensemble activity during rapid eye movement sleep. *Neuron* **29**:145–156.

Luczak, A., Bartho, P., Marguet, S. L., Buzsáki, G., and Harris, K. D. (2007). Sequential structure of neocortical spontaneous activity in vivo. *Proc Natl Acad Sci USA* **104**:347–352.

MacLean, J. N., Watson, B. O., Aaron, G. B., and Yuste, R. (2005). Internal dynamics determine the cortical response to thalamic stimulation. *Neuron* **48**:811–823.

MacLeod, K., Backer, A., and Laurent, G. (1998). Who reads temporal information contained across synchronized and oscillatory spike trains? *Nature* **395**:693–698.

Magee, J. C. and Johnston, D. (1997). A synaptically controlled, associative signal for Hebbian plasticity in hippocampal neurons. *Science* **275**:209–213.

Mainen, Z. F. and Sejnowski, T. J. (1995). Reliability of spike timing in neocortical neurons. *Science* **268**:1503–1506.

Mao, B. Q., Hamzei-Sichani, F., Aronov, D., Froemke, R. C., and Yuste, R. (2001). Dynamics of spontaneous activity in neocortical slices. *Neuron* **32**:883–898.

Marr, D. (1971). Simple memory: a theory for archicortex. *Phil Trans R Soc Lond B* **262**:23–81.

Marshall, L., Henze, D. A., Hirase, H., *et al.* (2002). Hippocampal pyramidal cell-interneuron spike transmission is frequency dependent and responsible for place modulation of interneuron discharge. *J Neurosci* **22**:RC197.

Mazurek, M. E. and Shadlen, M. N. (2002). Limits to the temporal fidelity of cortical spike rate signals. *Nat Neurosci* **5**:463–471.

Mehta, M. R., Lee, A. K., and Wilson, M. A. (2002). Role of experience and oscillations in transforming a rate code into a temporal code. *Nature* **417**:741–746.

Middlebrooks, J. C., Clock, A. E., Xu, L., and Green, D. M. (1994). A panoramic code for sound location by cortical neurons. *Science* **264**:842–844.

Mokeichev, A., Okun, M., Barak, O., *et al.* (2007). Stochastic emergence of repeating cortical motifs in spontaneous membrane potential fluctuations in vivo. *Neuron* **53**:413–425.

Nadasdy, Z., Hirase, H., Czurko, A., Csicsvari, J., and Buzsáki, G. (1999). Replay and time compression of recurring spike sequences in the hippocampus. *J Neurosci* **19**:9497–9507.

O'Keefe, J. and Recce, M. L. (1993). Phase relationship between hippocampal place units and the EEG theta rhythm. *Hippocampus* **3**:317–330.

Optican, L. M. and Richmond, B. J. (1987). Temporal encoding of two-dimensional patterns by single units in primate inferior temporal cortex. III. Information theoretic analysis. *J Neurophysiol* **57**:162–178.

Oram, M. W., Foldiak, P., Perrett, D. I., and Sengpiel, F. (1998). The 'Ideal Homunculus': decoding neural population signals. *Trends Neurosci* **21**:259–265.

Oram, M. W., Wiener, M. C., Lestienne, R., and Richmond, B. J. (1999). Stochastic nature of precisely timed spike patterns in visual system neuronal responses. *J Neurophysiol* **81**:3021–3033.

Oram, M. W., Hatsopoulos, N. G., Richmond, B. J., and Donoghue, J. P. (2001). Excess synchrony in motor cortical neurons provides redundant direction information with that from coarse temporal measures. *J Neurophysiol* **86**:1700–1716.

Panzeri, S., Petersen, R. S., Schultz, S. R., Lebedev, M., and Diamond, M. E. (2001). The role of spike timing in the coding of stimulus location in rat somatosensory cortex. *Neuron* **29**:769–777.

Perez-Orive, J., Bazhenov, M., and Laurent, G. (2004). Intrinsic and circuit properties favor coincidence detection for decoding oscillatory input. *J Neurosci* **24**:6037–6047.

Prut, Y., Vaadia, E., Bergman, H., *et al.* (1998). Spatiotemporal structure of cortical activity: properties and behavioral relevance. *J Neurophysiol* **79**:2857–2874.

Riehle, A., Grun, S., Diesmann, M., and Aertsen, A. (1997). Spike synchronization and rate modulation differentially involved in motor cortical function. *Science* **278**:1950–1953.

Romo, R., Hernandez, A., Zainos, A., Lemus, L., and Brody, C. D. (2002). Neuronal correlates of decision-making in secondary somatosensory cortex. *Nat Neurosci* **5**:1217–1225.

Shadlen, M. N. and Newsome, W. T. (1998). The variable discharge of cortical neurons: implications for connectivity, computation, and information coding. *J Neurosci* **18**:3870–3896.

Singer, W. (1993). Synchronization of cortical activity and its putative role in information processing and learning. *Annu Rev Physiol* **55**:349–374.

Skaggs, W. E. and McNaughton, B. L. (1996). Replay of neuronal firing sequences in rat hippocampus during sleep following spatial experience. *Science* **271**:1870–1873.

Skaggs, W. E., McNaughton, B. L., Wilson, M. A., and Barnes, C. A. (1996). Theta phase precession in hippocampal neuronal populations and the compression of temporal sequences. *Hippocampus* **6**:149–172.

Skinner, B. F. (1938). *The Behavior of Organisms: An Experimental Analysis*. New York: Appleton-Century.

Softky, W. R. and Koch, C. (1993). The highly irregular firing of cortical cells is inconsistent with temporal integration of random EPSPs. *J Neurosci* **13**:334–350.

Spruston, N. and Johnston, D. (1992). Perforated patch-clamp analysis of the passive membrane properties of three classes of hippocampal neurons. *J Neurophysiol* **67**:508–529.

Stevens, C. F. and Zador, A. M. (1998). Input synchrony and the irregular firing of cortical neurons. *Nat Neurosci* **1**:210–217.

Stopfer, M., Bhagavan, S., Smith, B. H., and Laurent, G. (1997). Impaired odour discrimination on desynchronization of odour-encoding neural assemblies. *Nature* **390**:70–74.

Stopfer, M., Jayaraman, V. and Laurent, G. (2003). Intensity versus identity coding in an olfactory system. *Neuron* **39**:991–1004.

Teich, M. C. (1989). Fractal character of the auditory neural spike train. *IEEE Trans Biomed Eng* **36**:150–160.

Tolman, E. C. (1932). *Purposive Behavior in Animals and Men*. New York: Century.

Usrey, W. M. and Reid, R. C. (1999). Synchronous activity in the visual system. *Annu Rev Physiol* **61**:435–456.

Vaadia, E., Haalman, I., Abeles, M., *et al.* (1995). Dynamics of neuronal interactions in monkey cortex in relation to behavioural events. *Nature* **373**:515–518.

von Stein, A., Chiang, C., and Konig, P. (2000). Top-down processing mediated by interareal synchronization. *Proc Natl Acad Sci USA* **97**:14748–14753.

Wallenstein, G. V. and Hasselmo, M. E. (1997). GABAergic modulation of hippocampal population activity: sequence learning, place field development, and the phase precession effect. *J Neurophysiol* **78**:393–408.

Wehr, M. and Zador, A. M. (2003). Balanced inhibition underlies tuning and sharpens spike timing in auditory cortex. *Nature* **426**:442–446.

Wilson, M. A. and McNaughton, B. L. (1994). Reactivation of hippocampal ensemble memories during sleep. *Science* **265**:676–679.

Wood, E. R., Dudchenko, P. A., Robitsek, R. J., and Eichenbaum, H. (2000). Hippocampal neurons encode information about different types of memory episodes occurring in the same location. *Neuron* **27**:623–633.

4

Neural population recording in behaving animals: constituents of a neural code for behavioral decisions

ROBERT E. HAMPSON AND SAM A. DEADWYLER

Introduction

A major advantage conferred by recording from populations of neurons from any brain area is the potential to determine how that population *encodes* or represents information about a sensory input, behavioral task, motor movement, or cognitive decision. The ultimate purpose of populations of neural ensemble, recording and analysis can then be characterized as understanding: (1) what does the ensemble encode? (2) how does the ensemble encode it? and finally, (3) how do brain structures use that ensemble code?

In the hippocampus, the anatomy has been studied extensively such that connections between the major principal cell groups are well characterized and the local "functional" circuitry is currently under intense investigation. Neurons have been recorded in all major subfields in the hippocampus, and cell identification via firing signature or local analysis is not a problem in most cases. In the same manner, anatomical connections between subfields are also known; therefore, it is possible to position recording electrodes along specific anatomic projections to record ensembles of neurons with known anatomic connectivity. Given these factors, we have used multineuron recording techniques to determine how neural activity within hippocampal circuits is integrated with behavioral and cognitive events. However, as in many brain systems, the make-up of the ensembles is at least as critical as the techniques used to analyze the

Information Processing by Neuronal Populations, ed. Christian Hölscher and Matthias Munk.
Published by Cambridge University Press. © Cambridge University Press 2009.

ensemble data, or "codes." In addition, the functional connectivity that gives rise to such codes may not be constant; in fact variations in functional connectivity may produce different codes for different cognitive events. Finally there is the question of relationship to empirical events and behavior, i.e. which is responsible for the other? – is the neural code the basis for the behavior or is it a reflection of it?

Neural codes in brain

A major goal in neuroscience research is to identify patterns of neural activity that directly correlate with external actions (behavior, motor movements, etc.). A desired outcome of identifying codes is then to employ them to guide or alter performance, thus confirming that the neural patterns encode information essential to performance of the task, be it motor, sensory, or cognitive. Analysis techniques designed to extract neural coding of external events have been proposed for nearly 40 years (Gerstein and Perkel, 1969; Abeles and Gerstein, 1988; Gochin et al., 1994). In the case of cortical motor function, early work by Georgopoulos et al. (Georgopoulos et al., 1986; Lee et al., 1998; Georgopoulos, 2000) and Schwartz et al. (Moran and Schwartz, 1999; Schwartz and Moran, 1999) derived "population vectors" in which specific arm movements in monkeys were predicted by precise firing in populations of cortical neurons. As proof of the functional relevance of these neural codes, this work has evolved in recent years into a potential means for controlling artificial limbs (Chapin et al., 1999; Donoghue et al., 2004; Friehs et al., 2004; Wessberg and Nicolelis, 2004). However, with some exceptions (Talwar et al., 2002) it has proven more difficult to identify neural patterns that predict behaviors dependent on cognition (Freedman et al., 2002; Wallis and Miller, 2003; Ergorul and Eichenbaum, 2004; Stepniewska et al., 2005).

What constitutes a neural code?

There are two likely models by which information is encoded by neural ensembles: (a) "sparse, distributed" models assume that (1) recording neurons from a specific area will reveal few neurons that encode information relevant to the code, (2) many neurons within a given brain area will be silent or have no correlation to the information, and (3) information theoretic analysis reveal that the information encoded by each neuron is novel, with little redundancy across neurons (Georgopoulos et al., 1986; Georgopoulos et al., 1989; Moser and Moser, 1998; Moran and Schwartz, 1999; Schwartz and Moran, 1999; Frank et al., 2000; Laubach et al., 2000; Hampson et al., 2004; Bilkey and Clearwater, 2005; Nicolelis, 2005).

Examples of sparse encoding are prevalent in hippocampus: patterns of neural ensemble firing have been shown to map the spatial environment (Huxter et al., 2003; Bilkey and Clearwater, 2005; Leutgeb et al., 2005a; Touretzky et al., 2005; Wilson et al., 2005), replay that map for memory consolidation (Louie and Wilson, 2001; Sirota et al., 2003; Ribeiro and Nicolelis, 2004), indicate the familiarity or novelty of the environment (Fenton et al., 2000; Buzsáki, 2005; Leutgeb et al., 2005a), register distal and proximal cues (Huxter et al., 2003; Knierim and Rao, 2003; Battaglia et al., 2004), and identify locations in the environment where behaviorally relevant events occur (Wood et al., 2000; de Hoz et al., 2004; Ergorul and Eichenbaum, 2004). Adjacent neurons typically do not encode the same or adjacent places (Jensen and Lisman, 2000; Redish et al., 2001; Shapiro, 2001; Lee et al., 2004), and a random sampling of neurons may not encode all possible places in the environment (Zhang et al., 1998; Lee et al., 2004; Leutgeb et al., 2005b).

On the other hand, (b) "dense" models require that (1) most neurons within a given region will encode information relevant to the code, (2) information theoretic analyses reveal a high degree of redundancy between neurons with respect to the information encoded by each neuron, and (3) specific neural correlates exhibit some form of topography or "map" of the information encoded. A typical example of dense encoding is the topographical maps of stimulus information in auditory, visual, and sensory cortex. Information is densely encoded, since the majority of neurons in a given section of cortex respond to the same (or similar) features or characteristics of the stimulus. Encoding of relevant events seems to be a feature of individual cortical neurons, whereby one can randomly sample a single neuron for each specific stimulus feature (i.e. auditory frequency) by recording at intervals along the topographic representation within that brain region, and construct a complete code for the stimulus from such arbitrary "ensembles." Encoding of a specific stimulus feature by dense networks is typically involves modulation of firing rate – i.e. a neuron will fire at a high rate if the stimulus contains the feature for which the neuron is "tuned," but will fire at a low rate if that feature is absent (Pouget et al., 2003). Motor cortex also consists of a dense, topographic representation of the various muscle systems required for body and limb movement. However, random sampling of neurons along that topography does not necessarily yield a functional code for specific limb movement. An analysis termed "population vector" by Georgopoulos (Georgopoulos et al., 1986, 1989) and extensively applied to prediction of motor movement (Georgopoulos, 1994; Schwartz and Moran, 1999; Laubach et al., 2000; Nicolelis, 2001; Chapin, 2004) constructs ensembles by recording neurons and determining the "tuning" of each neuron with respect to movement direction and distance. Once a sufficient number of neurons have been recorded that encode the majority of possible limb movements, one or two neurons for each representation are combined into a single ensemble. Such an ensemble is highly successful in predicting motor movements, and have become a powerful basis for the development of neutrally activated prosthetics (Nicolelis and Chapin, 2002; Paninski et al., 2004; Lebedev et al., 2005).

Identifying neural population codes

If the information within neural ensembles is both *distributed* and *locally encoded*, it may be difficult to identify the specific aspects of single neuron firing that critically correspond to the encoded information. In addition to place fields, hippocampus and connected structures also encode "associations" – between temporal events (Dusek and Eichenbaum, 1997; Wallenstein et al., 1998; Agster et al., 2002; Eichenbaum, 2004), between places and behaviorally relevant information (Hampson et al., 1999b; Wood et al., 2000, 2001; Ergorul and Eichenbaum, 2004), or object recognition and categorization (Suzuki et al., 1997; Rolls, 1999; Fortin et al., 2004; Hampson et al., 2004; Rolls et al., 2005). Such codes are not simply constructed of individual neurons with specific correlates, but may be a part of a larger more complex code transmitted by many different neurons that only encode portions of an overall *pattern*. In this case information transmitted by the population reflects the functional code resident within the entire ensemble that is critical to the transference of information to the next stage.

Hippocampal ensemble encoding and the delayed-nonmatch-to-sample task

This laboratory has extensively examined the role of rat hippocampal neurons in encoding working memory initially in the delayed-match-to-sample (DMS) task, then in the delayed-nonmatch-to-sample (DNMS) task, for nearly 15 years (Hampson et al., 1993, 1999a, 1999b; Hampson and Deadwyler, 1994, 2000b, 2003; Deadwyler and Hampson, 1995, 1997, 2004; Deadwyler et al., 1996). These studies have determined that not only do hippocampal neural ensembles encode representation of critical behavioral events, but the neural encoding can actually *predict* subsequent behavioral responses in the task (Hampson and Deadwyler, 1996a, 1996b; Deadwyler and Hampson, 2004; Hampson et al., 2005; Simeral et al., 2006).

Single neuron firing in the DNMS task

Briefly, the DNMS task consists of Sample, Delay, and Nonmatch Phases. In the Sample Phase, one of two levers (Left or Right) is selected at random by a computer program, and extended into the behavioral chamber. The animal presses the lever, causing the lever to retract and start the Delay Phase (signaled by illumination of a cue light on the opposite wall of the experimental chamber). During the Delay the animal must break a photobeam with its nose ("nosepoke") at least once, within a randomly selected 1–30s delay interval. Once the Delay times out, the cue light is extinguished, and both levers are extended into the chamber signaling start of the Nonmatch Phase. As in the Sample Phase, the Nonmatch response consists of the animal pressing one of the two levers; however, only a response to the lever *opposite* to the Sample lever (i.e. a "non" match) is reinforced with a drop of water. Incorrect (match) responses are not reinforced and are signaled by a time-out in which the chamber lighting is extinguished for 5s. Subsequent DNMS trials commence after a 10s intertrial interval (ITI), with presentation of a randomly selected Sample lever. Animals typically perform sessions of 100–200 DNMS trials per day and reach a criterion of >90% correct responses on trials with ≤5s delay, decaying to approximately 60% correct responses on trials with >20s delays (Deadwyler et al., 1996).

A key finding with respect to single neuron firing recorded on the hippocampal array was that the majority consisted of a specific "functional cell types" (FCTs) with different behavioral correlates of DNMS task-related events such as lever position and/or phase of the task. The combined "ensemble firing pattern" is reflected as a transient increase in firing rate during the Sample or Nonmatch Phase and slowly increased firing in the Delay Phase if the delay interval exceeds 10s (Deadwyler et al., 1996; Deadwyler and Hampson, 2004). However, the defining characteristic of an FCT in this context is that the neuron responds only to a specific combination of events within the trial. Less complex FCTs respond only in terms of one dimension or feature such as lever position (Left or Right) or trial phase (Sample or Match) but the firing is specific to only one of the two possible occurrences on a given trial within the session (Hampson et al., 1999b). More complex FCTs respond to "conjunctions" (Eichenbaum et al., 1989; Otto and Eichenbaum, 1992) or combinations of position and phase (Hampson et al., 1999b). Figure 4.1 shows four FCTs tuned to fire to different combinations of lever position and DNMS task phase as evidenced by their differential firing patterns (peri-event histograms and single-trial rastergrams) on different trial types within the session. FCTs which encoded only lever position (Fig. 4.1a, b) increased firing whenever the lever response irrespective of task phase (Sample and Nonmatch), was either Left or Right, irrespective of task phase while the Phase FCT (Fig. 4.1c, d) showed increased firing during the Nonmatch Phase irrespective to the lever response position (Left or Right).

Figure 4.2 shows "conjunctive" FCTs (first demonstrated by Otto and Eichenbaum, 1992) in which the neural firing rate increased for this cell only at specific combinations of position *and phase of the task*. Thus the coincidence

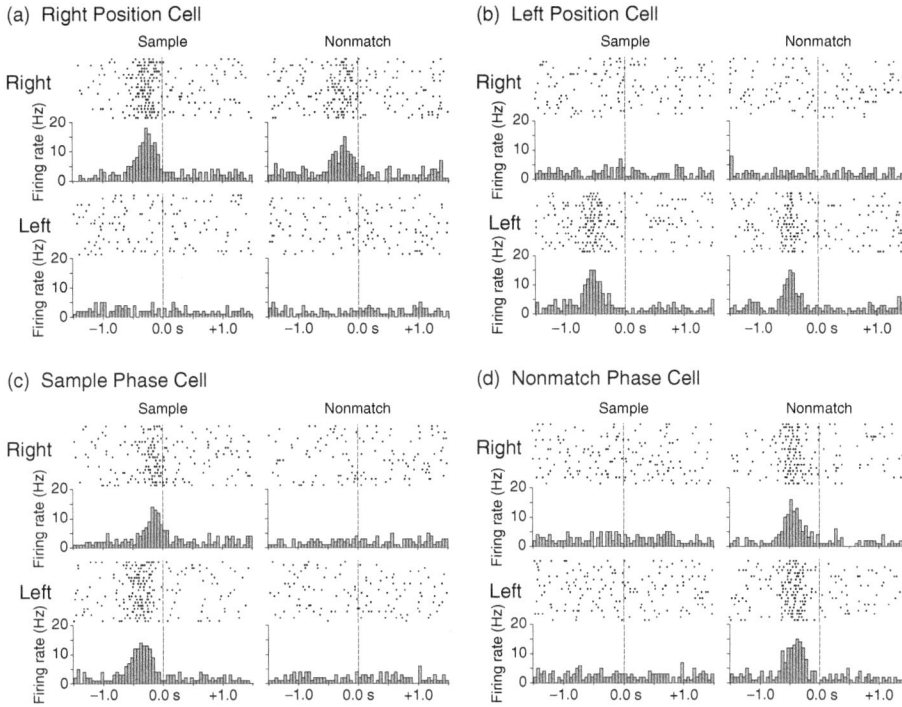

Figure 4.1 Peri-event histograms and rastergrams illustrating different types of functional cell types (FCTs, i.e. behavioral correlates) of hippocampal cells. (a) Right Position Cell. Peri-event histograms were constructed ±1.5s around the Right and Left Sample and Nonmatch lever responses. Peri-event histograms (below) are averages over 25 trials, single-trial records are depicted in the rastergrams above, with each dot representing a single extracellular action potential for the identified cell. Each row of the rastergram represents single trial. Histogram bins = 50ms. Note increased firing on Right responses irrespective of Sample or Nonmatch resulting in identification as a Right Position Cell. (b) Left Position Cell. Peri-event histograms and rastergrams constructed as above for a Left cell complementary to the Right cell in (a). (c) Sample Phase Cell. Peri-event histograms and rastergrams constructed as in (a) of a cell that increased firing rate on Sample responses, irrespective of Left or Right Position. (d) Nonmatch Phase Cell. Firing patterns complementary to the Sample cell in (c).

occurrence of the position and phase dimensions were encoded. Figure 4.2a illustrates a Right Sample which fired at the "conjunction" of Right position and Sample phase. The specificity of this coincident firing is shown by the failure of the cell to fire at a similar rate during any of the other events (Left or Nonmatch) in the task. Figure 4.2b shows a similar Left Nonmatch FCT; classes of these "event-specific" conjunctive cells have also been identified that for

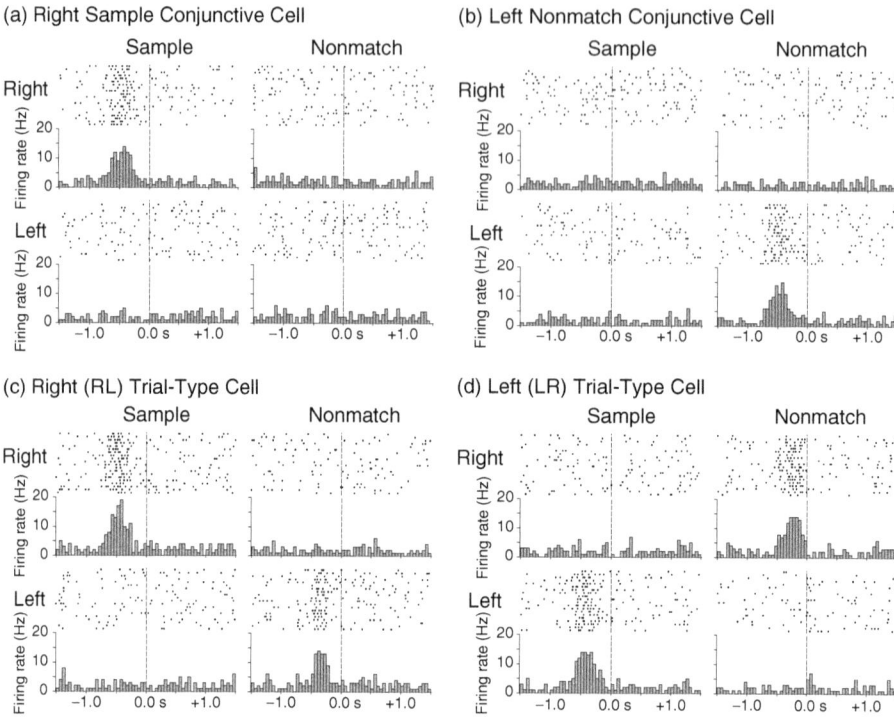

Figure 4.2 Peri-event histograms and rastergrams for "conjunctive" FCTs that fire only in response to specific combinations of position and phase. (a–b) Conjunctive Cell. Peri-event histograms and rastergrams (constructed as described in Fig. 4.1a) of neurons that increase firing to the specific "conjunction" (combination) of phase and lever position. There are four possible conjunctive cells: Right Sample (a), Left Sample (not shown), Right Nonmatch (not shown) and Left Nonmatch (b). (c) Right Trial Type Cell. Peri-event histograms and rastergrams of a neuron with a firing pattern that combines two lever responses appropriate to a Right DNMS trial (Right Sample/Left Nonmatch). Note that the Trial-Type FCT essentially combines the Right Sample and Left Nonmatch Conjunctive FCTs shown in (a) and (b). The complementary Left Trial-Type FCT encoding Left Sample/Right Nonmatch is shown in (d).

Left Sample and Right Nonmatch FCTs (Hampson *et al.*, 1999b). A third type of FCT represents an even higher-order conjunction, namely the encoding of an entire trial (Trial-Type cells) based on the type of response sequence required to satisfy the DNMS reward contingency when a given lever is presented (Left or Right) in the Sample phase. Trial-Type cells appear to get their inputs from event specific FCTs as shown in Fig. 4.2c and d in a manner that combines the two *appropriate* event specific conjunctive firing patterns for a given trial (i.e. the

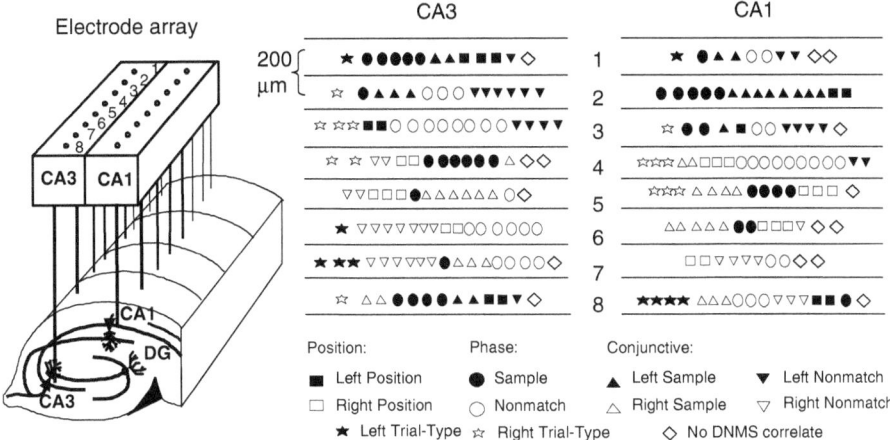

Figure 4.3 Distribution of 231 categorized hippocampal cells from 16 animals (i.e. 16 ensembles) plotted with respect to the position each cell was recorded with respect to dorsal hippocampal CA3 and CA1 subfields. Electrode positions (inset, left) are indicated by numbers 1–8 at center and represent successive 200-μm placements along the longitudinal axis of the hippocampus. Each cell is identified by symbol representing each cells behavioral correlate (FCT) in the DNMS task. Only 15 neurons were recorded that did not exhibit FCTs and are indicated by open diamonds.

Right Trial-Type cell depicted in Fig. 4.2c combines the conjunctive firing of the Right Sample and Left Nonmatch cells in Fig. 4.2a and b, respectively). It is important to note that in well-trained animals there are no Trial-Type FCTs that are *not* appropriate to the nonmatch rule, further validating the potential hierarchical encoding scheme as a means of contending with the trial-to-trial variability in Sample lever position which defines the type of contingent response in the nonmatch phase.

Classification of hippocampal CA1 and CA3 neurons into the above described 10 different types of FCTs has accounted for more than 90% of recorded neurons in the DNMS task. Figure 4.3 shows the segregation into FCTs of 231 hippocampal CA3 and CA1 principal cells recorded from 16 animals during performance of the DNMS task (Hampson et al., 1999b). Neurons were selected for firing rate (0.25–5.0 Hz baseline firing rate) and consistency of firing (waveform, firing rate, and behavioral correlate) over 5 or more days. Two important features of the neurons recorded during DNMS performance are illustrated in Fig. 4.3: (1) only 15 of 231 neurons recorded (6.5%) showed no behavioral correlate (unfilled diamond symbols), and (2) the FCTs were not randomly distributed throughout the hippocampus; neurons of similar or "compatible" FCTs were clustered along

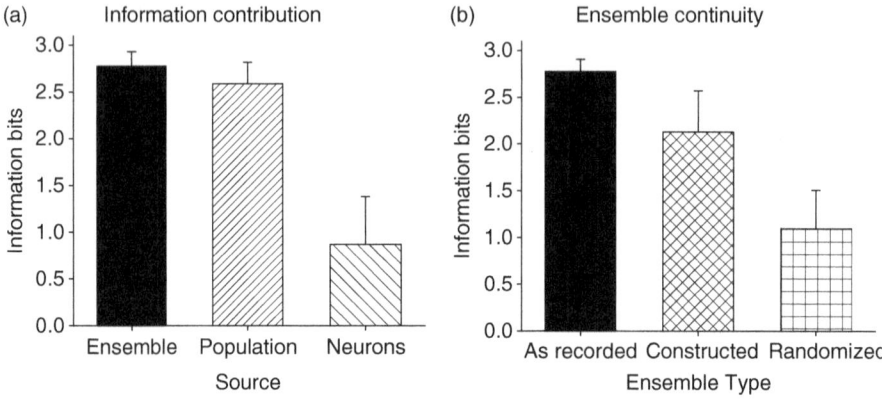

Figure 4.4 Information content analysis of the 16 ensembles illustrated in Fig. 4.3 with respect to information encoded in the DNMS task. (a) Mean information content (in bits) encoded by the 16 ensembles, by the entire population of 231 neurons, and as mean single neuron encoding. Ensembles and single neurons were scored according to accuracy of representation of DNMS lever response position (Left vs. Right), phase (Sample vs. Nonmatch) and behavioral outcome (correct vs. error) thus comprising 3 orthogonal bits of DNMS task information. Each ensemble typically included sufficient FCTs to encode all 3 bits, differences in actual score indicate variance in accuracy of encoding. Single FCTs could encode at most 1 bit. (b) Comparison of information content across ensemble. Ensembles consisted of simultaneously recorded neurons ("As recorded"), constructed by selecting 15 neurons from the population including at least one of each FCT ("Constructed"), or from simultaneously recorded neurons in which the ensemble coherence was disrupted ("Randomized"). The Randomized spike trains retained original neuron identity and interspike intervals from the original ensemble, but reordered the sequence of spike firing within each individual neuron to disrupt coherent patterns across neurons. Mean (and SEM) information bits are plotted.

the longitudinal axis of the hippocampus in the same longitudinal locations in CA3 and CA1, possibly suggesting a dense, rather than a sparse encoding of the DNMS task by the hippocampus.

Information content of hippocampal ensembles

If hippocampal ensembles form a dense, highly redundant code to represent DNMS task-specific information, analysis of the information content of hippocampal ensembles should reveal the same information content can be represented by any combination of individually recorded hippocampal cells (cf. Hampson and Deadwyler, 1996b). Figure 4.4 shows the mean information content of DNMS task-specific information encoded by simultaneously recorded ensembles as well as single neurons in hippocampus. Information content is calculated as "bits" of information encoded, and each "bit" thus represents a possible dimension of encoding: i.e. 1 bit for position (Left vs. Right), 1 bit for phase (Sample vs. Nonmatch), and 1 bit for outcome (correct vs. error). The mean information was then calculated for a population of 231 neurons (16 ensembles, 12–21 neurons each) as (1) average per ensemble, (2) the population of 231 neurons as a whole, and (3) average per neuron. Figure 4.4a shows that average information content per ensemble was 2.78 ± 0.15 bits, which was not significantly different from the information encoded by the complete population (2.59 ± 0.23 bits, $F_{(1,52)} = 0.97$, N.S.). Given the FCTs shown in Fig. 4.1 and 4.2 it was not surprising that the average single neuron information content was significantly lower at 0.87 ± 0.51 bits ($F_{(1,52)} = 13.44$, $p < 0.001$). Figure 4.4b further shows ensemble information content for three different methods of ensemble construction: (1) "As recorded," which consisted of the 16 original ensembles containing 12–21 simultaneously recorded neurons; (2) "Constructed" ensembles that consisted of neuron spike trains selected at random from the population of 231 neurons (16 ensembles of 15 neurons each were constructed and tested in Fig. 4.4b); and (3) "Randomized" ensembles which were identical to the original 16 simultaneously recorded ensembles, except that the sequence of DNMS trials within each spike train were reordered to disrupt coherent firing patterns across neurons while maintaining the firing correlation to DNMS task events. Figure 4.4b shows that mean information content (2.78 ± 0.15 bits) of simultaneously recorded ensembles (Fig. 4.4b, "As recorded") was greater than that of the "Constructed" ensembles (2.13 ± 0.44 bits, $F_{(1,52)} = 8.93$, $p < 0.01$), while the mean information content of "Randomized" ensembles was even lower (1.10 ± 0.41 bits, $F_{(1,52)} = 15.39$, $p < 0.001$). In fact, the information content of Randomized ensembles was not significantly different from that of single neurons shown in Fig. 4.4a ($F_{(1,52)} = 1.26$, N.S.).

Given the results shown in Fig. 4.4, does the hippocampal encoding of the DNMS task represent a sparse, distributed, or a dense, redundant code? The code appears to be highly redundant, with individual neurons encoding a specific element of the information (Fig. 4.1–4.3 and Hampson et al. (1999b)) thus appearing similar to the dense codes present in sensorimotor cortical areas (Georgopoulos et al., 1986; Georgopoulos, 1994; Schwartz and Moran, 1999; Chapin, 2004). However, the fact that simultaneously recorded ensembles encoded additional information that was lost when ensembles were simply assembled by selecting spike trains (or FCTs) indicates the importance of the spatiotemporal pattern (i.e. across neurons and time) of firing, consistent with a sparse code (Hampson and Deadwyler, 1996b). In addition, randomizing the temporal sequence of neural firing while maintaining the same FCTs in an ensemble disrupted information encoding. This suggests that ensemble

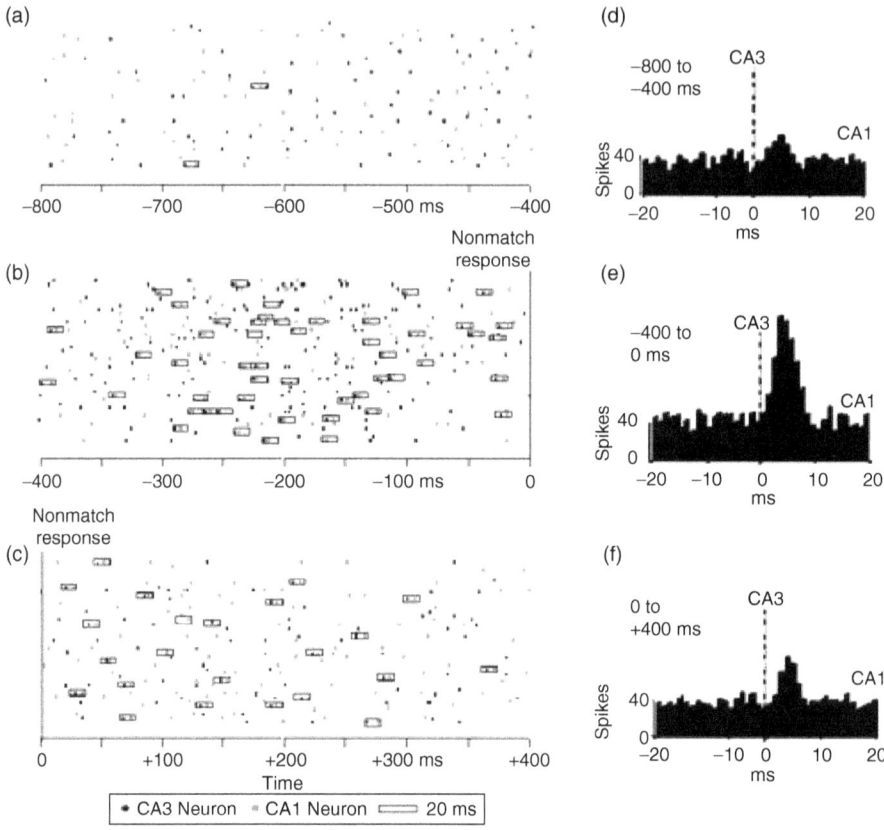

Figure 4.5 Transient cross-correlation between simultaneously recorded FCTs encoding the same DNMS behavioral event. (a–c) Rastergrams of single trial firing divided into three epochs during the Nonmatch Phase. (a) −800 to −400 ms; (b) −400 to 0 ms; (c) 0 to +400 ms relative to Nonmatch response. Black dots: CA3 Nonmatch FCT; crosses: CA1 Nonmatch FCT. Each row is a single trial. Rectangles indicate when both neurons fired within 20 ms. (d–f) Cross-correlation histograms of CA3 spike-triggered CA1 firing, constructed over 100 trials within each of the three indicated time epochs represented by the rastergrams at left (a–c): (d) −800 to −400 ms; (e) −400 to 0 ms; (f) 0 to +400 ms. Amplitude of correlation histograms is number of correlated spikes per bin (1 ms) over 100 trials.

encoding is dynamic, and *requires* the spatiotemporal pattern to fully represent DNMS task-relevant information.

Distributed encoding of information in hippocampal ensembles

Hippocampal neurons thus encode DNMS task-specific information via a code that shares the features of both dense and sparsely distributed models. Figure 4.5 illustrates a specific example of such temporal patterning between a CA3 and CA1 neuron that both encode the Nonmatch response. The peri-event histograms in Fig. 4.1 show that peak firing rates of FCTs occur approximately 0.0–0.5 s prior to the Sample or Nonmatch response. Therefore, the neural firing in Fig. 4.5 was divided into 3 epochs: 0.8 to 0.4 s before the Nonmatch (Fig. 4.5a, d), 0.4 to 0.0 s before the Nonmatch (Fig. 5b, e), and 0.0 to 0.4 s after the Nonmatch response (Fig. 5c, f). The rastergrams in Fig. 4.5(a–c) depict the firing of a pair of simultaneously recorded CA3 (crosses) and CA1 (black dots) neurons. The rectangles indicate instances in which the two neurons' firing was correlated within 20 ms. The number of correlated firings is minimal from −0.8 to −0.4 s, and maximal from −0.4 to 0.0 s as shown by the cross-correlograms corresponding to those intervals in Fig. 4.5(d–f) (Hampson et al., 2002). Similar transient correlations between CA3–CA3, CA3–CA1, and CA1–CA1 have been identified for Sample encoding. Of the 231 neurons represented in Fig. 4.3, there were 78 pairs of neurons with FCTs that encoded the same DNMS event. Of those, 56 pairs exhibited a significant increase in correlated firing (mean 36.7 ± 9.5 to greater than 64.1 ± 11.8 correlated spikes, $t_{(22)} = 4.7$, $p < 0.001$) within ± 0.5 s of the Sample or Nonmatch lever press events.

Figure 4.6 illustrates another dynamic feature of transient correlation between neurons within an ensemble. The three-dimensional surface in Fig. 4.6 is constructed from cross-correlograms (similar to Fig. 4.5d–f) of two CA1 Sample FCTs computed at 100-ms intervals from 1.5 s before to 1.5 s after the Sample response. Each correlogram can then be considered a "slice" of correlation (horizontal axis of Fig. 4.6) at a given time interval, and can be stacked in temporal order (vertical axis of Fig. 4.6) to create the surface plot shown in Fig. 4.6. As in Fig. 4.5, the maximal correlation occurs between 0.5 and 0.0 s before the response; however, following the Sample response, the temporal peak of the correlograms shifts from a time lag of 5 ms between firing of the paired neurons to a lag of 0 s. Similar transient correlation and temporal shifts of correlation have been implicated in the sequential processing of DNMS task-relevant information between hippocampus and dorsal subiculum (Deadwyler and Hampson, 2006).

Neural encoding and cognitive processing

Neural encoding in brain regions directly connected to sensory inputs or motor outputs frequently exhibit a dense, topographic code (Georgopoulos, 1994; Schwartz and Moran, 1999; Laubach et al., 2000; Nicolelis, 2001; Chapin, 2004), whereas encoding in brain areas further removed from sensory input and motor output often incorporate a sparse, distributed code (Wallenstein et al., 1998; Wood et al., 2001; Agster et al., 2002; Freedman et al., 2002; Hampson et al., 2004; Rolls et al., 2005). Hippocampal encoding of task-relevant DNMS

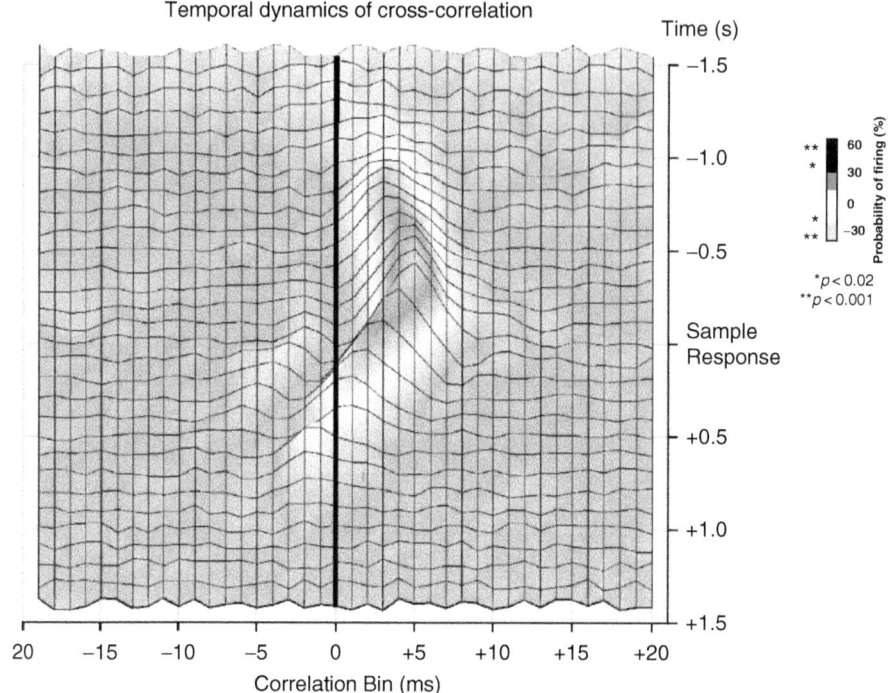

Figure 4.6 Three-dimensional cross-correlogram surfaces for two CA1 neurons encoding Sample response. Correlations were computed in 100-ms increments from 1.5s before to 1.5s after the Sample response over 100 DNMS trials and illustrate frequency of correlated firing in 1-ms bins. Correlation surface is shown in "overhead" view with beginning of the Sample epoch (−1.5s) at the upper edge. Inset bar indicates probability of correlated firing, asterisks ($^*p < 0.02$, $^{**}p < 0.001$) indicate significant correlation (peaks) or inverse correlations (troughs).

information as shown here incorporates features of both sparse and dense encoding models – with redundant information encoding among neurons, but dynamic information encoded by the ensemble as a whole that exceeds the information encoded by individual neurons. With some exceptions (Talwar et al., 2002) it has proven more difficult to identify neural encoding that can be used to predict cognitive processing, or behaviors dependent on cognition (Freedman et al., 2002; Wallis and Miller, 2003; Ergorul and Eichenbaum, 2004; Stepniewska et al., 2005).

Neural encoding as the determinant of behavioral decisions

Hippocampal neural activity has been thoroughly characterized with respect to behavioral performance in the DNMS task described here (Deadwyler and Hampson, 2004). As indicated in Fig. 4.1 and 4.2 CA1 and CA3 neuron ensembles exhibit task-specific firing correlates (FCTs), with individual neurons encoding response position, phase of the trial, and appropriate conjunctions of those two events (Deadwyler et al., 1996; Hampson et al., 1999b). Single neurons exhibit FCTs, yet the combined encoding of the ensemble is greater than that provided by the sum of the single neurons (see Fig. 4.4 and Hampson and Deadwyler, 1996b). This leads to the important observation that the DNMS "code" can be derived *on single trials* from ensembles of 15–30 hippocampal neurons (Hampson et al., 2001, 2005; Simeral et al., 2006). These single-trial codes can then be used to examine the question of whether the hippocampal code merely represents the behavioral events as they happen, or can predict behavioral decisions subsequent to the appearance of a given code within the ensemble.

To compute single trial codes, ensemble firing was analyzed using a canonical discriminant analysis (CDA), described extensively elsewhere (Hampson et al., 2001, 2005b; Simeral et al., 2006). Briefly, the analysis considers the ensemble firing of single DNMS trials to be a multidimensional vector of instantaneous firing rate across neurons and time intervals (around the Sample and Nonmatch responses). The CDA uses canonical correlation and linear discriminant analysis (Stevens, 2002) to reduce those dimensions and maximally discriminate between lever position, DNMS task phase, and behavioral outcome (correct vs. error). The resulting discriminant functions and scores can then be computed on single trials to identify the DNMS trial-specific information encoded by the ensemble (Deadwyler et al., 1996; Simeral et al., 2006). This technique previously demonstrated that across trials, "strong" codes resulted in correct DNMS performance, while "weak" codes resulted in errors (Hampson and Deadwyler, 1996a, 2000a; Deadwyler and Hampson, 2004). Figure 4.7 however demonstrates an important confirmation based on single-trial ensemble coding of Left vs. Right DNMS trials. Figure 4.7a shows single trial codes for correct "Left Trials" (Left Sample/Right Nonmatch) and "Right Trials" (Right Sample/Left Nonmatch) plotted across the entire trial including the interpolated delay periods. Discriminant scores are color-coded to correspond to encoding of Left vs. Right response position, and trials are plotted in order of increasing Delay duration to illustrate change in the code across the Delay. In Fig. 4.7a, it can be seen that for correct trials, the appropriate position code is present at the respective Sample and Nonmatch responses, with a gradual transition in the ensemble code from Left to Right response position during the Delay prior to when the Nonmatch response occurred. The codes yield a consistent pattern of transition across the temporal extent of the DNMS trials irrespective of duration of the Delay interval. However, if the ensemble encodes the incorrect position of the lever press during the Sample phase and the ensemble phase (Fig. 4.7b) a similar transition to the

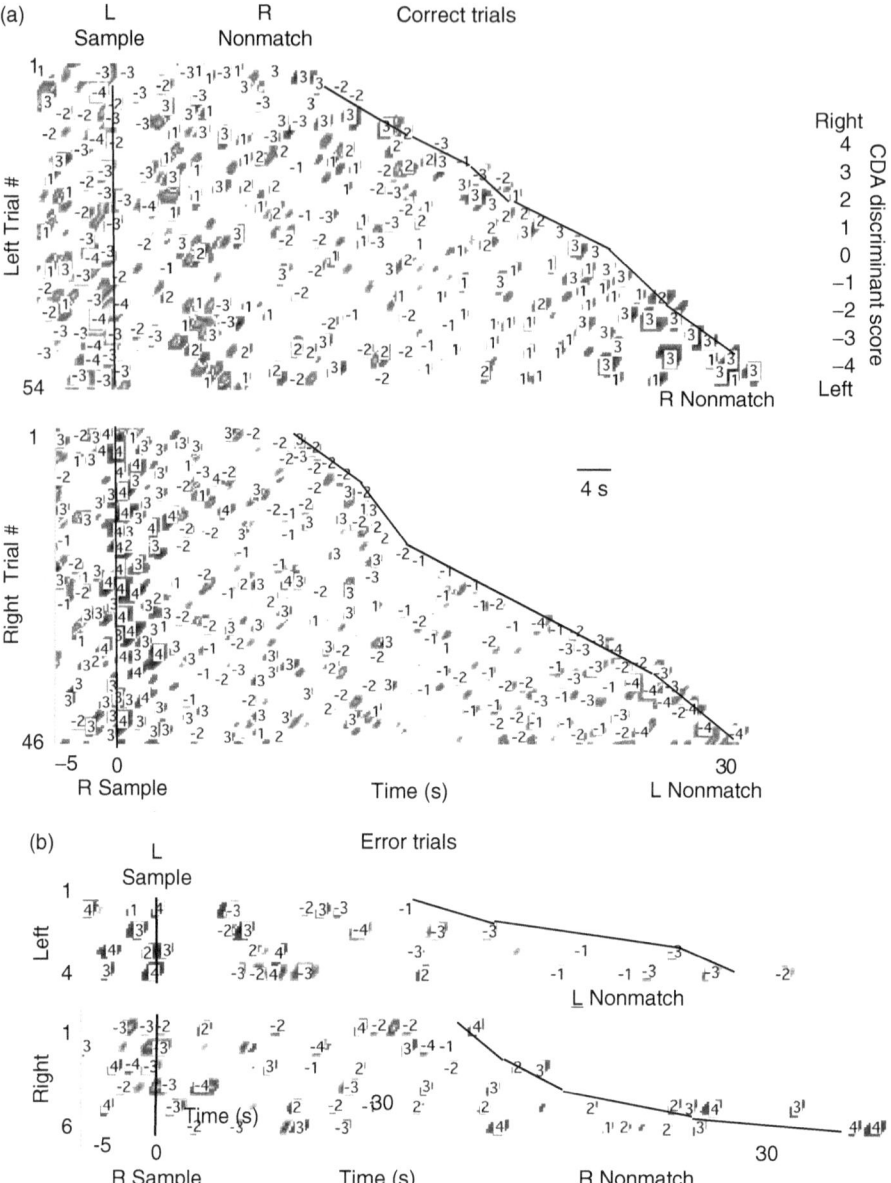

Figure 4.7 Dynamic ensemble representation of position within DNMS trials.
(a) Example of multineuron (ensemble) activity for a left (upper) and right (lower) single Correct DNMS trials represented as number values spanning 5s before the sample (S) to 5s after the Nonmatch (NM) responses. Respective left (L) and right (R) response positions are indicated. Ensemble phase and position discriminant scores were derived by "sliding" the CDA analysis in 250ms across the. Correct trials for one animal (single session) have been sorted by trial type and length of delay, and with numbers representing the corresponding position discriminant scores (scale at right).

opposite position code developed across the delay interval as on correct trials, even though the resulting code was for an incorrect response. Since Fig. 4.7b depicts error trials, the ensemble code was inappropriate to the actual behavior at the Sample, but the transition to the opposite pattern (erroneous) response at the Nonmatch phase was appropriate for the nonmatch rule applied to fallacious information.

Thus, the question of whether hippocampal encoding of DNMS information can be considered the determinant of the behavioral performance (i.e. the code determines the behavior) rather than the result of a prior behavioral decision (the behavior determines the code) is resolved by the single trial data. Figure 4.7b demonstrates that codes generated in the Sample phase actually *predicted* the animals' decision in the subsequent Nonmatch (recall) phase after an imposed 1–30-s delay interval (Deadwyler et al., 1996; Hampson and Deadwyler, 1996a), while at the same time, those codes misrepresented the actual lever pressed. Coupled with the finding that strength as well as accuracy of hippocampal encoding correlates with likelihood of correct trial-to-trial DNMS performance indicates that is it possible to "read" hippocampal encoding online to determine whether an error is likely to occur and make associated adjustments in task parameters.

Summary

Ensemble analyses applied to deriving neural codes by which the brain represents external events are typically divided into sparse (or distributed), and dense network models. Dense models are often topographic, and are exemplified by sensory representation or motor cortical encoding of movement. Sparse models include hippocampal place cell mapping and memory association. Hippocampal encoding of a Delayed-Nonmatch-to-Sample task, as shown here, exhibits features of both dense and sparse network models. The network is dense

Caption for Figure 4.7 (cont.)
Sample lever press is indicated by the vertical black line to the left, Nonmatch is indicated by the black line to the right, oblique due to different delay lengths. White space within trials indicates non-significant position scores (−0.5 to +0.5). White border space between numbers falls outside trial start and end points and does not contain data. (b) Error DNMS trials from the same animal and session. Trials were sorted in the same order as (a); however, the number code for position indicates that the ensemble code for Sample on these trials did not correspond to the actual lever pressed. Ensemble codes for Nonmatch on Error trials were consistent with the erroneous response position.

because >90% of hippocampal principal cells recorded encoded at least some DNMS task-relevant information, and multiple cells encode the same information. The network is sparse/distributed because the ensemble as a whole encodes more information than the sum of individual neurons. One reason for the dual nature of hippocampal encoding of the DNMS task is that correlation (and hence functional connectivity) between neurons within ensembles is dynamic, and varies throughout performance of a trial. This dynamic contributes to significant single-trial encoding which can be shown to both represent as well as predict behavioral responses in the DNMS task. Since the single-trial encoding actually indicates what behavioral decision will be made, this type of encoding is unique with respect to brain regions not directly connected to motor outputs and provides potential insight into how the brain functions to strategically encode and retrieve information during cognitive processing of complex behaviors.

Acknowledgments

This work was supported by National Institutes of Health grants DA03502, DA00119, DARPA contract (SPAWAR) N66001-02-C-8058 to SAD and National Institutes of Health grants DA08549, MH61397 to REH. The authors thank Vernell Collins, Jason Locke, D. Lynn Johnson, David B. King, Wilson Davis, Joanne Konstantopoulos, and John Simeral for technical support.

References

Abeles, M. and Gerstein, G. L. (1988). Detecting spatiotemporal firing patterns among simultaneously recorded single neurons. *J Neurosci* **60**:909–924.

Agster, K. L., Fortin, N. J., and Eichenbaum, H. (2002). The hippocampus and disambiguation of overlapping sequences. *J Neurosci* **22**:5760–5768.

Battaglia, F. P., Sutherland, G. R., and McNaughton, B. L. (2004). Local sensory cues and place cell directionality: additional evidence of prospective coding in the hippocampus. *J Neurosci* **24**:4541–4550.

Bilkey, D. K. and Clearwater, J. M. (2005). The dynamic nature of spatial encoding in the hippocampus. *Behav Neurosci* **119**:1533–1545.

Buzsáki, G. (2005). Theta rhythm of navigation: link between path integration and landmark navigation, episodic and semantic memory. *Hippocampus* **15**:827–840.

Chapin, J. K. (2004). Using multi-neuron population recordings for neural prosthetics. *Nat Neurosci* **7**:452–455.

Chapin, J. K., Moxon, K. A., Markowitz, R. S., and Nicolelis, M. A. (1999). Real-time control of a robot arm using simultaneously recorded neurons in the motor cortex. *Nat Neurosci* **2**:664–670.

de Hoz, L., Martin, S. J., and Morris, R. G. (2004). Forgetting, reminding, and remembering: the retrieval of lost spatial memory. *PLoS Biol* **2**:E225.

Deadwyler, S. A. and Hampson, R. E. (1995). Ensemble activity and behavior: what's the code? *Science* **270**:1316–1318.

Deadwyler, S. A. and Hampson, R. E. (1997). The significance of neural ensemble codes during behavior and cognition. *Annu Rev Neurosci* **20**:217–244.

Deadwyler, S. A. and Hampson, R. E. (2004). Differential but complementary mnemonic functions of the hippocampus and subiculum. *Neuron* **42**:465–476.

Deadwyler, S. A. and Hampson, R. E. (2006). Temporal coupling between subicular and hippocampal neurons underlies retention of trial-specific events. *Behav Brain Res* **174**:272–280.

Deadwyler, S. A., Bunn, T., and Hampson, R. E. (1996). Hippocampal ensemble activity during spatial delayed-nonmatch-to-sample performance in rats. *J Neurosci* **16**:354–372.

Donoghue, J. P., Nurmikko, A., Friehs, G., and Black, M. (2004). Development of neuromotor prostheses for humans. *Suppl Clin Neurophysiol* **57**:592–606.

Dusek, J. A. and Eichenbaum, H. (1997). The hippocampus and memory for orderly stimulus relations. *Proc Natl Acad Sci USA* **94**:7109–7114.

Eichenbaum, H. (2004). Hippocampus: cognitive processes and neural representations that underlie declarative memory. *Neuron* **44**:109–120.

Eichenbaum, H., Wiener, S. I., Shapiro, M. L., and Cohen, N. J. (1989). The organization of spatial coding in the hippocampus: a study of neural ensemble activity. *J Neurosci* **9**:2764–2775.

Ergorul, C. and Eichenbaum, H. (2004). The hippocampus and memory for "what," "where," and "when". *Learn Mem* **11**:397–405.

Fenton, A. A., Csizmadia, G., and Muller, R. U. (2000). Conjoint control of hippocampal place cell firing by two visual stimuli. I. The effects of moving the stimuli on firing field positions. *J Gen Physiol* **116**:191–209.

Fortin, N. J., Wright, S. P., and Eichenbaum, H. (2004). Recollection-like memory retrieval in rats is dependent on the hippocampus. *Nature* **431**:188–191.

Frank, L. M., Brown, E. N., and Wilson, M. (2000). Trajectory encoding in the hippocampus and entorhinal cortex. *Neuron* **27**:169–178.

Freedman, D. J., Riesenhuber, M., Poggio, T., and Miller, E. K. (2002). Visual categorization and the primate prefrontal cortex: neurophysiology and behavior. *J Neurophysiol* **88**:929–941.

Friehs, G. M., Zerris, V. A., Ojakangas, C. L., Fellows, M. R., and Donoghue, J. P. (2004). Brain-machine and brain-computer interfaces. *Stroke* **35**:2702–2705.

Georgopoulos, A. P. (1994). Population activity in the control of movement. *Int Rev Neurobiol* **37**:103–119.

Georgopoulos, A. P. (2000). Neural aspects of cognitive motor control. *Curr Opin Neurobiol* **10**:238–241.

Georgopoulos, A. P., Schwartz, A. B., and Kettner, R. E. (1986). Neuronal population encoding of movement direction. *Science* **233**:1416–1419.

Georgopoulos, A. P., Lurito, J. T., Petrides, M., Schwartz, A. B., and Massey, J. T. (1989). Mental rotation of the neuronal population vector. *Science* **243**:234–236.

Gerstein, G. L. and Perkel, D. H. (1969). Simultaneously recorded trains of action potentials: analysis and functional interpretation. *Science* **164**:828–830.

Gochin, P. M., Colombo, M., Dorfman, G. A., Gerstein, G. L., and Gross, C. G. (1994). Neural ensemble coding in inferior temporal cortex. *J Neurophysiol* **71**:2325–2337.

Hampson, R. E. and Deadwyler, S. A. (1994). Hippocampal representations of DMS/DNMS in the rat. *Behav Brain Sci* **17**:480–482.

Hampson, R. E. and Deadwyler, S. A. (1996a). Ensemble codes involving hippocampal neurons are at risk during delayed performance tests. *Proc Natl Acad Sci USA* **93**:13487–13493.

Hampson, R. E., Deadwyler, S. A. (1996b). LTP and LTD and the encoding of memory in small ensembles of hippocampal neurons. In: *Long-Term Potentiation*, vol. 3, ed. Baudry, M. and Davis, J., pp. 199–214. Cambridge, MA: MIT Press.

Hampson, R. E. and Deadwyler, S. A. (2000a). Cannabinoids reveal the necessity of hippocampal neural encoding for short-term memory in rats. *J Neurosci* **20**:8932–8942.

Hampson, R. E. and Deadwyler, S. A. (2000b). Differential information processing by hippocampal and subicular neurons. In: *The Parahippocampal Region: Special Supplement to the Annals of the New York Academy of Sciences*, ed. Witter, M. P., pp. 143–175. New York: New York Academy of Sciences.

Hampson, R. E. and Deadwyler, S. A. (2003). Temporal firing characteristics and the strategic role of subicular neurons in short-term memory. *Hippocampus* **13**:529–541.

Hampson, R. E., Heyser, C. J., and Deadwyler, S. A. (1993). Hippocampal cell firing correlates of delayed-match-to-sample performance in the rat. *Behav Neurosci* **107**:715–739.

Hampson, R. E., Jarrard, L. E., and Deadwyler, S. A. (1999a). Effects of ibotenate hippocampal and extrahippocampal destruction on delayed-match and -nonmatch-to-sample behavior in rats. *J Neurosci* **19**:1492–1507.

Hampson, R. E., Simeral, J. D., and Deadwyler, S. A. (1999b). Distribution of spatial and nonspatial Information in dorsal hippocampus. *Nature* **402**:610–614.

Hampson, R. E., Simeral, J. D., and Deadwyler, S. A. (2001). What ensemble recordings reveal about functional hippocampal cell encoding. *Prog Brain Res* **130**:345–357.

Hampson, R. E., Simeral, J. D., and Deadwyler, S. A. (2002). "Keeping on track": firing of hippocampal neurons during delayed-nonmatch-to-sample performance. *J Neurosci* **22**:RC198.

Hampson, R. E., Pons, T. P., Stanford, T. R., and Deadwyler, S. A. (2004). Categorization in the monkey hippocampus: a possible mechanism for encoding information into memory. *Proc Natl Acad Sci USA* **101**:3184–3189.

Hampson, R. E., Simeral, J. D., and Deadwyler, S. A. (2005). Cognitive processes in replacement brain parts: a code for all reasons. In: *Toward Replacement Parts for the Brain: Implantable Biomimetic Electronics as Neural Prosthesis*, ed. Berger, T. W. and Glanzman D. L., pp. 111–128. Cambridge, MA: MIT Press.

Huxter, J., Burgess, N., and O'Keefe, J. (2003). Independent rate and temporal coding in hippocampal pyramidal cells. *Nature* **425**:828–832.

Jensen, O. and Lisman, J. E. (2000). Position reconstruction from an ensemble of hippocampal place cells: contribution of theta phase coding. *J Neurophysiol* **83**:2602–2609.

Knierim, J. J. and Rao, G. (2003). Distal landmarks and hippocampal place cells: effects of relative translation versus rotation. *Hippocampus* **13**:604–617.

Laubach, M., Wessberg, J., and Nicolelis, M. A. (2000). Cortical ensemble activity increasingly predicts behaviour outcomes during learning of a motor task. *Nature* **405**:567–571.

Lebedev, M. A., Carmena, J. M., O'Doherty, J. E., et al. (2005). Cortical ensemble adaptation to represent velocity of an artificial actuator controlled by a brain–machine interface. *J Neurosci* **25**:4681–4693.

Lee, D., Port, N. P., Kruse, W., and Georgopoulos, A. P. (1998). Neuronal population coding: multielectrode recording in primate cerebral cortex. In: *Neuronal Ensembles: Strategies for Recording and Decoding*, ed. Eichenbaum, H. and Davis, J., pp. 239–254. New York: John Wiley.

Lee, I., Yoganarasimha, D., Rao, G., and Knierim, J. J. (2004). Comparison of population coherence of place cells in hippocampal subfields CA1 and CA3. *Nature* **430**:456–459.

Leutgeb, S., Leutgeb, J. K., Barnes, C. A., et al. (2005a). Independent codes for spatial and episodic memory in hippocampal neuronal ensembles. *Science* **309**:619–623.

Leutgeb, S., Leutgeb, J. K., Moser, M. B., and Moser, E. I. (2005b). Place cells, spatial maps and the population code for memory. *Curr Opin Neurobiol* **15**:738–746.

Louie, K. and Wilson, M. A. (2001). Temporally structured replay of awake hippocampal ensemble activity during rapid eye movement sleep. *Neuron* **29**:145–156.

Moran, D. W. and Schwartz, A. B. (1999). Motor cortical activity during drawing movements: population representation during spiral tracing. *J Neurophysiol* **82**:2693–2704.

Moser, M. B. and Moser, E. I. (1998). Distributed encoding and retrieval of spatial memory in the hippocampus. *J Neurosci* **18**:7535–7542.

Nicolelis, M. A. (2001). Actions from thoughts. *Nature* **409**(Suppl):403–407.

Nicolelis, M. A. (2005). Computing with thalamocortical ensembles during different behavioural states. *J Physiol* **566**:37–47.

Nicolelis, M. A. and Chapin, J. K. (2002). Controlling robots with the mind. *Sci Am* **287**:46–53.

Otto, T. and Eichenbaum, H. (1992). Neuronal activity in the hippocampus during delayed non-match to sample performance in rats: evidence for hippocampal processing in recognition memory. *Hippocampus* **2**:323–334.

Paninski, L., Shoham, S., Fellows, M. R., Hatsopoulos, N. G., and Donoghue, J. P. (2004). Superlinear population encoding of dynamic hand trajectory in primary motor cortex. *J Neurosci* **24**:8551–8561.

Pouget, A., Dayan, P., and Zemel, R. S. (2003). Inference and computation with population codes. *Annu Rev Neurosci* **26**:381–410.

Redish, A. D., Battaglia, F. P., Chawla, M. K., et al. (2001). Independence of firing correlates of anatomically proximate hippocampal pyramidal cells. *J Neurosci* **21**:1–6.

Ribeiro, S. and Nicolelis, M. A. (2004). Reverberation, storage, and postsynaptic propagation of memories during sleep. *Learn Mem* **11**:686–696.

Rolls, E. T. (1999). Spatial view cells and the representation of place in the primate hippocampus. *Hippocampus* 9:467–480.

Rolls, E. T., Xiang, J., and Franco, L. (2005). Object, space, and object–space representations in the primate hippocampus. *J Neurophysiol* 94:833–844.

Schwartz, A. B. and Moran, D. W. (1999). Motor cortical activity during drawing movements: population representation during lemniscate tracing. *J Neurophysiol* 82:2705–2718.

Shapiro, M. (2001). Plasticity, hippocampal place cells, and cognitive maps. *Arch Neurol* 58:874–881.

Simeral, J. D., Hampson, R. E., and Deadwyler, S. A. (2006). Behaviorally relevant neural codes in hippocampal ensembles: detection on single trials. In: *Synaptic Plasticity: From Basic Mechanisms to Clinical Applications*, ed. Baudry, M., Bi, X., and Schreiber, S., pp. 278–291. Cambridge, MA: MIT Press.

Sirota, A., Csicsvari, J., Buhl, D., and Buzsáki, G. (2003). Communication between neocortex and hippocampus during sleep in rodents. *Proc Natl Acad Sci USA* 100:2065–2069.

Stepniewska, I., Fang, P. C., and Kaas, J. H. (2005). Microstimulation reveals specialized subregions for different complex movements in posterior parietal cortex of prosimian galagos. *Proc Natl Acad Sci USA* 102:4878–4883.

Stevens, J. (2002). *Applied Multivariate Statistics for the Social Sciences*. Hillsdale, NJ: Lawrence Erlbaum.

Suzuki, W. A., Miller, E. K., and Desimone, R. (1997). Object and place memory in the macaque entorhinal cortex. *J Neurophysiol* 78:1062–1081.

Talwar, S. K., Xu, S., Hawley, E. S., et al. (2002). Rat navigation guided by remote control. *Nature* 417:37–38.

Touretzky, D. S., Weisman, W. E., Fuhs, M. C., et al. (2005). Deforming the hippocampal map. *Hippocampus* 15:41–55.

Wallenstein, G. V., Eichenbaum, H., and Hasselmo, M. E. (1998). The hippocampus as an associator of discontiguous events. *Trends Neurosci* 21:317–323.

Wallis, J. D. and Miller, E. K. (2003). Neuronal activity in primate dorsolateral and orbital prefrontal cortex during performance of a reward preference task. *Eur J Neurosci* 18:2069–2081.

Wessberg, J. and Nicolelis, M. A. (2004). Optimizing a linear algorithm for real-time robotic control using chronic cortical ensemble recordings in monkeys. *J Cogn Neurosci* 16:1022–1035.

Wilson, I. A., Ikonen, S., Gurevicius, K., et al. (2005). Place cells of aged rats in two visually identical compartments. *Neurobiol Aging* 26:1099–1106.

Wood, E. R., Dudchenko, P. A., Robitsek, R. J., and Eichenbaum, H. (2000). Hippocampal neurons encode information about different types of memory episodes occurring in the same location. *Neuron* 27:623–633.

Wood, E. R., Dudchenko, P. A., and Eichenbaum, H. (2001). Cellular correlates of behavior. *Int Rev Neurobiol* 45:293–312.

Zhang, K., Ginzburg, I., McNaughton, B. L., and Sejnowski, T. J. (1998). Interpreting neuronal population activity by reconstruction: unified framework with application to hippocampal place cells. *J Neurophysiol* 79:1017–1044.

5

Measuring distributed properties of neural representations beyond the decoding of local variables: implications for cognition

ADAM JOHNSON, JADIN C. JACKSON, AND A. DAVID REDISH

Introduction

Neural representations are distributed. This means that more information can be gleaned from neural ensembles than from single cells. Modern recording technology allows the simultaneous recording of large neural ensembles (of more than 100 cells simultaneously) from awake behaving animals. Historically, the principal means of analyzing representations encoded within large ensembles has been to measure the immediate accuracy of the encoding of behavioral variables ("reconstruction"). In this chapter, we will argue that measuring immediate reconstruction only touches the surface of what can be gleaned from these ensembles. We will discuss the implications of distributed representation, in particular, the usefulness of measuring self-consistency of the representation within neural ensembles. Because representations are distributed, neurons in a population can agree or disagree on the value being represented. Measuring the extent to which a firing pattern matches expectations can provide an accurate assessment of the self-consistency of a representation. Dynamic changes in the self-consistency of a representation are potentially indicative of cognitive processes. We will also discuss the implications of representation of non-local (non-immediate) values for cognitive processes. Because cognition occurs at fast timescales, changes must be detectable at fast (millisecond, tens of milliseconds) timescales.

Information Processing by Neuronal Populations, ed. Christian Hölscher and Matthias Munk. Published by Cambridge University Press. © Cambridge University Press 2009.

Representation

As an animal interacts with the world, it encounters various problems for which it must find a solution. The description of the world and the problems encountered within it play a fundamental role in how an animal behaves and finds a solution. Sensory and memory processes within the brain provide a description of the world and within that description the brain's decision-making processes must select some course of action or behavior.[1]

The resulting open question is how the information about the world is represented and organized across these brain areas. The use of information from the world in behavior involves two critical processes. The first process is appropriate transformation of information about the world into a representation that is relevant and useful for behavior. The second process is the projection of that representation onto a behavior that allows the animal to interact with its world.

We will call the representation of the world within the brain or any transformation of that representation toward behavior (even if the behavior is not executed) a *neural representation*. This definition is intentionally broad such that the operations underlying directly observable behavior *and* covert mental activities can be considered.

Encoding (tuning curves)

What makes a neuron fire? The question can be asked with respect to the neuron's immediate environment – its afferents and ion channels – and with respect to the world beyond. Answering the latter question requires knowing what information is *encoded* by the neuron. An *encoding model* describes a hypothesis relating the information represented by the cell (sensory, perceptual, motivational, motor, etc.) to the activity of a single neuron. The hypothesized relationship between the encoded information x and the neural activity, typically considered in terms of spikes, s can be written as the function $p(s(t)) = T(x(t))$ where $p(s(t))$ is the probability of a spike at time s at time (t).[2] This definition is easily extended to include both preceding experience and planned future behaviors in the encoded information x. For simplicity, the present discussion neglects identification of the precise temporal offset in describing the relationship between $s(t)$ and $x(t)$.

---

[1] The decision to not act is still a decision made.
[2] Of course, the actual activity is also a function of the history of spiking of the neuron (e.g. neurons show refractory periods, bursting, and other history-dependent processes).

These encoding models have classically been found by standard *tuning curves*. More recently, these encoding models have been stated in terms of Shannon information theory, identifying the mutual information between behavioral variables and spike timing (Rieke *et al.*, 1997; Dayan and Abbott, 2001). Other encoding definitions have been based on *linear filter kernels*, which reflect the recent history of variable x in the firing of a cell's spikes (Bialek *et al.*, 1991) or bursts (Kepecs and Lisman, 2003).[3] These encoding models can be measured relative to any available behavioral variable, whether it be immediate sensory input such as the frequency of an auditory tone, an immediate motor output such as the target of a saccade or the direction of a reach, or a cognitive variable such as the location of an animal in the environment.

Decoding (reconstruction)

Because the variability of a single cell is usually insufficient to fully describe the entire space of encoded information, information is generally encoded across a population of neurons that differ in their tuning curves (often described by a *family of tuning curves*), such as retinotopic Gaussians or place fields. If information is consistently represented across a population of neurons, then it should be possible to infer the expectation of the variable x by examining the neural activity across the population. This inference can be made using Bayes' rule

$$p(x,s) = p(x|s)p(s) = p(s|x)p(x) \tag{5.1}$$

where $p(s|x)$ is the probability of observing some set of neural activities given the variable of interest and $p(x|s)$ is the probability of the variable of interest given some set of neural activity. This means that the variable x can be decoded from the neural activity across the population s by

$$p(x|s) = p(s|x)p(x)/p(s). \tag{5.2}$$

The probability $p(x|s)$ describes how information can be read out or *decoded* from the network. What should be clear from this simple account is that decoding critically depends on the *encoding model*, $p(s|x)$.

The term s in Eq. (5.2) reflects the pattern of neural activity across the entire population of cells at time t. This analysis thus requires sufficient data to infer

---

[3] Kernel-based methods explain the observed neural activity in terms of both the represented information and the neuronal dynamics of the cell itself. The generative method below effectively extends this perspective to include unobserved variables that can only be determined by examining ensembles with decoding.

the probability density function across an n-dimensional space (where n is the number of cells in the ensemble). For even moderately sized ensembles, appropriate sampling of s thus requires an inordinate amount of data due to the curse of dimensionality. In many situations, it is convenient to assume that the activity of each cell is conditionally independent, relative to the represented variable x (Brown et al., 1998; Zhang et al., 1998), so that

$$p(x|s) = p(x) \prod_{i \in \text{cells}} \frac{p(s_i|x)}{p(s_i)}. \tag{5.3}$$

However, the validity of this assumption is still controversial (Nirenberg and Latham, 2003; Schneiderman et al., 2003; Averbeck et al., 2006).

Although Bayes' rule (Eq. 5.2) provides an optimal solution for decoding, even the simplified version (Eq. 5.3) is often not computationally tractable. As a result, several other non-probability-based methods have been developed for decoding (e.g. template matching: Wilson and McNaughton, 1993; Zhang et al., 1998; Averbeck et al., 2003a, linearly weighted averaging: Georgopoulos et al., 1983; Salinas and Abbott, 1994; Zhang et al., 1998). These can be considered as reduced forms of Bayes' rule (Dayan and Abbott, 2001).

Non-local reconstruction (memory and cognition)

While neural activity is typically measured and discussed in terms of an observable external variable $x(t)$: $p(s(t)) = T(x(t))$, a more inclusive statement is that neural activity reflects an internal representation of this variable. That internal representation can potentially deviate from the external world. This point is particularly important when investigating processes in which cognition potentially plays a role; one of the hallmarks of cognitive processing is the connection of the observable world with the animal's externally invisible goals or motivations (Tulving, 1983, 2001, 2002; Suddendorf and Busby, 2003; Gray, 2004; Ferbinteanu et al., 2006; Johnson and Redish, 2007).

During normal navigation, as rats perform active behavioral tasks on an environment, the first-order information encoded within hippocampal pyramidal cells is the location of the animal (O'Keefe and Nadel, 1978; Redish, 1999). However, when rats are sleeping or when they pause at feeder sites to eat or groom, the hippocampus changes state, and the hippocampal firing reflects internal dynamics rather than its primary inputs (O'Keefe and Nadel, 1978; Wilson and McNaughton, 1994; Chrobak and Buzsáki, 1994, 1996; Ylinen et al., 1995; Csicsvari et al., 1999; Chrobak et al., 2000). Cell firing during subsequent sleep states reflects recently experienced memories rather than the current location of the animal (Wilson and McNaughton, 1994; Kudrimoti et al., 1999;

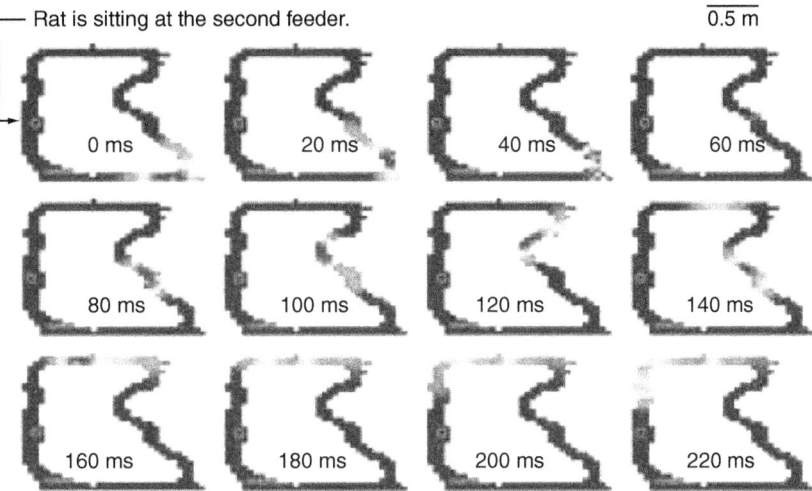

Figure 5.1 Replay of experience on the maze during an awake sharp wave. The rat is sitting at the second feeder throughout the event. The distribution starts at the base of the first T and moves through the full maze in 220 ms (typical behavioral run times through this maze = 10–12 s). The reconstructed location is indicated by color (light gray high probability, dark gray low probability). Panels arranged from left to right, top to bottom in 20 ms intervals. Note the coherent but non-local reconstruction of the representation during the sharp wave.

Nadasdy et al., 1999; Sutherland and McNaughton, 2000; Hoffmann and McNaughton, 2002; Lee and Wilson, 2002). Reconstruction during non-attentive waking states reveals representations of *non-local information* (Jensen and Lisman, 2000; Jackson et al., 2006) (see Fig. 5.1).

The slow dynamics in which reconstruction tracks behavior (Wilson and McNaughton, 1994; Brown et al., 1998; Zhang et al., 1998) and the fast dynamics of replay (e.g. Fig. 5.1) are examples of different *information processing modes* (Buzsáki, 1989; Harris-Warrick and Marder, 1991; Hasselmo and Bower, 1993; Redish, 1999). These two modes occur in recognizably distinct brain states, characterized by distinct local field potential frequencies. The neural firing patterns of both projection (pyramidal) cells and interneurons change between modes, as do the neuromodulators present (Vanderwolf, 1971; O'Keefe and Nadel, 1978; Hasselmo and Bower, 1993; Somogyi and Klausberger, 2005). These modes are thought to be differentially involved in learning, storage, and recall (Buzsáki, 1989; Hasselmo and Bower, 1993; Redish, 1999). Later in this chapter, we will discuss the implications of multiple generative models $p_M(s|x)$ for understanding these multiple information processing modes. In order to differentiate between models, we first need to address the question of *self-consistency*.

Self-consistency (coherency)

While the development of ensemble-based reconstruction methods such as those described above has allowed us to probe more deeply into the brain's processing of behavioral information, we run the risk of assuming that an animal's brain rigidly adheres to representing the present behavioral status of the animal. In doing so, reconstruction errors are viewed as "noise in the system," ignoring the cognitive questions of memory and recall that are fundamental to the brain's inner workings. For instance, what is recall or confusion and how does the brain represent competing values in ambiguous situations? To answer these questions, we need to consider how units within a network function together to form a coherent representation, i.e. one that is internally consistent across all units.

A *coherent* or *self-consistent* representation is one in which the firing of all neurons in a network conforms to some pattern expected from observations during normal (local, baseline) encoding. For instance, if one records from an ensemble of motor cortical cells, one possible model of the network would be to assume that the firing of each neuron is tuned to the direction of movement. This tuning, if it exists, should dictate the distributed pattern of activity across the neurons in the network. If the network is representing a particular direction, all neurons with any tuning to that direction should be firing to some degree specified by their respective tuning curves and neurons that are tuned to very different directions should be firing rarely if at all. In other words, neurons with similar preferred directions should respond similarly if the network is responding in a manner consistent with the data set used to construct the neuronal tuning curves. If this is not true, there is a fundamental difference between the model and the current status of the network. This principle allows for the formulation of a measure of the *coherency* or *self-consistency* of a neural ensemble (Redish et al., 2000; Jackson and Redish, 2003; Jackson, 2006) (see Fig. 5.2).

Figure 5.2b, c, and d show three hypothetical states for a network made up of neurons with tuning curves shaped like the one depicted in Fig. 5.2a, but centered at even intervals along x. The behavioral variables $\hat{x}_1$ and $\hat{x}_2$ are shown for reference. The pattern in (b) is consistent with behavioral variable $\hat{x}_2$ but not with $\hat{x}_1$. A reconstruction algorithm would yield value $\hat{x}_2$. If the actual value was $\hat{x}_1$, then reconstruction error $|\hat{x}_2 - \hat{x}_1|$ would be high even as the network state is internally consistent. The left mode of the pattern in (c) is consistent with behavioral variable $\hat{x}_1$ but neither mode is consistent with $\hat{x}_2$. A vector-based reconstruction algorithm would yield value $\hat{x}_2$, while template-matching or Bayesian methods would yield $\hat{x}_1$ or the right peak depending

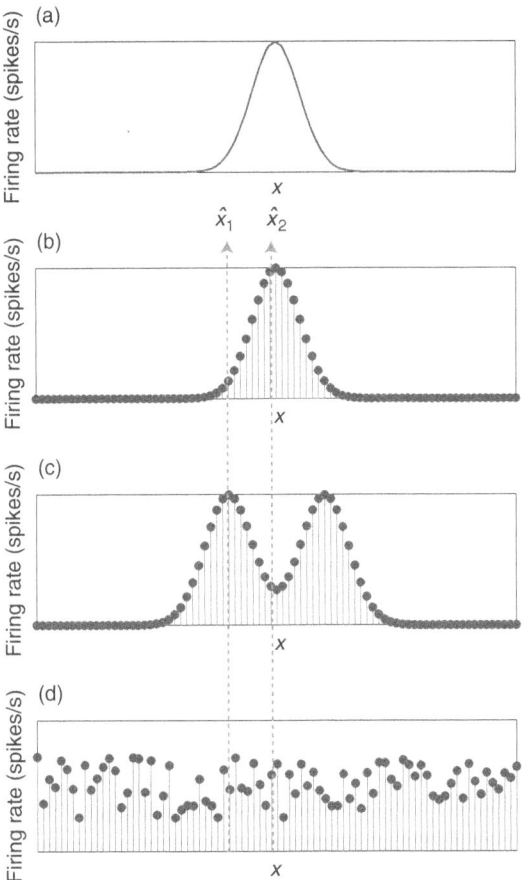

Figure 5.2 Self-consistency. (a) An example unimodal tuning curve. The stimulus or behavioral variable is on the x-axis with firing rate represented along the y-axis. (b) A "coherent" network firing pattern. The stimulus or behavioral variable is on the x-axis with firing rate of each neuron represented along the y-axis. Each line represents the location of a neuron's preferred stimulus, with height equal to the neuron's firing rate. If each neuron in a network had unimodal tuning curves identical to the neuron represented in (a) but with the peak firing occurring at a different preferred stimulus x, then when the preferred stimulus of the neuron in (a) is presented, this is the expected network firing pattern. This pattern is consistent with behavioral variable $\hat{x}_2$ but not with $\hat{x}_1$. (c) A bimodal representation would represent an ambiguous or incoherent state of the network described in (b), since the unimodal tuning curves would predict only one mode of activity should be possible for a single stimulus x and that outside this mode neurons should be silent. One mode is consistent with behavioral variable $\hat{x}_1$ but neither mode is consistent with $\hat{x}_2$. (d) As in (c) this representation would represent a confused or incoherent state of the network described in (b), since the unimodal tuning curves would predict a prominent mode of activity and that outside this mode neurons should be silent. This state is not consistent with either behavioral variable $\hat{x}_1$ or $\hat{x}_2$. (From Jackson, 2006.)

on the noise in the system. If the actual value was $\hat{x}_1$, then reconstruction error $|\hat{x}_2 - \hat{x}_1|$ would either be low or high depending on the reconstruction method and the noise in neuronal activity. However, neither reconstruction measure would reveal the underlying representational ambiguity. The state in (d) is not consistent with either behavioral variable $\hat{x}_1$ or $\hat{x}_2$. However, each reconstruction method would yield a value such as $\hat{x}_1$ or $\hat{x}_2$ even though the underlying state is completely random. Each of these scenarios suggests very different cognitive processes in this brain network; accessing these processes through an appropriate measure of internal consistency is one primary aim of this chapter.

Redish et al. (2000) first suggested that a mathematical comparison between expected and actual activity patterns could provide useful information about the dynamics of neural processes. They used such a comparison to identify when the hippocampal ensemble transitioned between two spatial representations.

Averbeck and colleagues (Averbeck, 2001; Averbeck et al., 2002, 2003b) recorded from frontal and parietal neural ensembles in a shape-copying task and compared the neural activity patterns during the time monkeys were actually drawing the shapes with the neural activity patterns during the preparatory period. They first calculated the n-dimensional tuple of firing rates during each segment of the copying process (e.g. $F_{\text{bottom line of square}} = (f_1, f_2, \ldots, f_n)$, where f_i is the firing rate of cell i, and n is number of cells in ensemble). They then measured the Euclidean distance between the firing rate patterns of the ensembles in each 25 ms bin of the preparatory period and the firing rate pattern of the ensemble during each segment of the copying process. They found that at the end of the preparatory period, the first component was most likely to be closest in this distance metric, while the second component was next closest, the third component next, and so forth.

In the limited condition of cosine-tuned neurons (Georgopoulos et al., 1983), one can measure self-consistency by measuring the length of the population vector (Smyrnis et al., 1992), which is a measure of the variance of the circular distribution (Mardia, 1972; Batschelet, 1981). Georgopoulos et al. (1988) used this to measure development of a motor plan during mental rotation tasks. As the animal developed a motor plan, the length of the population vector increased.

While the comparison process laid out by Jackson and Redish (2003) (and see below) requires a hypothesized behavioral variable $\hat{x}$ in order to define the expected activity packet $\hat{A}$, the hypothesized behavioral variable does not have to reflect anything about the outside world. It can be based on the estimated representation of a given variable $\hat{x}$. This estimated variable can be found by an internal decoding process (i.e. *reconstruction*). For example, Johnson et al.

(2005) recorded neural ensembles of head direction cells from the postsubiculum of rats in a cylinder-foraging task and calculated the coherency of the head direction representation relative to the reconstructed head direction $\hat{\phi}$. Populations that were highly self-consistent were more likely to provide an accurate representation of the world. In other words, actual reconstruction error (e.g. $|\hat{\phi} - \text{actual } \hat{\phi}|$) was reflected in the self-consistency of the representation, even though the self-consistency could be determined from entirely internal signals. Thus, if downstream structures used only self-consistent representations for making decisions, then the animal would be more likely to be using accurate representations of the outside world.

Comparing actual and expected activity patterns

In the following section, we review the results of Jackson and Redish (2003) and show the generality of the results. Further details can be found in the original paper.

Activity packets were defined as the weighted sum of the tuning curves. (Jackson and Redish (2003) showed that normalizing by the average tuning curve made this a linear calculation and simplified subsequent analyses.)

$$A(x,t) = \frac{\sum_k T_k(x) \cdot F_k(t)}{\sum_k T_k(x)} \tag{5.4}$$

where k ranges over the available cells in the ensemble, $T_k(x)$ is the tuning curve of cell k relative to variable x, and $F_k(t)$ is the firing rate of cell k at time t. The activity packet is thus a function over both the behavioral (possibly multidimensional) variable x and time. The expected activity packet is then defined as the weighted sum of the tuning curves, weighted, not by the actual firing rate of the cells, but rather by the *expected* firing rate of the cells.

$$\hat{A}(x,t) = \frac{\sum_k T_k(x) \cdot E(F_k(t))}{\sum_k T_k(x)} \tag{5.5}$$

$$= \frac{\sum_k T_k(x) \cdot T_k(\hat{x}(t))}{\sum_k T_k(x)} \tag{5.6}$$

where $\hat{x}(t)$ is the hypothesized value of variable x at time t. Once the actual activity packet $A(x, t)$ and the expected activity packet $\hat{A}(x, t)$ have been defined, then the self-consistency of the population can be measured by comparing the two packets. We have explored multiple comparison methods, including dot-product (DP: Redish et al., 2000), root-mean-squared error (RMS: Jackson and Redish, 2003), variance (VAR: Johnson et al., 2005), which can be brought into the same units as RMS by taking the square root and using the standard deviation instead (STD: Jackson, 2006) (see Jackson (2006) for a review).

$$C_{\text{DP}}(t) = \hat{A}(x,t) \cdot A(x,t) \tag{5.7}$$

$$I_{\text{RMS}}(t) = \frac{\sqrt{\int_x (A(x,t) - \hat{A}(x,t))^2 dx}}{\int_x \hat{A}(x,t) dx} \tag{5.8}$$

$$I_{\text{VAR}}(t) = \frac{\text{var}_x(A(x,t) - \hat{A}(x,t))}{\int_x \hat{A}(x,t) dx} \tag{5.9}$$

$$I_{\text{STD}}(t) = \frac{\text{stdev}_x(A(x,t) - \hat{A}(x,t))}{\int_x \hat{A}(x,t) dx}. \tag{5.10}$$

Here, we use a C to denote that the measure C_{DP} measures the consistency, or similarity, between the actual and expected representations; we use I to denote the other measures, which measure inconsistency, or dissimilarity, between the actual and expected activity packets. The integration is done over the entire representational space. We include C_{DP} for completeness, but we have found that I_{RMS} and I_{STD} are the most sensitive and recommend their use for experimental analyses (Jackson, 2006). Statistically, the self-consistency of the ensemble relative to hypothesized behavioral variable $\hat{x} \in x$ can be defined as the probability of accepting the null hypothesis that the actual and expected activity packets are the same:

$$H_0 : \forall_x \forall_t A(x,t) = \hat{A}(x,t). \tag{5.11}$$

In practice, we expect the validity of this hypothesis to vary over time and generally measure it as a function of time. For a given time t,

$$H_0(t) : \forall_x A(x,t) = \hat{A}(x,t). \tag{5.12}$$

The probability of accepting the null hypothesis can be found by empirically determining the probability distribution of the measurement of choice under conditions of stability. *Self-consistency* can then be defined as the deviation from this expected probability distribution. If the measure implemented detects differences between the activity packets, this probability is equal to the probability of seeing a larger difference between the actual and expected representation given the data in the training set. If this probability is sufficiently small, the actual and expected activity packets are more different than a large majority of the samples in our training set and we can reject the null hypothesis that the actual representation is the same as the expected representation.

Validation: simulations

Simulations provide a fast, efficient, and, most importantly, controlled means of generating data for the purposes of characterizing ensemble measures. The attractor network used for the simulations was a standard local-excitatory/global-inhibitory network which has been extensively studied (Wilson and Cowan, 1973; Amari, 1977; Ermentrout and Cowan, 1979; Kohonen, 1982, 1984; Redish, 1999; Eliasmith and Anderson, 2003; Jackson and Redish, 2003) and has been used to model numerous systems in the brain (Droulez and Berthoz, 1991; Munoz et al., 1991; Arai et al., 1994; Skaggs et al., 1995; Redish et al., 1996; Zhang, 1996; Samsonovich and McNaughton, 1997; Redish and Touretzky, 1998; Redish, 1999; Tsodyks, 1999, 2005; Doboli et al., 2000; Goodridge and Touretzky, 2000; Laing and Chow, 2001; Wills et al., 2005). Briefly, this network employed symmetric local excitatory connections between neurons with similar preferred directions and global inhibition with periodic boundary conditions. Thus, this network can be thought of as a circular ring of neurons with a stable attractor state consisting of a single mode of active neurons, which could be located anywhere on the ring. This local-excitation/global-inhbition, ring-based attractor network has several useful properties for the study of self-consistency; however, it is important to note that the self-consistency equations above (Eqs. 5.4–5.12) make no assumptions about the shape or structure of the tuning curves or the network connectivity (Jackson and Redish, 2003). The only assumption made is that tuning curves are stable over the course of the training and test sets.

Issue 1: Random network firing vs. stable activity mode

When started from random noise, neurons in a ring attractor will compete until a group of neighbors wins and the network settles to a stable mode of activity at that location (i.e. representing one direction). Neurons with preferred directions near this direction will have higher firing rates than those distant from this direction. Thus, the final mode will be randomly selected given a random input (Wilson and Cowan, 1973; Kohonen, 1977).

In this situation, reconstruction techniques such as the vector mean (Mardia, 1972; also known as the *population vector*: Georgopoulos et al., 1983) always provide an answer and cannot be used to differentiate the random and settled states. In contrast, self-consistency measures differentiate the random and settled states (see Fig. 5.3). Because of the non-linearities of the measure, I_{RMS} detected the time of settling accurately, displaying a stark difference between the two states. While in the random state, the self-consistency measurement showed that the random state was significantly different from the expected "bump" of

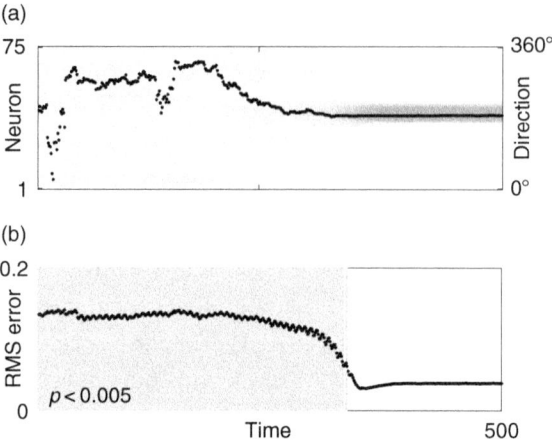

Figure 5.3 A simulation started with random input to the network settles to a stable state. (a) The neural activity. Time is shown in time-steps on the x-axis. Neurons ordered by their preferred direction (0°–360°) along the y-axis, shaded according to their firing rate. Black dots indicate the direction extracted from the population activity using population-vector reconstruction. Note that the reconstruction algorithm yields a position whether or not there is an actual mode of activity present at that location. (b) The I_{RMS} measure of inconsistency between actual and expected activity packets. During the random state, the discrepancy between the actual and expected activity packets is high ($p < 0.005$, gray zone). Upon reaching the stable state at time-step 342, the difference drops ($p > 0.005$, white zone). (From Jackson, 2006; see also Jackson and Redish, 2003.)

activity ($p < 0.005$). After the network transitioned to a stable representational state, the self-consistency measurement showed a higher probability of match.

Issue 2: Rotation vs. jump

When this system is in a stable state (i.e. representing one direction), and network inputs drive neurons with preferred directions near the represented direction (within 60° in our network), the represented direction will shift toward the input (Redish *et al.*, 1996; Zhang, 1996; Samsonovich and McNaughton, 1997; Redish, 1999). Chaining this extra-network excitation to the represented direction forces the network to rotate continuously. In contrast, when the network inputs drive neurons with preferred directions far from the represented direction (greater than 60° in our network), the system will non-linearly jump to a new direction if the strength of the drive is large enough to overcome the global inhibition (Zhang, 1996; Samsonovich and McNaughton, 1997; Redish, 1999). Reconstruction showed a smooth transition through intermediate orientations in both the rotation and jump conditions (Fig. 5.4). Reconstruction thus suggested that both of these transitions were simple rotations, yet the dynamics

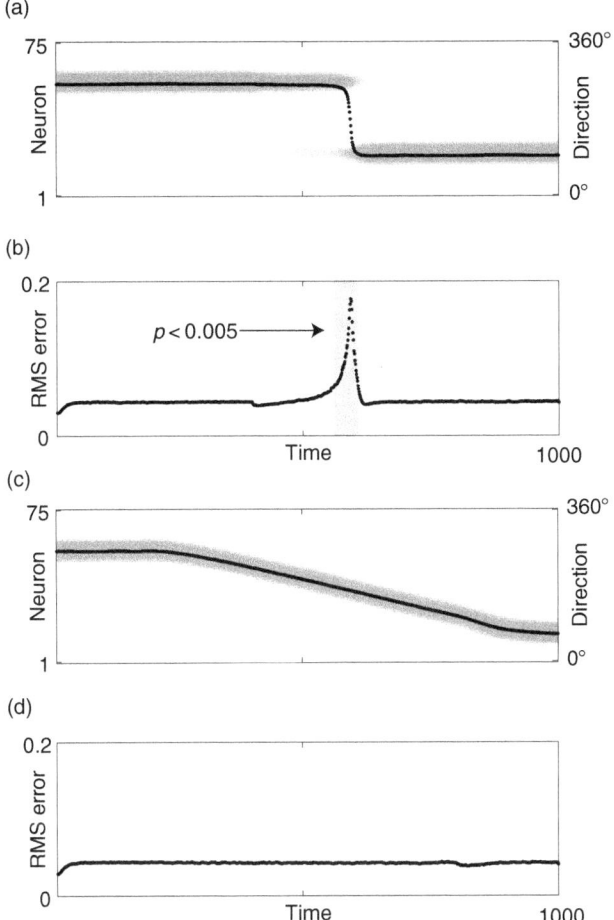

Figure 5.4 (a) Offset activity produces a jump in the representation. Layout as in Fig. 5.3. Note that the reconstructed position shows a smooth rotation from the initial position of activity before the jump, through positions where there is no network activity, to the final location of activity after the jump. (b) The I_{RMS} measure of inconsistency between actual and expected activity packets. The discrepancy between the actual and expected activity packets is low during the stable state, before and after the jump ($p > 0.005$, white zone), but high during the transient bimodal activity state at the moment of the jump from time-steps 562–609 ($p < 0.005$, gray zone). (c) A smooth rotation induced in the network yields stable results. Layout as above. The reconstructed position follows the activity of the network faithfully. (d) The I_{RMS} measure of inconsistency between actual and expected activity packets. Throughout the rotation, the network maintains a stable state with a small difference between the actual and expected activity packets ($p > 0.005$). (From Jackson, 2006; see also Jackson and Redish, 2003.)

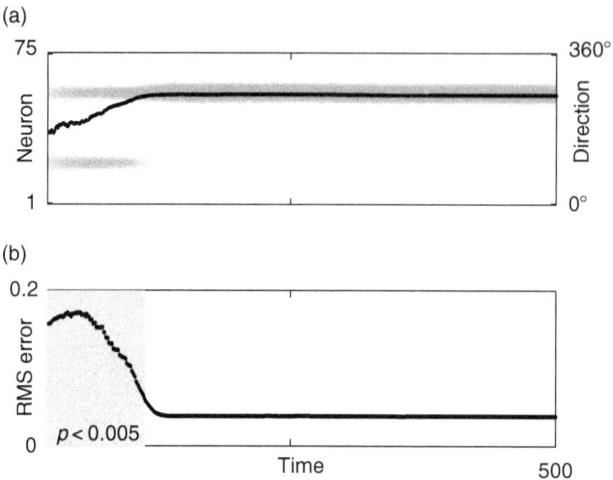

Figure 5.5 (a) A simulation started with competing inputs settles to a single mode of activity. Layout as in Fig. 5.3. Note that the reconstructed position shows a smooth rotation from the mean position, where there is no network activity to the winning location. (b) The I_{RMS} measure of inconsistency between actual and expected activity packets. The discrepancy between the actual and expected activity packets is high during the initial bimodal state before the competition is resolved ($p < 0.005$, gray zone), but low afterwards ($p > 0.005$, white zone). (From Jackson, 2006; see also Jackson and Redish, 2003.)

of these two transitions were fundamentally different. I_{RMS}, however, detected the difference. In the jump condition, I_{RMS} showed a strong transient increase at the time of transition (I_{RMS}: time-steps 562–609, $p < 0.005$), but no corresponding increase during the rotation (time-steps 200–800, $p > 0.005$).

Issue 3: Ambiguous vs. single-valued representations

If the network is started from a bimodal state (i.e. with inputs at two different directions), the population of neurons representing each input location will compete until the network settles on a single "bump" (Kohonen, 1977, 1982, 1984; Redish, 1999). This can serve as a selection process to resolve ambiguity (Wilson and Cowan, 1973; Kohonen, 1977; Redish and Touretzky, 1998). The location of the specific result will depend on the noise in the network and the difference in direction between the candidate inputs (Redish, 1997). During the settling process, there will temporarily be more than one bump of activity, one at the center of each input, essentially representing multiple values. Classical reconstruction techniques will be unable to determine whether or not the network has resolved the ambiguity. As shown in Fig. 5.5, self-consistency measures readily identify the resolution of the ambiguity.

Self-consistency in a Bayesian framework

The self-consistency measures reviewed above are based on observed changes in expected distributions, allowing the generation of statistical p-values identifying times when the representation significantly differs from the expected distribution (null hypothesis H_0: Eq. (5.12), above). Recent reconstruction analyses have been based on Bayesian and information measures (Rieke et al., 1997; Brown et al., 1998; Zemel et al., 1998; Zhang et al., 1998; Dayan and Abbott, 2001). It is possible to reinterpret the self-consistency question in terms of Bayesian reconstruction. In the methods above, the activity packet (Eq. (5.4)) measures the expected distribution of variable x given the firing rate at time t, F(t). In Bayesian terms, the posterior distribution $p(x|s(t))$ (Eq. 5.2) provides the term analogous to the expected activity packet above. Like the activity packet, this term is a function over the variable x and time. In the methods above, the self-consistency equations (Eqs. 5.7–5.10) measure the consistency of this reconstruction process relative to the (implied) model defined by the tuning curves T(x) (defined as the *expected activity packet*: Eq. 5.6). In the Bayesian formula, after decoding the neural representation, the validity can be determined by estimating how consistent the decoded representation is with the observed neural activity. One of the advantages of using a probabilistic approach in comparison to non-probabilistic estimate based methods is that rather that deriving a single estimate and using that in tandem with the tuning curves to develop the expected activities, the entire posterior distribution $p(x|s(t))$ can be used to generate the expected activity. Recall that the probability distribution (or the derived estimate) is a mapping over the space of x. Basic probability theory shows that the product of this probability with the encoding model produces a joint distribution over the spiking activity of the ensemble and the decoded variable. Substituting $p(x|s(t))$ for $p(x)$ in Eq. (5.2) gives

$$p(\hat{s}, \hat{x}) = p(\hat{s}|\hat{x})p(\hat{x}|s(t)) \tag{5.13}$$

where $\hat{x}$ reminds us that $p(\hat{x}|s(t))$ is estimating the distribution of the variable x given the observed spiking patterns, and $\hat{s}$ reminds us that $p(\hat{s}|\hat{x})$ is estimating the firing pattern s given our estimate of the variable x. Taking the marginal distribution with respect to the neural activity s (integrating across the variable x) provides the probability of a given neural activity set.

$$p_{\text{consistency}}(s) = \int_x p(s|x)p(\hat{x}|s)dx. \tag{5.14}$$

This formulation of consistency indicates that the probability that the observed neural activity was generated by the decoded neural representation provides a normative method for assessment for competing models, and makes clear that we need to make explicit the generative model used.

$$p_{\text{consistency}}(s|M) = \int_x p_M(s|x) p(\hat{x}|s, M) dx \tag{5.15}$$

where M indicates the generative model used. A generative model is *consistent* with the observed neural activity when $p_{\text{consistency}}(s|M)$ is high and *inconsistent* when this probability is low. Generative models can be compared using standard methods (such as odds ratios) and measured in decibans (Jaynes, 2003).

Multiple models in hippocampus

The previous sections have developed the idea that neural activity can represent many types of information – from sensory descriptions of the world to motor planning for behavior and even to the cognitive processes in between – and that the organization of this information is a critical aspect for both interpreting that information from within the system (as downstream neurons) or from outside the system (as experimenters). The suggestion of this approach is that rather than examining a decoded representation with respect to how well it matches an experimentally observed or controlled variable, a decoded representation should be examined on the basis of its intrinsic organization or consistency. In some instances a neural representation may match an observed variable and be well organized. In other instances, the neural representation may be disorganized. However, in other instances, the neural representation may completely disagree with an observed variable, but remain relatively well organized. Spatial representations within the hippocampus provide such an example.

Spatial representations in the hippocampus have been explored using a variety of decoding methods (Wilson and McNaughton, 1994; Brown et al., 1998; Zhang et al., 1998; Jensen and Lisman, 2000). Generally, the neural activity of place cells and the decoded spatial representation very well predicts the animal's position within the environment; however, recent studies have shown that place cell activity can remain well organized even when the decoded representation does not match the animal's position (Skaggs et al., 1996; Lee and Wilson, 2002; Johnson and Redish, 2005, 2007; Foster and Wilson, 2006; see also Fig. 5.1).

Within the hippocampus, multiple brain states have been identified based on characteristic local field potential activity (Vanderwolf, 1971; O'Keefe and Nadel, 1978). The hippocampal neural representations of space show different representational dynamics during these multiple brain states. In the theta regime, *phase precession* describes a dynamic that occurs during each theta cycle in which the spatial representation sweeps through positions recently occupied by the animal to positions that will likely be occupied by the animal (O'Keefe and Recce, 1993; Skaggs et al., 1996). The neural representation during this sweep is time-compressed approximately 10–15 times relative to animal behavior during task performance (Skaggs et al., 1996). In the large irregular activity (LIA) regime, *route replay* describes a dynamic that occurs during slow wave sleep, following task performance in which neuronal activity present during task performance is replayed (Kudrimoti et al., 1999; Nadasdy et al., 1999; Lee and Wilson, 2002). Spiking activity in sharp-wave replay is time-compressed 40 times relative to animal behavior during the task (Nadasdy et al., 1999; Lee and Wilson, 2002). The observation of both phase precession and sharp-wave ripples during awake states (O'Keefe and Nadel, 1978; O'Keefe and Recce, 1993; Foster and Wilson, 2006; Jackson et al., 2006; O'Neill et al., 2006), suggests that the hippocampal representation of space may operate with multiple spatiotemporal dynamics, even during awake behaviors.

Application of self-consistency measures provides a method for examining spatial representations in the hippocampus with respect to multiple spatiotemporal dynamics. Explicit comparison of multiple models of hypothesized representation dynamics allows identification of the underlying dynamical state of the neural representation. A model of the dynamics of a neural representation can be most simply described as a Markov process $p(\hat{x}_t|\hat{x}_{t-1})$, which gives the probability of the representation transitioning from the estimate $\hat{x}_{t-1}$ to a new estimate $\hat{x}_t$. These models can be as simple as a Brownian walk or as complex as a rigidly specified directional flow. Models of representational dynamics are easily added to the Bayes' decoding equation shown above (Eq. 5.2) and can be written in terms of a predictive filter.

The term predictive filter refers to the recursive application of a *prediction step* which predicts the temporal evolution of the neural representation given the previous prediction and the proposed dynamical model and a *correction step* which corrects the prediction based on the spikes observed at time t. The prediction can be written as

$$p(\hat{x}_t|s_{t-1}) = \int p(\hat{x}_t|\hat{x}_{t-1})p(\hat{x}_{t-1}|s_{t-1})d\hat{x}_{t-1} \tag{5.16}$$

where $p(\hat{x}_t|\hat{x}_{t-1})$ describes the hypothesized model of representation dynamics and the term $p(\hat{x}_{t-1}|s_{t-1})$ represents the previously predicted neural representation. The correction step can be written as

$$p(\hat{x}_t|s_{t-1}) = \frac{p(s_t|\hat{x}_t)p(\hat{x}_t|s_{t-1})}{p(s_t|s_{t-1})} \tag{5.17}$$

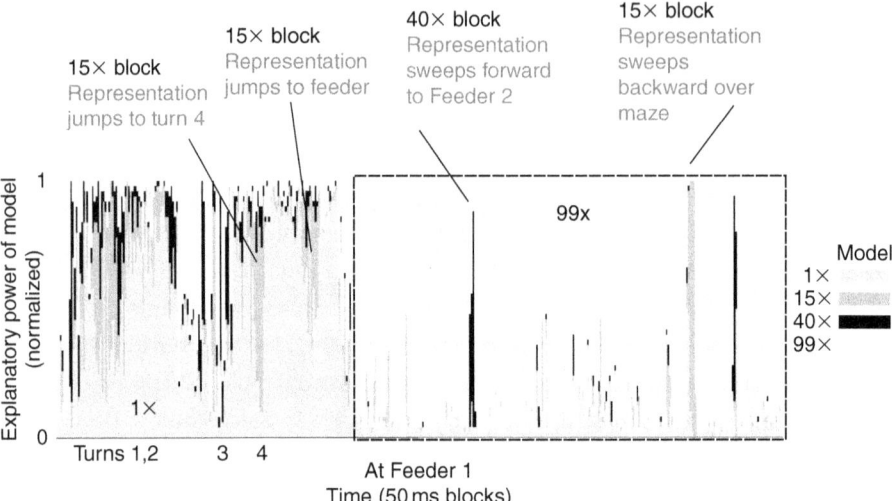

Figure 5.6 Multiple generative models in the hippocampus. Four generative models were examined: 1×, 15×, 40×, and 99×. During the first portion (turns 1–4), the animal was running through the maze. During the second portion, the animal paused at the first feeder to rest, groom, and eat.

where $p(s_t|s_{t-1})$ is the probability of the neural activity set s_t given the previous set of neural activity s_{t-1} and the term $p(\hat{x}_t|s_{t-1})$ represents a prediction from the previous equation. Predictive filters have been used for decoding neural activity within a variety of brain areas (e.g. Brown et al., 1998; Brockwell et al., 2004; Wu et al., 2006).

After decoding the neural representation for each of the proposed models, the validity of the hypothesized model can be determined by estimating how consistent the decoded representation is with the observed neural activity.

Four generative models were used to perform reconstruction from a hippocampal neural ensemble recorded from the CA1 and CA3 regions of an animal running on a 4-T multiple-T maze (Schmitzer-Torbert and Redish, 2002, 2004). Four generative models were examined, each of which allowed the probability distribution to spread at the prediction step with different rates: $1\times \equiv p(\hat{x}_t|\hat{x}_{t-1})^1$, $15\times \equiv p(\hat{x}_t|\hat{x}_{t-1})^{15}$, $40\times \equiv p(\hat{x}_t|\hat{x}_{t-1})^{40}$, and $99\times \equiv p(\hat{x}_t|\hat{x}_{t-1})^{99}$ where $p(\hat{x}_t|\hat{x}_{t-1})$ was a Gaussian function with σ proportional to the average velocity of the rat. The 99× model provided a nearly uniform distribution over the scale of the multiple-T maze. As can be seen in Fig. 5.6, different models were more consistent at different times, reflecting changes in the neural dynamics.

Model selection was accomplished by calculating an error between the expected spiking activity given the posterior distribution of each filter and observed spiking data as in Eq. (5.15). As noted above, the different generative

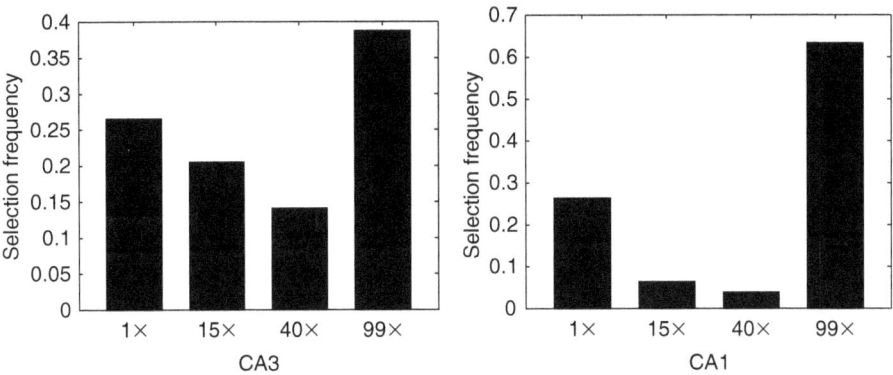

Figure 5.7 Percentage of samples in which each model was found to be the most consistent (Eq. 5.15). The 99× filter was often selected during jumps or intervals in which few spikes were fired.

models are hypothesized to reflect different information processing modes. In the hippocampus, these modes are reflected in local field potential signals (O'Keefe and Nadel, 1978; Buzsáki, 1989, 2006; Hasselmo and Bower, 1993; Redish, 1999), thus we hypothesized that the characteristic local field potential power spectrum for each spatiotemporal filter should show similar trends – specifically, the 1× and 15× filters should show increased power within theta frequencies (7–10 Hz) while the 40× filter should show increased power within slow wave delta (2–6 Hz) and sharp wave ripple (170–200 Hz) frequencies. Clear differences were found within slow wave and theta frequencies. Differences between the characteristic power spectra for each filter were similar between CA1 and CA3. Consistent with previous results (Lee et al., 2004; Leutgeb et al., 2004), subfield analysis found that more dynamic models (e.g. 99×) were more often selected in CA1 data sets, relative to the CA3 data sets (see Fig. 5.7).

While generative models have been broadly used to explain and decode neural activity (e.g. Brown et al., 1998; Rao and Ballard, 1999; Lee and Mumford, 2003; Brockwell et al., 2004; Serruya et al., 2004; Wu et al., 2006), one notable distinction should be made between the typical generative model formulation and the present formulation. Because we are concerned with the dynamical regulation of neural representations by cognitive processes, particularly explicit memory retrieval, we suggest that multiple generative models are necessary to explain observed neural activity. A single model is generally not enough because cognition requires the interactive use of dynamical information based on sensory or motor processes *and* planning, motivation or, for lack of another word, cognitive processes. Within each of these types of representation, cognition modulates an ongoing process. This is precisely the type of modulation that is sought when examining learning and memory or any cognitive processes and mathematically it can be identified as changes in the model prior $p_M(x)$. In terms of the generative model, this simply states that there exists a prior non-uniform distribution $p_1(x)$ which better describes the neural activity than a uniform distribution $p_2(x)$. The critical aspect of this formulation is that the goal is to completely generate the observed neural activity. Because of the probabilistic treatment, it becomes straightforward to integrate the contributions of both representation driven aspects of neural activity (e.g. above) and intrinsically driven neural dynamics such as refractory period (Frank et al., 2002).

Conclusions

A variety of experimental and theoretical results suggest the existence of cognitive processes requiring active memory use in decision-making. These processes are non-trivial to assess in human populations using such measures as self-report and are even more difficult to assess in non-human populations. Identifying such cognitive processes in non-human animals will require the development of measures to examine computations underlying these processes. Central to this approach is the development of statistical algorithms for decoding neural representations at multiple time scales and validation or error-assessment methods that allow characterization of cognitive processes related to, but not necessarily mirrored by, directly observable behavior. In this chapter, we have described a method for examining highly dynamic cognitive processes through observation of neural representations with multiple dynamics. Reconstruction alone cannot be used to infer internal states of an animal's sensory and cognitive networks such as the difference between random firing and well-represented variables. This is particularly important when considering issues of memory and recall. One function of memory is to appropriately link a current experience to a past experience; in the case of the hippocampus, this may mean using the same spatial map as was previously used in an environment. However, a primary usefulness of a memory is in its ability to influence disconnected experiences through recall of past events or episodes. In this case of recall, one would expect that neuronal firing would, by definition, be disconnected from the current behavioral state of the animal. Recall may be detected by reconstruction methods identifying values very different from the current behavioral value. Usually, these values are considered noise to be removed from a reconstruction algorithm. Using a self-consistency method like those presented here will allow an investigator to judge whether these aberrant reconstructions are truly valid representational events.

Acknowledgments

This work was primarily supported by R01-MH06829. Additional support was provided by NSF-IGERT training grant #9870633 (JCJ, AJ), by the Center for Cognitive Science at the University of Minnesota (AJ, T32HD007151) and a 3M Graduate Fellowship (AJ). We thank Paul Schrater for helpful discussions, particularly with the development of the generative models and model selection methods.

References

Amari, S. I. (1977). Dynamics of pattern formation in lateral-inhibition type neural fields. *Biol Cybernet* **27**:77–87.

Arai, K., Keller, E. L., and Edelman, J. A. (1994). Two-dimensional neural network model of the primate saccadic system. *Neur Networks* **7**:1115–1135.

Averbeck, B. B. (2001). Neural mechanisms of copying geometrical shapes. Ph.D. thesis, University of Minnesota.

Averbeck, B. B., Chafee, M. V., Crowe, D. A., and Georgopoulos, A. P. (2002). Parallel processing of serial movements in prefrontal cortex. *Proc Natl Acad Sci USA* **99**:13172–13177.

Averbeck, B. B., Chafee, M. V., Crowe, D. A., and Georgopoulos, A. P. (2003a). Neural activity in prefrontal cortex during copying geometrical shapes. I. Single cells encode shape, sequence, and metric parameters. *Exp Brain Res* **150**:127–141.

Averbeck, B. B., Crowe, D. A., Chafee, M. V., and Georgopoulos, A. P. (2003b). Neural activity in prefrontal cortex during copying geometrical shapes. II. Decoding shape segments from neural ensembles. *Exp Brain Res* **150**:142–153.

Averbeck, B. B., Latham, P. E., and Pouget, A. (2006). Neural correlations, population coding and computation. *Nat Rev Neurosci* **7**:358–366.

Batschelet, E. (1981). *Circular Statistics in Biology*. New York: Academic Press.

Bialek, W., Rieke, F., de Ruyter van Steveninck, R. R., and Warland, D. (1991). Reading a neural code. *Science* **252**:1854–1857.

Brockwell, A. E., Rojas, A. L., and Kass, R. E. (2004). Recursive bayesian decoding of motor cortical signals by particle filtering. *J Neurophysiol* **91**:1899–1907.

Brown, E. N., Frank, L. M., Tang, D., Quirk, M. C., and Wilson, M. A. (1998). A statistical paradigm for neural spike train decoding applied to position prediction from ensemble firing patterns of rat hippocampal place cells. *J Neurosci* **18**:7411–7425.

Buzsáki, G. (1989). Two-stage model of memory trace formation: a role for "noisy" brain states. *Neuroscience* **31**:551–570.

Buzsáki, G. (2006). *Rhythms of the Brain*. Oxford, UK: Oxford University Press.

Chrobak, J. J. and Buzsáki, G. (1994). Selective activation of deep layer (V–VI) retrohippocampal neurons during hippocampal sharp waves in the behaving rat. *J Neurosci* **14**:6160–6170.

Chrobak, J. J. and Buzsáki, G. (1996). High-frequency oscillations in the output networks of the hippocampal–entorhinal axis of the freely behaving rat. *J Neurosci* **16**:3056–3066.

Chrobak, J. J., Lörincz, A., and Buzsáki, G. (2000). Physiological patterns in the hippocampo-entorhinal cortex system. *Hippocampus* **10**:457–465.

Csicsvari, J., Hirase, H., Czurkó, A., and Buzsáki, G. (1999). Fast network oscillations in the hippocampal CA1 region of the behaving rat. *J Neurosci* **19**:1–4.

Dayan, P. and Abbott, L. F. (2001). *Theoretical Neuroscience.* Cambridge, MA: MIT Press.

Doboli, S., Minai, A. A., and Best, P. J. (2000). Latent attractors: a model for context-dependent place representations in the hippocampus. *Neur Comput* **12**:1009–1043.

Droulez, J. and Berthoz, A. (1991). A neural network model of sensoritopic maps with predictive short-term memory properties. *Proc Natl Acad Sci USA* **88**:9653–9657.

Eliasmith, C. and Anderson, C. H. (2003). *Neural Engineering.* Cambridge, MA: MIT Press.

Ermentrout, B. and Cowan, J. (1979). A mathematical theory of visual hallucination patterns. *Biol Cybernet* **34**:137–150.

Ferbinteanu, J., Kennedy, P. J., and Shapiro, M. L. (2006). Episodic memory: from brain to mind. *Hippocampus* **16**:704–715.

Foster, D. J. and Wilson, M. A. (2006). Reverse replay of behavioural sequences in hippocampal place cells during the awake state. *Nature* **440**:680–683.

Frank, L. M., Eden, U. T., Solo, V., Wilson, M. A., and Brown, E. N. (2002). Contrasting patterns of receptive field plasticity in the hippocampus and the entorhinal cortex: an adaptive filtering approach. *J Neurosci* **22**:3817–3830.

Georgopoulos, A. P., Caminiti, R., Kalaska, J. F., and Massey, J. T. (1983). Spatial coding of movement: a hypothesis concerning the coding of movement direction by motor cortical populations. *Exp Brain Res* **7(Suppl)**:327–336.

Georgopoulos, A. P., Kettner, R. E., and Schwartz, A. B. (1988). Primate motor cortex and free arm movements to visual targets in three-dimensional space. II. Coding of the direction of movement by a neuronal population. *J Neurosci* **8**:2928–2937.

Goodridge, J. P. and Touretzky, D. S. (2000). Modeling attractor deformation in the rodent head direction system. *J Neurophysiol* **83**:3402–3410.

Gray, J. (2004). *Consciousness: Creeping Up on the Hard Problem.* Oxford, UK: Oxford University Press.

Harris-Warrick, R. M. and Marder, E. (1991). Modulation of neural networks for behavior. *Annu Rev Neurosci* **14**:39–57.

Hasselmo, M. E. and Bower, J. M. (1993). Acetylcholine and memory. *Trends Neurosci* **16**:218–222.

Hoffmann, K. L. and McNaughton, B. L. (2002). Coordinated reactivation of distributed memory traces in primate neocortex. *Science* **297**:2070–2073.

Jackson, J. (2006). Network consistency and hippocampal dynamics: using the properties of cell assemblies to probe the hippocampal representation of space. Ph.D. thesis, University of Minnesota.

Jackson, J. C. and Redish, A. D. (2003). Detecting dynamical changes within a simulated neural ensemble using a measure of representational quality. *Network: Comput Neur Syst* **14**:629–645.

Jackson, J. C., Johnson, A., and Redish, A. D. (2006). Hippocampal sharp waves and reactivation during awake states depend on repeated sequential experience. *J Neurosci* **26**:12415–12426.

Jaynes, E. T. (2003). *Probability Theory*. Cambridge, UK: Cambridge University Press.

Jensen, O. and Lisman, J. E. (2000). Position reconstruction from an ensemble of hippocampal place cells: contribution of theta phase encoding. *J Neurophysiol* **83**:2602–2609.

Johnson, A. and Redish, A. D. (2005). Observation of transient neural dynamics in the rodent hippocampus during behavior of a sequential decision task using predictive filter methods. *Acta Neurobiol Exp* **65**(Suppl 2005):103. Poster 245.

Johnson, A. and Redish, A. D. (2007). Neural ensembles in CA3 transiently encode paths forward of the animal at a decision point. *J Neurosci* **27**:12176–12189.

Johnson, A., Seeland, K. D., and Redish, A. D. (2005). Reconstruction of the postsubiculum head direction signal from neural ensembles. *Hippocampus* **15**:86–96.

Kepecs, A. and Lisman, J. (2003). Information encoding and computation with spikes and bursts. *Network: Comput Neur Syst* **14**:103–118.

Kohonen, T. (1977). *Associative Memory: A System-Theoretical Approach*. New York: Springer.

Kohonen, T. (1982). Self-organized formation of topologically correct feature maps. *Biol Cybernet* **43**:59–69.

Kohonen, T. (1984). *Self-Organization and Associative Memory*. New York: Springer.

Kudrimoti, H. S., Barnes, C. A., and McNaughton, B. L. (1999). Reactivation of hippocampal cell assemblies: effects of behavioral state, experience, and EEG dynamics. *J Neurosci* **19**:4090–4101.

Laing, C. R. and Chow, C. C. (2001). Stationary bumps in networks of spiking neurons. *Neur Comput* **13**:1473–1494.

Lee, A. K. and Wilson, M. A. (2002). Memory of sequential experience in the hippocampus during slow wave sleep. *Neuron* **36**:1183–1194.

Lee, I., Yoganarasimha, D., Rao, G., and Knierim, J. J. (2004). Comparison of population coherence of place cells in hippocampal subfields of CA1 and CA3. *Nature* **430**:456–459.

Lee, T. S. and Mumford, D. (2003). Hierarchical Bayesian inference in the visual cortex. *J Optic Soc Am A* **20**:1434–1448.

Leutgeb, S., Leutgeb, J. K., Treves, A., Moser, M. B., and Moser, E. I. (2004). Distinct ensemble codes in hippocampal areas CA3 and CA1. *Science* **305**:1295–1298.

Mardia, K. V. (1972). *Statistics of Directional Data*. New York: Academic Press.

Munoz, D. P., Pélisson, D., and Guitton, D. (1991). Movement of neural activity on the superior colliculus motor map during gaze shifts. *Science* **251**:1358–1360.

Nadasdy, Z., Hirase, H., Czurkó, A., Csicsvari, J., and Buzsáki, G. (1999). Replay and time compression of recurring spike sequences in the hippocampus. *J Neurosci* **19**:9497–9507.

Nirenberg, S. and Latham, P. E. (2003). Decoding neuronal spike trains: how important are correlations? *Proc Natl Acad Sci USA* **100**:7348–7353.

O'Keefe, J. and Nadel, L. (1978). *The Hippocampus as a Cognitive Map*. Oxford, UK: Clarendon Press.

O'Keefe, J. and Recce, M. (1993). Phase relationship between hippocampal place units and the EEG theta rhythm. *Hippocampus* **3**:317–330.

O'Neill, J., Senior, T., and Csicsvari, J. (2006). Place-selective firing of CA1 pyramidal cells during sharp wave/ripple network patterns in exploratory behavior. *Neuron* **49**:143–155.

Rao, R. P. N. and Ballard, D. H. (1999). Predictive coding in the visual cortex: a functional interpretation of some extra-classical receptive-field effects. *Nat Neurosci* **2**:79–87.

Redish, A. D. (1997). Beyond the cognitive map: contributions to a computational neuroscience theory of rodent navigation. Ph.D. thesis, Carnegie Mellon University.

Redish, A. D. (1999). *Beyond the Cognitive Map: From Place Cells to Episodic Memory.* Cambridge, MA: MIT Press.

Redish, A. D. and Touretzky, D. S. (1998). The role of the hippocampus in solving the Morris water maze. *Neur Comput* **10**:73–111.

Redish, A. D., Elga, A. N., and Touretzky, D. S. (1996). A coupled attractor model of the rodent head direction system. *Network: Comput Neur Syst* **7**:671–685.

Redish, A. D., Rosenzweig, E. S., Bohanick, J. D., McNaughton, B. L., and Barnes, C. A. (2000). Dynamics of hippocampal ensemble realignment: time vs. space. *J Neurosci* **20**:9289–9309.

Rieke, F., Warland, D., de Ruyter van Steveninck, R., and Bialek, W. (1997). *Spikes.* Cambridge, MA: MIT Press.

Salinas, E. and Abbott, L. (1994). Vector reconstruction from firing rates. *J Comput Neurosci* **1**:89–107.

Samsonovich, A. V. and McNaughton, B. L. (1997). Path integration and cognitive mapping in a continuous attractor neural network model. *J Neurosci* **17**:5900–5920.

Schmitzer-Torbert, N. C. and Redish, A. D. (2002). Development of path stereotypy in a single day in rats on a multiple-T maze. *Arch Ital Biol* **140**:295–301.

Schmitzer-Torbert, N. C. and Redish, A. D. (2004). Neuronal activity in the rodent dorsal striatum in sequential navigation: separation of spatial and reward responses on the multiple-T task. *J Neurophysiol* **91**:2259–2272.

Schneiderman, E., Bialek, W., and Berry, M. J. (2003). Synergy, redundancy, and independence in population codes. *J Neurosci* **23**:11539–11553.

Serruya, M., Hatsopoulos, N., Fellows, M., Paninski, L., and Donoghue, J. (2004). Robustness of neuroprosthetic decoding algorithms. *Biol Cybernet* **88**:219–228.

Skaggs, W. E., Knierim, J. J., Kudrimoti, H. S., and McNaughton, B. L. (1995). A model of the neural basis of the rat's sense of direction. In: *Advances in Neural Information Processing Systems*, vol. 7, eds. Tesauro, G., Touretzky, D. S., and Leen, T. K., pp. 173–180. Cambridge, MA: MIT Press.

Skaggs, W. E., McNaughton, B. L., Wilson, M. A., and Barnes, C. A. (1996). Theta phase precession in hippocampal neuronal populations and the compression of temporal sequences. *Hippocampus* **6**:149–173.

Smyrnis, M., Taira, M., Ashe, J., and Georgopoulos, A. P. (1992). Motor cortical activity in a memorized delay task. *Exp Brain Res* **92**:139–151.

Somogyi, P. and Klausberger, T. (2005). Defined types of cortical interneurone structure space and spike timing in the hippocampus. *J Physiol (Lond)* **562**:9–26.

Suddendorf, T. and Busby, J. (2003). Mental time travel in animals? *Trend Cogni Sci* **7**:391–396.

Sutherland, G. R. and McNaughton, B. L. (2000). Memory trace reactivation in hippocampal and neocortical neuronal ensembles. *Curr Opin Neurobiol* **10**:180–186.

Tsodyks, M. (1999). Attractor network models of spatial maps in hippocampus. *Hippocampus* **9**:481–489.

Tsodyks, M. (2005). Attractor neural networks and spatial maps in hippocampus. *Neuron* **48**:168–169.

Tulving, E. (1983). *Elements of Episodic Memory*. New York: Oxford University Press.

Tulving, E. (2001). Episodic memory and common sense: how far apart? *Phil Trans R Soc Lond B* **356**:1505–1515.

Tulving, E. (2002). Episodic memory: from mind to brain. *Annu Rev Psychol* **53**:1–25.

Vanderwolf, C. H. (1971). Limbic-diencephalic mechanisms of voluntary movement. *Psychol Rev* **78**:83–113.

Wills, T. J., Lever, C., Cacucci, F., Burgess, N., and O'Keefe, J. (2005). Attractor dynamics in the hippocampal representation of the local environment. *Science* **308**:873–876.

Wilson, H. R. and Cowan, J. D. (1973). A mathematical theory of the functional dynamics of cortical and thalamic tissue. *Kybernetik* **13**:55–80.

Wilson, M. A. and McNaughton, B. L. (1993). Dynamics of the hippocampal ensemble code for space. *Science* **261**:1055–1058.

Wilson, M. A. and McNaughton, B. L. (1994). Reactivation of hippocampal ensemble memories during sleep. *Science* **265**:676–679.

Wu, W., Gao, Y., Bienenstock, E., Donoghue, J. P., and Black, M. J. (2006). Bayesian population decoding of motor cortical activity using a Kalman filter. *Neur Comput* **18**:80–118.

Ylinen, A., Bragin, A., Nadasdy, Z., *et al.* (1995). Sharp wave associated high-frequency oscillation (200Hz) in the intact hippocampus: network and intracellular mechanisms. *J Neurosci* **15**:30–46.

Zemel, R. S., Dayan, P., and Pouget, A. (1998). Probabilistic interpretation of population codes. *Neur Comput* **10**:403–430.

Zhang, K. (1996). Representation of spatial orientation by the intrinsic dynamics of the head direction cell ensemble: a theory. *J Neurosci* **16**:2112–2126.

Zhang, K., Ginzburg, I., McNaughton, B. L., and Sejnowski, T. J. (1998). Interpreting neuronal population activity by reconstruction: unified framework with application to hippocampal place cells. *J Neurophysiol* **79**:1017–1044.

6

Single-neuron and ensemble contributions to decoding simultaneously recorded spike trains

MARK LAUBACH, NANDAKUMAR S. NARAYANAN, AND EYAL Y. KIMCHI

Decoding simultaneously recorded spike trains

Pioneering studies of motor cortex by Georgopoulos and colleagues (e.g. Georgopoulos *et al.*, 1982) established that "population vectors," constructed from weighted averages of the responses of single neurons, can accurately predict behavioral variables, such as movement direction. This approach has been used to study population coding in a number of cortical systems and has led to the view that cortical neurons act as independent processors of information (e.g. Gochin *et al.*, 1994). However, some recent work has challenged this interpretation of neural population activity. For example, Schneidman *et al.* (2003) proposed interpreting neural ensemble activity by comparing ensemble information with information represented by the single neurons that comprise the ensemble. In a synergistic coding scheme, ensembles encode more than the sum of the component neurons. The advantage of synergy is that there can be a massive gain in information from the activity of multiple neurons. In a redundant coding scheme, the removal of individual neurons has little effect on encoding and thus the ensembles can be less noisy and less prone to errors. In Narayanan *et al.* (2005), we adapted the information-theoretical framework proposed by Schneidman *et al.* (2003) to measures of decoding of the performance

Information Processing by Neuronal Populations, ed. Christian Hölscher and Matthias Munk.
Published by Cambridge University Press. © Cambridge University Press 2009.

of a delayed response task with activity from the rodent motor cortex. The predictive relationship between neural firing rates and a categorical measure of behavior, e.g. correct vs. error performance of a reaction time task, was quantified using statistical classifiers. Estimates of decoding were made for single neurons and for all possible combinations of neurons in a given ensemble. Importantly, this approach is applicable to any type of data set. For example, if the behavioral measure of interest is represented continuously (e.g. arm movement: Wessberg et al., 2000; reaction times: M. Laubach, E.Y. Kimchi, N.S. Narayanan, and D.J. Woodward, unpublished data), a regression approach can be used and a continuous form of information theory (Kraskov et al., 2004) can be used to quantify interactions between neurons. The methods we used are general in nature and were developed in the field of machine learning (Hastie et al., 2003). An advantage of these methods over alternative approaches, such as point process models (Kass et al., 2005), is that we can apply the same methods to spike trains and simultaneously collected local field potentials (Narayanan and Laubach 2006b), membrane voltage from intracellular recordings (Sachdev et al., 2006), or signals collected with voltage-sensitive dyes (Davis et al., 2006). In Narayanan et al. (2005), we reported evidence for highly redundant interactions among neurons in the rodent motor cortex that underlie the decoding of behavioral outcomes in a lever-release delayed response task. The removal of individual neurons did little to degrade ensemble performance with regard to decoding behavioral outcome. Pairs and triplets of neurons were the exception, as many small ensembles were synergistic. The main novelty of our paper was showing that neural interactions depend on the size of the neural population. Redundant interactions among large ensembles of neurons have been reported for several brain systems: retina: Puchalla et al. (2005) and Schneidman et al. (2006); primate motor cortex: Carmena et al. (2005); primate prefrontal cortex: Averbeck et al. (2003) and Averbeck and Lee (2006); rodent taste cortex: Gutierrez et al. (2006). Synergistic interactions have also been reported in the literature (Gawne and Richmond, 1993; Maynard et al., 1999; Reich et al., 2001; Rolls et al., 2004; Samonds et al., 2004; Purushothaman and Bradley, 2005), but in each case the data sets that were investigated were rather small. Based on results in the Narayanan et al. (2005) study, we believe that resolving the issue of whether population coding in a given brain system is based on interactions between neurons will require that researchers record from relatively large groups of cells (>20). In this chapter, we will review our methods for using classifiers to study decoding. Our goal is to provide a "recipe" for others to use and to explain what we have learned by using a "pattern recognition approach" (Laubach, 2003).

Procedures for analyzing neural spike train data with statistical classifiers

We use classifiers to examine decoding because the methods allow us to examine interactions between large groups of neurons. Such data sets cannot be studied in practice using pure information theory. This might be due to not having enough trials to establish information theoretic metrics. However, in our view, the real problem is the size of the "codebook" (e.g. Strong et al., 1998) required to decode all possible patterns of spiking. If one tried to assess all possible interactions within an ensemble of 10 neurons (and using a single number representing firing rate, i.e. firing rate measured in one bin), this would require a look-up table, i.e. the "codebook," with 1023 elements. An ensemble with 15 neurons would require a table with 32 767 elements. With twice that number of neurons (30 neurons), there are 1 000 000 000 entries and this is with only a single measure of neural activity. There is no computer with sufficient memory to analyze all possible combinations of neural firing for an ensemble of 30 neurons measured over, for example, 100 10-ms bins. For this reason, dimension reduction and the use of classifiers (or regression) is critical. Classifiers are an alternative to using traditional statistical methods, such as the t-test, ANOVA, and their non-parametric equivalents. The strength of classifiers is that they measure *trial-to-trial variability* in neural firing. This is an aspect of neural function that needs to be measured if the goal is to understand how the brain processes information in real time. This issue is important for theoretical studies of neuronal activity in relation to behavior (e.g. Lo and Wang, 2006) and for the control of prosthetic devices (e.g. Wessberg et al., 2000, Taylor et al., 2002, Carmena et al., 2003). In practice, we have found many instances in which we obtain a significant difference in average firing rates across a set of behavioral conditions but this effect is not consistent from trial to trial (i.e. raster plots show that the cell does not fire at the same rate over all trials). As a result, some neurons might be considered to fire differently by a t-test and not produce high levels of classification. Classifiers have been developed by a large community of statisticians and applied mathematicians (see Hastie et al. (2003) for review) to deal with real-world data sets. For example, "modern" and flexible methods for classification, such as regularized discriminant analysis (Friedman, 1989), have been applied to data sets as in the detection of earthquakes and nuclear explosions (Anderson and Taylor, 2002) and analyzing gene microarrays (Guo et al., 2007). Such flexible methods have also been shown to be excellent at dealing with small data sets, for example, as in face recognition (Lu et al., 2003). Classifiers have been used in many studies of neural activity (see Deadwyler and Hampson (1997) for review). The use of classifiers was pioneered by Gochin *et al.* (1991), Miller *et al.* (1993), Schoenbaum and Eichenbaum (1995a 1995b), and Deadwyler *et al.* (1996). One of us, ML, has used *non-parametric* versions of these methods for more than 10 years to study relationships between neural firing and either sensory stimuli (Nicolelis *et al.*, 1998) or behavioral events (Laubach *et al.*, 2000, Krupa *et al.*, 2004, Narayanan *et al.*, 2005, Narayanan and Laubach, 2006a). The key advantage of non-parametric classifiers over classic methods, such as linear discriminant analysis (e.g. Briggman *et al.*, 2005), is that the non-parametric methods have limited assumptions about the predictor data. For example, predictors do not have to be normally distributed or have equal variance across task conditions. Such assumptions are rarely met by spike train data. Moreover, many algorithms for classification are freely available over the Internet, for example using software from the R or WEKA projects.[1] For these reasons, we believe that researchers need to make use of modern flexible methods for classification (see Hastie *et al.*, 2003). That said, we must admit that, despite using non-parametric methods, our approach is not entirely free of assumptions. For example, because we perform smoothing on our spike trains, when converting them to spike density functions, we eliminate the possibility of finding predictive relationships between behavior and high frequencies of spiking. Also, our estimates of decoding depend to some extent on the specific algorithm used for classification. It is possible (and probably even likely) that a classifier exists that will perform better on our data than the methods we have used here. For these reasons, we can only estimate a lower bound on the decoding of neuronal activity using classifiers. A critical issue for understanding the use of classifiers is the bias–variance trade-off (Geman *et al.*, 1992). That is, there are two sources of error, bias and variance, that impact the use of model-based methods such as classifiers. *Bias* refers to error between estimates of predictive information and true information. For statistical classifiers, there are three sources of bias (Witten and Frank, 2000). First, a statistical classifier might have a *language bias*. That is, some algorithms are better able to capture certain types of data than others based on how they represent information bearing aspects of the data set. Second, statistical classifiers have a *search bias* in the heuristic rules that they use to fit the data. If the rules are inappropriate, inaccurate conclusions may be reached. Finally, statistical classifiers can suffer from an *overfitting bias*, i.e. a tendency to find patterns that do not exist in the data. The second source of error in classification was *variance*, i.e. errors in classification due to the sample chosen from which mutual information was predicted. Variance in our data sets is accounted for by repeating classifications

---

[1] Their websites are R: http://cran.r-project.org/ and WEKA: http://www.cs.waikato.ac.nz/ml/weka/, respectively.

several times. In our research (e.g. Narayanan et al., 2005), we address the first two sources of bias by using multiple statistical classifiers with distinct internal representations and decision rules (e.g. regularized discriminant analysis and support vector machines). It is always possible that there exists another classifier that has an internal representation which better matches our data or that has more appropriate decision rules. The last source of bias, due to overfitting, is addressed in two ways. First, for every data set and every number of neurons, we apply classifiers to data in which the behavioral labels for all trials were randomly permuted. Second, we test our implementations of classifiers using benchmark data with known Bayesian error rates (the 'mlbench' library for R).

Overview: methods for decoding with classifiers

There are five main steps in our analysis procedures (Fig. 6.1a). First, we arrange neural firing rates into a "peri-event" format. For each neuron, a window is placed around a behavioral event of interest and spikes are recorded in 1-ms bins for each trial. This results in a matrix with rows representing time around the event of interest and columns representing each trial in a given behavioral session (Fig. 6.1b). Second, we *reduce the complexity of spike train data* using low-pass filtering and decimation. This compresses (or shortens) the spike train and converts the time series of spike times into an estimate of spike density over time. Third, we further reduce the dimensionality of the data, for each neuron, by using a *feature extraction* approach. Typically, a principal component or wavelet-based method is used to extract major components of variance in the spike density functions (Fig. 6.2). This procedure greatly reduced the dimensionality of our data sets. As described below, we typically go from a spike train with many 1-ms bins (e.g. 100) to a small set of features (e.g. three per neuron). Fourth, we use the extracted features to *train the classifier*. A subset of trials, called the training data, are used to define the parameters of the classifier. We usually use *leave-one-out cross-validation*, in which all but one trial is used as training data, or *10-fold cross-validation*, in which 90% of trials are used as training data (Fig. 6.3). We then *test the classifier*. Here, the trials not used to train the classifier are evaluated to measure how well the classifier performs on "new data." The fifth and final step is to *summarize the results*. We store the results of the classification of the training data into a *confusion matrix* and evaluate classifier performance in bits of information (Krippendorff, 1986). We also save *posterior probabilities*, i.e. how strongly the classifier associated a given trial with one categorical measure of behavior, and use these values to construct a curve for the receiver-operating characteristic (ROC) (Sing et al., 2005). This collection of methods is described in detail below. Two caveats are worth

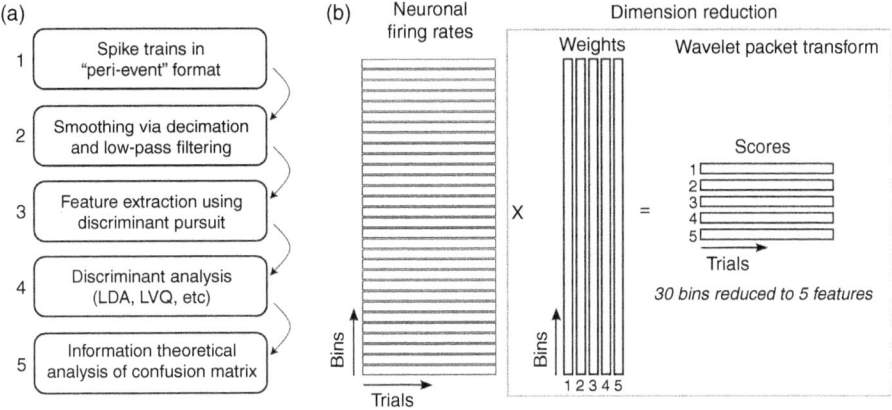

Figure 6.1 Procedures for decoding spike trains with classifiers. (a) Steps in the analysis. Spike trains are arranged in a "peri-event" format (see text for details). An estimate of spike density over peri-event time is then computed using decimation and low-pass filtering. Next, a wavelet-based discriminant pursuit is used to identify major components in the spike trains that are associated with differences in behavior. Scores for these "features" are used to train and test a classifier. Results from the classifier are expressed in terms of information theory by applying Shannon's equation for mutual information to the confusion matrix. (b) Dimension reduction due to use of wavelet-based feature extraction. Firing rates are stored in a matrix with rows representing trials and columns representing bins, 30 in this example. Discriminant pursuit identifies a series of filters that convert the 30-bin spike train into a single number called the score. The score reflects how strongly a given time-frequency filter was matched by the spike train on a single trial. If five scores are retained for training and testing the classifier, there is a sixfold reduction in the dimensionality of the data set.

mentioning here. First, it is important to assess effects with both leave-one-out and 10-fold cross-validation. Leave-one-out tends to work well given limited data but also is sensitive to serial correlations over trials, and this can give rise to spuriously high levels of noise correlations. Second, as is well known in regression analysis (Mosteller and Tukey, 1977), leave-one-out methods tend to give higher bias and 10-fold methods tend to give higher variance.

Reducing the complexity of spike train data

A major issue in multivariate statistics is the "curse of dimensionality" (Bellman 1961). That is, as one measures more and more variables, the multivariate space of the data grows at an exponential rate. This makes it difficult to specify stable and unique parameters for each of the many variables. There are several ways to avoid this issue when analyzing spike trains collected with multiple electrodes. One might simply measure firing rate using a relatively

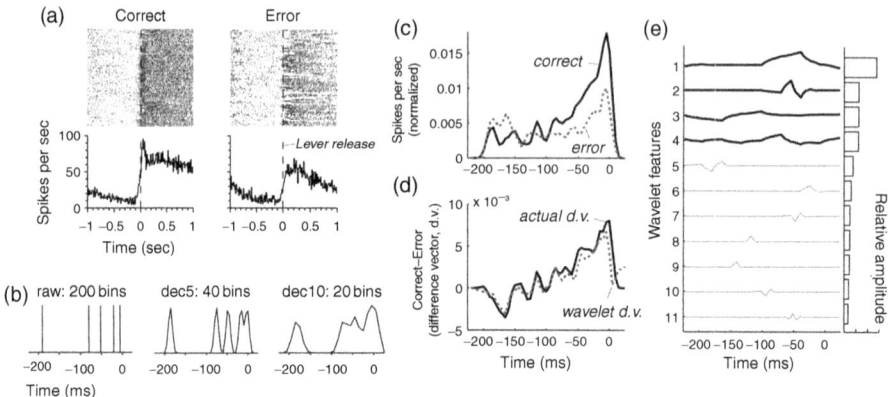

Figure 6.2 Example of spike density functions and feature extraction with discriminant pursuit. (a) Rasters and average peri-event histograms are shown for a neuron in rat motor cortex. Time 0 is when the lever was released in response to an auditory trigger stimulus. Bin size was 25 ms. The neuron fired at a reduced rate when the lever was released prematurely (error trial). (b) Conversion of spike train to spike density. On the left, spikes are shown as a point process. The two right plots show the results of decimating the spike train (1-ms bins) five and ten times to create estimates of spike density over 5- and 10-ms epochs. (c) Average firing rate for the neuron in (a) for the epoch from 200 ms before to 0 ms after lever release. (d) The "difference vector" used by the discriminant pursuit algorithm, defined as the average firing rate on correct trials minus the average firing rate on error trials. With 11 features, there is near-perfect reconstruction of the difference vector. (e) The time-course of the time-frequency features defined by discriminant pursuit. Comparisons between the variance accounted for by features from the actual data and from data with randomized group labels showed that only the top four features were larger than expected by chance. These features were used to train and test a classifier.

large bin during a task-epoch of interest and feed firing rate measures for an ensemble of neurons directly into a classifier. This would allow for examining hypotheses about how neural firing rates vary with a categorical aspect of behavior. However, neurons are not thought to process information in such a static manner, as in a look-up table. To address how firing rates covary with behavior requires measuring the rate of change, or slope, of the firing rate or using several large bins per neuron. Such additional measures can then be analyzed with a classifier. If one wishes to capture a more complete description of the time-varying nature of the firing rate, then it becomes a non-trivial task to define useful measures of neural firing. A pioneering effort on this topic was made by Richmond, Optican, and colleagues (e.g. Richmond and Optican, 1987). They used principal component analysis (PCA) to reduce a spike density functions to a small number of principal components (PCs) and found that

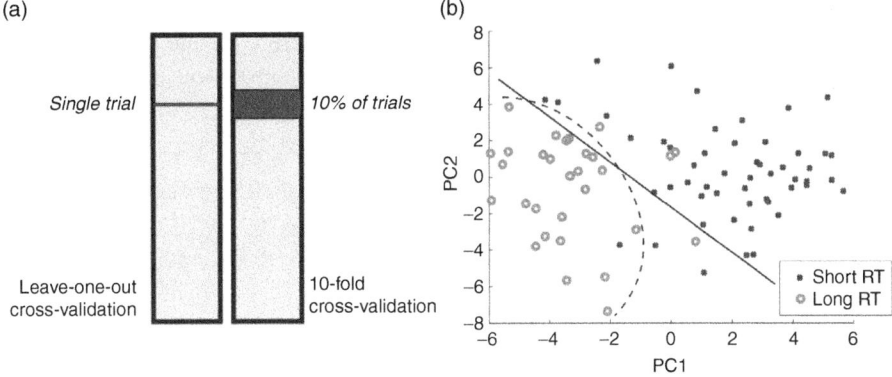

Figure 6.3 Use of classifiers for studying decoding. (a) Cross-validation is used either with "leave-one-out" partitioning or "10-fold" partitioning. In the "leave-one-out" procedure, a single trial is used as testing data and all other trials are used to train the classifier. In the "10-fold" procedure, 10% of trials are used as testing data and 90% of trials are used to train the classifier. (b) Example of a scatterplot of scores for an ensemble of neurons from motor cortex. There were 11 neurons with between two and seven features. Here, the first two principal components (PC1 and PC2) for the collection of features are shown. Clear clustering is revealed for trials with short and long reaction times (RT). Short RTs (filled circles) were defined as the lower tertile. Long RTs (open circles) were defined as the upper tertile. The linear decision boundary obtained with linear discriminant analysis is shown as a solid line. A curvilinear decision boundary obtained with regularized discriminant analysis, a non-parametric classifier, is shown as a dashed line. The identification of trials with long RTs was improved by using the non-parametric classifier.

they could represent most of the variance in spike trains from visual cortex in this manner. Each PC is formed by the weighted linear sum of the firing rate of the neuron over the time epoch of interest. The PCs are derived in an orthogonal manner such that each PC accounts for a unique portion of variance in the spike density function. The dot product of the coefficients for each PC and the spike density function gives a score for the PC for each trial in the experiment. Richmond and colleagues then evaluated how the scores for the PCs varied due to visual stimuli using information theory. They focused on single-neuron data and did not examine ensembles of neurons. Nevertheless, their approach is well suited for analysis of large numbers of neurons. Scores from PCs for each simultaneously recorded neuron can be combined and then decoding can be evaluated using a classifier. PCA is able to provide an efficient representation of temporal variations in a spike train. However, the method is known to have a limited ability to account for the fine details of local structure in signals and images. For this reason, we have used a wavelet-based method, discriminant pursuit (Buckheit and Donoho, 1995) in our research. This method is described in Laubach (2004). In the text below, we briefly describe how to use this methodology. There are two major steps in these procedures: (1) estimation of spike density functions and (2) application of the discriminant pursuit algorithm.

Spike density function

Spike trains are converted into continuous representations of firing rate prior to application of our wavelet method for dimension reduction. To do this, spikes are smoothed over 5-, 10-, or 20-ms intervals using a decimation method (Figure 6.2b). This procedure involves a low-pass filtering and resampling (see Fig. 6.2b) using standard routines (fir1, decimate, and filtfilt) from the Signal Processing toolbox for Matlab (The Mathworks, Natick, MA). Code for creating spike density functions is available on our laboratory's website.[2] (Note that we do not use filtfilt for preprocessing local field potentials, given the 0-phase nature of this filtering scheme.)

Discriminant pursuit

The difference between the mean spike density function for each neuron on different types of trials (e.g. conditioned vs. premature response) is computed and then decomposed with the wavelet packet transform. (Discriminant pursuit is not limited to two classes. In cases with many classes, c, the matrix of $c-1$ difference vectors is decomposed.) Each wavelet component is a feature of the spike train that can be ordered by the amount of between-groups variance it accounts for. By choosing to work with only the features that account for the largest amounts of between-groups variance (the "amplitude" of each feature), discriminant pursuit is able to dramatically reduce the number of data variables that are used for subsequent analysis. The number of significant features for each neuron was determined by comparing the amplitudes of features extracted from experimentally obtained data with those extracted from 1000 data sets in which the outcome of each trial was assigned pseudo-randomly. This procedure guarantees that the features obtained with discriminant pursuit are not due to differences in spike trains that could arise by chance alone. All procedures for feature extraction are carried out in Matlab (The Mathworks, Natick, MA) using the freely available "WaveLab" toolbox.[3] Computational details on the discriminant pursuit algorithm are given in Buckheit and Donoho (1995) and Laubach (2004).

---

[2] See http://spikelab.jbpierce.org/dp.
[3] WaveLab is available at http://www.stat.stanford.edu/~wavelab/. Matlab scripts for these feature extraction procedures are available at http://spikelab.jbpierce.org/dp or by request from the corresponding author.

Training a statistical classifier

Scores for each wavelet feature are used to train a classifier. As shown in Fig. 6.3a, part of the data is used as *training data* and part is used as *testing data*. The data set is split in this way to avoid *overfitting*. The parameters of the classifier are determined using the training data. The performance of the classifier is assessed using the testing data.[4] The overall idea of any classifier is to find a *decision boundary* in multivariate space. A classic method for classification, linear discriminant analysis (LDA), uses lines as boundaries between groups. In Fig. 6.3b, the solid line shows the type of boundary that LDA might use for classification. However, in many real-world data sets, performance improves if a curvilinear boundary is used (dashed line in Fig. 6.3b). This is the idea of regularized discriminant analysis (RDA) (Friedman 1989). There are several alternatives to this approach. One simple idea is to reduce the dimensionality of the predictor variables (e.g. the complete set of features from all neurons) and then use LDA on the scores from the resulting multivariate functions. This is the basis of penalized discriminant analysis (PDA) (Hastie *et al.*, 1995). Another idea, as used in support vector machines (SVM) (Hsu *et al.*, 2003), is to estimate the boundary based on points that are closest to an initial estimate of the boundary. Finally, a nearest-neighbor approach may be used, as in learning vector quantization (LVQ) (Kohonen 2000), in which there are multiple local boundaries estimated based on subsets of the training data. In practice, the choice of algorithm has little impact on evaluating neural coding (Narayanan *et al.*, 2005). However, we have found that some algorithms, such as LVQ, fail to show major gains in decoding if a very large number of predictors are used. As a result, LVQ has an inherent tendency to find redundant interactions within a given data set. For this reason, we have moved to using RDA, PDA, and SVM in our research. We use R for running classifiers on our data sets. Strengths of using R are that the classifiers are run using actual source code developed by the authors of each algorithm and that the program is open source and freely available.[5]

---

[4] It is well beyond the scope of this chapter to review the ideas that underlie the use of a classifier. Here, we only review essential points for understanding the methods we use for neural data analysis. The interested reader is referred to Hastie *et al.* (2003) for thorough coverage on these issue.

[5] RDA is available in the "klaR" library for R. PDA is available in the "mda" library. SVMs are available in the "e1071" library. LVQ is available in the "class" library. The R website is at http://www.R-project.org/.

Testing a statistical classifier

After the parameters of the classifier have been established, the "holdout" data are used to estimate the error rate of the classifier. In the case of leave-one-out cross-validation, a classifier is built for each trial in the data set. The parameters of the classifier are determined using all but the test trial. The test trial is then classified and the success and posterior probability of the classification is recorded. In the case of 10-fold cross-validation, the data set is broken into ten parts, with the order of trials randomized before assignment to each of the parts. Then, ten classifiers are trained from 90% of the total data set, using nine parts, and testing is done on 10% of the trials. That is, testing is done on the data that were "held out" of the training data set. As suggested by Witten and Frank (2000), we run ten such 10-fold cross-validations whenever we use this approach. In general, our experience has been that leave-one-out cross-validation works best for small data sets, with less than 100 trials per class (i.e. trial outcome), but is slow to use for larger data sets. In such cases, we use 10-fold cross-validation. As noted above, leave-one-out methods are biased by correlated noise and so we recommend that both leave-one-out and 10-fold cross-validation are used to confirm an important result.

Quantification of results using information theory

To summarize how well a given classifier discriminates between types of trials (e.g. different behavioral outcomes, different sensory stimuli), we use Shannon's equations for calculating mutual information, I_{AB}, from the confusion matrix (see below), as proposed by Krippendorff (1986). We prefer to use I_{AB} over the percentage of trials that are classified correctly because I_{AB} takes the underlying probabilities of the classes into account for evaluating a given classifier. As shown in Fig. 6.4a, the confusion matrix has four entries for a two-class discrimination problem:

> w is the number of trials from class 1 (e.g. conditioned lever releases in the delayed response task) that are classified correctly.
> x is the incorrectly classified trials from class 2 (e.g. premature lever releases classified as conditioned releases).
> y is the incorrectly classified trials from class 1 (e.g. conditioned lever releases classified as premature releases).
> z is for correctly classified trials from class 2 (e.g. premature releases).

I_{AB} is calculated as:

$$I_{AB} = H_A + H_B - H_{AB}. \tag{6.1}$$

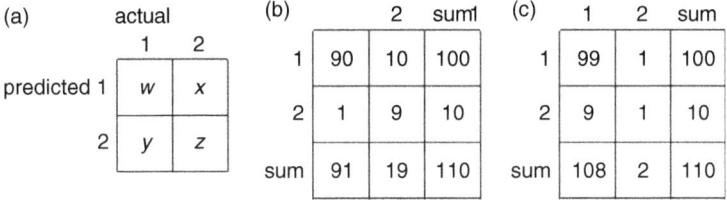

Figure 6.4 Application of mutual information to the confusion matrix obtained from testing a classifier. (a) Entries into the confusion matrix for correctly and incorrectly classified trials. See the text for details. (b) Example of a classifier that accurately discriminates between classes 1 and 2. Note that class 2 has many fewer trials than class 1. This would be a difficult problem for most methods in practice. (c) Example of a classifier that "lumps" all trials into the larger class. The classifiers in (b) and (c) can not be dissociated using the percentage of trials that are correctly classifier. Information theory is, however, able to dissociate between these results because the method accounts for the underlying probabilities of each class. See the text for details.

H_A is the "sender's entropy" that accounts for the probability distribution of the actual class assignments. This term is the sum of the following for each trial that was actually from a given class, i, and is determined from the actual class labels:

$$N_{\text{actual number for class } i} \div N_{\text{total}} \times \log_2(N_{\text{actual number for class } i} \div N_{\text{total}}) \times -1. \quad (6.2)$$

H_B is the "receiver's entropy" that accounts for the probability distribution of the classifier's predictions. This term is the sum of the following for each trial predicted to be a given class, i, and is determined from the list of trial classifications:

$$N_{\text{number classified as } i} \div N_{\text{total}} \times \log_2(N_{\text{number classified as } i} \div N_{\text{total}}) \times -1. \quad (6.3)$$

H_{AB} is the "total entropy" that accounts for the distribution of trials over the entire confusion matrix. This term is the sum of the following for each entry in the confusion matrix:

$$\left(N_{\text{entry in confusion matrix}} \div N_{\text{total}}\right) \times \log_2\left(N_{\text{entry in confusion matrix}} \div N_{\text{total}}\right) \times -1. \quad (6.4)$$

For example, if there are 110 trials in the data set, with 100 of class 1 and just 10 of class 2 and 90% of each class is classified correctly, then we have the confusion matrix in Fig. 6.4b. The sum along the diagonal divided by the total number of trials is the percentage correct: $99 \div 110$ is 90% correct. Here, H_A is 0.4395 bits, H_B is 0.6639 bits, and H_{AB} is 0.9085 bits. This gives I_{AB} as 0.1949 bits of information. This value is 44.4% of the total information needed for near-perfect classification (i.e. $0.1949 \div 0.4395$). By contrast, for the confusion matrix in Fig. 6.4c, we get about the same percentage correct, $100 \div 110$ or 90.9%. H_A is the same (0.4395 bits), H_B is just 0.1311 bits, and H_{AB} is 0.5555 bits. Now, this gives 0.015 bits of information, which is only 3.4% of the total information needed.

Determining what is expected by chance: the importance of looking at random data

The level of information that is expected to arise by chance may not always be equal to 0 bits. Classifiers can perform with a high level of bias (e.g. Geman et al., 1992). That is, a classifier can perform at levels better than expected by chance (i.e. 50% correct classification for a two-class problem) if given randomly generated data. Such a result can arise if a method is overly sensitive to small variances in the predictor variables (e.g. taking too many wavelet features), if cross-validation is not used, or if feature extraction is done without cross-validation. These situations should be avoided by using ideal procedures, that is, by performing feature extraction and classification using ten runs of 10-fold cross-validation. However, in some cases, as in the analysis of all-possible-combination of neurons (Narayanan et al., 2005), there are practical limitations for how long it takes to run a given analysis. It is then necessary to, for example, run classifiers on features defined without cross-validation. In such cases, it is critical to know how well a biased method performs if given random data. To examine significant interactions between an ensemble of 16 neurons, we must run 65353 classifiers for all possible combinations of neurons from 2 to 15. A typical cycle of feature extraction, training, and testing the classifier can take as long as 10 min on a single fast PC workstation. As such, we would need more than a week of computer time to complete the analysis. By taking only one set of features, the analysis time is cut by roughly one-third. To further reduce the time needed for these calculations, the analysis can be run in parallel over a cluster of workstations (Narayanan et al., 2005). To do this, we ran analyses across a set of eight dual-processor PCs (AMD Athlon) running a variant of Red Hat Linux (CentOS). We used the *snow* library for R, using a simple sockets-based approach, and with data sets and code stored locally on each identically configured PC. To date, we have not made use of methods for parallel programming, and so the time taken to complete analysis on a given data set was effectively equal to 1/16 that of running on a single PC.

As stated above, the consequence of not using cross-validation for feature extraction is that classifiers trained on random data (simulating random data for a "single neuron" 10000 times) can discriminate between two types of trials at levels that are significantly greater than 50% correct (or 0 bits of mutual information, I_{AB}, or area under the ROC curve, or AUC, of 0.5). The 95th percentile for accuracy values from a classifier given features from random data without

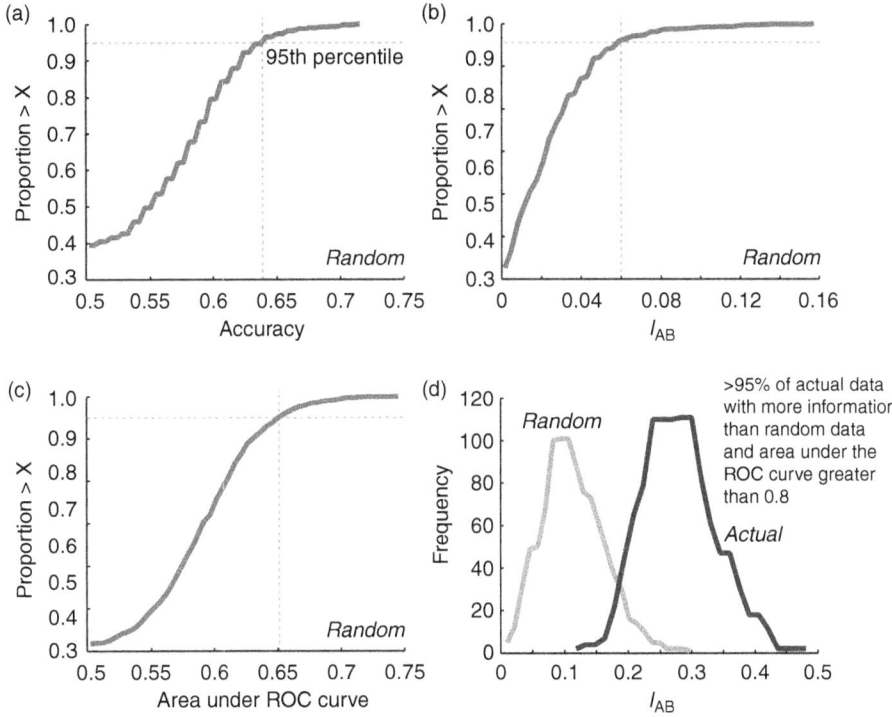

Figure 6.5 Results obtained from running classifiers on random data. (a) Percentage of trials that are classified correctly. (b) Mutual information (I_{AB}). (c) Area under the curve from a receiver-operating characteristic. (d) Comparison between mutual information for actual data and an equal number of randomly generated variables.

cross-validation can be greater than 0.6 (fraction of trials that are classified correctly) (Fig. 6.5a). The mutual information from such classifiers can be as large as 0.06 bits (Fig. 6.5b) and the area under the ROC curve can be as large as 0.65 (Fig. 6.5c). Nevertheless, these values are not as high as actual neural data if the data set contains neurons that are significantly modulated by behavior. For example, the data sets in Narayanan et al. (2005), less than 5% of classifiers from the actual data provided as little information as the classifiers run on random data (Fig. 6.5d). Based on these studies, we have revised our threshold to 0.06 bits of information (see Narayanan et al., 2005) for concluding that decoding by a neuron or ensemble was significantly better than expected by chance. We use this same value for reporting significant effects in analyses of neural interactions (i.e. synergy and redundancy calculations below). In addition, we have recently moved from using the discriminant pursuit algorithm to using a simple wavelet packet decomposition of the overall average spike density function, that is, the average response over all trials or the average response for one type of trial (Narayanan and Laubach, 2006a). By using features from a wavelet packet analysis on such an average response, the bias is only due to the classifier, and is reduced to ~0.03 bits for a two-class problem.

Case study: decoding correct and error responses based on motor cortical spike trains

Behavioral and electrophysiological methods

Here, we illustrate our approach to studying neural interactions using data from the rat motor cortex collected during a lever-release delayed response (Laubach et al., 2000; Narayanan et al., 2005; Narayanan and Laubach, 2006a). In the task, rats pressed a lever to initiate a trial and then maintained the lever-press over a variable delay period. At the end of the delay, they received a trigger stimulus (tone or vibration of the lever) and had to release the lever within 600 ms of stimulus onset to obtain a reward. This type of trial is called a "conditioned" response. "Premature" responses occurred if the lever was released before the trigger stimulus. Late responses (RT > 600 ms) were infrequent after initial training sessions (<10% of trials) and were not examined with classifiers. Neural ensemble recordings were made using a 64-channel Plexon recording system (MNAP). Single units were identified by the presence of a consistent spike waveform with an amplitude at least three times larger than background activity and by using interspike interval histograms that showed few, if any, spikes with short (<2 ms) intervals. We only analyzed units whose activity was consistently modulated by the behavioral task. A unit had to exhibit a modulation in firing rate in a window around the time of lever release (from 1 second before to 0.5 seconds after lever release). This was confirmed by comparing firing rates during the RT epoch (0–600 ms around trigger stimulus) with firing rates in pseudo-randomly chosen epochs throughout the behavioral session. Additionally, a unit had to have, on average, at least one spike per trial in the 200 ms prior to lever release.

Methods for data analysis

Features were extracted with discriminant pursuit and fed into a statistical classifier. The classifiers were trained on subsets of trials (i.e. training data) and made predictions for behavioral outcomes for trials not included in the training data (i.e. testing data). Leave-one-out cross-validation was used. Several different classifiers were used to study neural interactions in terms of synergy and redundancy (LVQ, SVM, PDA). As described above, predictive information for each data set was calculated via the Shannon entropy function applied to classification matrices (Krippendorff, 1986) constructed from the testing data.

Quantification of synergy and redundancy

To assess the contributions of each neuron to decoding using the neuron as part of an ensemble, we removed features for each neuron from each ensemble and re-estimated the predictive information available from the remaining neurons. The difference between the predictive information of the full ensemble and the predictive information of a subensemble without a given neuron was defined as I_c, the information contributed to the ensemble by that neuron. To measure the statistical relationship between a neuron and the rest of the ensemble, we defined the information contributed by single neurons to the ensemble, P_i, based on the difference between the individual predictive information of the neuron and its contribution to ensemble predictive information:

$$P_i = I_c - I_i \tag{6.5}$$

where I_c is defined as $I_e - I_e|i$. That is, I_c is the contribution of the kth neuron to the ensemble as measured by the predictive information lost after removing the kth neuron from the ensemble of N neurons (where $1 = k = N$). I_i is the predictive information (in bits) encoded by kth neuron, determined by classification of the features associated with the kth neuron. This equation answers the question: how does removing a neuron from an ensemble change ensemble predictive information? There are three possible results for removing a neuron from an ensemble that define the type of net contribution made by a neuron to information encoding at the ensemble level (Fig. 6.6). If P_i is approximately zero (if $I_c \approx I_i$), then the information lost by removing a neuron is equal to the individual information encoded by that neuron. In this case, the informational contribution to the ensemble is the same magnitude as its individual contribution. Such a neuron contributes *independent* information to the ensemble. Second, if P_i is negative (if $I_c < I_i$), the information lost by removing a neuron is less than the individual information encoded by that neuron. In this case, the informational contribution of that neuron to the ensemble is less than its individual information. Such a neuron contributes *redundant* information to the ensemble. Third, if P_i is positive (if $I_c > I_i$), the information lost by removing a neuron is more than the individual information encoded by that neuron. In this case, the informational contribution of that neuron to the ensemble is more than its individual information. Such a neuron contributes *synergistic* information to the ensemble. We defined interactions at the ensemble level as P_e, based on the ensemble information and the sum of the individual information of its component neurons (Schneidman et al., 2003):

$$P_e = I_e - \sum_1^n I_i \tag{6.6}$$

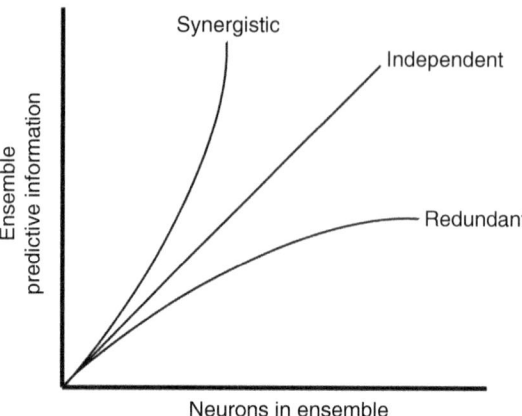

Figure 6.6 Synergy and redundancy in neuronal ensembles. As neurons are added to an independent ensemble, predictive information increases proportionally to number of neurons in the ensemble. If the ensemble is redundant, predictive information increases slower than number of neurons added to the ensemble. If the ensemble is synergistic, predictive information increases faster than number of neurons added to the ensemble.

where I_e is the predictive information of the ensemble, and I_i, as above, is the predictive information of individual neurons in the ensemble. This equation answers the question: how is predictive information decoded from the collective activity of neurons? That is, how do functional interactions between neurons influence decoding correlations between spike activity and behavior? As above, there are three possible coding schemes (Fig. 6.6). First, if P_e is zero, ensemble predictive information is the same as the sum of predictive information from neurons in the ensemble. As a whole, this ensemble is neither redundant nor synergistic. Second, if P_e is negative, ensemble predictive information is the less than the sum of predictive information from neurons in the ensemble. As a whole, this ensemble is *redundant*. Third, if P_e is positive, ensemble predictive information is more than the sum of predictive information from neurons in the ensemble. As a whole, this ensemble is *synergistic*. Importantly, there are limits on the wider applicability of these measures of functional interactions between neurons that are due to the fact that both the behavioral and neuronal data that are analyzed are from a limited sample of the animals' complete behavioral history and that it is possible that a given neuron will have non-stationary response properties over a prolonged period of time. This latter point is especially problematic and, in our opinion, has not received the attention it deserves in the literature on neural coding.

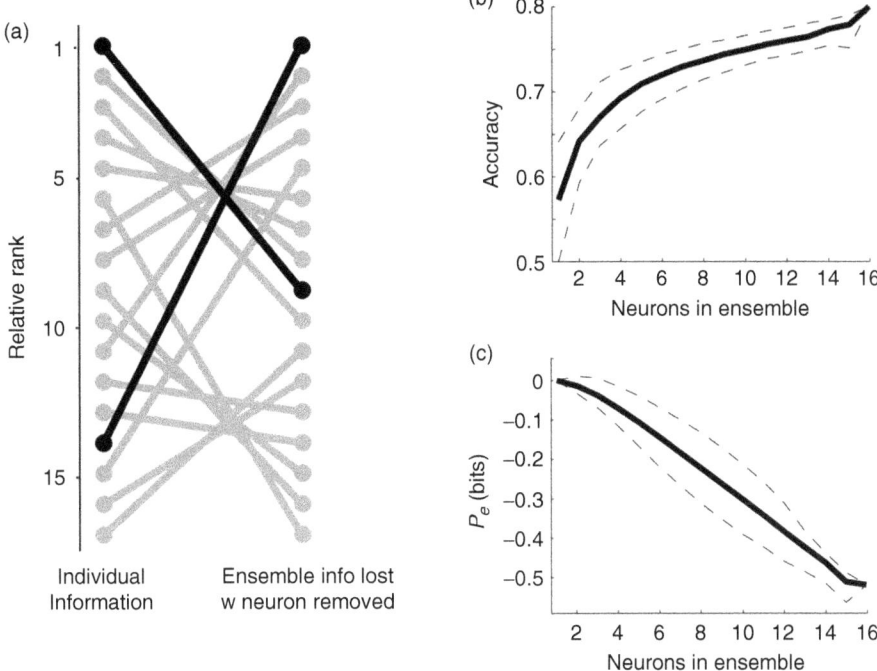

Figure 6.7 Neuronal interactions in decoding population activity. (a) Relative rank of neurons based on classification using only features from each neuron (left) and based on how much information is lost if the neuron is removed from the ensemble (right). (b) Increase in accuracy (percentage of trials classified correctly) for ensembles with two to 16 neurons. (c) Increase in negative P_e (redundancy) with increasing numbers of neurons in an ensemble. (Data are from Narayanan et al., 2005.)

Evidence for redundant neural interactions in rodent motor cortex

The application of the analysis framework described above led to three main results. First, the amount of information about the performance of a behavioral task that can be decoded with the activity of a single neuron was not predictive of how much information that neuron contributed to decoding with an ensemble of neurons (Fig. 6.7a). Second, almost all motor cortical neurons (96%) contributed redundant information, i.e. P_i was negative, to an ensemble (Fig. 6.7c). This redundancy led to a minimal loss of predictive information until most neurons were removed from a given ensemble (Fig. 6.7b), where information drops off once more than 10 of 16 neurons are removed. Interestingly, redundant pairs of neurons tended to produce misclassifications on the same trials (Fig. 6.8). A major factor driving redundancy appears to be correlated noise (Averbeck and Lee, 2006). As shown in Fig. 6.9, shuffling the trial

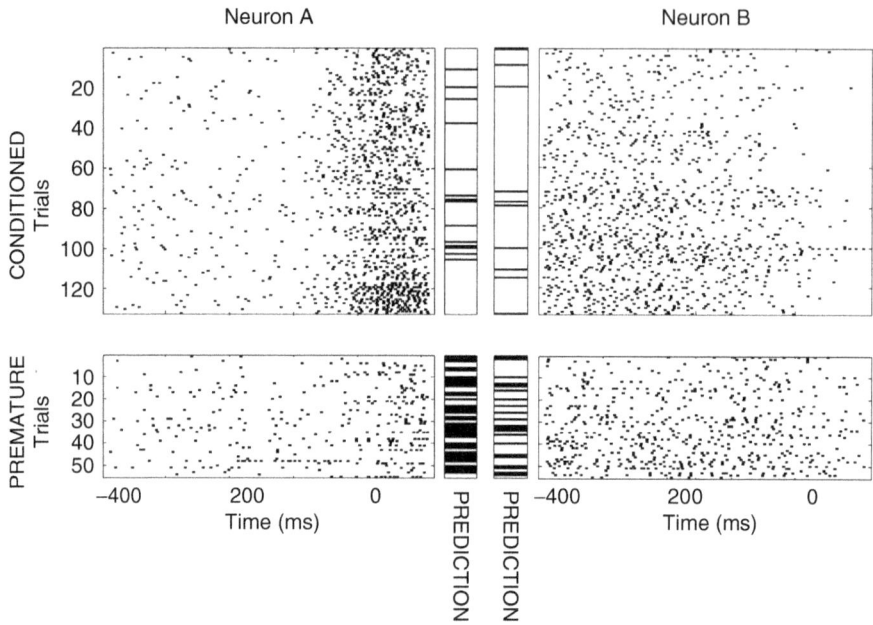

Figure 6.8 Covarying predictions of task performance by a pair of redundant neurons: rasters from neurons A (0.22 bits) and B (0.11 bits) are plotted next to their predicted class labels (two middle columns). White – classifier predicted conditioned response, black – classifier predicted premature response. Note that this pair, which has 0.17 bits of redundancy, often produces misclassification on the same trials.

orders for the neurons increased classification accuracy. This result shows that a major component of the redundant information between the neurons was due to correlations in the trial-by-trial variability of the spike trains. Third, the balance of synergistic and redundant interactions depended on the number of neurons in the ensemble. Small ensembles could exhibit synergistic interactions (e.g. $23 \pm 9\%$ of ensembles with two neurons were synergistic). By contrast, larger ensembles exhibited mostly redundant interactions (e.g. $99 \pm 0.1\%$ of ensembles with eight neurons were redundant).

These results suggest that single neurons in the rat motor cortex do not function independently of one another. Rather than gaining information about task performance by measuring the simultaneous activity of many motor cortical neurons, information was redundantly represented across neurons. Redundancy is a hallmark of distributed processing, as the removal of individual neurons from an ensemble results in a "graceful degradation" (McClelland et al., 1986) of information that does not affect overall system performance. In the motor cortex, as pointed out by Meijers and Eijkman (1974), redundancy might exist to overcome the large amount of variability in the onset latencies

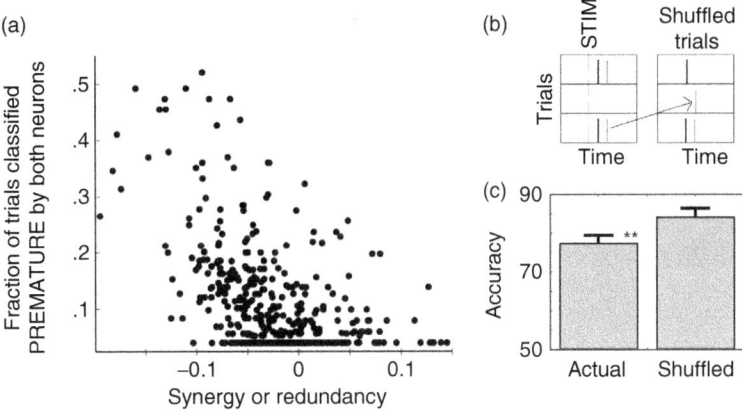

Figure 6.9 Role of correlated noise in redundant interactions between motor cortical neurons. (a) P_e for pairs of neurons plotted against the fraction of trials that both neurons classified as premature responses reveals a significant linear relationship. Synergistic interactions occurred when P_e was positive and redundant interactions occurred when P_e was negative. (b) An illustration trial-shuffling, where spikes occurring on certain trials are mixed randomly, and the spikes from two redundant neurons might end up on different trials. STIM, stimulation. (c) Ensemble accuracy compared with ensemble accuracy after trial-shuffling within groups (i.e. shared noise was decorrelated) reveals significant increase in classification accuracy, and hence a decrease in redundancy, after trials have been shuffled within each class of trials.

of corticospinal neurons relative to onset latencies in muscles. As such, redundancy may be an important organizing principle for information processing in the motor cortex, and perhaps also in other neural systems.

Potential mechanisms for synergy and redundancy: simulated ensembles

In order to understand the statistical relationships between neurons, we simulated a simple model network. In our simulations, we could vary the relationships of the simulated neurons to stimuli and to separate and common noise sources. By manipulating the influences of the stimuli and noise upon simulated neurons, we could readily create independent, redundant, and synergistic neural ensembles.

The simulations were run as follows. We constructed simple model ensembles with three types of elements: stimuli, noise sources, and simulated neurons (Fig. 6.10a). Stimuli were uniformly distributed random numbers between 0 and 1. These numbers were evenly thresholded into states (e.g. classes of 0 and 1)

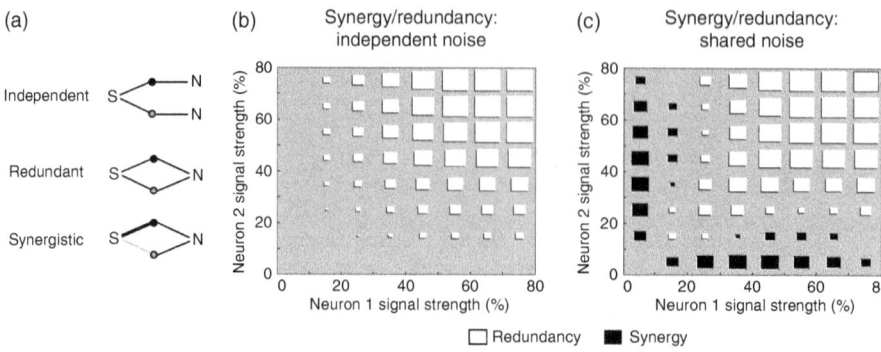

Figure 6.10 Example of simple networks with different informational interrelationships. (a) Top: an independent ensemble, with neurons that are driven by a signal, but each neuron has separate noise sources. Middle: a redundant ensemble, with neurons that are driven by a signal and have a shared noise source. Bottom: a synergistic ensemble, with neurons that are driven uniquely by a signal and a shared noise source that can provide information about the reliability of the stimulus. Note that these are canonical examples, and by changing the relative weights of stimulus and noise contributions, it is possible to create various coding schemes. (b) Synergy and redundancy among pairs of neurons with independent noise. Boxes show average value of synergy/redundancy: white = redundancy, black = synergy, gray = independent, size of box = magnitude of relationship. As signal strength increased for each neuron, redundancy increased. (c) Synergy and redundancy among pairs of neurons with shared noise. Colors as in panel (b). When one neuron has low signal strength, pairs were more synergistic. When both neurons have moderate signal strengths, pairs were redundant – compare values around 40% signal strength for both neurons in panel (c) with values around 40% signal strength in panel (b).

for information calculations. Noise sources were normally distributed random numbers generated separately from the stimuli (mean = 0, variance = 1). Noise sources could affect neurons individually or could be shared between neurons. An ensemble consisted of two or more neurons with predetermined stimulus and noise coefficients. The coefficients were varied between 0 and 1 in 0.1 increments to sample a large number of different coefficient combinations ($>10^6$). For each neuron, activity on a given trial was the sum of stimuli, individual noise, and shared noise values, each multiplied by their own coefficient. We generated 2000 trials of random stimuli and noise, from which we calculated neural activity by using the chosen coefficients. We also thresholded the activity of these simulated neurons to create discrete outputs (e.g. either 0 or 1). For any given neuron, signal strengths and shared noise fractions could be calculated by comparing the sizes of the coefficients.

$$\text{SignalStrength} = \frac{C_{Stimulus}}{C_{Stimulus} + C_{IndependentNoise} + C_{SharedNoise}} \qquad (6.7)$$

$$\text{SharedNoiseFraction} = \frac{C_{SharedNoise}}{C_{SharedNoise} + C_{IndependentNoise}}. \qquad (6.8)$$

We quantified the relationship and interactions between stimuli and neurons using several approaches. We used the thresholded stimuli and thresholded neural activity to calculate predictive information using several approaches. We used a statistical classifier (LVQ) to predict the stimulus of a given trial given the neural outputs on that trial. From the classification results we calculated predictive information and synergy/redundancy values. We also directly calculated information measures between stimuli and the neural responses for each ensemble. The mutual information (I) between a stimulus and a response can be calculated as (Cover and Thomas, 1991):

$$I(S;R) = \sum_{s \in S} \sum_{r \in R} p(s,r) \log_2 \left[\frac{p(s,r)}{p(s)p(r)} \right]. \qquad (6.9)$$

We calculated the mutual information between stimuli and individual neurons by using the thresholded stimuli as S and the thresholded neural responses as R. We also calculated the mutual information between stimuli and pairs of neurons by jointly considering the thresholded activity of two neurons R_1 and R_2. To calculate synergy and redundancy values, we adopted the approach of Schneidman *et al.* (2003), who calculated the difference between the mutual information of the stimulus and ensemble and the sum of the mutual information between the stimulus and individual neurons:

$$P(R_1, R_2) = I(S; R_1, R_2) - I(S; R_1) - I(S; R_2). \qquad (6.10)$$

We confirmed the results of these two approaches for our simple simulations using multiple linear regression on the unthresholded stimulus and neural data. This type of analysis achieves the ideal explanation of variance for continuous variables in a linear system. In this condition, we used percentage of variance explained (standard R_2 values), analogously to predictive information. P was calculated by comparing the variance explained by the pair compared to the sum of the variances explained by individual neurons. We made no attempt to scale these values to bits, since we were interested in the net relationships.

Effect of independent noise on neural statistical relationships

We first examined neuron pairs that only possessed independent noise (ensembles with neurons whose shared noise fractions were <0.1). At low levels of stimulus influence (both neurons' SignalStrength <0.2), the ensembles were neither synergistic nor redundant, and the information of the ensembles was nearly equivalent to the sum of the individual information of the neurons ($P = -0.00 \pm 0.00$). This is because the neurons shared neither stimulus nor noise relationships in common. However, *pairs of neurons with very high signal strength (>0.8) tended to be redundant* ($P = -0.96 \pm 0.01$). Redundancy increased rapidly as the stimulus effects began to outweigh the separate noise effects on the neurons (compare diagonal from bottom left to top right for Fig. 6.10b).

Effect of shared noise on neural statistical relationships

Although the activity of two neurons can be related because they are both related to a stimulus, they can also be related because they share a noise source. We next added shared noise influences onto our pairs of neurons to explore its effects on synergy and redundancy. We now specifically focused our attention on pairs of neurons that had most of their noise influence derived from shared noise (SharedNoise fractions >0.5 for both neurons). We found that even with shared noise, *pairs of neurons with very low signal strength (<0.2) tended to be independent* ($P = 0.03 \pm 0.13$), and *pairs of neurons with very high signal strength (>0.8) tended to be redundant* ($P = -0.96 \pm 0.01$), similar to the patterns observed with separate noise. However, *if either neuron had moderate signal strength (0.4–0.6), shared noise could cause either redundancy or synergy* ($P = -0.14 \pm 0.28$) (Fig. 6.10c). The effects of shared noise depended on relative signal strength. *If neurons had similar signal strength (0.2–0.4), shared noise increased redundancy* ($P = -0.17 \pm 0.13$ with shared noise, compared to $P = -0.07 \pm 0.03$ with independent noise, $p \ll 0.001$). *If neurons had different stimulus strengths (one neuron's less than 0.2, the other's greater than 0.3), shared noise created synergy* ($P = 0.15 \pm 0.13$ with shared noise, compared to $P = -0.01 \pm 0.02$ with independent noise, $p \ll 0.001$). This result suggests that if there is shared noise in a moderately predictive system, statistical interrelationships are complicated and independent ensembles are less likely to be encountered.

Ensemble size and neural statistical relationships

We were interested in what occurred as we added more neurons and more stimuli to our simulated ensembles. In general, *the larger the ensemble, the greater the level of redundancy* (R between number of neurons and ensemble $P = -0.38$, $p \ll 0.001$). As we increased the number of stimuli to be explained, the ensembles tended to become slightly more synergistic (R between number of stimuli and ensemble $P = 0.09$, $p < 0.01$). Thus with many neurons and a simple stimulus space, there is likely to be great overlap and much redundancy. However, with few neurons and a complex stimulus space, there is more likely to be less overlap between neurons and a greater opportunity for synergy.

This result predicts that *as ensembles grow, they may experience two regimes of statistical dependencies*. In small ensembles, synergy is favored as informational gains match or outstrip the independent information contributed by the component neurons. In larger ensembles, redundancy is favored once there is a relatively large amount of information represented by the ensemble.

Summary: simulated ensembles

Even within a simple linear system, we are able to explore conditions that give rise to synergistic or redundant statistical relationships. In our simulations, we demonstrated that direct neural interactions are not necessary to create synergy or redundancy. Rather, the mere presence of shared noise makes independent coding less likely. However, whether shared noise causes synergy or redundancy depends on other factors, such as the initial signal strength of cells within an ensemble. This suggests that pooling responses from neurons recorded separately can inappropriately underestimate or overestimate the total amount of information in the brain. Lastly, synergy is more likely to be found when analyzing a few cells in a more complex behavioral setting; whereas redundancy is more likely to be found when analyzing many cells in a more simple behavioral setting. One must consider these factors when comparing results collected in different brain areas and in different tasks.

Discussion

In this chapter, we have described how our group uses classifiers to analyze data from multiple electrode recordings in relation to behavioral measures of operant task performance. We explained how we reduce the complexity of spike trains from many simultaneously recorded neurons. Next, we discussed how to use the resulting low-dimensional representations of spike trains to train and test classifiers. We then described how to quantify synergistic and redundant interactions among an ensemble of neurons, the contributions of single neurons and neural ensembles to decoding, and the effects of correlated noise on classifier performance. Finally, we explored how synergy and redundancy might arise using simulations, and found that one major determinant of neuronal interactions is differences in event-related modulations in firing rates. By providing this summary of our research, we hope that other groups will be able to make use of these methods. We encourage any reader interested in our implementations of methods for feature extraction, classification, and quantification of neuronal interactions to contact us and to also take a look at our website (http://spikelab.jbpierce.org) for source code for Matlab and R. Our belief is that progress on understanding how populations of neurons represent information about behavior will advance only if researchers make their methods open to other groups. We close by raising three issues that motivate our work on population coding. First, we believe that a major issue is whether coding schemes are equivalent across brain areas. Differences in local circuit properties of many brain areas are known to exist (Shepherd, 2003). These issues are not usually addressed in computational and theoretical studies of population coding. However, we have found that there are major differences in how behavior is decoded by ensembles of motor cortical and striatal neurons (unpublished data). Second, we believe that studies using multiple electrode recording must be done in a "hypothesis driven" framework. Data from simultaneously recorded populations of neurons provide the opportunity to answer salient questions about how different brain networks process information. Ideally, the experimental approach should depend on processing in the target brain area (Narayanan and Laubach, 2006; Narayanan et al., 2006) and be guided by careful behavioral studies that involve lesions or reversible inactivations. Third, we believe that future challenges for our field include understanding functional interactions between brain areas (Miller and D'Esposito, 2005), and that the techniques discussed here will facilitate analysis of data from future experiments that record from multiple neurons across several brain systems.

Acknowledgments

Critical comments on the manuscript by Bruno Averbeck, Jose Carmena, and Nicole Horst are gratefully acknowledged.

References

Anderson, D. N. and Taylor, S. R. (2002). Application of regularized discrimination analysis to regional seismic event identification. *Bull Seism Soc Am* **92**:2391–2399.

Averbeck, B. B. and Lee, D. (2006). Effects of noise correlations on information encoding and decoding. *J Neurophysiol* **95**:3633–3644.

Averbeck, B. B., Crowe, D. A., Chafee, M. V., and Georgopoulos, A. P. (2003). Neural activity in prefrontal cortex during copying geometrical shapes. II. Decoding shape segments from neural ensembles. *Exp Brain Res* **150**:142–153.

Bellman, R. E. (1961). *Adaptive Control Processes*. Princeton, NJ: Princeton University Press.

Briggman, K. L., Abarbanel, H. D., and Kristan, W. B. (2005). Optical imaging of neuronal populations during decision-making. *Science* **307**:896–901.

Buckheit, J. and Donoho, D. L. (1995). Improved linear discrimination using time-frequency dictionaries. *Proc SPIE*:540–531.

Carmena, J. M., Lebedev, M. A., Crist, R. E., *et al.* (2003). Learning to control a brain–machine interface for reaching and grasping by primates. *PLoS Biol* **1**:E42.

Carmena, J. M., Lebedev, M. A., Henriquez, C. S., and Nicolelis, M. A. (2005). Stable ensemble performance with single-neuron variability during reaching movements in primates. *J Neurosci* **25**:10712–10716.

Cover, T. M. and Thomas, J. A. (1991). *Elements of Information Theory*. Hoboken, NJ: Wiley-Interscience.

Davis, D., Laubach, M., Cohen, L., and Pieribone, V. P. (2006). Voltage-sensitive dye imaging of whisker evoked responses in mouse barrel cortex. *Soci Neurosci Abstr* **134**:12.

Deadwyler, S. A. and Hampson, R. E. (1997). The significance of neural ensemble codes during behavior and cognition. *Annu Rev Neurosci* **20**:217–244.

Deadwyler, S. A., Bunn, T., and Hampson, R. E. (1996). Hippocampal ensemble activity during spatial delayed-nonmatch-to-sample performance in rats. *J Neurosci* **16**:354–372.

Friedman, J. H. (1989). Regularized discriminant analysis. *J Am Stat Assoc* **84**:165–175.

Gawne, T. J. and Richmond, B. J. (1993). How independent are the messages carried by adjacent inferior temporal cortical neurons? *J Neurosci* **13**:2758–2771.

Geman, S., Bienenstock, E., and Doursat, R. (1992). Neural networks and the bias/variance dilemma. *Neur Comput*. **4**:1–58.

Georgopoulos, A. P., Kalaska, J. F., Caminiti, R., and Massey, J. T. (1982). On the relations between the direction of two-dimensional arm movements and cell discharge in primate motor cortex. *J Neurosci* **2**:1527–1537.

Gochin, P. M., Colombo, M., Dorfman, G. A., Gerstein, G. L., and Gross, C. G. (1994). Neural ensemble coding in inferior temporal cortex. *J Neurophysiol* **71**:2325–2337.

Guo, Y., Hastie, T., and Tibshirani, R. (2007). Regularized linear discriminant analysis and its application in microarrays. *Biostatistics* **8**:86–100.

Gutierrez, R., Carmena, J. M., Nicolelis, M. A., and Simon, S. A. (2006). Orbitofrontal ensemble activity monitors licking and distinguishes among natural rewards. *J Neurophysiol* **95**:119–133.

Hastie, T., Buja, A., and Tibshirani, R. (1995). Penalized discriminant analysis. *Ann Stats* **23**:73–102.

Hastie, T., Tibshirani, R., and Friedman, J. (2003). *The Elements of Statistical Learning*. New York: Springer.

Hsu, C. W., Chang, C. C., and Lin, C. J. (2003). *A Practical Guide to Support Vector Classification*. Available online at www.csie.ntu.edu.tw/~cjlin/papers/guide/guide.pdf.

Kass, R. E., Ventura, V., and Brown, E. N. (2005). Statistical issues in the analysis of neuronal data. *J Neurophysiol* **94**:8–25.

Kohonen, T. (2000). *Self-Organizing Maps*. New York: Springer.

Kraskov, A., Stögbauer, H., and Grassberger, P. (2004). Estimating mutual information. *Phys Rev E* **69**:066138.

Krippendorff, K. (1986). *Information Theory: Structural Models for Qualitative Data*. Thousand Oaks, CA: Sage Publications.

Krupa, D. J., Wiest, Shuler, M. G., Laubach, M., and Nicolelis, M. A. (2004). Layer-specific somatosensory cortical activation during active tactile discrimination. *Science* **304**:1989–1992.

Laubach, M. (2003). Verifying neuronal codes: the statistical pattern recognition approach. *Annu Int Conf IEEE Engineering in Medicine and Biology* 3:2147–2150.

Laubach, M. (2004). Wavelet-based processing of neuronal spike trains prior to discriminant analysis. *J Neurosci Methods* 134:159–168.

Laubach, M., Wessberg, J., and Nicolelis, M. A. (2000). Cortical ensemble activity increasingly predicts behavior outcomes during learning of a motor task. *Nature* 405:567–571.

Lo, C. C. and Wang, X. J. (2006). Cortico-basal ganglia circuit mechanism for a decision threshold in reaction time tasks. *Nat Neurosci* 9:956–963.

Lu, J., Plataniotis, K. N., and Venetsanopoulos, A. N. (2003). Regularized discriminant analysis for the small sample size problem in face recognition. *Pattern Recogn Lett* 24:3079–3087.

Maynard, E. M., Hatsopoulos, N. G., Ojakangas, C. L., et al. (1999). Neuronal interactions improve cortical population coding of movement direction. *J Neurosci* 19:8083–8093.

McClelland, J., Rumelhart, D., and Hinton, G. (1986). The appeal of parallel distributed processing. In: *Computational Models of Cognition and Perception*, ed. McClelland, J., Rumelhart, D., and Hinton, G., vol. 1, pp. 3–44. Cambridge, MA: MIT Press.

Meijers, L. M. M. and Eijkman, E. G. J. (1974). The motor system in simple reaction time experiments. *Acta Psychol* 38:367–377.

Miller, B. T. and D'Esposito, M. (2005). Searching for the top in top-down control. *Neuron* 48:535–538.

Miller, E. K., Li, L., Desimore, D., et al. (1993). Activity of neurons in anterior inferior temporal cortex during a short-term memory task. *J Neurosci* 13:1460–1478.

Mosteller, D. E. and Tukey, J. W. (1977). *Data Analysis and Regression: A Second Course in Statistics*. Reading, MA: Addison-Wesley.

Narayanan, N. S. and Laubach, M. (2006a). Top-down control of motor cortex ensembles by dorsomedial prefrontal cortex. *Neuron* 52:921–931.

Narayanan, N. S. and Laubach, M. (2006b). Prefrontal control of low-frequency oscillations in motor cortex. Computational and Systems Neuroscience meeting abstr. I-32.

Narayanan, N. S., Kimchi, E. Y., and Laubach, M. (2005). Redundancy and synergy of neuronal ensembles in motor cortex. *J Neurosci* 25:4207–4216.

Narayanan, N. S., Horst, N. K., and Laubach, M. (2006). Reversible inactivations of rat medial prefrontal cortex impair the ability to wait for a stimulus. *Neuroscience* 139:865–876.

Nicolelis, M. A., Ghazanfar, A. A., Stambaugh, C. R., et al. (1998). Simultaneous encoding of tactile information by three primate cortical areas. *Nat Neurosci* 1:621–630.

Puchalla, J. L., Schneidman, E., Harris, R. A., and Berry, M. J. (2005). Redundancy in the population code of the retina. *Neuron* 46:493–504.

Purushothaman, G. and Bradley, D. C. (2005). Neural population code for fine perception in area MT. *Nat Neurosci* 8:99–106.

R Development Team T (2003). R: A language and environment for statistical computing. In: *R Foundation for Statistical Computing*.

Reich, D. S., Mechler, F., and Victor, J. D. (2001). Independent and redundant information in nearby cortical neurons. *Science* **294**:2566–2568.

Richmond, B. J. and Optican, L. M. (1987). Temporal encoding of two-dimensional patterns by single units in primate inferior temporal cortex. II. Quantification of response waveform. *J Neurophysiol* **57**:147–161.

Rolls, E. T., Aggelopoulos, N. C., Franco, L., and Treves, A. (2004). Information encoding in the inferior temporal visual cortex: contributions of the firing rates and the correlations between the firing of neurons. *Biol Cybernet* **90**:19–32.

Sachdev, R., Laubach, M., Mazer, J. A., and McCormick, D. A. (2006). Decoding stimulus preferences of rodent somatosensory cortical neurons. *Soci Neurosci Abstr* **109**:2.

Samonds, J. M., Allison, J. D., Brown, H. A., and Bonds, A. B. (2004). Cooperative synchronized assemblies enhance orientation discrimination. *Proc Natl Acad Sci USA* **101**:6722–6727.

Schoenbaum, G. and H. Eichenbaum (1995a). Information coding in the rodent prefrontal cortex. I. Single-neuron activity in orbitofrontal cortex compared with that in pyriform cortex. *J Neurophysiol* **74**:733–750.

Schoenbaum, G. and H. Eichenbaum (1995b). Information coding in the rodent prefrontal cortex. II. Ensemble activity in orbitofrontal cortex. *J Neurophysiol* **74**:751–762.

Schneidman, E., Bialek, W., and Berry, M. J. (2003). Synergy, redundancy, and independence in population codes. *J Neurosci* **23**:11539–11553.

Schneidman, E., Berry, M. J., Segev, R., and Bialek, W. (2006). Weak pairwise correlations imply strongly correlated network states in a neural population. *Nature* **440**:1007–1012.

Shepherd, G. M. (2003). *The Synaptic Organization of the Brain*, 5th edn. New York: Oxford University Press.

Sing, T., Sander, O., Beerenwinkel, N., and Lengauer, T. (2005). ROCR: visualizing classifier performance in R. *Bioinformatics* **21**:3940–3941.

Strong, S. P., Koberle, R., de Ruyter van Stevenick, R., and Bialek, W. (1998). Entropy and information in neural spike trains. *Phys Rev Lett* **80**:197–200.

Taylor, D. M., Tillery, S. I., and Schwartz, A. B. (2002). Direct cortical control of 3D neuroprosthetic devices. *Science* **296**:1829–1832.

Wessberg, J., Stambaugh, C. R., Kralik, J. D., *et al.* (2000). Real-time prediction of hand trajectory by ensembles of cortical neurons in primates. *Nature* **408**:361–365.

Witten, I. and Frank, E. (2000). *Data Mining*. San Diego, CA: Academic Press.

Part III NEURONAL POPULATION INFORMATION CODING AND PLASTICITY IN SPECIFIC BRAIN AREAS

7

Functional roles of theta and gamma oscillations in the association and dissociation of neuronal networks in primates and rodents

CHRISTIAN HÖLSCHER

Introduction

Hebb's postulates

As described as long as 70 years ago, Donald Hebb developed a theory of processes by which information could be encoded in the brain, and memory could be laid down.

He formulated three main postulates:

(1) Connections between neurons increase in efficacy in proportion to how successful the presynaptic neuron activates the postsynaptic activity.

"When an axon of cell A is near enough to excite B and repeatedly or persistently takes part in firing it, some growth process or metabolic change takes place in one or both cells such that A's efficiency, as one of the cells firing B, is increased" (Hebb, 1949, page 62). The first time that long-term potentiation of synaptic transmission (LTP) as induced by high-frequency stimulation (HFS) had been observed was in 1969 (published in Bliss and Lømo, 1973).

(2) Groups of neurons which tend to fire together form a cell assembly whose activity can persist after the triggering event and serve to represent information. Research on this issue of synchronous activity in

Information Processing by Neuronal Populations, ed. Christian Hölscher and Matthias Munk. Published by Cambridge University Press. © Cambridge University Press 2009.

the CNS has been ongoing for many decades (see e.g. Singer, 1994; Vanderwolf, 2000; and chapters in this book).

(3) Cognitive processes are based on the sequential activation of sets of such cell assemblies.

This set of rules has been paraphrased as "neurons that fire together, wire together, neurons out of sync fail to link."

If these postulates hold, it should be possible to analyze whether neurons that cooperate in the processing of information are firing in synchrony, and neurons that encode different information contents are separated and never fire in synchrony.

In order to be able to do this, there are several conditions that have to be met: (a) it is important to reliably separate single neurons during the recording – numerous studies have identified the dynamics of populations of neurons, but did not identify and isolate single neurons; (b) it has to be known which type of information the neurons encode. Again, this has not been done in many studies and it requires a testing system that offers different types of stimuli and that then can correlate this with the firing activity of the isolated neurons. It certainly requires recordings in awake and behaving animals.

I will present several examples of single-cell recording in the visual cortex of the primate and also in the hippocampus of the rat to illustrate the principles and to test whether the firing activity of neurons is in any way modulated in time.

The role of the perirhinal cortex in visual memory encoding in primates

Rate coding in the perirhinal cortex: encoding of specific information

In previous studies I have conducted experiments to test the role the perirhinal cortex plays in the establishment of complex scene memory. In these experiments, primates (*Macaca mulatta*) were shown novel and familiar images in a delayed match-to-sample task, and the activity of isolated neurons was correlated with the image presentation (see Figs. 7.1 and 7.2). This task design has a clear working memory component, and it is possible to analyze memory correlations in the firing activity of neurons. In another study, long-term memory was tested by presenting large data sets of visual images that were either new or familiar to the primate. It was found that a subset of neurons showed a dynamic range of firing modulation that correlates with working memory or long-term memory performance. Several subsets of neurons were identified. Some groups of neurons responded to specific images (Fig. 7.1), while others responded to the presentation of sample images independently of the identity of the image.

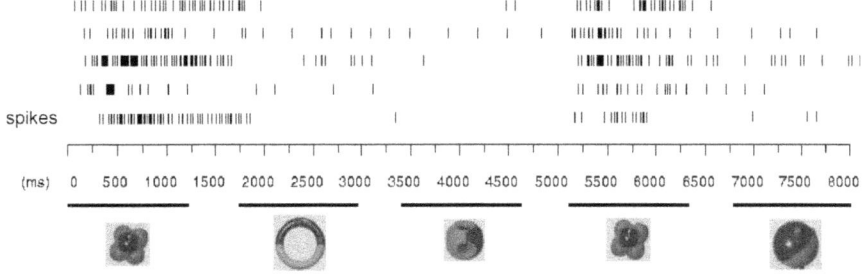

Figure 7.1 Example of the spike trains of one neuron recorded during the presentation of images in a delayed match-to-sample task. The firing rate of the neuron is increased during the presentation of the rewarded image. Responses of five different presentations are shown. This particular neuron responded to this specific image only. Some of the images shown to the monkey are shown here. The originals were in color but are shown here as 265 gray-level reproductions (see Hölscher and Rolls (2002) for details).

Yet other neurons showed responses only to combinations of conditions, e.g. to specific images that were rewarded, but not the same image shown in a non-rewarded position (as a distractor) (see Fig. 7.2) (Hölscher and Rolls, 2002; Hölscher et al., 2003b). These different types of neuronal responses are classic examples for rate coding (see Chapter 1). Each set of neurons appears to encode a specific part of the overall task information. Some neurons encode the beginning, some the end of the task. Others encode the rewarded image. This can be non-specific (all neurons fire for all rewarded images) or image-specific (only specific images activate the neuron, and only if this image is rewarded). Others encode the position of an image in the presentation (second, third, or fourth image) (see Fig. 7.2 for examples of such neuronal responses).

It is easy to imagine that the monkey will be able to perform the task if it has all of this information available. Furthermore, it is also clear that the information needs to be put in context in order to be useful. Interestingly enough, very few cells show "complex" responses, e.g. react to specific images *only* if they are rewarded. Most cells appear to encode single parts of the overall information only (this type of encoding has been observed many times in many other different cortical areas of the visual system; see, e.g. Baylis and Rolls, 1987; Jellema and Perrett, 2006; see also Chapter 4 for details on similar rate coding mechanisms in the hippocampus).

How are networks that encode different information "segments" of a task functionally associated?

It is clear that the individual elements of the delay-to-matching task that are encoded in the individual neurons (networks of neurons, more likely)

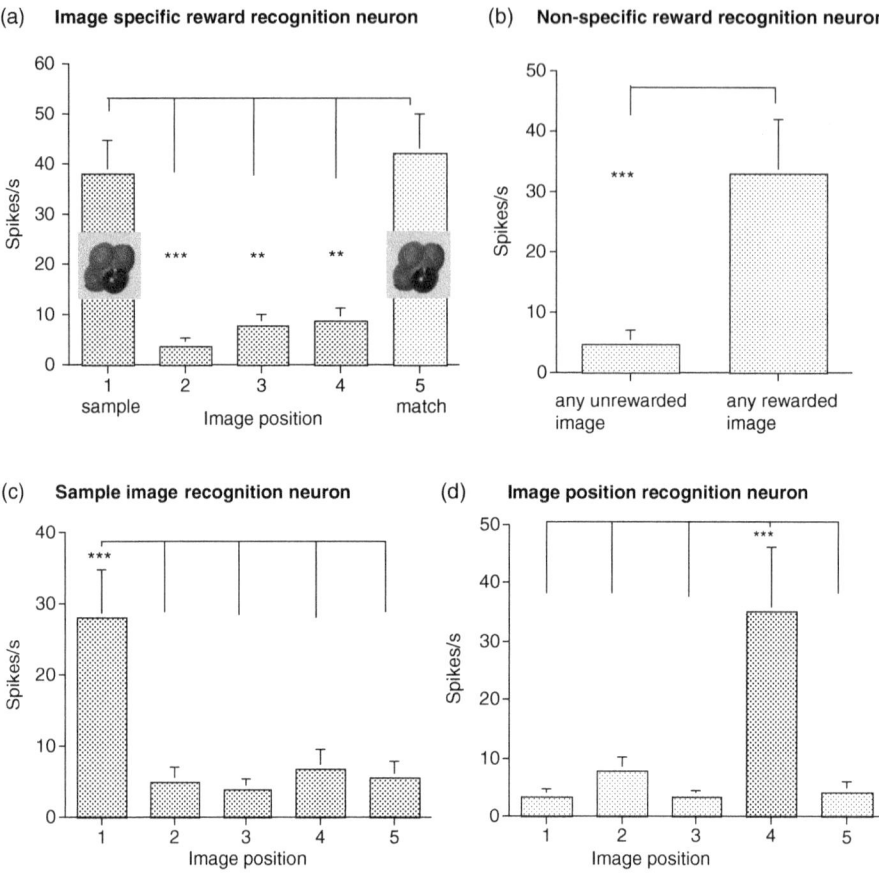

Figure 7.2 (a) A neuron that only fires if a specific image was shown in the rewarded position (as a sample and as a match image, at any position, but not as a non-rewarded distractor image). Some neurons only fired if the image was shown as a match image. The image presented to the monkey is shown. (b) Other cells were active if a rewarded image was shown, independent of the image identity or position in the task. (c) Yet other cells were active at any sample image presentation, independently of the image identity. (d) Some cells encoded specific positions within the task. Shown is a cell that was active when the fourth image was shown. ** $= p < 0.01$, *** $= p < 0.001$. For details on the methods and design, see Hölscher and Rolls (2002) and Hölscher et al. (2003).

have to be associated in a logical way in order to enable the primate to perform the task successfully. The association (e.g. reward with a specific image shown as the sample image) has to be flexible and fast. It should not take longer than a few seconds in order to be useful in such visual association tasks, and it should be fully reversible (if a different image is rewarded). Standard mechanisms of long-term potentiation of synaptic transmission would be of little use here, since they are slow to switch or reverse (Hölscher, 2001a). One mechanism could be that the individual networks are associated temporarily by phase-locking them into a larger network. This association could be easily and quickly formed and broken up again. This is a concept that can be easily tested with the available recorded single-cell data.

Synchronization of single cells/networks

This data set of neuronal activity that responds to defined stimuli then was available for further analysis. In a post-hoc analysis, I investigated whether neurons that showed similar firing profiles during stimulus presentation were in any way synchronized with each other, and whether neurons that showed clearly different firing profiles were not synchronized or even desynchronized. The advantage of this post-hoc test is that the neurons have not been selected for any oscillatory firing patterns, as only the firing intensity during stimulus presentation was a screening factor during the set-up of the experiment. Therefore, this data set is unbiased in regard to any oscillatory activity embedded in the firing activity.

Several neurons have been recorded simultaneously at the same electrode, and separated off-line using principal component analysis (Harris *et al.*, 2000). At first, it was analyzed whether the firing probability of neurons was modulated within gamma frequencies, as has been described many times previously (Munk *et al.*, 1996; Penttonen *et al.*, 1998; Fries *et al.*, 2001). It was found that 80% of neurons were clearly modulated in their firing probability in the range of 60–80 Hz (see Fig. 7.3a). The next step was to analyze how individual neurons interacted with each other, if at all. When comparing the spike timing intervals between different neurons in a cross-correlation, it became clear that neurons that were involved in the representation of associated stimuli, e.g. neurons that responded to specific images, and neurons that responded to specific positions of images within the delayed match-to-sample task sequence were time-locked and synchronized within the timescales of gamma activity (see Fig. 7.3b). Conversely, neurons that were not involved in encoding the same stimuli (during simultaneous presentation of several stimuli) were not correlated in any way (see Fig. 7.3c). This result suggests that neurons that are associated in their information processing job are also synchronized in their firing activity. Dissociated neurons are not synchronized, suggesting they are part of separate networks that are active within the cortex and that can be functionally and frequency decoupled, suggesting that there are different oscillators within the same cortex area that are not directly connected. This result supports several studies that showed that gamma activity becomes synchronized between cortical areas that are functionally coupled, while areas that are working on separate problems

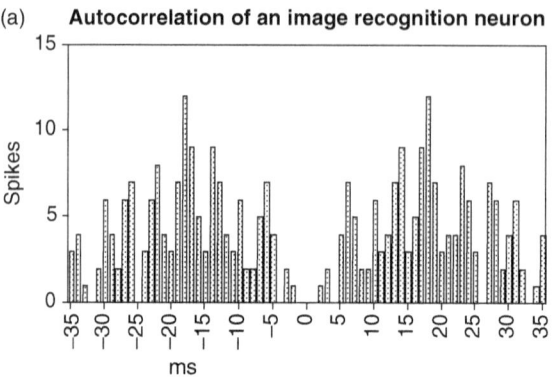

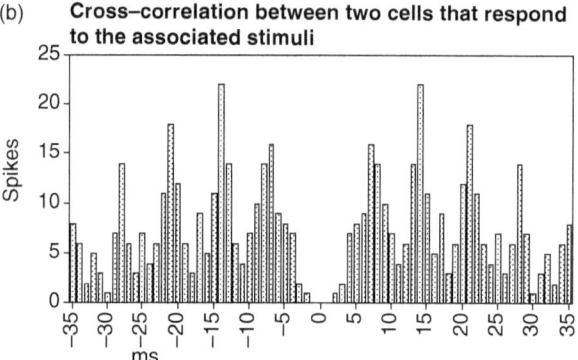

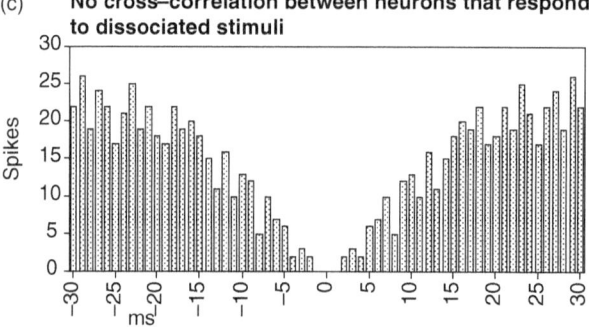

Figure 7.3 (a) In an autocorrelation analysis, the interval between subsequent spikes is analyzed. If the neuron is modulated in firing probability by gamma rhythm, a modulation of firing probabilities should be visible in the autocorrelation. This neuron that was driven by visual stimuli showed a modulation in the 85 Hz frequency range. (b) In a cross-correlation analysis, the timing between spikes of two neurons is analyzed. If the firing probability of the neurons is coupled in the gamma range, a modulation of firing probabilities for both neurons in the gamma range should be visible. Indeed, all neurons that were driven by associated stimuli (specific image and reward, or image and position in the delay-to-match sequence) were coupled in the gamma range (about 60–80 Hz). (c) Not all neurons are coupled in their firing

are not coupled or even are actively decoupled by anti-synchronization, most likely to keep separate information and networks apart and interference to a minimum. For example, in a study of neuronal activity in the areas 17 and 18 of the cat visual cortex, it was found that the firing probability of neurons, in response to the presentation of optimally aligned light bars within their receptive field, oscillated with a peak frequency of around 40 Hz. The neuronal firing pattern was tightly correlated with the phase and amplitude of oscillatory local field potentials recorded through the same electrode, and neurons with similar response properties fired together (Gray and Singer, 1989). Two further examples of neuronal recordings during the performance of an object recognition task or of a visual–motor coordination task that show synchronization of cortical areas on a large scale are given here. In a task that combined motor coordination and visual object recognition, EEG recordings showed episodes of increased broadband coherence from sensory, motor, and higher-order cortical sites of macaque monkeys. Cortical areas that are very far apart displayed coherent EEG activity without involving other intervening sites. The same regions were not synchronous when the monkey performed tasks that did not require the coordination of visual and motor cortex areas (Bressler et al., 1993). In a different study, people had to identify black-and-white drawings of faces. EEG recordings showed that the visual areas were synchronized during the presentation and in particular during the recognition of the faces. If the visual stimulus was not a face but a random black-and-white image, no large-area coherence was seen (Varela et al., 2001). Taken together, these results suggest that EEG oscillations, and the subsequent modulation and synchronization of neuronal firing activity, play a crucial role in the coordination of networks involved in information processing.

Gamma oscillations can bring neurons together that work together

This simple post-hoc analysis of the firing properties of neurons already shows that the firing properties of neurons is constrained in time, and both synchronous and asynchronous activity is indeed observed, as predicted by the theory. The neurons that are involved in processing the same type of information are synchronized in their firing activity, while other neurons that are not involved are either desynchronized or kept separate in their peak firing time.

Caption for Figure 7.3 (cont.)
probability. Neurons that did not respond to associated stimuli were not cross-correlated and modulated in the gamma range. This suggests that they are not part of the same network. For technical details see Gochin et al. (1991), Harris et al. (2000), Hölscher (2003), and Hölscher et al. (2003a).

"Place cells" in the hippocampus of the navigating rat

The observation that neurons in the hippocampus can be spatially tuned for local areas (place fields) within the environment that the rat explores was made almost 40 years ago (O'Keefe and Dostrovsky, 1971). In the modern experimental set-up, tetrodes (bundles of four electrodes) are implanted in the hippocampus of a rat, and allow the recording and separation of several neurons simultaneously. Principal component analysis of signals from four different channels enables a good separation of single units (Harris et al., 2000).

Gamma oscillations can keep neurons apart that do not work together

The place cell recording technique offers unique conditions for investigating the properties of different hippocampal neuronal networks that are active during a recording session. It has been found in the past that neurons that are recorded simultaneously almost never encode the same information (have the same place field) (O'Keefe, 1979; Redish et al., 2001; Jacob et al., 2002). The percentage of active neurons in the hippocampus at any given time is quite low, and the anatomical architecture resembles a sparse coding network (Braitenberg and Schütz, 1983; Rolls, 2001; Buzsáki, 2006). Therefore, it is assumed that active networks in the hippocampus contain only relatively few neurons that are spread apart in space, and the likelihood of recording two neurons of the same network simultaneously is therefore very low. Hence, all "place cell" neurons recorded at one session ordinarily will encode different information. This makes this recording set-up an ideal model for testing how independent neuronal networks are separated. Figure 7.4 shows such a neuron and where it fires in the environment that the animal explores.

In addition, recording neuronal activity in the hippocampus is a good brain area for studying phase coding. The spike timing of neurons is very well controlled, while the firing rate of neurons is low and rarely exceeds 10 Hz (Hölscher, 2003; Hölscher et al., 2003a). Neuronal activity is modulated in their firing rate to some degree, mostly by the running speed of the animal (Czurko et al., 1999). An inactive rat shows almost no place cell activity, underscoring the importance of movement to drive these cells (Foster et al., 1989). The low frequency (around 5 Hz) of neuronal activity makes it unlikely that all information is only encoded by a rate code (Hölscher, 2003). In addition, most excitatory neurons in the hippocampus fire in a burst mode (so-called complex cells). In this mode, neurons fire in a burst of around 5–7 spikes at around 200 Hz with virtually no spontaneous activity between bursts (see Fig. 7.5 for an example). This burst firing mode requires a high level of spike-timing control.

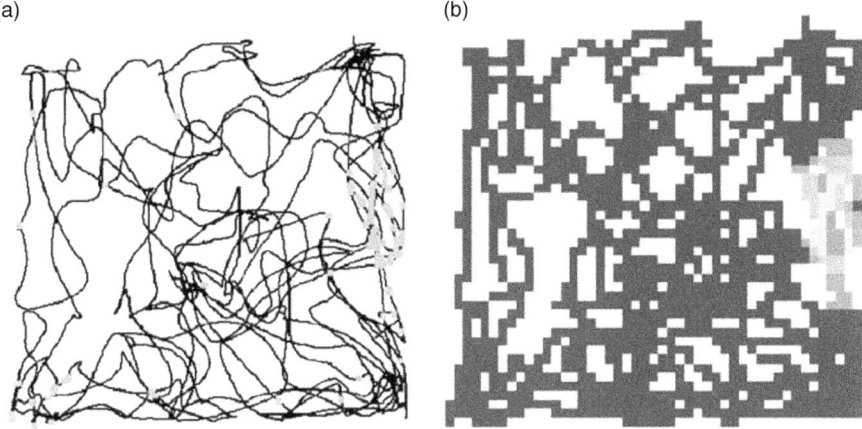

Figure 7.4 Example of a single-cell recording in the hippocampus of a rat during exploration of a box. (a) The spikes of the cell (gray dots) are shown and the track that the animal takes during the exploration. It is easy to see that the neuron has a clear firing preference for a particular area in the box. This encoding of spatial information gave rise to the term "place cell" (O'Keefe, 1979), though the cells do encode other qualities of information as well (Hölscher, 2003). (b) After normalization of the data (spikes per second per area explored) to avoid a bias by the difference in time spent in any give area, the firing probability is shown as a false-color map (gray scale in this image). Color versions of such images and further technical details can be found in Jacob *et al.* (2002) and Hölscher *et al.* (2003a).

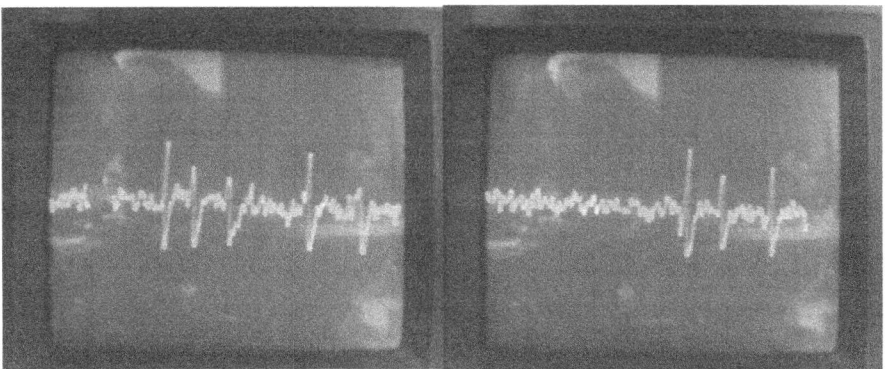

Figure 7.5 Two oscilloscope recordings of a complex spike. These cells fire in a very controlled way, about 3–7 spikes at around 200 Hz. The timing of the action potentials is under tight control by interneurons that fire in the gamma range (for details see Buzsáki, 1997; Penttonen *et al.*, 1998). In between the bursts, the cells are inhibited and can be silent for several seconds. This cell was recorded in the perirhinal cortex of a macaque, but hippocampal complex spike cells in the rat show very similar properties.

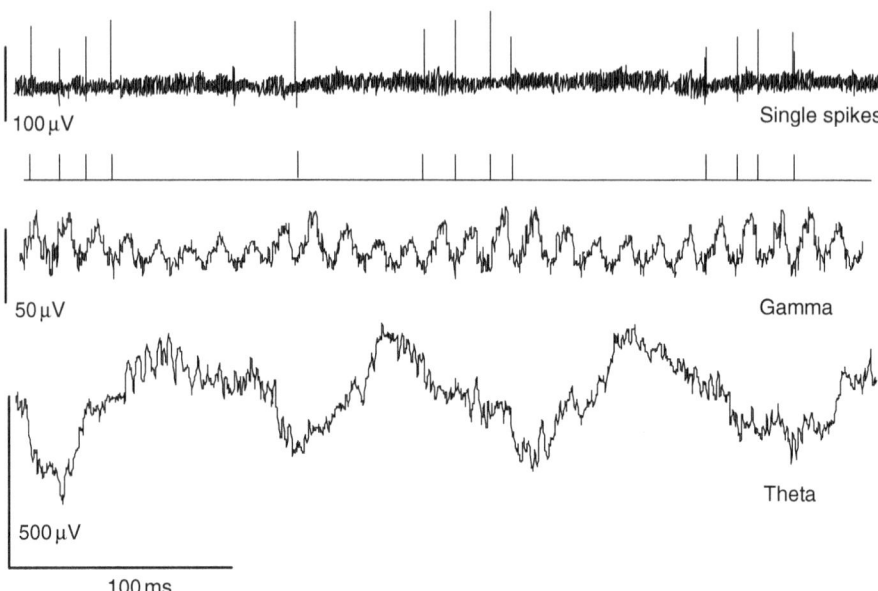

Figure 7.6 In vivo recording of hippocampal EEG and single neuron activity in vivo in the freely moving rat. The trace has been filtered using three different band pass filters to show action potentials of neurons and the correlation to gamma and theta EEG oscillations in the hippocampus. Note that the timing of action potentials is tightly controlled by gamma and theta rhythm. Firing activity is bundled into bursts of high-frequency activity.

The fact that neuronal spike timing in the hippocampus is highly controlled, however, makes it very likely that information is encoded in the precise timing of spikes, as suggested by several scientists (O'Keefe and Recce, 1993; Skaggs et al., 1996). Numerous studies have analyzed the timing of neuronal activity in relation to theta and gamma oscillations. It became clear very early on that neuronal spikes are phase-locked with theta activity (Fox and Ranck, 1975), and with gamma activity (Csicsvari et al., 1999). The phenomenon and functional significance of phase-locking and phase precession of single-cell activity during exploration has been under intense investigation ever since (O'Keefe and Recce, 1993; Skaggs et al., 1996; Tsodyks et al., 1996; Mehta et al., 2000).

Figure 7.6 shows a recording of neuronal activity in the freely moving rat, filtered for spikes, theta, or gamma activity. The recording shows that neuronal activity is tightly controlled by inhibitory field potentials in the gamma and theta range. The interneurons that fire in these frequencies (Buzsáki, 1997; Chrobak and Buzsáki, 1998) impose their firing patterns onto the excitatory pyramidal neurons. Figure 7.7 shows how several place cells are separated by gamma rhythm in their firing activity. This recording shows that different place cells never fire

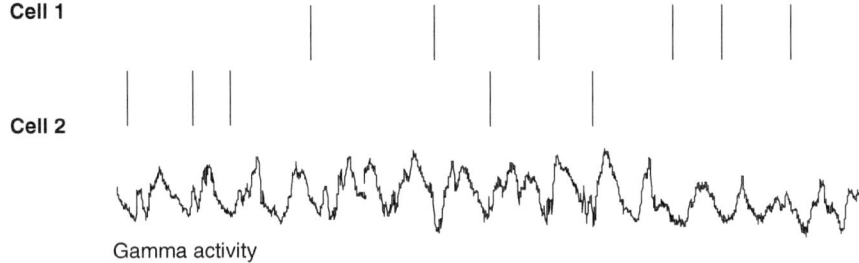

Figure 7.7 The firing activity of several place cells, and how they are separated by at least one gamma oscillation. Cell 1 is a place cell that fires after cell 2. There is some overlap in time of the firing activity of both cells, but they are never active simultaneously. This mechanisms could potentially avoid an overlap of neuronal networks that encode separate information units that are dissociated.

simultaneously, but are always separated by at least one gamma phase. The cells fire in a phase-coded fashion, with tightly controlled timing of action potentials.

How do EEG oscillations control excitatory neuronal activity?

Theta and gamma oscillations are basically modulations of local field potentials, predominately generated by interneuron–pyramidal neuron feedback loops that are tuned to specific frequencies (Csicsvari *et al.*, 1999) (see Chapter 2). The increase and decrease of local inhibition imposes a time pattern on the firing probability of pyramidal neurons. The neurons will fire at a much increased probability when the activity of the local interneurons that create the inhibitory oscillations are at their lowest. The modulation of firing probabilities by these local oscillations is the mechanism that controls spike timing, and therefore is the most likely candidate for the mechanism that can functionally "connect" neurons to work within the same network (as in the primate recordings), or to keep networks separate to avoid the merging of two separate networks, which would result in a complete loss of information of both networks (as in the place cell recordings in the rat) (Jensen *et al.*, 1996; Rolls, 2001). As will be discussed below in Chapters 16 and 17, a disturbance of these EEG oscillations has functional implications to cognitive processes (see also Uhlhaas and Singer (2006) for a review). Fenton (see Chapter 16) has shown that when theta and gamma rhythms are disturbed by pharmacological means, there is subsequently disturbance to the spike timing control of neurons in the hippocampus. As a direct effect, neurons that previously fired separately in time now may overlap in their firing timing and start to fire synchronously, while at the same time the ability of rats to navigate a spatial task was found to be impaired (Olypher *et al.*, 2006).

More functional roles of EEG oscillations: synchronous neuronal activity and synaptic plasticity

Long-term potentiation as a model for cellular learning mechanisms

So far, we have only described correlations between the occurrence of theta and gamma oscillations, neuronal spike timing, and information processing and performance in navigation and memory tasks. Correlations are of importance for making a case for potential roles of oscillations in the brain (see the excellent book by György Buzsáki (2006) on this topic). One role has already been suggested for oscillation activity – to order the firing activity in time and to connect neurons that interact in the processing of information while keeping neurons apart that do not process the same information. A second role that the coordination and synchronization of neuronal activity could play is that of inducing synaptic plasticity (Hebb's rule). If neurons are synchronized in their activity, one important side effect is that the excitatory input to target neurons also will be synchronous. It has been demonstrated in numerous in vitro studies in hippocampal slices that the timing of excitatory input is of central importance for the induction of LTP, and that synchronous input from several independent afferents is a very effective method of inducing LTP. Stanton and others have shown that stimulating independent inputs in a weak but synchronous pattern will induce LTP at the synapses of the target neuron, while desynchronous stimulation does not. In fact, desynchronous stimulation can induce long-term depression (LTD) or depotentiation of LTP if the conditions are right (Stanton and Sejnowski, 1989; Stanton, 1996). Recent work has shown that LTP can be induced by pairing pre- and postsynaptic stimulations within a time window of 15 ms, and LTD can be induced by mismatch of stimulation timing. The spike timing of spike bursts is also crucial, with the timing of the first spike of a burst being crucial for synaptic plasticity induction (Froemke and Dan, 2002; Froemke *et al.*, 2006).

The time windows for inducing LTP via synchronous stimulation are well within the realms of natural neuronal firing frequencies. The cellular mechanisms that underlie the control of synchronous activity and these synaptic changes have been analyzed in great detail (see e.g. Debanne, 1996; Paulsen and Sejnowski, 2000) (see also Chapter 1 for details). A summation of excitatory input depolarizes cell membranes to a larger extent than non-synchronous input, thereby activating more calcium channels and increasing intracellular calcium influx, an important second messenger which activates the cellular machinery that induces LTP (Artola *et al.*, 1990). When stimulating asynchronously, an optimal membrane depolarization will not be reached, and fewer ion channels are activated that only cause a moderate influx of calcium which either does not induce LTP or even induces LTD (Artola *et al.*, 1990; Artola and Singer, 1993). Therefore if excitatory inputs from independent efferent projections are synchronized in time, the effect will be an upregulation of synaptic activity, and only the synapses that are part of the active network and cooperate in processing information will be affected (Stanton, 1996; Xu *et al.*, 2006). Desynchronization of excitatory input, in contrast, can decrease synaptic input (Stanton and Sejnowski, 1989), thereby "disengaging" neuronal networks that are not directly functionally related.

Mechanisms for facilitating long-term potentiation

The modulation of local inhibition and the synchronization of excitatory input by theta and gamma oscillations have two effects: (1) the modulation of firing probability and synchronization of neurons when firing, and (2) reduced inhibition of target neurons by interneurons. With this mechanism in place, it is easy to postulate a mechanism by which synaptic weights are controlled in a context-dependent way. These mechanisms could explain how memory is formed during learning. It has been suggested by Hebb that only use-dependent synapses should be changed, and that neurons that fire within the same timeframe should be "wired together" to create a memory trace in the brain. The question has been how this mechanism works, and whether the oscillations would be effective enough to drive specific changes of synaptic weights.

Since Hebb's postulate, considerable research had been conducted in the area of EEG activity and single-cell recording, and in the area of how synaptic plasticity is induced in the brain. In the past, however, research had been divided into these two disciplines that did not always communicate with each other. With the development of the hippocampal slice technique to record LTP in vitro (Bliss and Collingridge, 1993), hundreds of scientists focused on the investigation of pharmacological and biochemical processes that underlie the upregulation of synaptic transmission. It had been assumed that this processes is the cellular mechanism of memory formation. In contrast, the Hebb postulate of synchronous network activity was not considered by many researchers in this area to be of importance. Instead, LTP was induced by trains of powerful high-frequency stimulation (HFS). This first indication that LTP indeed had some connection with memory formation came as late as 1986 (Morris *et al.*, 1986).

In parallel, many scientists were investigating the properties of neuronal activity in vivo and the role of synchronous activity in humans and in behaving animals (see Chapters 9, 14, and 15). These scientists, however, focused their work on the processes of synaptic plasticity in addition to their important contributions to information processing in the cortex, and did not research the mechanisms of memory formation and LTP in vivo. Therefore, Hebb's first and second postulates have not been analyzed or discussed in context for many years.

Problems with the theory of long-term potentiation as a mechanism or memory formation

Since the phenomenon of LTP fits the theoretical model suggested by Hebb very well, most scientists assumed that LTP is the cellular basis for memory formation. However, after a number of unexpected results that showed that HFS-induced LTP does not always correlate with memory formation (see Hölscher (1999) for a discussion), the focus shifted away from in vitro investigations of LTP to in vivo studies of behaving animals and studies of correlations of LTP with memory formation. From the same laboratory that published the original finding that a block of N-methyl-D-aspartate (NMDA) receptors impairs learning abilities and LTP induction in the hippocampus (Morris *et al.*, 1986), a second study was published that showed that this effect could well be an artefact (Bannerman *et al.*, 1995). Others have shown similar results, e.g. a lack of correlation between LTP in area CA1 of the hippocampus and spatial learning (Hölscher *et al.*, 1997c), a lack of correlation between the block of LTP by NMDA receptor antagonists and spatial learning (Saucier *et al.*, 1996), and many other cases (see Hölscher (1999) and Hölscher (2001a) for reviews).

One important aspect of the lack of LTP/memory correlation involved the fact that the stimulation technique used by most people for inducing LTP was not physiological in any way. LTP is usually induced by giving long trains of HFS, a protocol that had been developed to induce maximal LTP without any reference to physiological processes. This was not a concern for most people until the mismatch between LTP and learning became apparent.

Facilitation of long-term potentiation by taking advantage of naturally occurring disinhibition induced by theta activity

The physiological neuronal activity in the hippocampus that is observed in living brains does not include long trains of high-frequency activity, but rather includes low-frequency activity (1–10Hz), or "complex spike" cells that deliver short bursts with breaks of silence. Therefore, I decided to investigate to what degree the natural modulation of local inhibition during theta oscillations affects the induction of synaptic plasticity. When stimulating CA1 hippocampal neurons in vivo phase-locked with theta rhythm, I observed that as few as five spikes can induce reliable LTP, if the stimulation is delivered phase-locked with theta rhythm, at the point of lowest local inhibition (see Fig. 7.8). This shows that the modulation of local inhibition by theta oscillations is a crucial factor in controlling the induction of synaptic plasticity. When stimulating phase-locked

(a) Stimulation phase-locked with theta oscillations

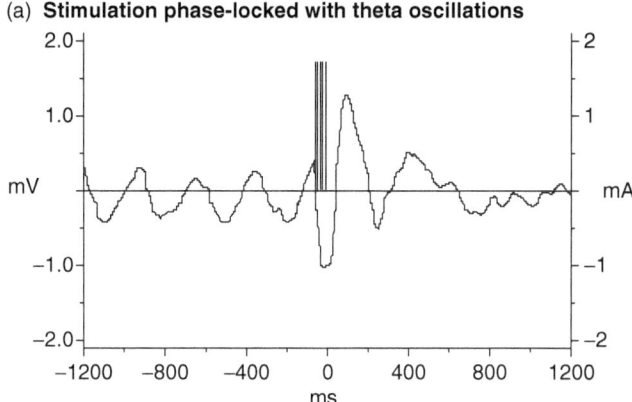

(b) Stimulation with 5 stimuli on the positive phase of theta

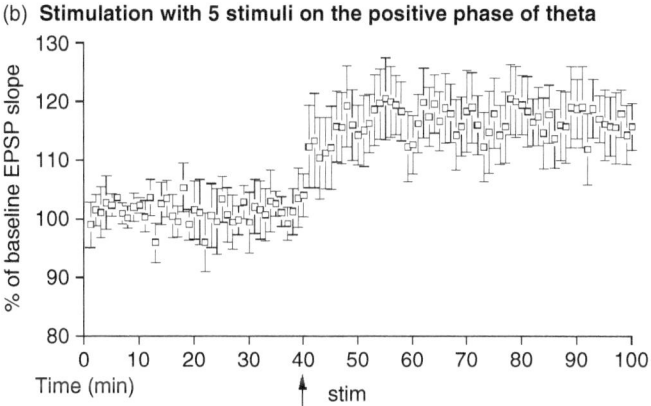

(c) Effect of stimulation with 3×5 stimuli on the positive theta wave

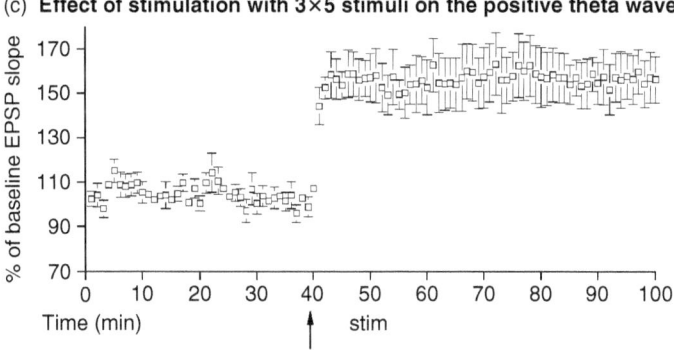

Figure 7.8 (a) Electrical stimulation of the hippocampus CA1 region by stimulating Shaffer collateral fibers phase-locked with endogenous theta rhythm in vivo. A train of five stimuli at 200 Hz is given at the time of lowest local inhibition by inhibitory activity of interneurons that fire in the theta range. This stimulation method imitates complex spike activity found in the hippocampus of awake rats. (b) Induction of synaptic plasticity, a potential cellular mechanism of memory formation, by giving

with the highest point of local inhibition of the theta phase, no such facilitating effect on synaptic transmission was seen. In fact, it was possible to depotentiate previously potentiated synapses using this method (Hölscher et al., 1997b).

As described before, in the brains of navigating rats or attentive primates, so-called complex spike cells fire with bursts of about five spikes, phase-locked with theta rhythm. Therefore, this novel stimulation protocol was able to show that complex spike activity can potentially induce synaptic plasticity in the working brain under in vivo conditions. Further studies even showed that a drug that previously was shown to block HFS-induced LTP, but not the learning of spatial tasks (Hölscher et al. 1997a, 1997c), did not block the type of LTP induced by the novel theta-phase locked type of stimulation (discussed in Hölscher 2001b, 2002). This finding could potentially resolve the apparent discrepancy between lack of LTP in the hippocampus and the preserved ability to learn spatial tasks.

Oscillations induce synchronous input which in turn can induce long-term potentiation

Theta and gamma oscillations can depolarize target neurons sufficiently to enable the controlled induction of synaptic plasticity within a time window that is equivalent to complex cell activity (e.g. place cells) in navigating rats. Furthermore, it has been demonstrated in hippocampal slices that the timing of excitatory input is of central importance, and that synchronous input from independent input projections is a very effective method of inducing LTP. Neuronal activity is synchronized by theta and gamma rhythm, as shown in Fig. 7.3. Therefore, the comparably simple mechanism of inducing such oscillations has very powerful effects on information processing and on memory formation. The synapses that are going to be changed in the process are exactly the ones that are involved in the processing of information, and therefore are the ones that are required to be changed in order to ensure a memory trace of the information that has been processed. The information has been "frozen" within the nervous system.

Caption for Figure 7.8 (cont.)
one train of five spikes phase-locked with theta rhythm. The increase of the slope of evoked field potentials signifies an increase in the efficacy of synapses in area CA1. Ordinarily, more than 100 stimuli using a high-frequency stimulation protocol would be required to induce synaptic plasticity in this area. (c) When giving three trains of five stimuli, the increase of synaptic activity is even greater. Two or three bursts of complex spike activity are commonly seen in the awake brain and could easily induce synaptic plasticity. (For technical details see Hölscher et al. (1997b) and for further discussions see Hölscher (1997, 2001b); see also Chapter 8.)

Sequence learning and high-order temporal coding in the hippocampus

Neuronal networks in the hippocampus develop sequences

I have described mechanisms by which neurons can be organized functionally into networks by the synchronization of their activity. The same mechanism can induce synaptic changes that can "freeze" such network activity in time. When taking this principle further, it follows that networks that drive secondary (follow-up) networks can establish sequences of network activation (first *a*, then *b*, then *c*). These sequences can be entrained and recalled if required.

There is ample evidence to show that the hippocampus is very much involved in the learning of complex sequences. For example, in a non-spatial task where rats have to learn both the correct associations between smell and food reward, and the correct sequence of smell presentation, it was found that the rats were not able to solve such a task without the hippocampus (Dusek and Eichenbaum, 1997; Fortin *et al.*, 2002). Another example is the ability to learn a classical conditioning task with a delay between the conditioned stimulus and the unconditioned stimulus (Kim *et al.*, 1995). This task is a sequential task, and requires the storage of timing information. The correct timing is reflected in the sequential firing activity of neurons in the hippocampus. These sequences that have been entrained can also be found in the neuronal activity of the hippocampus (see Fig. 7.9). When recording the firing properties of several neurons simultaneously, it has been shown that the neuronal activity of different neuronal populations is ordered sequentially. Using multi-electrode techniques, firing patterns can be recorded when the rat performs a sequential behavioral task, and the same firing pattern is reproduced when the animal returns to that section and repeats the same behavioral activity. Such patterns of neuronal activity can even reappear in REM sleep, suggesting that they are not the product of the temporal order of the specific sensory input that the animal receives at that section, but that these sequential patterns are truly stored in memory (Kudrimoti *et al.*, 1999; Louie and Wilson, 2001). This means that not only is the pattern of firing activity of *one* neuron reproducible (as shown in "place cell" activity), but the activity of groups of neurons that display very different firing patterns (e.g. burst activity, tonic activity, single spike firing, etc.) is preserved in memory and can be recalled when necessary. This patterned time order of neuronal activity could store information such as the sequential classic conditioning odor tasks described above (Dusek and Eichenbaum, 1997; Prut *et al.*, 1998; Fortin *et al.*, 2002) (see Fig. 7.9).

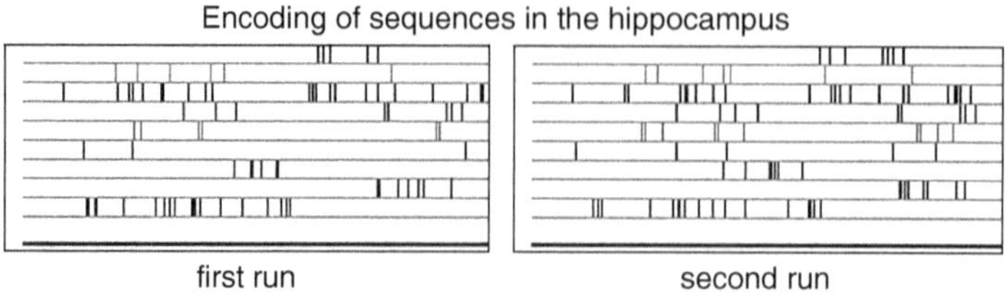

Encoding of sequences in the hippocampus

first run   second run

Figure 7.9 Encoding of sequences and high-order time coding in the hippocampus: each line represents the spike trains of a different neuron. (Left) Nine neurons recorded simultaneously during a spatial task in which the animal has to turn left in a square maze. (Right) When the animal has to turn left again at the same place in the second run, a very similar sequence of neuronal network activity is observed from the same nine neurons. Each neuron fires in a specific reproducible pattern, such as tonic or burst activity, and the relation of the firing activities of all neurons to each other is also very similar and reproducible. Some studies even show reproducible firing sequences during REM sleep. This indicates that networks encode time sequences and temporal network firing patterns that are stored and can be reproduced. (See Hölscher (2003) Dusek and Eichenbaum (1997), Prut *et al.* (1998), Fortin *et al.* (2002) and text for further details.)

The neural basis for episodic memory?

The role of the hippocampus in episodic learning has been discussed many times (Zola-Morgan *et al.*, 1986; Eichenbaum, 2001). In principle, episodic memory consists of the association of individual events with a time sequence in which these events took place. The embedding of discrete items such as faces or reward locations within a sequence of events requires that the neuronal networks involved must evolve a sequential activity pattern that can be stored and replayed. The specific sequences can store information of behavioral sequential tasks that cannot be stored in a single network. The observed firing characteristics fulfill the criteria required for such a system (Buzsáki and Chrobak, 1995; Jensen and Lisman, 1996; Skaggs *et al.*, 1996; Kamondi *et al.*, 1998; Singer, 1999; Bland and Oddie, 2001; Hölscher, 2001b; Lisman *et al.*, 2001).

Conclusion

When taking the information presented in the chapter together, we can conclude that Hebb's postulates turned out to be a far-sighted and most modern concept.

(1) The proposed rules for changing synaptic gain (pre- and postsynaptic sites must be depolarized simultaneously to induce a potentiation of transmission) turned out to be correct for the control of synaptic

plasticity, and are under the control of theta and gamma oscillations. Synchronous input indeed increases synaptic gain in the CNS, while asynchronous input does not, or even induces the opposite.

(2) The proposal for neuronal networks that are connected by synchronous firing turned out to be true in many studies of neuronal activity in the brains of behaving animals or humans. Also, single-cell recording studies showed synchronous activity of groups of neurons in the cortex during the performance of cognitive tasks. Furthermore, neurons that encode separate information are kept separate in their firing timing by gamma activity.

(3) Separate networks are activated in sequences and could be the basis for memory formation of sequential tasks or of episodic memory in general. Furthermore, as the last two chapters of this book describe, a disturbance of the temporal order of neuronal activity correlate with disturbances of mental processes.

References

Artola, A. and Singer, W. (1993). Long-term depression of excitatory synaptic transmission and its relationship to long-term potentiation. *Trends Neurosci* **16**:480–487.

Artola, A., Brocher, S., and Singer, W. (1990). Different voltage-dependent thresholds for inducing long-term depression and long-term potentiation in slices of rat visual cortex. *Nature* **347**:69–72.

Bannerman, D.M., Good, M.A., Butcher, S.P., Ramsay, M., and Morris, R.G.M. (1995). Distinct components of spatial learning revealed by prior training and NMDA receptor blockade. *Nature* **378**:182–186.

Baylis, G. and Rolls, E. (1987). Responses of neurons in the inferior temporal cortex in short term and serial recognition memory tasks. *Exp Brain Res* **65**:614–622.

Bland, B.H. and Oddie, S.D. (2001). Theta band oscillation and synchrony in the hippocampal formation and associated structures: the case for its role in sensorimotor integration. *Behav Brain Res* **127**:119–136.

Bliss, T.V.P. and Collingridge, G.L. (1993). A synaptic model of memory: long-term potentiation in the hippocampus. *Nature* **361**:31–39.

Bliss, T. and Lømo, T. (1973). Long-lasting potentiation of synaptic transmission in the dentate area of the anaesthetised rabbit following stimulation of the perforant path. *J Physiol (Lond)* **232**:331–356.

Braitenberg, V. and Schütz, A. (1983). Some anatomical comments on the hippocampus. In: *Neurobiology of the Hippocampus*, ed. Seifert, W., pp. 21–37. London: Academic Press.

Bressler, S.L., Coppola, R., and Nakamura, R. (1993). Episodic multiregional cortical coherence at multiple frequencies during visual task performance. *Nature* **366**:153–156.

Buzsáki, G. (1997). Functions for interneuronal nets in the hippocampus. *Can J Physiol Pharmacol* **75**:508–515.

Buzsáki, G. (2006). *Rhythms of the Brain*. Oxford, UK: Oxford University Press.

Buzsáki, G. and Chrobak, J.J. (1995). Temporal structure in spatially organized neuronal ensembles: a role for interneuronal networks. *Curr Opin Neurobiol* **5**:504–510.

Chrobak, J.J. and Buzsáki, G. (1998). Operational dynamics in the hippocampal-entorhinal axis. *Neurosci Biobehav Rev* **22**:303–310.

Csicsvari, J., Hirase, H., Czurko, A., Mamiya, A., and Buzsáki, G. (1999). Oscillatory coupling of hippocampal pyramidal cells and interneurons in the behaving rat. *J Neurosci* **19**:274–287.

Czurko, A., Hirase, H., Csicsvari, J., and Buzsáki, G. (1999). Sustained activation of hippocampal pyramidal cells by "space clamping" in a running wheel. *Eur J Neurosci* **11**:344–352.

Debanne, D. (1996). Associative synaptic plasticity in hippocampus and visual cortex: cellular mechanisms and functional implications. *Rev Neurosci* **7**:29–46.

Dusek, J. and Eichenbaum, H. (1997). The hippocampus and memory for orderly stimulus relations. *Proc Natl Acad Sci USA* **94**:7109–7114.

Eichenbaum, H. (2001). The hippocampus and declarative memory: cognitive mechanisms and neural codes. *Behav Brain Res* **127**:199–207.

Fortin, N.J., Agster, K.L., and Eichenbaum, H.B. (2002). Critical role of the hippocampus in memory for sequences of events. *Nat Neurosci* **5**:458–462.

Foster, T.C., Castro, C.A., and McNaughton, B.L. (1989). Spatial selectivity of rat hippocampal neurons: dependence on preparedness for movement. *Science* **244**:1580–1582.

Fox, S.E. and Ranck, J.B.J. (1975). Localization and anatomical identification of theta and complex spike cells in dorsal hippocampal formation of rats. *Exp Neurol* **49**:299–313.

Fries, P., Reynolds, J.H., Rorie, A.E., and Desimone, R. (2001). Modulation of oscillatory neuronal synchronization by selective visual attention. *Science* **291**:1560–1563.

Froemke, R.C. and Dan, Y. (2002). Spike-timing-dependent synaptic modification induced by natural spike trains. *Nature* **416**:433–438.

Froemke, R.C., Tsay, I.A., Raad, M., Long, J.D., and Dan, Y. (2006). Contribution of individual spikes in burst-induced long-term synaptic modification. *J Neurophysiol* **95**:1620–1629.

Gochin, P., Miller, E., Gross, C., and Gerstein, G. (1991). Functional interactions among neurons in inferior temporal cortex of the awake macaque. *Exp Brain Res* **84**:505–516.

Gray, C.M. and Singer, W. (1989). Stimulus-specific neuronal oscillations in orientation columns of cat visual cortex. *Proc Natl Acad Sci USA* **86**:1698–1702.

Harris, K., Henze, D., Csicsvari, J., Hirase, H., and Buzsáki, G. (2000). Accuracy of tetrode spike separation as determined by simultaneous intracellular and extracellular measurements. *J Neurophysiol* **84**:401–414.

Hebb, D. O. (1949). *The Organization of Behavior*. New York: John Wiley.
Hölscher, C. (1997). Long-term potentiation: a good model for learning and memory? *Pro Neuro-Psychopharmacol Biol Psychiat* **21**:47–68.
Hölscher, C. (1999). Synaptic plasticity and learning and memory: LTP and beyond. *J Neurosci Res* **58**:62–75.
Hölscher, C. (2001a). Long-term potentiation as a model for memory mechanisms: the story so far. In: *Neural Mechanisms of Memory Formation*, ed. Hölscher, C., pp. 1–36. Cambridge, UK: Cambridge University Press.
Hölscher, C. (2001b). Long-term potentiation induced by stimulation on the positive phase of theta rhythm: a better model for learning and memory? In: *Neural Mechanisms of Memory Formation*, ed. Hölscher, C., pp. 146–166. Cambridge, UK: Cambridge University Press.
Hölscher, C. (2002). Metabotropic glutamate receptors control gating of spike transmission in the hippocampus area CA1. *Pharmacol Biochem Behav* **73**:307–316.
Hölscher, C. (2003). Time, space, and hippocampal functions. *Rev Neurosci* **14**:253–284.
Hölscher, C. and Rolls, E. T. (2002). Perirhinal cortex neuronal activity is actively related to working memory in the macaque. *Neur Plast* **9**:41–51.
Hölscher, C., Anwyl, R., and Rowan, M. (1997a). Block of HFS-induced LTP in the dentate gyrus by 1S,3S-ACPD: further evidence against LTP as a model for learning. *Neuroreport* **8**:451–454.
Hölscher, C., Anwyl, R., and Rowan, M. (1997b). Stimulation on the positive phase of hippocampal theta rhythm induces long-term potentiation which can be depotentiated by stimulation on the negative phase in area CA1 *in vivo*. *J Neurosci* **17**:6470–6477.
Hölscher, C., McGlinchey, L., Anwyl, R., and Rowan, M. J. (1997c). HFS-induced long-term potentiation and depotentiation in area CA1 of the hippocampus are not good models for learning. *Psychopharmacology* **130**:174–182.
Hölscher, C., Jacob, W., and Mallot, H. (2003a). Reward modulates neuronal activity in the hippocampus of the rat. *Behav Brain Res* **142**:181–191.
Hölscher, C., Rolls, E. T., and Xiang, J. (2003b). Perirhinal cortex neuronal activity related to long-term familiarity memory in the macaque. *Eur J Neurosci* **18**:2037–2046.
Jacob, W., Mallot, H., and Hölscher, C. (2002). Broad-range place fields of hippocampal pyramidal cells: hints for motor representation in the hippocampal formation? *Proc Society for Neuroscience*, vol. 34, abstract 477.471.
Jellema, T. and Perrett, D. I. (2006). Neural representations of perceived bodily actions using a categorical frame of reference. *Neuropsychologia* **44**:1535–1546.
Jensen, O. and Lisman, J. (1996). Hippocampal CA3 region predicts memory sequences: accounting for the phase precession of place cells. *Learn Mem* **3**:279–287.
Jensen, O., Idiart, M. A. P., and Lisman, J. E. (1996). Physiologically realistic formation of autoassociative memory in networks with theta/gamma oscillations: role of fast NMDA channels. *Learn Mem* **3**:243–256.

Kamondi, A., Acsady, L., Wang, X. J., and Buzsáki, G. (1998). Theta oscillations in somata and dendrites of hippocampal pyramidal cells in vivo: activity-dependent phase-precession of action potentials. *Hippocampus* **8**:244–261.

Kim, J., Clark, R., and Thompson, R. (1995). Hippocampectomy impairs the memory of recently, but not remotely, acquired trace eyeblink conditioned responses. *Behav Neurosci* **109**:195–203.

Kudrimoti, H. S., Barnes, C. A., and McNaughton, B. L. (1999). Reactivation of hippocampal cell assemblies: effects of behavioral state, experience, and EEG dynamics. *J Neurosci* **19**:4090–4101.

Lisman, J., Jensen, O., and Kahana, M. (2001). Towards a physiologic explanation of behavioral data on human memory: the role of theta-gamma oscillations and NMDAR-dependent LTP. In: *Neuronal Mechanisms of Memory Formation*, ed. Hölscher, C., pp. 195–223. Cambridge, UK: Cambridge University Press.

Louie, K. and Wilson, M. A. (2001). Temporally structured replay of awake hippocampal ensemble activity during rapid eye movement sleep. *Neuron* **29**:145–156.

Mehta, M. R., Quirk, M. C., and Wilson, M. A. (2000). Experience-dependent asymmetric shape of hippocampal receptive fields. *Neuron* **25**:707–715.

Morris, R. G. M., Anderson, E., Lynch, G. S., and Baudry, M. (1986). Selective impairment of learning and blockade of LTP by a NMDA receptor antagonist, AP5. *Nature* **319**:774–776.

Munk, M., Roelfsema, P., König, P., Engel, A., and Singer, W. (1996). Role of reticular activation in the modulation of intracortical synchronization. *Science* **272**:271–274.

O'Keefe, J. (1979). A review of the hippocampal place cells. *Prog Neurobiol* **13**:419–439.

O'Keefe, J. and Dostrovsky, J. (1971). The hippocampus as a spatial map: preliminary evidence from unit activity in the freely-moving rat. *Brain Res* **34**:171–175.

O'Keefe, J. and Recce, M. L. (1993). Phase relationship between hippocampal place units and the EEG theta rhythm. *Hippocampus* **3**:317–330.

Olypher, A. V., Klement, D., and Fenton, A. A. (2006). Cognitive disorganization in hippocampus: a physiological model of the disorganization in psychosis. *J Neurosci* **26**:158–168.

Paulsen, O. and Sejnowski, T. J. (2000). Natural patterns of activity and long-term synaptic plasticity. *Curr Opin Neurobiol* **10**:172–179.

Penttonen, M., Kamondi, A., Acsady, L., and Buzsáki, G. (1998). Gamma frequency oscillation in the hippocampus of the rat: intracellular analysis in vivo. *Eur J Neurosci* **10**:718–728.

Prut, Y., Vaadia, E., Bergman, H., et al. (1998). Spatiotemporal structure of cortical activity: properties and behavioral relevance. *J Neurophysiol* **79**:2857–2874.

Redish, A. D., Battaglia, F. P., Chawla, M. K., et al. (2001). Independence of firing correlates of anatomically proximate hippocampal pyramidal cells. *J Neurosci* **21**: RC134.

Rolls, E. (2001). Neuronal networks, synaptic plasticity, and memory systems in primates. In: *Neuronal Mechanisms of Memory Formation*, ed. Hölscher, C., pp. 224–262. Cambridge, UK: Cambridge University Press.

Saucier, D., Hargreaves, E. L., Boon, F., Vanderwolf, C. H., and Cain, D. P. (1996). Detailed behavioral analysis of water maze acquisition under systemic NMDA or muscarinic antagonism: nonspatial pretraining eliminates spatial learning deficits. *Behav Neurosci* **110**:103–116.

Singer, W. (1994). Time as coding space in neocortical processing: a hypothesis. In: *Temporal Coding in the Brain*, ed. Buzsáki, G., pp. 51–79. Berlin: Springer-Verlag.

Singer, W. (1999). Time as coding space? *Curr Opin Neurobiol* **9**:189–194.

Skaggs, W. E., McNaughton, B. L., Wilson, M. A., and Barnes, C. A. (1996). Theta phase precession in hippocampal neuronal populations and the compression of temporal sequences. *Hippocampus* **6**:149–172.

Stanton, P. K. (1996). LTD, LTP, and the sliding threshold for long-term synaptic plasticity. *Hippocampus* **6**:35–42.

Stanton, P. K. and Sejnowski, T. J. (1989). Associative long-term depression in the hippocampus induced by Hebbian covariance. *Nature* **339**:215–218.

Tsodyks, M. V., Skaggs, W. E., Sejnowski, T. J., and McNaughton, B. L. (1996). Population dynamics and theta rhythm phase precession of hippocampal place cell firing: a spiking neuron model. *Hippocampus* **6**:271–280.

Uhlhaas, P. J. and Singer, W. (2006). Neural synchrony in brain disorders: relevance for cognitive dysfunctions and pathophysiology. *Neuron* **52**:155–168.

Vanderwolf, C. H. (2000). What is the significance of gamma wave activity in the pyriform cortex? *Brain Res* **877**:125–133.

Varela, F., Lachaux, J. P., Rodriguez, E., and Martinerie, J. (2001). The brainweb: phase synchronization and large-scale integration. *Nat Rev Neurosci* **2**:229–239.

Xu, N., Ye, C., Poo, M., and Zhang, X. (2006). Coincidence detection of synaptic inputs is facilitated at the distal dendrites after long-term potentiation induction. *J Neurosci* **26**:3002–3009.

Zola-Morgan, S., Squire, L. R., and Amaral, D. G. (1986). Human amnesia and the medial temporal region: enduring memory impairment following a bilateral lesion limited to CA1. *J Neurosci* **6**:2950–2967.

8

Theta rhythm and bidirectional plasticity in the hippocampus

JAMES M. HYMAN AND MICHAEL E. HASSELMO

Introduction

The local field potential from the hippocampus of an awake, stationary animal is filled with seemingly random low-amplitude high-frequency activity and arrhythmic high-amplitude low-frequency activity. To the untrained eye it is nearly impossible to discern any clear patterns or relationships of the signal to behavior, as the local field potential appears to be only noise. Then something striking happens, as the animal starts to move around, an extremely rhythmic high-amplitude 6–10-Hz sinusoidal waveform appears and instantly offers a window into the relationship between the hippocampal field potential and the animal's behavior. When Green and Arduini (1954) first recorded local field potentials activity from the hippocampus of behaving animals they saw this rhythmic activity (theta rhythm; 6–10-Hz high-amplitude sinusoidal activity: Buzsáki et al., 1986) and it was clear to them that theta rhythm played a significant role in hippocampal function. Theta appeared whenever the animal walked, ran, sniffed, oriented, reared, or went into rapid eye movement (REM) sleep, and theta was notably absent during consummatory behaviors (eating and drinking), grooming, and slow wave sleep.

The nature of theta's role in hippocampal processing was somewhat clarified when studies of the effect of hippocampectomy in humans and rodents were performed, and the hippocampus's central role in associative memory was suggested (Scoville and Milner, 1957; Morris et al., 1982). Since those original studies, further work has shown memory impairments in a wide range of behavioral and cognitive tasks that utilize multiple sensory modalities and behavioral responses (Eichenbaum, 2000). Based on those and other data Mishkin (1978) proposed a cortical–hippocampal memory network in which all major association cortices relay sensory, cognitive, and emotional information into the hippocampus (via the parahippocampal areas), and as output the hippocampus in turn provides remembered information back to these association areas.

In 1983, Buzsáki *et al.* lesioned the fornix and recorded a muted and almost absent theta rhythm in the hippocampus (Buzsáki *et al.*, 1983), and many studies have revealed learning and memory deficits following fornix transection that are often similar to the deficits seen in hippocampectomized animals (M'Harzi *et al.*, 1987; Numan and Quaranta, 1990; Whishaw and Tomie, 1997). These results point to theta rhythm playing an integral role in proper hippocampal memory operation. It is clear from the 40 years of research since Green and Arduini that the hippocampus is a central component in the higher-order association long-term memory circuit and that a robust theta rhythm is necessary in the hippocampus for this circuit to function correctly.

In this chapter, theta's role in memory processing, and specifically in synaptic modification, will be discussed. First, we will present a brief review of the role of the hippocampus in memory processing, concentrating on theta rhythm's role in memory process, unit activity, and synaptic modification. Then, a new model for in vivo plasticity will be presented that fuses intracellular Ca^{2+} level bidirectional plasticity with in vivo bidirectional theta phasic plasticity. This section discusses neuroplastic changes within the hippocampus, including: data from population plasticity studies, looking at theta burst and theta locked stimulation. Also, theta related intracellular and extracellular single neuron plasticity data and models are discussed.

Hippocampal function: lessons from theta rhythm and place cells

Case Vanderwolf (Vanderwolf, 1968, 1971, 1975) conducted a series of explorations of hippocampal rhythmic slow activity (RSA) (or "theta rhythm" as it would later become known). Using a range of different mammalian species, Vanderwolf found theta was closely correlated with many different behaviors (including running, rearing, digging, walking, and manipulation of objects with forelimbs). Further he found that in a different set of largely immobile behaviors (including eating, grooming, chewing, and vocalizing), the theta rhythm was noticeably absent from hippocampal field potential activity. Vanderwolf centered his work on theta's role in motor functioning, and envisioned the interactions of the hippocampus and neocortex as part of a sensory and motor integration system. Green and Arduini had envisioned theta being involved in dynamic emotional processing, while Vanderwolf saw a more automatic form of motor processing. Future research would help support both ideas, but would mostly support an alternative idea for hippocampal–cortical interactions based upon mnemonic processing.

Concrete evidence of the role of the rodent hippocampus had not yet been established in 1971, when O'Keefe and Dostrovsky first characterized the tendency of hippocampus neurons to fire in relation to an animal's current position in space. As the animal moves through an environment a series of different place cells are active, so that time spent in different areas provokes spatially related firing rate increases in different hippocampus principal cells (O'Keefe and Dostrovsky, 1971). These findings led O'Keefe and Conway (1978) to resurrect an old theory of learning originally put forward by Tolman (Hebb, 1949), termed the cognitive mapping hypothesis. O'Keefe hypothesized that a Cartesian map was created within the hippocampus through the spatial firing properties of hippocampus cells. This map is based upon an animal's location in space and the animal's current visual viewpoint. O'Keefe argued that the hippocampus in rodents serves to construct these maps in all environments the animal encounters and these maps are used for successful navigation through those environments. However, even in their initial report the authors pointed out that the firing of hippocampus principal cells seemed to be affected by other behavioral factors (besides location), and that some gross behaviors in multiple different locations were encoded by hippocampus cells like grooming, sniffing, and orienting (O'Keefe and Dostrovsky, 1971).

Hippocampal cellular activity and theta rhythm

Our understanding of the role of theta was furthered by discoveries that ablation of the fornix or medial septum greatly reduces the magnitude and occurrence of theta activity in the hippocampus (Andersen et al., 1979; Buzsáki et al., 1983). Considerable research has focused on the correlation of theta rhythm with learning and memory (Berry and Thompson, 1978; Winson, 1978; Givens and Olton, 1990; Vertes and Kocsis, 1997; Berry and Seager, 2001). An early study showed that the impairment in a spatial memory task caused by lesions of the medial septum was correlated with the amount of reduction of the hippocampal theta rhythm (Winson, 1978). Lesions of the medial septum and fornix reduce hippocampal theta power (Rawlins et al., 1979) and cause impairments in a number of memory-guided tasks, including spatial alternation (Givens and Olton, 1990; Aggleton et al., 1995; Ennaceur et al., 1996), delayed nonmatch to position (Markowska et al., 1989), operant delayed alternation

(Numan and Quaranta, 1990), and spatial reversal (M'Harzi et al., 1987). The impairments appear specific to recent, episodic memory, as fornix lesions do not impair the initial learning of a goal location, but impair the learning of reversal (M'Harzi et al., 1987), and medial septal inactivation does not impair reference memory, but impairs recent episodic memory in continuous conditional discrimination (Givens and Olton, 1994).

The role in learning is supported by extensive data in rabbits. The rate of learning is faster in individual rabbits when the hippocampal EEG has the highest amount of theta power (Berry and Thompson, 1978). When delivery of the conditioned stimulus is timed to appear during periods of theta rhythm, the rate of conditioning to the stimulus is enhanced in both delay conditioning (Seager et al., 2002) and trace conditioning (Griffin et al., 2004). Theta rhythm appears to reset its phase for encoding new stimuli during presentation of visual stimuli in a delayed match-to-sample task (Givens, 1996) but not during a reference memory task, and this phase resetting allows enhanced induction of long-term potentiation (McCartney et al., 2004). Phase resetting shows specificity for item encoding versus retrieval probe phases in human memory tasks (Rizzuto et al., 2003), suggesting a role for phase reset in determining appropriate dynamics for encoding and retrieval.

Taken together these results indicate that not only is theta intrinsically involved in mnemonic processing, but that it is a necessary component of proper hippocampus functioning. The question remained as to how theta affected hippocampus processing and what role this had in the formation of associations that are the heart of memories (Eichenbaum et al., 1987). To examine this idea, researchers began to study theta rhythm and hippocampus cellular activity together.

In 1986, Fox, Wolfson, and Ranck recorded principal cells, or complex spike cells, in the CA fields and examined the temporal dynamics of actions potentials to theta rhythm. They found that most cells fired with a clear phasic relationship to the ongoing theta rhythm in both awake and anesthetized conditions. This finding was not very surprising since it had been thought that the oscillations of theta must be generated by the firing of hippocampal neurons (Buzsáki et al., 1983); however, this was the first comprehensive presentation of this effect.

In 1993, O'Keefe's group reported the tendency of hippocampal place cells to precess relative to the theta wave as the animal traversed through the place field (O'Keefe and Recce, 1993). In 1996, Skaggs et al. published a report that detailed this phase precession phenomenom (Skaggs et al., 1996). They found that principal cells in CA1 would begin to fire at earlier phases of the theta wave as the animal went through the place field. Skaggs found principal cells show peak firing at a relatively small phase of the theta rhythm ($\sim 120°$) and that interneurons fire phasically to different points over the entire 360° theta cycle, though each individual interneuron fired to a smaller phase than most principal cells (Skaggs and McNaughton, 1996).

In the same study Skaggs *et al.* (1996) showed that spikes from principal cells with sequential place fields (as the animal moved in one direction on a thin platform) would fire sequentially within the same theta cycle. So within the theta cycle there appeared to be a "readout" of the sequential place cells, and the temporal order of the place fields was preserved in relation to the hippocampus theta rhythm. Skaggs used these data to argue that the compressed "readout" within a theta cycle allowed the hippocampus to order events temporally, and the phase precession effect is either a product of or the cause of this "readout." These results imply that theta enables and encourages associations to form between place cells with adjoining fields, and moreover theta rhythm may be the mechanism that allows principal cells to fire in relation to all types of behaviors, locations, and states (one example being the formation of place cell firing: Skaggs & McNaughton, 1996). Huxter *et al.* (2003) found that theta phase by position gave more precise predictions about an animal's current location than did traditional analysis of firing rates by position. They argued that hippocampus principal cells encode information about location within a place field via the temporal code of the spike times in relation to theta and information about speed of travel is encoded via the firing rate of these cells.

Theta and long-term potentiation

In 1949, Donald Hebb theorized that rapid changes in synaptic efficacy could allow a circuit of interconnected excitatory neurons to encode associations (Hebb, 1949). Later work would show that these changes could either potentiate or depress synapses and both of these changes can contribute to the encoding of learnt information. Long-term potentiation (LTP) is a long-lasting enhancement of synaptic excitability that can be elicited by larger-than-normal current stimulation in a high-frequency burst or with theta frequency stimulation. LTP can be observed in several pathways in the hippocampus and in many other brain areas (Bliss and Lomo, 1973; Bliss and Collingridge, 1993). In contrast, long-term depression (LTD) is characterized by a similarly long-lasting decrease in the size of evoked potentials (Malenka and Bear, 2004). The most common occurrences of LTD appear after prolonged periods of low-frequency stimulation (usually 1–2 Hz for 5–20 min) and it has been recorded in many cortical and subcortical areas (Dudek and Bear, 1993). Both LTP and LTD can be detected as instantaneous changes in the strength of synapses (via alterations in synaptic transmission); however, both also lead to lasting morphological changes at the synapse that alter synapses in somewhat permanent ways.

The perforant pathway of the hippocampus is an optimal area to observe both LTP and LTD due to the dense array of fibers from glutamatergic neurons in the entorhinal cortex entering the hippocampus (Stanton, 1996). The connection between neuroplasticity and hippocampal memory functioning was quite clear and intuitive, but how theta rhythm fit in with plasticity was less intuitive. One of the earliest links between theta activity and synaptic alteration was the observation that trains of stimuli delivered at intervals equal to the frequency of theta more readily induced LTP than similar stimulation at other frequencies (Larson and Lynch, 1986; Greenstein *et al.*, 1988). The most vigorous and reliable LTP is typically produced by delivering repeated high-frequency, high-amplitude bursts of stimulation, but comparable amplitude and duration of LTP can be provoked by simply delivering relatively low-amplitude theta frequency stimulation (8–10 Hz).

These findings lead Pavlides *et al.* (1988) to examine if LTP could be generated with a lower number of high-frequency stimuli. They performed an experiment where tetanic bursts were applied at specific temporal points of the ongoing hippocampal theta rhythm. They found in urethane-anesthetized rats that LTP is more effectively induced in the dentate gyrus when a tetanus was delivered on positive phases of theta (peaks) than when identical stimulation was given on the negative phases (troughs). In this study theta rhythm was provoked by stimulation of subcortical regions and it remained to be seen whether naturally driven theta rhythm produced similar plastic results. Huerta and Lisman (1995) extended these findings into the CA fields, using the slice protocol, when they timed single tetanic burst to different phases of cholinergically driven theta oscillations. They found a stimulation burst delivered on the peak elicited LTP, while a burst on the trough lead to LTD. Hölscher, Anwyl, and Rowan (Hölscher *et al.*, 1997) replicated these slice results in vivo, when they elicited LTP by delivering tetanic bursts to the peak of the theta wave in region CA1 with spontaneous theta in anesthetized animals. In this study it was also found that LTP could be depotentiated by delivery of a high-frequency stimulus to the troughs of theta. Unlike Huerta and Lisman (1995), they found that similar tetanic bursts delivered to the negative or zero phase of theta in an unpotentiated animal did not change the evoked potential. Orr *et al.* (2001) found that tetanic stimulation to the perforant path at the peak of dentate fissure theta produced LTP in fascia dentata in freely moving animals. When they stimulated on the trough of theta they found a significant increase in evoked potentials 24 hours after high-frequency stimulation, and by 48 hours after high-frequency burst stimulation, evoked potentials showed a slight decrease from baseline. This study was the first demonstration in awake animals that peak-locked tetanic bursts produce more potentiation than stimulation to other temporal points of theta. It still was not clear whether the depotentiation exhibited in earlier studies when bursts were applied during the trough of theta was related to LTD. The importance of this exhibition is integral to the idea of LTD being involved in real-time learning, because there are not many instances, outside the laboratory, that neurons will fire preferentially at low frequencies for long periods of time. From a psychological perspective it is not common for a learning stimulus to be presented repeatedly for long periods of time (i.e. 1–2 Hz presentations for 5–20 min: Malenka & Bear, 2004). It remained to be seen whether the theta phase-locked LTD recorded in the slice would also exist in vivo (Huerta and Lisman, 1995).

In 2003, workers in our laboratory showed that tetanic bursts delivered to the trough of theta did produce depression in previously unpotentiated synapses (Hyman et al., 2003). They found that when stimulation was delivered to the exact trough (most negative point on the wave) evoked potentials decreased over 10% in amplitude (Fig. 8.1). Identical stimulation delivered to opposite phases of the ongoing theta wave produced opposing changes in synaptic efficacy. In this study negative phase stimulations were limited to the apex of the trough, while the previous studies had included data when tetanic stimuli had been timed to any negative part of the theta wave. This effectively limited the window of stimulation to ∼20 ms in this study as compared to ∼60–80 ms in previous studies (this idea will be discussed more fully later in this chapter). This was the first time that stimulation more similar to endogenous neuronal activity (tetanic bursts) had produced LTD without any pharmaceutical manipulations. This result confirmed the slice results of Huerta and Lisman (1995), and they show that theta rhythm provides hippocampal neurons with the means to selectively decrease or increase synaptic efficacy in real time if a given neuron fires in a burst at one phase of theta or the other.

Ca^{2+} level dependent bidirectional hippocampal synaptic plasticity

Hippocampal LTP can be induced without a tetanus or theta-frequency stimulation, if the pre- and postsynaptic neurons are sequentially stimulated within ∼40 ms repeatedly (Bi and Poo, 1998). This spike timing dependent plasticity (STDP), like tetanic burst or theta frequency LTP, is dependent on the activation of N-methyl-D-aspartate (NMDA) channels (Collingridge et al., 1983), which leads to an influx of Ca^{2+} (Jahr & Stevens, 1987). While LTD can be induced by prolonged low-frequency stimulation or theta phasic tetanic burst stimulation, it can also be induced if the postsynaptic neuron is stimulated to spike just before the presynaptic neuron spikes (Bi and Poo, 1998). STDP is a more natural model of plasticity than tetanic burst, low-frequency LTD protocols,

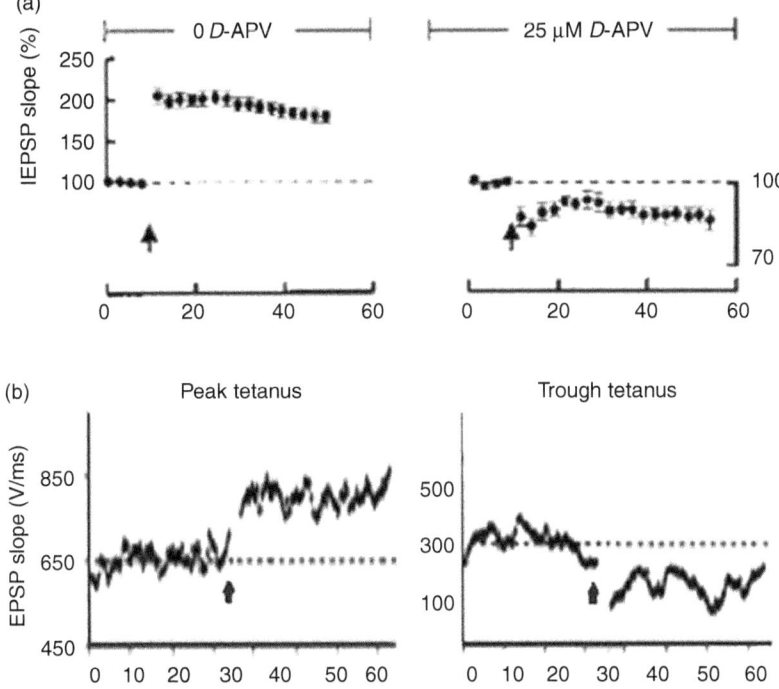

Figure 8.1 Bidirectional plastic changes in vitro and in vivo. (a) (Left) The effects of four 100-Hz tetani applied in the absence of D-APV (an NMDA antagonist: high Ca^{2+} levels). LTP was elicited (slope of the evoked potential increased by more than 50% of baseline levels) for 40 min after the tetani application. (Right) Results from when the same tetanus protocol was delivered to a slice with 25 μM D-APV. LTD was provoked (decrease in slope of ~10–15%) for 40 min following tetanization. (From Cummings et al., 1996.) (b) (Left) Test potential slopes before and after application of three 200-Hz tetani to the peak of local theta in stratum radiatum of CA1, in a behaving rat. LTP was elicited (~10–15% increase in evoked potential slope) lasting for 30 min after tetanization. (Right) The results of the same tetanus protocol delivered to the trough of theta in a behaving animal. LTD was provoked (10–15% decrease in slope) that lasted for at least 30 min following tetanization. In all four graphs an arrow shows the time of tetanus delivery, and the dashed line represents the mean of baseline test potential slope. (From Hyman et al., 2003.)

or theta frequency protocols, because the changes are occurring, presumably, at only one synapse, thus offering an appropriate model for discussion of in vivo plasticity (Wittenberg and Wang, 2006). However, while LTP and LTD can be induced via many different types of stimulation patterns the induced plastic changes are all similar and all changes rely on a postsynaptic intracellular Ca^{2+} influx (Lynch et al., 1983; Mulkey and Malenka, 1992).

In 1989, Lisman put forward a model of hippocampal bidirectional plasticity that postulated that differing intracellular Ca^{2+} levels would lead to either LTP or LTD given identical presynaptic stimulation (Lisman, 1989). The model argued LTD will result from moderately elevated levels of intracellular Ca^{2+}, while LTP will result from larger increases in intracellular Ca^{2+}. This idea was extended in the experimental findings of Cummings et al. (1996), where the authors controlled Ca^{2+} influx during tetanic burst stimulation (via a postsynaptic voltage clamp) leading to LTD in the voltage-clamped cell while the evoked field potential showed LTP. Cummings et al. (1996) showed that identical stimulation would yield postsynaptic LTP when the cell was allowed to function freely. This idea was fully quantified when Cormier, Greenwood, and Connor (2001) delivered identical stimulation to cells with moderate levels of Ca^{2+} (inducing LTD) or high levels of Ca^{2+} (inducing LTP), and established the threshold at which the direction of plasticity changes.

LTP changes are dependent upon NMDA receptor activation, which leads to increased intracellular Ca^{2+}. High levels of Ca^{2+} lead to autophosporylation of CaM-kinase II which is the beginning of a series of changes that result in increased AMPA receptor efficacy (Lisman, 2001). Lisman proposed that when low levels of Ca^{2+} are present, phosphatase 1 is activated which dephosphorylates CaM-kinase II, which leads to decreased postsynaptic receptor efficacy. In fact, if a pharmacological block of CaM-kinase II is present, even presynaptic then postsynaptic STDP stimulation will lead to LTD (Wang et al., 2005). It is thought that Ca^{2+} level changes are also at the heart of the different effects that changing the relative timing of pre- vs. post-synaptic stimulation has in STDP protocols (Rubin et al., 2005). Cho et al., (2001) discovered a third plasticity state, when Ca^{2+} levels are between the thresholds for LTD and LTP, called "no man's land" where no plastic changes occur. In this "no man's land" the autophosphorylation of CaM-kinase II appears to be counteracted by dephosphorylation by phosphatase 1, balancing each other out (Lisman, 2001). The role of Ca^{2+} in plasticity is so great, that manipulation of Ca^{2+} levels alone can trigger both LTP and LTD (Yang et al., 1999). This body of literature has shown that postsynaptic intracellular levels of Ca^{2+} are the determining factor for whether a synapse will be depressed or potentiated. Specifically, moderate levels of Ca^{2+} will lead to LTD, and high levels of Ca^{2+} will lead to LTP.

An integrative model of theta rhythmic stimulation and Ca^{2+} levels

Before examining plastic effects of theta, it is important to understand the relationship of theta rhythm to intracellular Ca^{2+} levels in hippocampal neurons. Unfortunately it is not yet possible to measure intracellular Ca^{2+} accurately in vivo, yet there are data that indirectly examine this idea. The recorded hippocampal theta signal is created by the interplay between rhythmically firing hippocampal cells, afferents from the entorhinal cortex (excitatory), and afferents from the medial septum (inhibitory). When discussing theta rhythm it is important to keep in mind that theta recorded in different areas of the hippocampus will appear different. For instance if you are recording from two electrodes, one near the hippocampal fissure, and the other near the CA1 pyramidal layer, these two signals will be 180° out of phase. CA1 and CA3 neurons fire at the peak of CA1 (stratum pyramidale) theta (thus the trough of fissure theta: Fox et al., 1986; Skaggs et al., 1996) and the membrane potential of the apical dendrites of these neurons is 180° out of phase with somatic oscillations and spiking activity (Kamondi et al., 1998). Intracellular membrane potentials are the opposite of extracellular field potential oscillations (Kamondi et al., 1998), so the apical dendrites are hyperpolarized at the peak of pyramidal theta, and at their most depolarized at the trough of pyramidal theta. The somata of CA1 pyramidals will be depolarized at the peak of pyramidal theta (peak spiking activity at this phase: Skaggs et al., 1996), and they will be hyperpolarized at the trough of pyramidal theta. So at the peak of pyramidal theta apical dendrites are depolarized and the somata are hyperpolarized, while at the trough of pyramidal theta the apical dendrites are hyperpolarized and the somata will be depolarized (Kamondi et al., 1998). This relationship is due in large part to the movement of Ca^{2+} across the cellular membrane, controlled by periodic inhibition and excitatory afferent stimulation via entorhinal cortex neurons, which are firing at this phase of theta (Stewart et al., 1992).

Ca^{2+} may also play a role in the manifestation of theta rhythm. Bonansco and Buno (2003) showed in vitro that blocking of Ca^{2+} channels eliminates rhythmic bursts induced through NMDA microiontophoresis, thus blocking the theta-like oscillation created in their slices of hippocampus. Furthermore, they showed that these oscillations of field potentials are produced by dendritic Ca^{2+} spiking, which supports previous findings that show Ca^{2+} channel effects alone (when Na^+ and K^+ channels are muted) can produce intracellular phase-locked "theta" oscillations (Ylinen et al., 1995). It is reasonable to argue that intracellular Ca^{2+} levels are fluctuating along with membrane potential oscillations during endogenous theta rhythm.

In 2003, an experiment conducted by our laboratory applied tetanic stimulation timed to opposite phases of hippocampal theta rhythm led to opposing changes in the evoked potential (Hyman et al., 2003). In this study theta rhythm was recorded locally in stratum radiatum near the hippocampal fissure (fissure theta is 180° out of phase with CA field pyramidal layer theta: Buzsáki et al.,

1986). So the dendrites of the postsynaptic neurons in stratum radiatum are hyperpolarized during pyramidal layer spiking activity and Ca^{2+} levels in the dendrites are quite low. The effects of induction of LTP vs. LTD may depend upon relative depolarizations of dendrites even though this dendritic depolarization is out of phase with soma depolarization. Thus, in Hyman *et al.* (2003) when tetanic stimulation was delivered during the trough of theta recorded in stratum radiatum (close to trough of pyramidal theta), the apical dendrites were at their most hyperpolarized (even though the somata would be depolarized at this time and this is the time of peak spiking). This leads to the low postsynaptic Ca^{2+} condition in the stimulation protocol Lisman (1989) described in which LTD occurs, due to phosphatase 1 activation, and subsequent dephosphorylation of CaM-kinase II. At this phase of theta intracellular Ca^{2+} in the dendrite should be at its lowest concentrations during the theta cycle. In our study tetanic stimulation at this phase of theta did lead to LTD, and it also explains why previous studies had recorded depotentiation when tetanic stimulation was delivered to the negative phase of theta (Pavlides *et al.*, 1988; Hölscher *et al.*, 1997; Orr *et al.*, 2001). Unlike those other studies we limited trough stimulations to the most negative point of the theta cycle, and it is possible that only at this most negative point do the levels of intracellular Ca^{2+} cross the threshold from "no man's land" to the LTD zone.

At the opposite phase of theta, at the peak of theta recorded in pyramidale (when somatic membrane potentials are hyperpolarized) dendritic potentials are at their most depolarized (Kamondi *et al.*, 1998). When tetanic stimulation was delivered at this phase the high levels of Ca^{2+} in the postsynaptic dendrites could cause LTP to occur (Pavlides *et al.*, 1988; Huerta and Lisman, 1995; Hölscher *et al.*, 1997; Orr *et al.*, 2001; Hyman *et al.*, 2003). Tetanic stimulation during periods of high Ca^{2+} levels leads to Ca^{2+}-dependent autophosphorylation of CaM-kinase II, and eventually enhanced AMPA receptor functioning (Lisman, 2001).

The framework shown in Fig. 8.2 provides functional dynamics that could be interpreted as hippocampal theta rhythm providing separate phases for encoding versus retrieval dynamics (Hasselmo *et al.*, 2002). As shown in Fig. 8.2, the long-term potentiation of synapses in stratum radiatum occurs when entorhinal cortex input causes depolarization of the apical dendrite, but when the soma is hyperpolarized. This corresponds to the encoding phase proposed in previous models (Hasselmo *et al.*, 2002; Hasselmo and Eichenbaum, 2005). This allows associations to be formed with new information arriving from entorhinal cortex, without the new input causing spiking activity in the CA1 pyramidal cells. This prevents spiking activity caused by retrieval from interfering with encoding of new memories. This also prevents new information from being interpreted as if it were retrieved from memory. (In fact, spiking in response to entorhinal

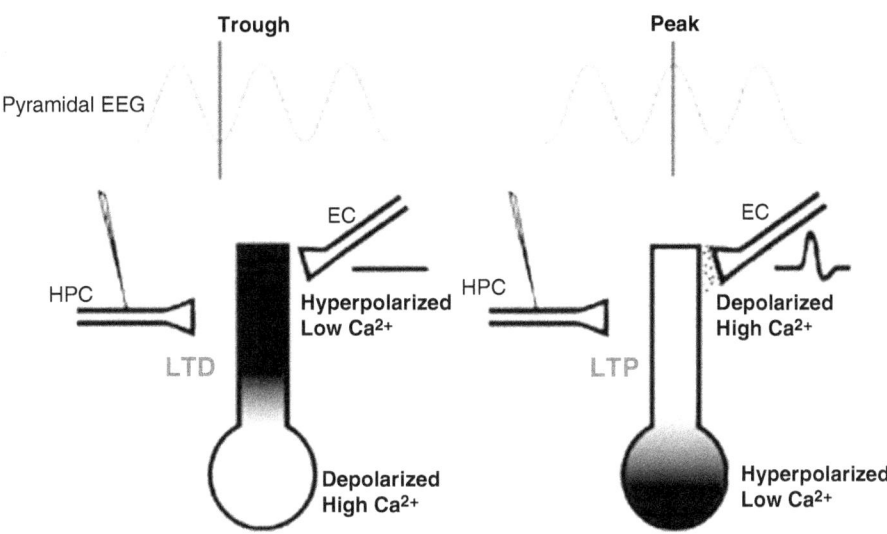

Figure 8.2 Ca^{2+} and plasticity phase differences of hippocampal theta rhythm. The central neuron is a CA1 pyramidal cell. This neuron receives afferent input from entorhinal cortex (EC), and also from other parts of the hippocampus (HPC). The HPC inputs could be Schaeffer collateral (from CA3) or from recurrent CA1 connections (as in Hyman et al., 2003). At the top are traces representing the EEG from the pyramidal layer of CA1; vertical lines show which phase in the theta cycle (trough on left, peak on right). When theta is at the trough, the soma of the CA1 neuron is depolarized and at its peak firing rate, while the apical dendrites are hyperpolarized, and EC input is weak. At this time Ca^{2+} is at low levels in the apical dendrite. When strong stimulation comes from the HPC input, LTD will results at this synapse. When the theta cycle is at the peak, the soma of the CA1 neuron is hyperpolarized, and the apical dendrites are depolarized, as excitatory EC input is strong. Now dendritic Ca^{2+} levels are high, and when input arrives from HPC, LTP will occur at that synapse.

input could underlie the phenomenon of déjà vu, in which new information is perceived as if it were being retrieved from memory.) The strengthening of synapses dependent on dendritic depolarization during hyperpolarization of the soma specifically differs from the standard view of STDP depending on back-propagation of a spike induced in the soma. However, this framework is supported by data showing that LTP can be induced with dendritic spikes, even when the soma is hyperpolarized (Golding et al., 2002).

In contrast, the opposite phase of theta shown in Fig. 8.2 provides dynamics appropriate for retrieval. During this phase, the soma is depolarized, allowing previously modified synapses to cause spiking activity in CA1, but the dendrites do not have sufficient calcium for LTP, thereby preventing the old retrieved information from being encoded as new. In fact, the induction of LTD could allow gradual decay of older associations within hippocampal circuits, which may be useful for some behaviors (Hasselmo et al., 2002).

The Ca^{2+} model can be applied to other plasticity protocols such as STDP and theta frequency stimulation. Ca^{2+} levels are supralinear when presynaptic stimulation precedes a postsynaptic spike, and sublinear when postsynaptic stimulation precedes presynaptic stimulation protocol is used (Koester and Sakmann, 1998). Although there is some discrepancy as to whether Ca^{2+} levels alone can explain STDP, or whether there is a need to integrate models relying on after-depolarization or some coincidence detector (Karmarkar et al., 2002), or because of after-hyperpolarization (Gorchetchnikov et al., 2005). The situation is different for theta frequency stimulation protocols. In these protocols multiple tetanic stimulations are delivered at theta frequencies, and optimal LTP is recorded as compared with identical numbers of tetani delivered at different interstimulus intervals (Larson and Lynch, 1986). This framework could also account for pulse-primed potentiation (Dunwiddie and Lynch, 1979; Diamond and Rose, 1994). During in vitro preparations it is most likely that these test pulses are manipulating intracellular currents, relying upon the natural theta rhythmicity of hippocampal neurons and thus Ca^{2+} levels are fluctuating accordingly. During in vivo theta frequency LTP protocols it is most likely that the theta rhythm is reset by the incoming tetanus and thus subsequent tetani will fall on the positive phase. Electrically driven or behaviorally driven theta reset both result in an alignment of the first positive phase occurring ~200 ms following stimulation (Buno et al., 1978; Givens, 1996). Furthermore tetani delivered after a behaviorally driven theta reset lead to robust LTP (McCartney et al., 2004).

If we assume Ca^{2+} level oscillations are opposite from the spiking activity of hippocampal principal cells, thus the dendrites are out of phase with soma spiking, then the phase of the current theta rhythm will determine the level of Ca^{2+} in the dendrites of hippocampal cells. In turn the levels of Ca^{2+} determine the directionality of plastic changes, so the phase of theta rhythm controls the directionality of any possible changes within the entire hippocampus.

Conclusion

In examining theta's role in synaptic modification, the extent and nature of theta's impact on the hippocampus can be seen. The recorded local field potentials of theta are the product of the entire hippocampus, and the phase of this sinusoidal pattern determines the direction of plastic changes for the entire population. These affects can be seen when populations of hippocampal cells are stimulated in awake, behaving animals (Hölscher et al., 1997;

Orr et al., 2001; Hyman et al., 2003); when small populations are stimulated in hippocampal slice field recordings (Larson and Lynch, 1986; Huerta and Lisman, 1995); and when individual neurons are stimulated and individual synapses are isolated in whole cell patch clamp and current clamp experiments (Cummings et al., 1996; Bi and Poo, 1998).

Theta rhythm is one of the most vital components for proper hippocampal functioning. Without it, animals seem to be left with a less-functional hippocampus. Hippocampal spiking activity is tightly controlled by theta, at the same time sensory, behavioral, cognitive, or affective stimuli are controlling that activity. Synaptic modification is dependent on, or rather determined by, theta. Intracellular potentials and ion levels are regulated by theta. The impact of theta rhythm on the hippocampus can be seen in both macroscopic and microscopic scales, and all levels in between.

References

Aggleton, J.P., Neave, N., Nagle, S., and Hunt, P.R. (1995). A comparison of the effects of anterior thalamic, mamillary body and fornix lesions on reinforced spatial alternation. *Behav Brain Res* **68**:91–101.

Andersen, P., Bland, H.B., Myhrer, T., and Schwartzkroin, P.A. (1979). Septo-hippocampal pathway necessary for dentate theta production. *Brain Res* **165**:13–22.

Berry, S.D. and Seager, M.A. (2001). Hippocampal theta oscillations and classical conditioning. *Neurobiol Learn Mem* **76**:298–313.

Berry, S.D. and Thompson, R.F. (1978). Prediction of learning rate from the hippocampal electroencephalogram. *Science* **200**:1298–1300.

Bi, G.Q. and Poo, M.M. (1998). Synaptic modifications in cultured hippocampal neurons: dependence on spike timing, synaptic strength, and postsynaptic cell type. *J Neurosci* **18**:10464–10472.

Bliss, T.V. and Collingridge, G.L. (1993). A synaptic model of memory: long-term potentiation in the hippocampus. *Nature* **361**:31–39.

Bliss, T.V. and Lomo, T. (1973). Long-lasting potentiation of synaptic transmission in the dentate area of the anaesthetized rabbit following stimulation of the perforant path. *J Physiol* **232**:331–356.

Bonansco, C. and Buno, W. (2003). Cellular mechanisms underlying the rhythmic bursts induced by NMDA microiontophoresis at the apical dendrites of CA1 pyramidal neurons. *Hippocampus* **13**:150–163.

Buno, W., Jr., Garcia-Sanchez, J.L., and Garcia-Austt, E. (1978). Reset of hippocampal rhythmical activities by afferent stimulation. *Brain Res Bull* **3**:21–28.

Buzsáki, G., Leung, L.W., and Vanderwolf, C.H. (1983). Cellular bases of hippocampal EEG in the behaving rat. *Brain Res* **287**:139–171.

Buzsáki, G., Czopf, J., Kondakor, I., and Kellenyi, L. (1986). Laminar distribution of hippocampal rhythmic slow activity (RSA) in the behaving rat: current-source density analysis, effects of urethane and atropine. *Brain Res* **365**:125–137.

Cho, K., Aggleton, J. P., Brown, M. W., and Bashir, Z. I. (2001). An experimental test of the role of postsynaptic calcium levels in determining synaptic strength using perirhinal cortex of rat. *J Physiol* **532**:459–466.

Collingridge, G. L., Kehl, S. J., and McLennan, H. (1983). Excitatory amino acids in synaptic transmission in the Schaffer collateral-commissural pathway of the rat hippocampus. *J Physiol* **3**:33–46.

Cormier, R. J., Greenwood, A. C., and Connor, J. A. (2001). Bidirectional synaptic plasticity correlated with the magnitude of dendritic calcium transients above a threshold. *J Neurophysiol* **85**:399–406.

Cummings, D. D., Wilcox, K. S., and Dichter, M. A. (1996). Calcium-dependent paired-pulse facilitation of miniature EPSC frequency accompanies depression of EPSCs at hippocampal synapses in culture. *J Neurosci* **16**:5312–5322.

Diamond, D. M. and Rose, G. M. (1994). Stress impairs LTP and hippocampal-dependent memory. *Ann N Y Acad Sci* **746**:411–414.

Dudek, S. M. and Bear, M. F. (1993). Bidirectional long-term modification of synaptic effectiveness of adult and immature hippocampus. *J Neurosci* **13**:2190–2198.

Dunwiddie, T. V. and Lynch, G. (1979). The relationship between extracellular calcium concentrations and the induction of hippocampal long-term potentiation. *Brain Res* **169**:103–110.

Eichenbaum, H. (2000). A cortical–hippocampal system for declarative memory. *Nat Rev Neurosci* **1**:41–50.

Eichenbaum, H., Kuperstein, M., Fagan, A., and Nagode, J. (1987). Cue-sampling and goal-approach correlates of hippocampal unit activity in rats performing an odor-discrimination task. *J Neurosci* **7**:716–732.

Ennaceur, A., Neave, N., and Aggleton, J. P. (1996). Neurotoxic lesions of the perirhinal cortex do not mimic the behavioural effects of fornix transection in the rat. *Behav Brain Res* **80**:9–25.

Fox, S. E., Wolfson, S., and Ranck, J. B., Jr. (1986). Hippocampal theta rhythm and the firing of neurons in walking and urethane anesthetized rats. *Brain Res* **62**:495–508.

Givens, B. (1996). Stimulus-evoked resetting of the dentate theta rhythm: relation to working memory. *Neuroreport* **8**:159–163.

Givens, B. S. and Olton, D. S. (1990). Cholinergic and GABAergic modulation of the medial septal area: effect on working memory. *Behav Neurosci* **104**:849–855.

Givens, B. and Olton, D. S. (1994). Local modulation of basal forebrain: effects on working and reference memory. *J Neurosci* **14**:3578–3587.

Golding, N., Staff, N., and Spruston, N. (2002). Dendritic spikes as a mechanism for cooperative long-term potentiation. *Nature* **418**:326–331.

Gorchetchnikov, A., Versace, M., and Hasselmo, M. E. (2005). A model of STDP based on spatially and temporally local information: derivation and combination with gated decay. *Neur Networks* **18**:458–466.

Green, J. D. and Arduini, A. A. (1954). Hippocampal electrical activity and arousal. *J Neurophysiol* **17**:533–557.

Greenstein, Y. J., Pavlides, C., and Winson, J. (1988). Long-term potentiation in the dentate gyrus is preferentially induced at theta rhythm periodicity. *Brain Res* **438**:331–334.

Griffin, A. L., Asaka, Y., Darling, R. D., and Berry, S. D. (2004). Theta-contingent trial presentation accelerates learning rate and enhances hippocampal plasticity during trace eyeblink conditioning. *Behav Neurosci* **118**:403–411.

Hasselmo, M. E. and Eichenbaum, H. (2005). Hippocampal mechanisms for the context-dependent retrieval of episodes. *Neur Networks* **15**:689–707.

Hasselmo, M. E., Bodelon, C., and Wyble, B. P. (2002). A proposed function for hippocampal theta rhythm: separate phases of encoding and retrieval enhance reversal of prior learning. *Neur Comput* **14**:793–817.

Hebb, D. O. (1949). *The Organization of Behavior: A Neuropsychological Theory*. New York: John Wiley.

Hölscher, C., Anwyl, R., and Rowan, M. J. (1997). Stimulation on the positive phase of hippocampal theta rhythm induces long-term potentiation that can be depotentiated by stimulation on the negative phase in area CA1 in vivo. *J Neurosci* **17**:6470–6477.

Huerta, P. T. and Lisman, J. E. (1995). Bidirectional synaptic plasticity induced by a single burst during cholinergic theta oscillation in CA1 in vitro. *Neuron* **15**:1053–1063.

Huxter, J., Burgess, N., and O'Keefe, J. (2003). Independent rate and temporal coding in hippocampal pyramidal cells. *Nature* **425**:828–832.

Hyman, J. M., Wyble, B. P., Goyal, V., Rossi, C. A., and Hasselmo, M. E. (2003). Stimulation in hippocampal region CA1 in behaving rats yields long-term potentiation when delivered to the peak of theta and long-term depression when delivered to the trough. *J Neurosci* **23**:11725–11731.

Jahr, C. E. and Stevens, C. F. (1987). Glutamate activates multiple single channel conductances in hippocampal neurons. *Nature* **325**:522–525.

Kamondi, A., Acsady, L., Wang, X. J., and Buzsáki, G. (1998). Theta oscillations in somata and dendrites of hippocampal pyramidal cells in vivo: activity-dependent phase-precession of action potentials. *Hippocampus* **8**:244–261.

Karmarkar, U. R., Najarian, M. T., and Buonomano, D. V. (2002). Mechanisms and significance of spike-timing dependent plasticity. *Biol Cybern* **87**:373–382.

Koester, H. J. and Sakmann, B. (1998). Calcium dynamics in single spines during coincident pre- and postsynaptic activity depend on relative timing of back-propagating action potentials and subthreshold excitatory postsynaptic potentials. *Proc Natl Acad Sci USA* **95**:9596–9601.

Larson, J. and Lynch, G. (1986). Induction of synaptic potentiation in hippocampus by patterned stimulation involves two events. *Science* **232**:985–988.

Lisman, J. (1989). A mechanism for the Hebb and the anti-Hebb processes underlying learning and memory. *Proc Natl Acad Sci USA* **86**:9574–9578.

Lisman, J. E. (2001). Three Ca^{2+} levels affect plasticity differently: the LTP zone, the LTD zone and no man's land. *J Physiol* **532**:285.

Lynch, G., Larson, J., Kelso, S., Barrionuevo, G., and Schottler, F. (1983). Intracellular injections of EGTA block induction of hippocampal long-term potentiation. *Nature* **305**:719–721.

Malenka, R. C. and Bear, M. F. (2004). LTP and LTD: an embarrassment of riches. *Neuron* **44**:5–21.

Markowska, A. L., Olton, D. S., Murray, E. A., and Gaffan, D. (1989). A comparative analysis of the role of fornix and cingulate cortex in memory: rats. *Exp Brain Res* **74**:187–201.

McCartney, H., Johnson, A. D., Weil, Z. M., and Givens, B. (2004). Theta reset produces optimal conditions for long-term potentiation. *Hippocampus* **14**:684–687.

M'Harzi, M., Palacios, A., Monmaur, P., *et al.* (1987). Effects of selective lesions of fimbria-fornix on learning set in the rat. *Physiol Behav* **40**:181–188.

Mishkin, M. (1978). Memory in monkeys severely impaired by combined but not by separate removal of amygdala and hippocampus. *Nature* **273**:297–298.

Morris, R. G., Garrud, P., Rawlins, J. N., and O'Keefe, J. (1982). Place navigation impaired in rats with hippocampal lesions. *Nature* **297**:681–683.

Mulkey, R. M. and Malenka, R. C. (1992). Mechanisms underlying induction of homosynaptic long-term depression in area CA1 of the hippocampus. *Neuron* **9**:967–975.

Numan, R. and Quaranta, J. R., Jr. (1990). Effects of medial septal lesions on operant delayed alternation in rats. *Brain Res* **531**:232–241.

O'Keefe, J. and Conway, D. H. (1978). Hippocampal place units in the freely moving rat: why they fire where they fire. *Exp Brain Res* **31**:573–590.

O'Keefe, J. and Dostrovsky, J. (1971). The hippocampus as a spatial map: preliminary evidence from unit activity in the freely-moving rat. *Brain Res* **34**:171–175.

O'Keefe, J. and Recce, M. L. (1993). Phase relationship between hippocampal place units and the EEG theta rhythm. *Hippocampus* **3**:317–330.

Orr, G., Rao, G., Houston, F. P., McNaughton, B. L., and Barnes, C. A. (2001). Hippocampal synaptic plasticity is modulated by theta rhythm in the fascia dentata of adult and aged freely behaving rats. *Hippocampus* **11**:647–654.

Pavlides, C., Greenstein, Y. J., Grudman, M., and Winson, J. (1988). Long-term potentiation in the dentate gyrus is induced preferentially on the positive phase of theta-rhythm. *Brain Res* **439**:383–387.

Rawlins, J. N., Feldon, J., and Gray, J. A. (1979). Septo-hippocampal connections and the hippocampal theta rhythm. *Exp Brain Res* **37**:49–63.

Rizzuto, D. S., Madsen, J. R., Bromfield, E. B., *et al.* (2003). Reset of human neocortical oscillations during a working memory task. *Proc Natl Acad Sci USA* **100**:7931–7936.

Rubin, J. E., Gerkin, R. C., Bi, G. Q., and Chow, C. C. (2005). Calcium time course as a signal for spike-timing-dependent plasticity. *J Neurophysiol* **93**:2600–2613.

Scoville, W. B. and Milner, B. (1957). Loss of recent memory after bilateral hippocampal lesions. *J Neurol Neurosurg Psychiat* **20**:11–21.

Seager, M. A., Johnson, L. D., Chabot, E. S., Asaka, Y., and Berry, S. D. (2002). Oscillatory brain states and learning: impact of hippocampal theta-contingent training. *Proc Natl Acad Sci USA* **99**:1616–1620.

Skaggs, W. E. and McNaughton, B. L. (1996). Replay of neuronal firing sequences in rat hippocampus during sleep following spatial experience. *Science* **271**:1870–1873.

Skaggs, W. E., McNaughton, B. L., Wilson, M. A., and Barnes, C. A. (1996). Theta phase precession in hippocampal neuronal populations and the compression of temporal sequences. *Hippocampus* **6**:149–172.

Stanton, P. K. (1996). LTD, LTP, and the sliding threshold for long-term synaptic plasticity. *Hippocampus* **6**:35–42.

Stewart, M., Quirk, G. J., Barry, M., and Fox, S. E. (1992). Firing relations of medial entorhinal neurons to the hippocampal theta rhythm in urethane anesthetized and walking rats. *Exp Brain Res* **90**:21–28.

Vanderwolf, C. H. (1968). Recovery from large medial thalamic lesions as a result of electroconvulsive therapy. *J Neurol Neurosurg Psychiat* **31**:67–72.

Vanderwolf, C. H. (1971). Limbic–diencephalic mechanisms of voluntary movement. *Psychol Rev* **78**:83–113.

Vanderwolf, C. H. (1975). Neocortical and hippocampal activation relation to behavior: effects of atropine, eserine, phenothiazines, and amphetamine. *J Comp Physiol Psychol* **88**:300–323.

Vertes, R. P. and Kocsis, B. (1997). Brainstem–diencephalo-septohippocampal systems controlling the theta rhythm of the hippocampus. *Neuroscience* **81**:893–926.

Wang, H. X., Gerkin, R. C., Nauen, D. W., and Bi, G. Q. (2005). Coactivation and timing-dependent integration of synaptic potentiation and depression. *Nat Neurosci* **8**:187–193.

Whishaw, I. Q. and Tomie, J. A. (1997). Perseveration on place reversals in spatial swimming pool tasks: further evidence for place learning in hippocampal rats. *Hippocampus* **7**:361–370.

Winson, J. (1978). Loss of hippocampal theta rhythm results in spatial memory deficit in the rat. *Science* **201**:160–163.

Wittenberg, G. M. and Wang, S. S. (2006). Malleability of spike-timing-dependent plasticity at the CA3–CA1 synapse. *J Neurosci* **26**:6610–6617.

Yang, S. N., Tang, Y. G., and Zucker, R. S. (1999). Selective induction of LTP and LTD by postsynaptic $[Ca^{2+}]i$ elevation. *J Neurophysiol* **81**:781–787.

Ylinen, A., Soltesz, I., Bragin, A., *et al.* (1995). Intracellular correlates of hippocampal theta rhythm in identified pyramidal cells, granule cells, and basket cells. *Hippocampus* **5**:78–90.

9

Distributed population codes in sensory and memory representations of the neocortex

MATTHIAS MUNK

Distributed representations are the inevitable consequence of devoting large neuronal circuits to a detailed and adaptive analysis of complex information. The neocortical sheet with its extensive cortico-cortical connectivity is characterized by ubiquitous massive divergence and convergence, sparseness and reciprocity of the vast majority of connections. It therefore appears as an optimized dynamical structure for detailed and adaptive analysis on the one hand and the operation of multiple parallel neuronal processes required for optimizing speed and accuracy of information processing on the other hand. The established concepts of information coding in the cortex are based on tuning functions of many individual neurons thought to express their stimulus specificity in an independent way. A second level of organization is usually attributed to the spatial relations of neurons in topographically organized representations like those in sensory areas and the convergence of neuronal signals carrying information from different modalities into "higher" areas which are more involved in executive functions or the formation of complex memory representations. It has been argued that the collection of neuronal signals from consecutive recording sessions can be used to reconstruct population codes as it has been done with the "population vector" analysis in motor, sensory, and memory areas of the cortex. It is clear that the success of this method relies on fixed neuronal response properties which have been consolidated in cortical circuits over long periods of time such that the spatial pattern and the mixture

Information Processing by Neuronal Populations, ed. Christian Hölscher and Matthias Munk.
Published by Cambridge University Press. © Cambridge University Press 2009.

of neurons contributing to the population response is reasonably stable. There is also evidence both from micro-electrode experiments and from more recent neuroimaging studies that activity patterns in populations of temporal cortex are stimulus selective and allow for category specific identification of the stimulus material. The spatial resolution of measurements is crucial to draw the correct conclusions about the neuronal composition of a representation. However, beyond such long-term representations, it is essential for the brain to process rapidly changing relations among content and context for which more dynamic neuronal representations are required. Here we will explore the mesoscopic range of cortical signals that might contribute to distributed codes in the cortex with particular emphasis on the malleability of representations as the underlying mechanism of learning processes.

Why is neuronal coding a difficult issue?

How the brain and in particular its most complex part, the neocortex, codes information and how the signals carrying information are used by neuronal processes for the control of behavior remains the central problem of neuroscience despite more than half a century of intensive research. This may in part be explained by the fact that the neocortex is the structure with by far the largest proportion of self-referencing connections (Salin and Bullier, 1995; Braitenberg and Schüz, 1998; Douglas and Martin, 2004, 2007) and, probably as a consequence, expresses more "ongoing" self-maintaining activity than activity related to stimuli or any other external events (Arieli et al., 1996). Of course then, the widely used approach of treating the brain as an input–output system can have only limited success. Still, we have learnt lots of useful details about a huge range of neuronal procedures and mechanisms that can explain well-defined functions like for example perceiving the movement direction of a visual stimulus (Salzman et al., 1992; Salzman and Newsome, 1994; Parker and Newsome, 1998) or other more complex partial functions like visually guided reaching (Goodale et al., 1986; Whitney et al., 2003; Goodale and Westwood, 2004). What is nevertheless missing is a comprehensive understanding of the embedding of these many partial functions into ongoing processes or, with a wider perspective, into a "brain's life" which means that all prior experience and memories may influence processing. In terms of neuronal coding, the embedding of specific functions requires that representations which use a particular neuronal code are sensitive to the context, but at the same time allow for a sufficient degree of invariance in order to make perceptual processes reliable and therefore support recognition.

The sensitivity of neuronal processing to context has mostly been studied with respect to external factors like additional sensory information e.g. in the surround of receptive fields (Phillips and Singer, 1997) or the effect of concurrent processing in a different modality (Calvert, 2001; Macaluso and Driver, 2005; Schroeder and Foxe, 2005; Ghazanfar and Schroeder, 2006). A different form of context in the time domain is provided by the influence of memory representations on sensory processing which determine familiarity or novelty (Xiang and Brown, 2004). But there are many other less specific factors which contribute to the context of processing a particular entity like a stimulus. For example attention and expectation have profound influence on the behavioral effect of perceptual processes. These factors are at least in part linked to internal and ongoing processes. Other brain processes like thinking and imagination are entirely internal but are apparently able to activate neuronal representations in a fashion such that their results are compatible with those derived from processing input (Golland *et al.*, 2006) or operating output (Llinas *et al.*, 1998). This requires that codes are employed which support congruent processing of the content, but differential processing of the context. Therefore the neuronal representations have to share certain features, but not others.

In the case of processing sensory information, the actual neuronal implementation needs to change depending on the context and the state of the system. But what does this mean for the employed neuronal codes? Different states do not necessarily need to be as different as sleep and waking in order to affect for example sensory processing, but may simply reflect the transition from one processing mode like the identification of an expected sensory stimulus to a subsequent mode during which information about an identified stimulus needs to be maintained and updated. Although most of the previous work suggests that a single neuronal code is used to conserve information through such transitions, it appears as rather unlikely that this would allow for successful retrieval and utilization of the relevant information in particular if the context keeps changing. To some extent, there needs to be constancy across states and context in order to be able to detect and make use of common content. However, other elements of the neuronal representation need to differ significantly, at least to avoid confusion at the behavioral level. For processing of sensory information, the constant elements might be represented in earlier areas which are much less susceptible to plastic changes (Hochstein and Ahissar, 2002) and have a larger capacity for detailed information while the more variable context elements might be processed in higher areas such as in the parietal and frontal lobes. However, this division of labor is not useful if content and context require a similar degree of spatial resolution and therefore need to be processed in the same areas. In such cases, a simple distributed code would not work and the system needs to organize distributed activity in such a way that different processes can run simultaneously without interference in the same network. Several proposals have been made (Delage, 1919; Hebb, 1949; Singer *et al.*, 1997) for how ensembles of neurons can self-organize for this purpose (see also Chapters 3 and 7).

The situation is worse if the context and its impact on the processing of the content keep changing, which is usually the case in natural environments. Then not only relations among ensembles of neurons need to be flexible, but the code they employ for the representation of content cannot work like a simple look-up table, no matter how complex it might be. It needs to adapt to the new requirements such that the combinatorial complexity of content and context can still be decoded by the central executive. This means that codes probably evolve depending on the requirements of the individual history of representations. In such a procedural context, it appears as rather unlikely that the neuronal mechanism for maintaining information for example in short-term memory is based on sustained spike firing in prefrontal areas which is involved in planning and deciding on behavioral responses rather than serving as a passive storage device. Planning means monitoring, updating, and initiating adaptations of processing strategy which all depends on the behavioral context. As a consequence, neuronal coding has to involve constant dynamical restructuring of activity which has to incorporate contextual information and therefore has to include many distributed circuits if not the entire brain without loosing the specificity of concurrent processing.

What properties does a neuronal code need to have?

Before exploring which features of neuronal activity have so far been identified as putative codes, it might be useful to consider in a more general framework what properties a neuronal code needs to have, given that brains are causally involved in controlling behavior. It may seem trivial, but the simple fact that experimenters can extract information from neuronal signals which appears to be relevant in a certain experiment does not mean that these signals are actually used by the brain under investigation to compute the observed behavioral response. In general, each neuronal signal that systematically changes depending on variable external and/or internal conditions may contain features that have the potential to qualify as a code. In that sense, signals with systematically covarying features certainly carry information about the conditions, but how much variance do they need to explain before we can qualify particular signal features as a code?

What is a code really? By definition, a code is an instruction or rule with which a signal that contains information is transformed into another signal that can serve as a message or command which is transmitted to a receiver or target system. The main historical reasons for using a code was to hide information from unwanted receivers, or in order to transmit more efficiently, to use fewer signals and thus save energy and gain speed. The latter reason is at the heart of bio-economical constraints and is certainly a criterion for all biological codes that nervous systems have developed under selection pressure throughout evolution. Implicit in the term "code," as it is normally used, is that the rules have been developed by somebody or at least by some external instance and have been applied to a specific transmission system. This is definitively not the case for a self-organizing system like the brain. In self-organizing systems, the constant feedback from the behavioral outcome onto the processing architecture requires that representations are constantly adapted (see Chapter 7 for improved memory representations in perirhinal cortex). The fact that brains are self-organizing is one of the reasons why it is so difficult to unravel neuronal codes, at least in more complex nervous systems.

But independently from their origin, any successful use of a code depends as much on the receiver as on the sender of messages. Therefore, a neuronal code can only be deciphered if, in addition to the transmitted neuronal activity patterns that have been generated by applying the (so far unknown) rules of a particular code, the neuronal readout mechanism has been identified and used to interpret the transmitted signals. This is only feasible if the activity of target neurons and transmitting afferents are monitored simultaneously, i.e. the result of postsynaptic integration has been captured by recordings such that the behavior of the postsynaptic neuron can be causally related to features of the observed synaptic input. Causality can be inferred if the activity patterns of the target population can be reconstructed from the activity patterns of the sender populations after applying the transformation (Laurent, 2002), or experimentally, by reversible and selective manipulation of the essential feature of the transmitted activity (Laurent, 2002; Miyashita, 2004; Wilson and Laurent, 2005).

Attempts to determine the contribution of presynaptic neurons have been successful for identified retinogeniculate (Cleland *et al.*, 1971; Levick *et al.*, 1972) and thalamocortical connections (Alonso *et al.*, 1996) as well as for binaural inputs to brainstem nuclei responsible for directional hearing (Carney, 1999; Konishi, 2003). In these systems, the anatomical relations among pre- and postsynaptic cells are clear. However, the problem with identified connections in neocortex is that there is almost always a huge number of often widely distributed afferents which feed into each target neuron. It is therefore very difficult if not impossible to derive the transformation rules to predict correctly the effect on the activity patterns of target neurons. Nevertheless, with massive parallel monitoring of neighboring neuron populations in primary visual cortex, as it has been possible to predict the firing pattern of individual cells from the surrounding activity patterns (Tsodyks *et al.*, 1999). If larger groups of neurons coordinate their activity such that they form an assembly (as comprehensively discussed in Chapters 3 and 7), deciphering their code becomes more complicated, because there are several possibilities how assemblies can code information and the decoded signal is no longer a simple scalar-measure-like firing rate (Pouget *et al.*, 2000; Averbeck *et al.*, 2006).

Assemblies are defined by the relative timing of spike discharge across the member neurons. Therefore, an assembly can only be detected by identifying and proving a statistically reliable fixed temporal relation of firing times among two or more simultaneously recorded neurons. One possible – and probably the simplest – type of assembly code consists of the simple combination of functional properties the individual members express as a consequence of co-activation by e.g. sensory stimulation. Although the information represented by the assembly is defined by response properties of the participating neurons, it is not only available as a consequence of sensory stimulation, but whenever the assembly is active. However, as Harris has nicely described in this book (Chapter 3), assemblies follow each other in sequences which suggests that assemblies may need to pass on information among each other, in particular if they are engaged in the same process. For deciphering such an inter-assembly code, it is not only necessary to identify large enough parts of the participating assemblies, but also to know how subsequent assemblies could interact. Only then is it possible to find out how one distributed activity pattern is translated into a subsequent pattern which is the basis for deriving a transformation rule.

In order to unravel the transformation rules of an unknown code by reverse engineering of activity patterns, it is necessary to investigate the transformations for a whole range of conditions. The typical approach of a neurophysiologist would be to analyze a reasonably large range of input–output relations to derive the transformation rule which unfortunately suffers from practical and conceptional limitations. First, the number of stimuli and corresponding behavioral and neuronal responses that can be recorded and analyzed in a reasonable amount of time is very limited compared to the real parameter space most brains have evolved for. Second, a lot of the information a brain receives and processes will never be expressed in any measurable output but will change representations, influence internal states and therefore create variance of the output which will result in more complicated or even ambiguous rules. Of course, the typical experimental strategy is to keep all context parameters as constant as possible, but if context is ignored, the generality of the results one could obtain is rather limited. One way of reducing this variance is to include contextual information, and if possible, study its influence in a parametric way. In addition, there is the problem of the history of a certain brain like its specific memory content that might influence even simple rules. If for example a perceptual process is facilitated by the existence of a memory trace, then the individual experience with memories relevant for the perceptual process under study will cause inter-individual variability that might compromise the derivation of generalized rules. Finally, if the development of a brain function during ontogeny differs among subjects with respect to the availability and level of maturation of circuits, then the same function might employ different codes and maybe even different circuits. So, in practice there are numerous difficulties and caveats with deciphering neuronal codes which can only be overcome by very careful experimental designs and systematic quantitative assessment of data.

Established and proposed codes

Which features of brain signals have in the past been considered as a code for information processing? The most fundamental signal that has been associated with information processing is the activation of neurons, circuits, or entire brain systems upon the occurrence of sensory stimuli. Although the observed activity changes are generally only correlated with the occurrence of a stimulus or movement, as opposed to proving their causal influence, a change in net activity has always been considered to represent an important signal. Therefore it was not easily accepted when visual neurophysiologists investigated neuronal responses under more natural conditions, i.e. during free viewing, and found that there are no more obvious rate changes when a stimulus crossed the receptive field during an exploratory eye movement (Gallant et al., 1998). Along these lines, strong increases of activity as observed under highly artificial laboratory conditions using extremely reduced stimulus features in the early days of visual neuroscience (Hubel and Wiesel, 1962) were considered an important signal change while neuronal responses which are expressed for example as reduction of activity (Quintana et al., 1988; Nowak et al., 1995a) have mostly been ignored. Silent neurons may be as important for a code like zeros in a binary code. The same or even more lack of interest occurred towards other more subtle activity features like changes in variability or embedded fine time structure. Even if features of a signal are subtle, they may be very important for subsequent signal integration or the reorganization of the system. Minor differences in the time structure of spike trains may appear negligible at the level of single units, while in a network of such units, the precise timing of spikes converging onto common target cells will make an important difference for their efficiency to integrate and hence their output (Munk, 2001).

Beyond activity patterns observed in single neurons, codes have been proposed which make use of activity coordinated among distributed populations of neurons. Based on the finding that neocortical circuits express oscillations in the gamma frequency range (Gray and Singer, 1987, 1989; Eckhorn et al., 1988) synchronization of remote neuron populations (Engel et al., 1991a; Munk et al., 1995) was shown to code for stimulus continuity and common fate (Gray and Singer, 1989; Engel et al., 1991b; Nowak et al., 1995b; Kreiter and Singer, 1996) (for reviews see Singer (1999) and Munk and Neuenschwander (2000)). Evidence for binding-by-synchrony in the non-human primate cortex is provided by experiments in which dynamically changing shapes have to be integrated and successfully processed such that the monkey can give correct responses (Kreiter and Singer, 1996; Tallon-Baudry et al., 2004; Hirabayashi and Miyashita, 2005; Mandon and Kreiter, 2005) while attempts to qualify stimulus dependent synchrony as information carrier have unfortunately failed (Aggelopoulos et al., 2005). Although synchronization of high-frequency oscillations is potentially a very powerful code for highly distributed processing of related information (Engel et al., 1992; Herrmann et al., 2004), there are many open questions concerning the generality of this mechanism. For example, if synchronization should serve a larger number of parallel processes instead of constituting a single global solution, the degrees of freedom for large functional networks to engage in different processes is rather limited if the only parameter for separating different ensembles would be oscillation frequencies.

These difficulties can in principle be overcome if synchrony does not only depend on rhythms at a particular frequency, but requires that a particular connectivity pattern like converging connections with matched conduction delays, which may also be specific for different processes, can be responsible for the generation and maintenance of synchrony (Diesmann et al., 1999). Although such a connectivity-dependent mechanism may at first glance appear as too constrained and mutually exclusive to phase-coupled oscillations as synchronizing mechanism, the coordination of spatiotemporal spike patterns, which constitute the physiological correlate of synfire chains (Abeles, 1991), could take advantage of oscillatory patterns at each synfire stage, but may also operate independently. The real charm of the synfire mechanism is that it might operate on any combination of connections, in any direction and can probably travel through parts of the network that are involved in many other processes without too much interference, because each stage requires only a single well-timed spike per cell. Therefore, synchrony based on the synfire mechanism could constitute many independent processes with very little interference. Most importantly, synfire processes are self-organizing and probably depend only very little on external events, because each individual process constitutes, continues and ceases depending on the state of the contributing connections and not on the state of the entire system. All of these properties make neuronal synchrony a much more sophisticated principle of cortical processing than a single global state and despite the diversity of processes and functions it can serve, it is always directly contributing to the integration and distribution of information at a very global level.

Activation at a more global scale involving entire networks of areas as revealed by imaging methods can certainly not unravel any code, but reflects general processing strategies of the brain which by their pattern can provide information about the circuits actually used, their interactions, and by their statistical relation to stimuli and other external events can give an account of the level or degree of processing that has occurred up to a certain point within a network. With imaging techniques like functional magnetic resonance imaging (fMRI), hemodynamic responses are triggered by increased energy demand which allows for the identification of brain regions that seem to be important for concurrent processing. Unfortunately, nobody knows today the exact nature of these hemodynamic responses or what changes in neuronal activity patterns are required to cause the observed hemodynamic changes. However, there is no simple relation between hemodynamic and neuronal responses in the sense of a linear correlation, although the sign of BOLD and multi-unit responses seems to correlate on a trial-by-trial basis (Shmuel et al., 2006). If we turn our perspective and ask what kind of neuronal processes are energy demanding so that they could potentially trigger the expensive luxury perfusion known as the BOLD response, there are a number of options that need to be sorted out in the future. One interesting link between changes in neuronal activity patterns and a hemodynamic response has more recently been proposed by uncovering a positive correlation between the strength of neuronal oscillations and the strength of a very localized hemodynamic response in superficial layers of cat visual cortex (Niessing et al., 2005), the correlation with firing rates being significantly weaker. Already the finding that coordinated network activity can be correlated with the hemodynamic response rather than net activity levels of representative neurons suggests that neuronal processing itself is in general highly distributed and may cause high energy demand due to the need for coordination among remote neuron populations.

The fact that patterned activity at the macroscopic level of hemodynamic signals carries information about stimulus categories (Haxby et al., 2001) brings us back to the question how the brain codes information. Of course, there are other reasons for distributed representations as we will see later, but if the brain including the human brain expresses widespread stimulus-category-specific activity patterns that allow for a reliable reconstruction of what the tested subjects saw (Haxby et al., 2000, 2001; Kamitani and Tong, 2005, 2006) then distributed spatial patterns of activity seem to reflect a coding strategy. So far, all studies employing techniques not measuring neuronal activity directly were thought to reveal only the localization of regions in the brain that express high energy demand and therefore supposedly play an important role for active processing. However, as spatial and temporal resolution of functional imaging techniques improves, they can indeed provide more and more detailed information about the macroscopic and mesoscopic aspects of distributed representations (see Chapter 13) and thus contribute important information to the understanding of neuronal codes which micro-electrode experiments cannot reveal, because in most cases they are bound to deliver very local information. Nevertheless, results based solely on hemodynamic techniques will never answer the question how neurons compute behavior or what the *neuronal algorithms* of particular cognitive functions are (N.K. Logotheis, personal communication).

Beyond conceptually simple codes

Most examples of sensory codes show a simple intuitive relation between stimulus properties like their intensity and features of neuronal activity such as discharge rate. However, there are activity features that could qualify as neuronal codes which certainly do not show such a simple relation, but may have more effect on synaptic integration at a subsequent processing stage. For example, the coding of absolute temperature and rate of temperature change in peripheral nerves seems to be achieved by changes in oscillation frequency and a noise component (Braun *et al.*, 1994) which influences the probability of spike discharge. This way signals even on single fibers can be used for the coding of multiple-stimulus parameters. Although not independent or incompatible with a rate code, burst patterns during sensory responses have been shown to carry information about the stimulus (Tsodyks and Markram, 1997; Snider *et al.*, 1998; Oram *et al.*, 1999; Reinagel and Reid, 2000; Lu *et al.*, 2001; Krahe and Gabbiani, 2004; Wang *et al.*, 2005) which implies yet another decoding mechanism, probably being able to rely mostly on temporal summation during very short intervals. For distributed codes, the relative timing of activity is crucial (deCharms, 1998). The most explicit form of response timing is reflected in latency codes (Gawne *et al.*, 1996; VanRullen *et al.*, 2005; Chase and Young, 2007), which may in some cases provide only partial information (Nelken *et al.*, 2005). Although first spikes may represent a strong and effective signal for subsequent processing in systems like the auditory and somatosensory systems, they might not necessarily carry all aspects of stimulus information.

Higher-order information like the global configuration of several stimuli can be reflected in patterns of neuronal synchronization, which in most cases develops during the course of sensory responses (Singer, 1999). Such a coding mechanism might be particularly useful if distributed representations express oscillatory activity patterns which can self-organize by coordinating the phase relations of oscillations. The latter mechanism of neuronal synchronization has the interesting feature that underlying oscillations can not only sharpen a sensory representation (Schaefer et al., 2006), but allow for more difficult perceptual operations like detecting noisy or ambiguous stimuli (Rodriguez et al., 1999). The formation of synchronized ensembles occurs at the population level, however, population codes have been divided (deCharms, 1998) into those which assume independent coding of huge numbers of cells (Georgopoulos et al., 1988), or which depend on neuronal interactions to organize ensembles (Singer et al., 1997; Harris et al., 2003). It is, however, difficult to believe – and I don't know of any evidence – that cortical neurons can operate as independent units, given that most cortical neurons outside layer 4 receive almost exclusively cortico-cortical input and their output is tightly correlated to activity patterns of large cortical ensembles which certainly have a strong bias for nearby populations of cells (Tsodyks et al., 1999), but have also been shown to draw from more remote populations depending on the influence they can exert (Munk et al., 1995).

Cortical network operations at the mesoscopic level

Which options does the cortical network provide as carrier signals for a code? If we want to understand in a more general framework how the system architecture of the neocortical network can support complex neuronal operations we have basically two options. Either we look straight at the anatomical patterns of cortico-cortical connections and try to derive theoretically possible neuronal operations and test them in a computer simulation as has been done for synchronizing axons (Houzel et al., 1994; Innocenti et al., 1994) and synfire chains (Abeles, 1991). The other option is to measure neuronal activity patterns and to correlate them to the underlying connectivity, the difficulty of course being to achieve a correct and meaningful match, as for example signs of synaptic connections between neighboring neurons in vivo (Hata et al., 1993; Constantinidis et al., 2001, 2002) raise the question what kind of neuronal computation they really support. A very useful extension of the latter approach is to manipulate either the connectivity patterns by lesion (Miyashita, 2004) or reversible inactivation and study the function of the respective subsystem (Fuster and Alexander, 1970; Fuster et al., 1981; Payne et al., 1996). These studies have revealed driving input in the case of feedforward connections (Girard and Bullier, 1989; Girard et al., 1991a, 1991b, 1992) or modulatory input mediated through feedback connections (Hupe et al., 1998; Bullier et al., 2001), but rarely interactions among neurons have been studied in vivo which could be related to self-organizing processes like cortical network oscillations (Engel *et al.*, 1991a; Munk *et al.*, 1995). The latter two studies revealed that direct cortico-cortical connections connecting primary visual areas across the corpus callosum play a crucial role for stimulus-dependent cortical synchronization. No general rules could be derived yet for how functional connectivity determines neuronal operations that are required for processing higher degrees of complexity as they occur in real-world situations. Although complexity can formally be measured and interpreted as an increase in mutual information (Tononi *et al.*, 1996), there is no simple relationship between connectivity and complexity (Sporns *et al.*, 2000).

If we now try to identify connectivity features that can in principle support the coding of complex information by neuronal processes, we may consider in the first place the extraordinary degree of divergence at almost any point in the cortical network. The conventional understanding of diverging functional connections is the distribution of activity for parallel processing which can improve the speed and reliability of processing (Rousselet *et al.*, 2003). If neurons, in particular in higher areas, in which no simple relations exist to stimulus dimensions, can each be considered as representing a new dimension, diverging connections could in principle support an expansion of activity patterns into a high-dimensional space. It is known from anatomical studies (Rockland, 1992, 2002) that cortico-cortical connections do not only spread locally and into neighboring columns which may serve to strengthen the signal and improve redundancy, but also spread out over several millimeters which in the case of feedforward connections may serve increases of receptive field size. However, as divergence increases more and more in anterior visual areas, it suggests that divergence serves the generation of higher-order feature selectivity and lateral interactions at this level, e.g. for comparative processing rather than a simple spread of visual signals in areas TEO and TE (Saleem *et al.*, 1993; Tanigawa *et al.*, 2005).

From a functional point of view, little is known about the topography of cortical interactions in inferotemporal areas except that pairs of neurons which show signs of synaptic connections have more similar stimulus selectivity (Gochin *et al.*, 1991) and that correlated spike firing is more prevalent if stimuli can be more readily identified as faces than control stimuli (Hirabayashi and Miyashita, 2005). However, evidence for high-dimensional codes does not require in the first place that those neuronal interactions are identified which constitute the activity patterns carrying high-dimensional information. It would be much more desirable to demonstrate that the information-carrying patterns are read by subsequent circuits and utilized for generating behavior. Evidence for high-dimensional coding has been gained in the motor system, beginning with the simple combination of feature selective activity, representing firing rates from non-simultaneously recorded neurons by the so-called "population vector" (Georgopoulos et al., 1988). Later, similar principles were applied to analyzing activity in inferotemporal cortex (Rolls et al., 1997). More recently, in both the visual and motor system, the activity of simultaneously recorded neurons contained more (Taylor et al., 2002) or less (Aggelopoulos et al., 2005) useful ensemble information. However, none of the studies on higher-dimensional codes provides a proposal for a neuronal mechanism for setting up the information-carrying activity patterns nor how these patterns are read out and the information used for generating behavior.

But how can we read a high-dimensional signal, understand how the brain uses these signals, and therefore retrieve a code? If we want to use the ultimate output, i.e. behavior, to understand the readout mechanism, we cannot assume that imposing complex activity patterns onto the executive system solves the problem, because there are many intermediate processing steps which will all contribute a large variety of modifications to the transformation of distributed patterns such that the outcome will not be predictable as a whole. This leaves us with the conclusion that we have to identify all the intermediate processing stages and understand the transformations performed at each stage and integrate the contributions of each stage into the entire process. This then requires that we need to understand the readout at each stage, which could be quite different, and decipher the respective codes. For example, codes for sensory processing and codes for memory representations might be different and would then require that signals undergo major transformations. From our current technical perspective, physiologists can only unravel some, probably more local, rules, but an entire series of transformations as proposed above will most probably be only tractable in sufficiently elaborate neuronal network models integrating transformation rules from several stages. In addition, I think that we have not yet identified all possible levels of functional organization of activity in large networks like the neocortex.

For this reason, I want to discuss in the following section three types of activity patterns which are promising candidate signals for serving neuronal computations, all of which depend on intact cortical connectivity and, each of which is characterized by its own dynamical properties. First, the most recently discovered patterns are cortical *avalanches* which represent waves of transmitted synchronous activity (Plenz and Thiagarajan, 2007). Second, I will discuss spatio-temporal spike patterns as a correlate of *synfire chains* which had originally been postulated as a results of a well-thought-through concept of how cortical connectivity might determine brain activity (Abeles, 1982). Last, but not least, the oldest form of electrical brain activity which can in many instances assigned to the activity of cortical circuits are *oscillations*. They were theoretically predicted (Delage, 1919) before their discovery (Berger, 1929) and since then have become an important constituent of our modern concepts of brain function (Salinas and Sejnowski, 2001; Buzsáki and Draguhn, 2004; Herrmann *et al.*, 2004).

Avalanches

Neuronal avalanches are the description for propagating synchronized signals of large neuronal populations as they can be observed as population spikes in cortical slices and slice cultures (Beggs and Plenz, 2003, 2004) and more recently as oscillatory activity in cortical field potentials of ongoing activity in resting monkeys (Petermann *et al.*, 2006). The attractive aspect of investigating avalanches in vivo is that they may solve the problem of the needle in the haystack by transforming functional connectivity to the mesoscopic level. Finding monosynaptically connected neurons in neocortex in vivo over slightly larger distances than local connections within the same column is technically at least extremely difficult if not impossible. If, however, the transmission of signals requires cooperative processes like synchrony of a few dozens or more neurons to overcome sparse divergent connectivity (Abeles, 1982), then avalanches (Fig. 9.1a) would reflect successful transmission processes at the population level.

Depending on the spatiotemporal dynamics of these processes, the neuronal network can be in a "critical" state (Fig. 9.1b) which essentially means that the diversity of processes with respect to their size and duration is scale-free (Plenz and Thiagarajan, 2007). Translated into the terminology of branching processes (Fig. 9.1c), which is a more concrete model for the propagation of synchronous cortical activity, a critical state reflects stable dynamics: on average over large samples, the number of subsequently active sites remains constant. This stability has been shown to depend on a balance between excitation and inhibition (Beggs and Plenz, 2003) and an optimal concentration of dopamine (Stewart and Plenz, 2006). Future studies will have to determine how specific these state estimators are for predicting behavior and whether different states may support or even require different types of neuronal coding. Depending on when avalanches occur during a task, they could either reflect feedforward dissemination of sensory information, lateral interactions that might stabilize or reorganize activity patterns of a distributed representation, or might reflect some kind of a readout process which for example a higher area could use to scan lower-level representations before generating a behavioral response. Above all, patterned neuronal activity is not confined to the stimulus or behavior-related processes, but is as relevant for intrinsic and ongoing processes (Arieli *et al.*, 1996; Tsodyks *et al.*, 1999) which have more recently been observed in

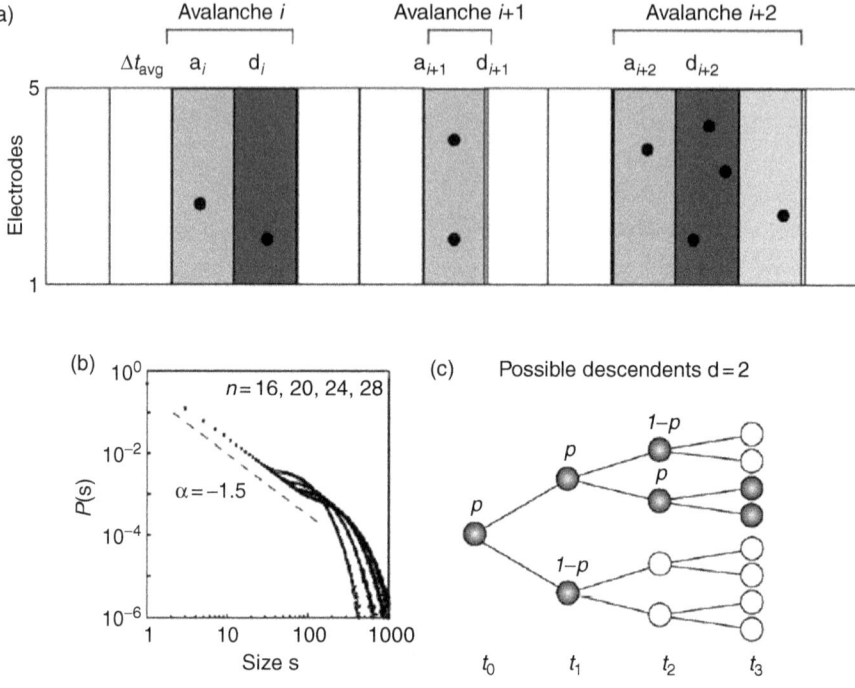

Figure 9.1 Neuronal avalanches. (a) Avalanches of different length and size. Each dot represents an event which in most studied cases reflects synchronous population activity. Events can occur sequentially (left), synchronously (middle), or in any combination of the latter (right). (b) Size distribution of avalanches: the frequency P(s) is a function of event size, i.e. how many sites participate follows a power law with an exponent of −1.5 to indicate critical behavior. (c) Branching pattern of a self-organizing critical process. (Adapted from Plenz and Thiagarajan, 2007.)

the human brain (Nir et al., 2006). In this context it is worth noting that in the human brain, entire territories seem to be devoted to processing intrinsic signals, in contrast to the well-known sensory domains (Golland et al., 2006) (Fig. 9.2).

Synfire chains

Originally, the concept of synchronous transmission was developed on the basis of the underlying structural mechanisms of brain signals (Abeles, 1982), the key connectivity feature on which synfire chains were based being the ubiquitous divergence and convergence of cortico-cortical connectivity. Network simulations have shown that volleys of synchronous spikes can propagate in a stable fashion through many stages of a model network (Abeles, 1982; Diesmann et al., 1999) and can group into separate chains or even regroup (Diesmann et al., 1999), depending on the external input and the network's

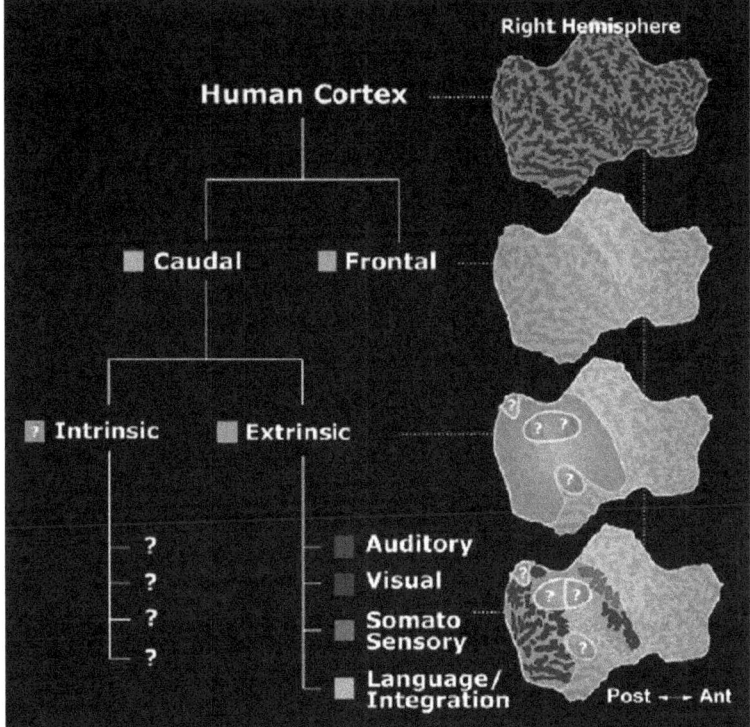

Figure 9.2 Intrinsic and extrinsic cortical systems related to natural scene stimulation. The extrinsic system is characterized by widespread activations driven by sensory stimulation in different modalities. In contrast, territories within the gaps of the extrinsic system were identified by highly correlated activity among each other and are referred to as intrinsic system. (Modified with permission from Golland *et al.*, 2006.)

dynamical state. The neurophysiological correlate of synfire chains are precise "spatiotemporal spike patterns" (as demonstrated in Fig. 9.3) which have by now been shown to occur mostly in frontal areas in close correlation with the behavioral task (Prut *et al.*, 1998; Shmiel *et al.*, 2005), but are still much under debate whether the statistical tests are valid (Baker and Lemon, 2000; Mokeichev *et al.*, 2007).

The fact that so far very few research laboratories have been able to reproduce the observation of precise spatiotemporal patterns both in vitro (Ikegaya *et al.*, 2004; Yuste *et al.*, 2005) and in vivo (Villa *et al.*, 1999) does not mean that synfire chains do not exist; as has once been put forward for gamma frequency oscillations (Tovee and Rolls, 1992). It is obviously difficult to find patterns in a reliable way. If synfire chains serve to organize sparse and distributed representations

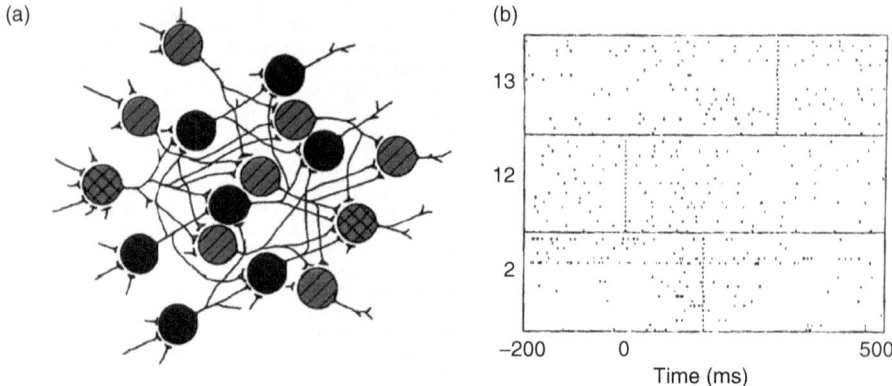

Figure 9.3 Spatiotemporal spike patterns. (a) Schematic drawing of a network supporting synfire activity. (b) Spatiotemporal patterns recorded from frontal cortex in behaving monkeys. The synfire chain first propagates through cell 12, then cell 2 and finally cell 13. (Adapted from Abeles *et al.*, 1994.)

then the probability of detecting spatiotemporal patterns in a small set of arbitrary recordings is simply very small. One of the interesting features of spatiotemporal patterns is that theoretically a very large number of more or less independent chains could coexist and, as simulations have shown, they could fuse or part. Much more easily than signals that need to disperse in phase space of oscillatory activity patterns in order to separate or differentiate, synfire chains can evolve into a high-dimensional coding space defined by functional neuronal pathways and, depending on the availability of recurrent pathways, activity patterns can persist for longer durations. If expansion and contraction of a representation were to be related to subsequent stages of synfire chains, then changes of distributed representations could be expected to occur in very rapid sequences. Such rapid sequences are reminiscent of the "phase sequences" discussed elsewhere in this book (see Chapter 3).

Oscillations

It is far beyond the scope of this chapter to give a comprehensive review on oscillatory brain activity. However, there are some aspects of the structure–function relationship of cortical oscillations which might be useful for understanding how large populations of cortical neurons may be coordinated such that they can directly code information or support complex neuronal operations. The first and probably most fundamental functional property of brain oscillations is their state dependence as reported in the pioneer study on human EEG for the alpha rhythm (Berger, 1929) which provides among other important properties contextual information. Although alpha is often addressed as an idling rhythm, the phase relation between alpha rhythm and the time of stimulus presentation determines how fast and reliably a subject will respond (Klimesch *et al.*, 1996). This appears as a pure attentional mechanism; however, in the time domain, global repetitive excitability changes also provide a temporal context for sensory coding. The theta rhythm, well known for its role in place cell coding, is another important example where the time structure of an oscillation is directly used as reference signal for coding spatial information (see Chapter 7). In a similar way, rhythms at other frequency ranges are to some extent state dependent, but may participate in or support neuronal coding of information. Beta and gamma rhythms are both highly state dependent and probably play similar important roles for attentional modulation of neuronal processing. Beta rhythms are more prevalent in the motor system while gamma oscillations have up to now been mostly studied in sensory systems.

The most profound feature of oscillations at any frequency is their inherent time structure which means that they provide a very regular signal that neurons can use to define time relations either between a global rhythm, e.g. reflecting a state, and local events like spike firing in response to sensory input, as employed in place coding, or between many local rhythms to define relations among remote neuron populations in global networks. The latter could in principle support long-distance transmission of spikes (Salinas and Sejnowski, 2001), because pre- and postsynaptic neurons with coordinated fluctuations of excitability are more prepared to exchange signals when postsynaptic integration is facilitated by coincidence of presynaptic spikes with the depolarized phase of postsynaptic potentials (see inset in Fig. 9.4). As outlined in Fig. 9.4, phase relations differ among simultaneously recorded field potentials in remote cortical areas of monkeys performing a difficult visuomotor task. None of the signal pairs synchronize with zero phase lag (A7–A4 is 360°) and some (V4–A6) exhibit phase shifts around 90° which correspond to 8–9 ms at a frequency around 30 Hz. This shift is well in the range of latency differences between cortical areas and therefore coupled oscillatory signals with such a phase shift could facilitate serial spike transmission between areas. This needs to be tested in the future by approaches as described elsewhere in this book (see Chapter 14).

Population codes and distributed representations

What can examples of complex representations tell us about sensory processing and memory? Population codes were originally analyzed in non-simultaneous recordings of multiple single neurons by combining their response properties and comparing the results to stimulus properties (Perrett *et al.*, 1982; Rolls *et al.*, 1997; for review, see Pouget *et al.*, 2000) or motor behavior

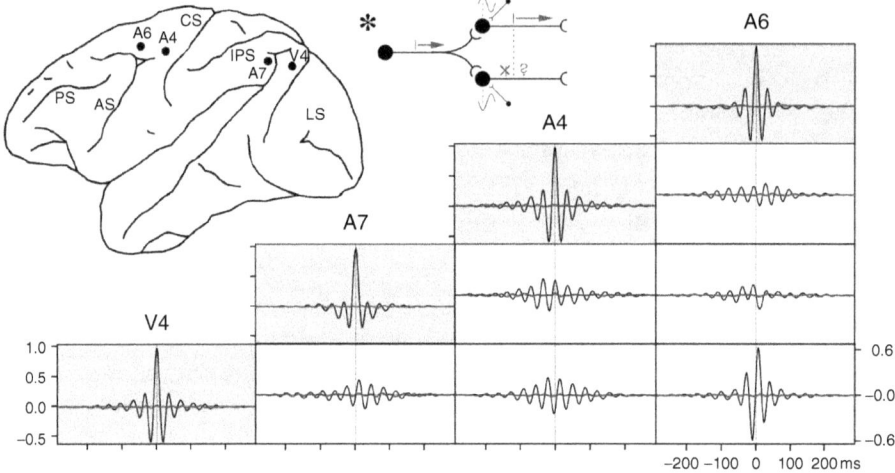

Figure 9.4 Oscillations. Scheme of lateral aspect of a macaque brain with recording positions (upper left) and averaged autocorrelation (gray) and cross-correlation (white) functions with shift predictors (lower right) computed from transcortical field potentials recorded simultaneously in four cortical areas. Note that the phase relations among different oscillatory processes as provided by the cross-correlations are very different. The inset (*) illustrates the basic idea of how oscillations – depending on their phase relation to the presynaptic signal – can modulate synaptic transmission of spike signals sent across multiple branches of the same axon.

(Georgopoulos et al., 1988; Kruse et al., 2002). Most of the information about stimuli is usually extracted from response amplitudes like the discharge rate of neurons and some studies even tried to show that correlated neuronal signals like precise spike synchrony added very little information (Aggelopoulos et al., 2005). However, signal correlations are more frequently considered (Averbeck et al., 2006) which suggests that the structural organization and intrinsic circuit dynamics appear to have gained ground for concepts of neuronal processes underlying perceptual and cognitive functions. Also, concepts including structural aspects and circuit dynamics are more compatible with the concept of assemblies which have a much longer tradition (Delage, 1919; Hebb, 1949) in particular in the field of memory research (for review, see Harris (2005) and Chapter 3). In this section I try to relate evidence for population codes, distributed representations and assemblies as investigated at the neocortical level in primates. Nevertheless, the first step of analyzing complex representations requires a description of functional aspects like tuning profiles, i.e. the preferences of neurons for certain stimuli or some of their features.

One of the first examples of a distributed representation of complex stimuli like faces was based on recordings from inferotemporal cortex (Young and

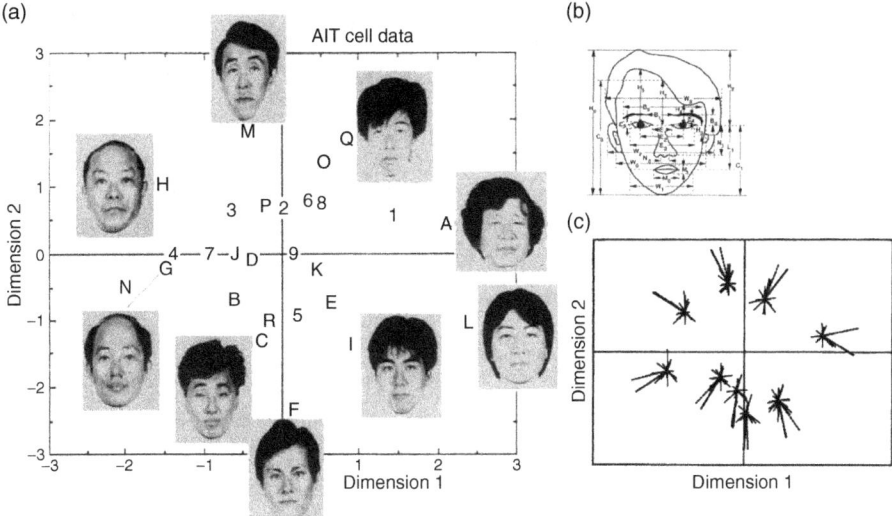

Figure 9.5 Distributed and sparse representation: an example is shown from a study in primate inferior temporal cortex using human faces. (a) Neuronal population data (Young and Yamane, 1992) after multidimensional scaling. (b) Parameters used to generate a physical model of the faces. (c) Population vectors indicating the match between model and neuronal data.

Yamane, 1992) in which the authors showed that the tuning of individual cells was broad, but that the acuity of the population of cells was close to identifying individual faces (Fig. 9.5). Another example directly demonstrating the more general case of distributed representations for multiple objects from different categories based on a quite different technique is illustrated in the introductory chapter. Inferotemporal cortex was investigated with optical recording in order to visualize the spatial layout of activity patterns and how these patterns change when the complexity of visual objects was manipulated (Tsunoda et al., 2001; Yamane et al., 2006). Using both optical and micro-electrode recordings, Tanifuji and colleagues showed that patterns of multiple activity blobs represent various visual objects in which neurons express corresponding object selectivity (Tsunoda et al., 2001). However, the complexity of the object did not simply express in the number and arrangement of blobs (Fig. 1.4b, in Chapter 1) suggesting that the spatial layout of activity at the level of the cortical area was not a set of convergence zones in which simpler feature combinations arriving from more posterior areas were simply integrated. The authors also concluded that distributed representations might be more complex by drawing from active and inactive columns. These representations were investigated in anesthetized monkeys suggesting that the activity patterns were most likely caused by feedforward activation. Unfortunately, no compatible data for domains in temporal cortex are yet available for monkeys that need to generate a behavioral response based on successful processing of complex objects.

These examples show that representations for complex visual objects are distributed over large populations of neurons in inferotemporal cortex. There are no studies that have searched for or investigated a neuronal readout mechanism for such highly distributed representations. The conventional idea of how such distributed representations could be integrated assumes converging connections which carry signals coding for the same object or related objects to subsequent processing levels. This concept implies that the knowledge about relations is implemented in the wiring of connections for example to prefrontal cortex. Although we have no data on the physiology of signal transmission from temporal to prefrontal cortex, we know that a rigid set of feedforward connections is very unlikely to solve the problem of selecting and providing the behaviorally relevant information to the next stage. If selective routing (Olshausen et al., 1993; Constantinidis and Steinmetz, 2001) of neuronal signals is based on oscillatory processes as recent work in the dorsal visual stream (Buschman and Miller, 2007; Saalmann et al., 2007) suggests, then the question remains how feedback processes can structure oscillatory activities in inferotemporal cortex such that spike transmission to prefrontal cortex is dynamically and selectively optimized in accord with the task demands. It is rather unlikely that the mere existence of coherence in a broad frequency band between two brain structures is a sufficient explanation for directed neuronal interactions as some authors have suggested (Schoffelen et al., 2005).

Visual representations during short-term memory: units versus populations

What needs to happen in the brain if sensory information is to be maintained? The established concept consists of a transfer of preprocessed neuronal activity to areas like the prefrontal and parietal cortex which can switch neuronal activity into a sustained mode of firing. The transition of firing mode is thought to stabilize the stimulus related activity and thereby hold information as an "engram." There are quite different concepts for the stabilization of activity patterns that serve the maintenance of information, suggestions ranging from simple membrane states (Marder et al., 1996), to reverberating or recurrent activity (Lau and Bi, 2005; Machens et al., 2005; Wang et al., 2005). Since the content of short-term memory needs to be constantly updated in order to serve working memory, no sluggish or highly local mechanisms are useful to control maintenance. A mechanism that can follow instructions from systems that control behavioral rules (Wallis *et al.*, 2001; Miller *et al.*, 2002; Wallis and Miller, 2003), probably in a top–down fashion, and at the same time select among diverse sensory input that may be relevant for achieving a certain behavioral goal needs to be able to integrate many different dynamical aspects in an adaptive fashion and at the same time provide sufficient stability of representations such that maintenance remains a reliable process. This is unlikely to be achieved by a single dynamical principle like sustained spike firing or more or less local reverberations of activity. This does not exclude that neuronal activity patterns like sustained spike firing may be a reliable indicator that neurons participate in an active short-term memory process, but sustained firing alone does not qualify as the only relevant mechanism, causally related to behavioral performance.

The observation that multiple intricate mechanisms have to cooperate in order to achieve adaptive memory performance suggests that processes occur at multiple levels involving multiple circuits and different levels of organization. Very local processes expressed at the level of single or multiple units have been shown to carry information about different stimuli; however, the degree of selectivity of single-unit activity was found to be too low to allow for a reliable discrimination of different stimuli (Miller *et al.*, 1996). Selectivity has been quantified by using an index that relates responses to best and worst stimuli such that a value of 1 means perfect selectivity and a value of 0 means no selectivity. As can be seen on the upper left of Fig. 9.6, showing the distribution of selectivity indices of single units replotted from Miller *et al.* (1996), most units have very low indices, the average being 0.19. In contrast, analyses of multi-unit recordings, which could be expected to express even less selectivity, because signals from several neurons are mixed, revealed, if anything, higher indices (Fig. 9.6 lower left). This finding already indicates that populations of neurons may be a useful level of organization for a better understanding of memory mechanisms.

If individual units do not provide sufficiently precise information about memorized items like visual objects, then accuracy of the coding can only be achieved by coordinated activities of neuronal populations. In the example shown in Fig. 9.6, activity of 15 simultaneously recorded prefrontal sites is patterned (Fig. 9.6c) depending on what stimulus (Fig. 9.6b) the monkey saw. Of course, such patterns could be random noise and only rigorous testing based on resampling can reveal which fraction of patterns is stable. In Fig. 9.7, the fraction of stable patterns changes over time and as a function of the number of contributing sites (Staedtler *et al.*, 2006), revealing that representations are more transient than expected. Comparing the time course of stable patterns with the time course of firing rates (Fig. 9.7) reveals that different groups of sites

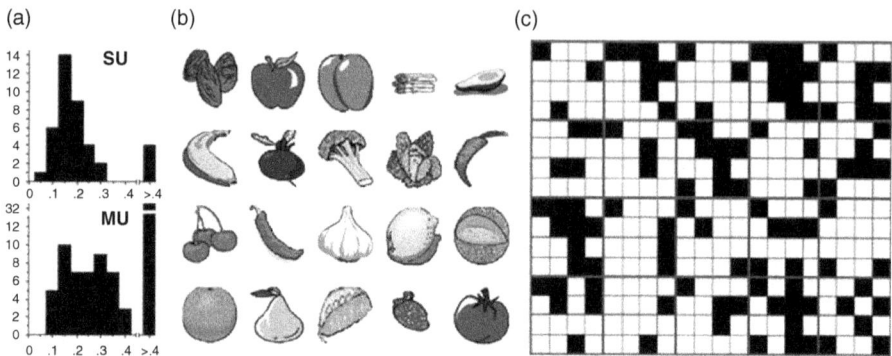

Figure 9.6 Stimulus selectivity and distributed activity patterns in lateral prefrontal cortex of monkeys performing a simple visual memory task. (a) Selectivity index for single units (SU) and multi-units (MU) (Miller et al., 1996). (b) The 20 familiar objects the monkeys had to memorize. (c) Patterns of activity observed during early delay. Each pattern is plotted into the grid of 4 × 4 electrodes and each black square represents a significant (1% criterion) deviation of activity from the pre-stimulus baseline.

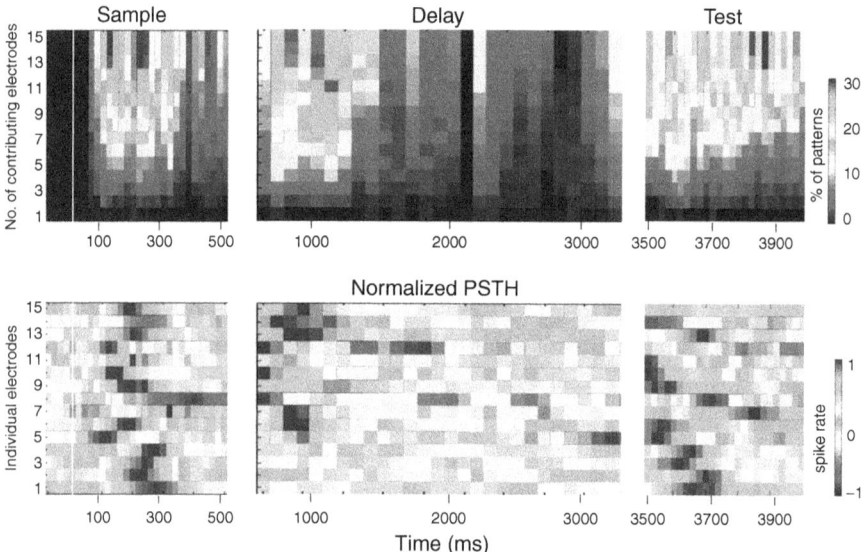

Figure 9.7 Strength of ensemble selectivity and simultaneous modulation of firing rates. (Top) Stimulus selectivity of distributed activity patterns expressed as percentage of stable non-random population patterns (z-axis) plotted as a function of number of contributing sites (y-axis) and time (x-axis). Sliding windows are 100, 300, and 100 ms for Sample, Delay, and Test epochs, respectively. (Bottom) Sliding window Peri Stimulus Time Histogram (PSTH) normalized for each channel's minimal and maximal rates.

are co-activated during different epochs with increased frequency of stable patterns. This suggests that different groups of neurons cooperate in order to constitute the distributed activity patterns at the population level. To compare quantitatively what these different levels, population patterns and samples of single or multiple units, contribute to the actual processes that lead to behavioral performance is work in progress. Given that multiple cortical and subcortical areas participate in all complex cognitive functions, the next more global levels of organization as they can be investigated with mass signals and imaging will add new and certainly more complex types of processes. The challenge is to determine how different neuronal processes as discussed above (and in Chapters 3, 4, and 7) combine and complement each other in order to achieve appropriate behavior.

Conclusion

Compared to the research on hippocampus, our understanding of neuronal coding and integrative processes in neocortex is a long way behind. Although neocortical organization is considerable more complex, there are some advantages of studying neocortex. The fact that cortex has all these different modalities makes it easier to test whether processes can be found in more than one subsystem and may therefore constitute a general principle. Activity features like inhibitory response components or precise spike timing have to be studied very carefully, because at the level of ensembles, they might be of more profound importance than at the single-unit level. In order to establish causal relations between neuronal information coding and behavior, it will be necessary to perform experiments in which neuronal processes are reversibly manipulated with the consequence that behavior covaries in a predictable way, and that the corresponding transformations of neuronal signals between the input and the manipulated processing stage, or the latter and its output, are so well understood that they can be reproduced in a model. In order to achieve this goal, we need to know a lot more about the actual neuronal interactions which cause neuronal representations and their transformations, if possible, while the brains under study are dealing with real-world problems.

References

Abeles, M. (1982). *Local Cortical Circuits: An Electrophysiological Study*. Heidelberg, Germany: Springer-Verlag.

Abeles, M. (1991). *Corticonics: Neural Circuits of the Cerebral Cortex*. Cambridge, UK: Cambridge University Press.

Abeles, M., Prut, Y., Bergman, H., and Vaadia, E. (1994). Synchronization in neuronal transmission and its importance for information processing. *Prog Brain Res* **102**:395–404.

Aggelopoulos, N. C., Franco, L., and Rolls, E. T. (2005). Object perception in natural scenes: encoding by inferior temporal cortex simultaneously recorded neurons. *J Neurophysiol* **93**:1342–1357.

Alonso, J. M., Usrey, W. M., and Reid, R. C. (1996). Precisely correlated firing in cells of the lateral geniculate nucleus. *Nature* **383**:815–819.

Arieli, A., Sterkin, A., Grinvald, A., and Aertsen, A. (1996). Dynamics of ongoing activity: explanation of the large variability in evoked cortical responses. *Science* **273**:1868–1871.

Averbeck, B. B., Latham, P. E., and Pouget, A. (2006). Neural correlations, population coding and computation. *Nat Rev Neurosci* **7**:358–366.

Baker, S. N. and Lemon, R. N. (2000). Precise spatiotemporal repeating patterns in monkey primary and supplementary motor areas occur at chance levels. *J Neurophysiol* **84**:1770–1780.

Beggs, J. M. and Plenz, D. (2003). Neuronal avalanches in neocortical circuits. *J Neurosci* **23**:11167–11177.

Beggs, J. M. and Plenz, D. (2004). Neuronal avalanches are diverse and precise activity patterns that are stable for many hours in cortical slice cultures. *J Neurosci* **24**:5216–5229.

Berger, H. (1929). Über das Elektrencephalogramm des Menschen. *Arch Psychiat Nervenkr* **87**:527–570.

Braitenberg, V. and Schüz, A. (1998). *Cortex: Statistics and Geometry of Neuronal Connections*. Heidelberg, Germany: Springer-Verlag.

Braun, H. A., Wissing, H., Schafer, K., and Hirsch, M. C. (1994). Oscillation and noise determine signal transduction in shark multimodal sensory cells. *Nature* **367**:270–273.

Bullier, J., Hupe, J. M., James, A. C., and Girard, P. (2001). The role of feedback connections in shaping the responses of visual cortical neurons. *Prog Brain Res* **134**:193–204.

Buschman, T. J. and Miller, E. K. (2007). Top–down versus bottom–up control of attention in the prefrontal and posterior parietal cortices. *Science* **315**:1860–1862.

Buzsáki, G. and Draguhn, A. (2004). Neuronal oscillations in cortical networks. *Science* **304**:1926–1929.

Calvert, G. A. (2001). Crossmodal processing in the human brain: insights from functional neuroimaging studies. *Cereb Cortex* **11**:1110–1123.

Carney, L. H. (1999). Temporal response properties of neurons in the auditory pathway. *Curr Opin Neurobiol* **9**:442–446.

Chase, S. M. and Young, E. D. (2007). First-spike latency information in single neurons increases when referenced to population onset. *Proc Natl Acad Sci USA* **104**:5175–5180.

Cleland, B. G., Dubin, M. W., and Levick, W. R. (1971). Simultaneous recording of input and output of lateral geniculate neurones. *Nat New Biol* **231**:191–192.

Constantinidis, C. and Steinmetz, M. A. (2001). Neuronal responses in area 7a to multiple-stimulus displays. I. Neurons encode the location of the salient stimulus. *Cereb Cortex* **11**:581–591.

Constantinidis, C., Franowicz, M. N., and Goldman-Rakic, P. S. (2001). Coding specificity in cortical microcircuits: a multiple-electrode analysis of primate prefrontal cortex. *J Neurosci* **21**:3646–3655.

Constantinidis, C., Williams, G. V., and Goldman-Rakic, P. S. (2002). A role for inhibition in shaping the temporal flow of information in prefrontal cortex. *Nat Neurosci* **5**:175–180.

deCharms, R. C. (1998). Information coding in the cortex by independent or coordinated populations. *Proc Natl Acad Sci USA* **95**:15166–15168.

Delage, Y. (1919). *Le Rêve: Étude psychologique, philosophique et littéraire*. Paris: Presses Universitaires de France.

Diesmann, M., Gewaltig, M. O., and Aertsen, A. (1999). Stable propagation of synchronous spiking in cortical neural networks. *Nature* **402**:529–533.

Douglas, R. J. and Martin, K. A. (2004). Neuronal circuits of the neocortex. *Annu Rev Neurosci* **27**:419–451.

Douglas, R. J. and Martin, K. A. (2007). Recurrent neuronal circuits in the neocortex. *Curr Biol* **17**:R496–R500.

Eckhorn, R., Bauer, R., Jordan, W., *et al.* (1988). Coherent oscillations: a mechanism of feature linking in the visual cortex? Multiple electrode and correlation analyses in the cat. *Biol Cybernet* **60**:121–130.

Engel, A. K., Konig, P., Kreiter, A. K., and Singer, W. (1991a). Interhemispheric synchronization of oscillatory neuronal responses in cat visual cortex. *Science* **252**:1177–1179.

Engel, A. K., Konig, P., and Singer, W. (1991b). Direct physiological evidence for scene segmentation by temporal coding. *Proc Natl Acad Sci USA* **88**:9136–9140.

Engel, A. K., Konig, P., Kreiter, A. K., Schillen, T. B., and Singer, W. (1992). Temporal coding in the visual cortex: new vistas on integration in the nervous system. *Trends Neurosci* **15**:218–226.

Fuster, J. M. and Alexander, G. E. (1970). Delayed response deficit by cryogenic depression of frontal cortex. *Brain Res* **20**:85–90.

Fuster, J. M., Bauer, R. H. and Jervey, J. P. (1981). Effects of cooling inferotemporal cortex on performance of visual memory tasks. *Exp Neurol* **71**:398–409.

Gallant, J. L., Connor, C. E., and Van, E. (1998). Neural activity in areas V1, V2 and V4 during free viewing of natural scenes compared to controlled viewing. *Neuroreport* **9**:85–90.

Gawne, T. J., Kjaer, T. W., and Richmond, B. J. (1996). Latency: another potential code for feature binding in striate cortex. *J Neurophysiol* **76**:1356–1360.

Georgopoulos, A. P., Kettner, R. E., and Schwartz, A. B. (1988). Primate motor cortex and free arm movements to visual targets in three-dimensional space. II. Coding of the direction of movement by a neuronal population. *J Neurosci* **8**:2928–2937.

Ghazanfar, A. A. and Schroeder, C. E. (2006). Is neocortex essentially multisensory? *Trends Cogn Sci* **10**:278–285.

Girard, P. and Bullier, J. (1989). Visual activity in area V2 during reversible inactivation of area 17 in the macaque monkey. *J Neurophysiol* **62**:1287–1302.

Girard, P., Salin, P. A., and Bullier, J. (1991a). Visual activity in areas V3a and V3 during reversible inactivation of area V1 in the macaque monkey. *J Neurophysiol* **66**:1493–1503.

Girard, P., Salin, P. A., and Bullier, J. (1991b). Visual activity in macaque area V4 depends on area 17 input. *Neuroreport* **2**:81–84.

Girard, P., Salin, P. A., and Bullier, J. (1992). Response selectivity of neurons in area MT of the macaque monkey during reversible inactivation of area V1. *J Neurophysiol* **67**:1437–1446.

Gochin, P. M., Miller, E. K., Gross, C. G., and Gerstein, G. L. (1991). Functional interactions among neurons in inferior temporal cortex of the awake macaque. *Exp Brain Res* **84**:505–516.

Golland, Y., Bentin, S., Gelbard, H., et al. (2006). Extrinsic and intrinsic systems in the posterior cortex of the human brain revealed during natural sensory stimulation. *Cereb Cortex* **17**:766–777.

Goodale, M. A. and Westwood, D. A. (2004). An evolving view of duplex vision: separate but interacting cortical pathways for perception and action. *Curr Opin Neurobiol* **14**:203–211.

Goodale, M. A., Pelisson, D., and Prablanc, C. (1986). Large adjustments in visually guided reaching do not depend on vision of the hand or perception of target displacement. *Nature* **320**:748–750.

Gray, C. M. and Singer, W. (1987). Stimulus-specific neuronal oscillations in the cat visual cortex: a cortical functional unit. *Soc Neurosci Abstr* **13**:404.3.

Gray, C. M. and Singer, W. (1989). Stimulus-specific neuronal oscillations in orientation columns of cat visual cortex. *Proc Natl Acad Sci USA* **86**:1698–1702.

Harris, K. D., Csicsvari, J., Hirase, H., Dragoi, G., and Buzsáki, G. (2003). Organization of cell assemblies in the hippocampus. *Nature* **424**:552–556.

Hata, Y., Tsumoto, T., Sato, H., Hagihara, K., and Tamura, H. (1993). Development of local horizontal interactions in cat visual cortex studied by cross-correlation analysis. *J Neurophysiol* **69**:40–56.

Haxby, J. V., Hoffman, E. A., and Gobbini, M. I. (2000). The distributed human neural system for face perception. *Trends Cogn Sci* **4**:223–233.

Haxby, J. V., Gobbini, M. I., Furey, M. L., et al. (2001). Distributed and overlapping representations of faces and objects in ventral temporal cortex. *Science* **293**:2425–2430.

Hebb, D. O. (1949). *The Organization of Behavior*. New York: John Wiley.

Herrmann, C. S., Munk, M. H., and Engel, A. K. (2004). Cognitive functions of gamma-band activity: memory match and utilization. *Trends Cogn Sci* **8**:347–355.

Hirabayashi, T. and Miyashita, Y. (2005). Dynamically modulated spike correlation in monkey inferior temporal cortex depending on the feature configuration within a whole object. *J Neurosci* **25**:10299–10307.

Hochstein, S. and Ahissar, M. (2002). View from the top: hierarchies and reverse hierarchies in the visual system. *Neuron* **36**:791–804.

Houzel, J.C., Milleret, C., and Innocenti, G. (1994). Morphology of callosal axons interconnecting areas 17 and 18 of the cat. *Eur J Neurosci* **6**:898–917.

Hubel, D.H. and Wiesel, T.N. (1962). Receptive fields, binocular interaction and functional architecture in the cat's visual cortex. *J Physiol* **160**:106–154.

Hupe, J.M., James, A.C., Payne, B.R., et al. (1998). Cortical feedback improves discrimination between figure and background by V1, V2 and V3 neurons. *Nature* **394**:784–787.

Ikegaya, Y., Aaron, G., Cossart, R., et al. (2004). Synfire chains and cortical songs: temporal modules of cortical activity. *Science* **304**:559–564.

Innocenti, G.M., Lehmann, P., and Houzel, J.C. (1994). Computational structure of visual callosal axons. *Eur J Neurosci* **6**:918–935.

Kamitani, Y. and Tong, F. (2005). Decoding the visual and subjective contents of the human brain. *Nat Neurosci* **8**:679–685.

Kamitani, Y. and Tong, F. (2006). Decoding seen and attended motion directions from activity in the human visual cortex. *Curr Biol* **16**:1096–1102.

Klimesch, W., Doppelmayr, M., Schimke, H., and Pachinger, T. (1996). Alpha frequency, reaction time, and the speed of processing information. *J Clin Neurophysiol* **13**:511–518.

Konishi, M. (2003). Coding of auditory space. *Annu Rev Neurosci* **26**:31–55.

Krahe, R. and Gabbiani, F. (2004). Burst firing in sensory systems. *Nat Rev Neurosci* **5**:13–23.

Kreiter, A.K. and Singer, W. (1996). Stimulus-dependent synchronization of neuronal responses in the visual cortex of the awake macaque monkey. *J Neurosci* **16**:2381–2396.

Kruse, W., Dannenberg, S., Kleiser, R., and Hoffmann, K.P. (2002). Temporal relation of population activity in visual areas MT/MST and in primary motor cortex during visually guided tracking movements. *Cereb Cortex* **12**:466–476.

Lau, P.M. and Bi, G.Q. (2005). Synaptic mechanisms of persistent reverberatory activity in neuronal networks. *Proc Natl Acad Sci USA* **102**:10333–10338.

Laurent, G. (2002). Olfactory network dynamics and the coding of multidimensional signals. *Nat Rev Neurosci* **3**:884–895.

Levick, W.R., Cleland, B.G., and Dubin, M.W. (1972). Lateral geniculate neurons of the cat: retinal inputs and physiology. *Invest Ophthalmol* **11**:302–311.

Llinas, R., Ribary, U., Contreras, D., and Pedroarena, C. (1998). The neuronal basis for consciousness. *Phil Trans R Soc Lond B* **353**:1841–1849.

Lu, T., Liang, L., and Wang, X. (2001). Temporal and rate representations of time-varying signals in the auditory cortex of awake primates. *Nat Neurosci* **4**:1131–1138.

Macaluso, E. and Driver, J. (2005). Multisensory spatial interactions: a window onto functional integration in the human brain. *Trends Neurosci* **28**:264–271.

Machens, C.K., Romo, R., and Brody, C.D. (2005). Flexible control of mutual inhibition: a neural model of two-interval discrimination. *Science* **307**:1121–1124.

Mandon, S. and Kreiter, A. K. (2005). Rapid contour integration in macaque monkeys. *Vision Res* **45**:291–300.

Marder, E., Abbott, L. F., Turrigiano, G. G., Liu, Z., and Golowasch, J. (1996). Memory from the dynamics of intrinsic membrane currents. *Proc Natl Acad Sci USA* **93**:13481–13486.

Miller, E. K., Erickson, C. A., and Desimone, R. (1996). Neural mechanisms of visual working memory in prefrontal cortex of the macaque. *J Neurosci* **16**:5154–5167.

Miller, E. K., Freedman, D. J., and Wallis, J. D. (2002). The prefrontal cortex: categories, concepts and cognition. *Phil Trans R Soc Lond B* **357**:1123–1136.

Miyashita, Y. (2004). Cognitive memory: cellular and network machineries and their top-down control. *Science* **306**:435–440.

Mokeichev, A., Okun, M., Barak, O., et al. (2007). Stochastic emergence of repeating cortical motifs in spontaneous membrane potential fluctuations in vivo. *Neuron* **53**:413–425.

Munk, M. H. (2001). Role of gamma oscillations for information processing and memory formation in the neocortex. In: *Neuronal Mechanisms of Memory Formation*, ed. Hölscher, C., pp. 167–194. Cambridge, UK: Cambridge University Press.

Munk, M. H. and Neuenschwander, S. (2000). High-frequency oscillations (20 to 120 Hz) and their role in visual processing. *J Clin Neurophysiol* **17**:341–360.

Munk, M. H., Nowak, L. G., Nelson, J. I., and Bullier, J. (1995). Structural basis of cortical synchronization. II. Effects of cortical lesions. *J Neurophysiol* **74**:2401–2414.

Nelken, I., Chechik, G., Mrsic-Flogel, T. D., King, A. J., and Schnupp, J. W. (2005). Encoding stimulus information by spike numbers and mean response time in primary auditory cortex. *J Comput Neurosci* **19**:199–221.

Niessing, J., Ebisch, B., Schmidt, K. E., et al. (2005). Hemodynamic signals correlate tightly with synchronized gamma oscillations. *Science* **309**:948–951.

Nir, Y., Hasson, U., Levy, I., Yeshurun, Y., and Malach, R. (2006). Widespread functional connectivity and fMRI fluctuations in human visual cortex in the absence of visual stimulation. *Neuroimage* **30**:1313–1324.

Nowak, L. G., Munk, M. H., Girard, P., and Bullier, J. (1995a). Visual latencies in areas V1 and V2 of the macaque monkey. *Visual Neurosci* **12**:371–384.

Nowak, L. G., Munk, M. H., Nelson, J. I., James, A. C., and Bullier, J. (1995b). Structural basis of cortical synchronization. I. Three types of interhemispheric coupling. *J Neurophysiol* **74**:2379–2400.

Olshausen, B. A., Anderson, C. H., and Van, E. (1993). A neurobiological model of visual attention and invariant pattern recognition based on dynamic routing of information. *J Neurosci* **13**:4700–4719.

Oram, M. W., Wiener, M. C., Lestienne, R., and Richmond, B. J. (1999). Stochastic nature of precisely timed spike patterns in visual system neuronal responses. *J Neurophysiol* **81**:3021–3033.

Parker, A. J. and Newsome, W. T. (1998). Sense and the single neuron: probing the physiology of perception. *Annu Rev Neurosci* **21**:227–277.

Payne, B. R., Lomber, S. G., Villa, A. E., and Bullier, J. (1996). Reversible deactivation of cerebral network components. *Trends Neurosci* **19**:535–542.

Perrett, D. I., Rolls, E. T., and Caan, W. (1982). Visual neurones responsive to faces in the monkey temporal cortex. *Exp Brain Res* **47**:329–342.

Petermann, T., Lebedev, M. A., Nicolelis, M., and Plenz, D. (2006). Neuronal avalanches in vivo. *Soc Neurosci Abstr* **32**:539.1.

Phillips, W. A. and Singer, W. (1997). In search of common foundations for cortical computation. *Behav Brain Sci* **20**:657–683.

Plenz, D. and Thiagarajan, T. C. (2007). The organizing principles of neuronal avalanches: cell assemblies in the cortex? *Trends Neurosci* **30**:101–110.

Pouget, A., Dayan, P., and Zemel, R. (2000). Information processing with population codes. *Nat Rev Neurosci* **1**:125–132.

Prut, Y., Vaadia, E., Bergman, H., et al. (1998). Spatiotemporal structure of cortical activity: properties and behavioral relevance. *J Neurophysiol* **79**:2857–2874.

Quintana, J., Yajeya, J., and Fuster, J. M. (1988). Prefrontal representation of stimulus attributes during delay tasks. I. Unit activity in cross-temporal integration of sensory and sensory-motor information. *Brain Res* **474**:211–221.

Reinagel, P. and Reid, R. C. (2000). Temporal coding of visual information in the thalamus. *J Neurosci* **20**:5392–5400.

Rockland, K. S. (1992). Configuration, in serial reconstruction, of individual axons projecting from area V2 to V4 in the macaque monkey. *Cereb Cortex* **2**:353–374.

Rockland, K. S. (2002). Non-uniformity of extrinsic connections and columnar organization. *J Neurocytol* **31**:247–253.

Rodriguez, E., George, N., Lachaux, J. P., et al. (1999). Perception's shadow: long-distance synchronization of human brain activity. *Nature* **397**:430–433.

Rolls, E. T., Treves, A., and Tovee, M. J. (1997). The representational capacity of the distributed encoding of information provided by populations of neurons in primate temporal visual cortex. *Exp Brain Res* **114**:149–162.

Rousselet, G. A., Thorpe, S. J., and Fabre-Thorpe, M. (2003). Taking the MAX from neuronal responses. *Trends Cogn Sci* **7**:99–102.

Saalmann, Y. B., Pigarev, I. N., and Vidyasagar, T. R. (2007). Neural mechanisms of visual attention: how top-down feedback highlights relevant locations. *Science* **316**:1612–1615.

Saleem, K. S., Tanaka, K., and Rockland, K. S. (1993). Specific and columnar projection from area TEO to TE in the macaque inferotemporal cortex. *Cereb Cortex* **3**:454–464.

Salin, P. A. and Bullier, J. (1995). Corticocortical connections in the visual system: structure and function. *Physiol Rev* **75**:107–154.

Salinas, E. and Sejnowski, T. J. (2001). Correlated neuronal activity and the flow of neural information. *Nat Rev Neurosci* **2**:539–550.

Salzman, C. D. and Newsome, W. T. (1994). Neural mechanisms for forming a perceptual decision. *Science* **264**:231–237.

Salzman, C. D., Murasugi, C. M., Britten, K. H., and Newsome, W. T. (1992). Microstimulation in visual area MT: effects on direction discrimination performance. *J Neurosci* **12**:2331–2355.

Schaefer, A.T., Angelo, K., Spors, H., and Margrie, T.W. (2006). Neuronal oscillations enhance stimulus discrimination by ensuring action potential precision. *PLoS Biol* **4**:e163.

Schoffelen, J.M., Oostenveld, R., and Fries, P. (2005). Neuronal coherence as a mechanism of effective corticospinal interaction. *Science* **308**:111–113.

Schroeder, C.E. and Foxe, J. (2005). Multisensory contributions to low-level, "unisensory" processing. *Curr Opin Neurobiol* **15**:454–458.

Shmiel, T., Drori, R., Shmiel, O., et al. (2005). Neurons of the cerebral cortex exhibit precise interspike timing in correspondence to behavior. *Proc Natl Acad Sci USA* **102**:18655–18657.

Shmuel, A., Augath, M., Oeltermann, A., and Logothetis, N.K. (2006). Negative functional MRI response correlates with decreases in neuronal activity in monkey visual area V1. *Nat Neurosci* **9**:569–577.

Singer, W. (1999). Neuronal synchrony: a versatile code for the definition of relations? *Neuron* **24**:49–25.

Singer, W., Engel, A.K., Kreiter, A.K., et al. (1997). Neuronal assemblies: necessity, signature and detectability. *Trends Cogn Sci* **1**:252–261.

Snider, R.K., Kabara, J.F., Roig, B.R., and Bonds, A.B. (1998). Burst firing and modulation of functional connectivity in cat striate cortex. *J Neurophysiol* **80**:730–744.

Sporns, O., Tononi, G., and Edelman, G.M. (2000). Connectivity and complexity: the relationship between neuroanatomy and brain dynamics. *Neural Networks* **13**:909–922.

Staedtler, E.S., Pipa, G., Muckli, L.F., Goebel, R., and Munk, M.H. (2006). Dissimilarity of firing rate patterns suggest population coding of visual objects in prefrontal cortex during short-term memory. *Soc Neurosci Abstr* **32**:812.14.

Stewart, C.V. and Plenz, D. (2006). Inverted-U profile of dopamine-NMDA-mediated spontaneous avalanche recurrence in superficial layers of rat prefrontal cortex. *J Neurosci* **26**:8148–8159.

Tallon-Baudry, C., Mandon, S., Freiwald, W.A., and Kreiter, A.K. (2004). Oscillatory synchrony in the monkey temporal lobe correlates with performance in a visual short-term memory task. *Cereb Cortex* **14**:713–720.

Tanigawa, H., Wang, Q., and Fujita, I. (2005). Organization of horizontal axons in the inferior temporal cortex and primary visual cortex of the macaque monkey. *Cereb Cortex* **15**:1887–1899.

Taylor, D.M., Tillery, S.I., and Schwartz, A.B. (2002). Direct cortical control of 3D neuroprosthetic devices. *Science* **296**:1829–1832.

Tononi, G., Sporns, O., and Edelman, G.M. (1996). A complexity measure for selective matching of signals by the brain. *Proc Natl Acad Sci USA* **93**:3422–3427.

Tovee, M.J. and Rolls, E.T. (1992). Oscillatory activity is not evident in the primate temporal visual cortex with static stimuli. *Neuroreport* **3**:369–372.

Tsodyks, M.V. and Markram, H. (1997). The neural code between neocortical pyramidal neurons depends on neurotransmitter release probability. *Proc Natl Acad Sci USA* **94**:719–723.

Tsodyks, M., Kenet, T., Grinvald, A., and Arieli, A. (1999). Linking spontaneous activity of single cortical neurons and the underlying functional architecture. *Science* **286**:1943–1946.

Tsunoda, K., Yamane, Y., Nishizaki, M., and Tanifuji, M. (2001). Complex objects are represented in macaque inferotemporal cortex by the combination of feature columns. *Nat Neurosci* **4**:832–838.

VanRullen, R., Guyonneau, R., and Thorpe, S. J. (2005). Spike times make sense. *Trends Neurosci* **28**:1–4.

Villa, A. E., Tetko, I. V., Hyland, B., and Najem, A. (1999). Spatiotemporal activity patterns of rat cortical neurons predict responses in a conditioned task. *Proc Natl Acad Sci USA* **96**:1106–1111.

Wallis, J. D. and Miller, E. K. (2003). From rule to response: neuronal processes in the premotor and prefrontal cortex. *J Neurophysiol* **90**:1790–1806.

Wallis, J. D., Anderson, K. C., and Miller, E. K. (2001). Single neurons in prefrontal cortex encode abstract rules. *Nature* **411**:953–956.

Wang, X., Lu, T., Snider, R. K., and Liang, L. (2005). Sustained firing in auditory cortex evoked by preferred stimuli. *Nature* **435**:341–346.

Whitney, D., Westwood, D. A., and Goodale, M. A. (2003). The influence of visual motion on fast reaching movements to a stationary object. *Nature* **423**:869–873.

Wilson, R. I. and Laurent, G. (2005). Role of GABAergic inhibition in shaping odor-evoked spatiotemporal patterns in the *Drosophila* antennal lobe. *J Neurosci* **25**:9069–9079.

Xiang, J. Z. and Brown, M. W. (2004). Neuronal responses related to long-term recognition memory processes in prefrontal cortex. *Neuron* **42**:817–829.

Yamane, Y., Tsunoda, K., Matsumoto, M., Phillips, A. N., and Tanifuji, M. (2006). Representation of the spatial relationship among object parts by neurons in macaque inferotemporal cortex. *J Neurophysiol* **96**:3147–3156.

Young, M. P. and Yamane, S. (1992). Sparse population coding of faces in the inferotemporal cortex. *Science* **256**:1327–1331.

Yuste, R., MacLean, J. N., Smith, J., and Lansner, A. (2005). The cortex as a central pattern generator. *Nat Rev Neurosci* **6**:477–483.

10

The role of neuronal populations in auditory cortex for category learning

FRANK W. OHL AND HENNING SCHEICH

Introduction

Auditory cortex function beyond bottom–up feature detection

Until the 1980s the auditory cortex was mainly conceptualized as the neuronal structure implementing the top hierarchy level of bottom–up processing of physical characteristics (*features*) of auditory stimuli. In that respect, plastic changes in anatomical and functional principles were only considered relevant for developmental processes towards an otherwise stable adult brain. Presently, this view has been replaced by a conceptualization of auditory cortex as a structure holding a strategic position in the interaction between bottom–up and top–down processing (for review see Irvine, 2007; Scheich *et al.*, 2007), in particular auditory learning (for review see Weinberger, 2004; Irvine and Wright, 2005; Ohl and Scheich, 2005).

In this chapter we review experimental evidence from gerbil and macaque auditory cortex that has led to this change of view about auditory cortex function. It will be argued that a fundamental understanding of the role of auditory cortex in learning has required to move beyond the study of simple classical conditioning and feature detection learning, for which auditory cortex does not seem to be a generally necessary structure (see below). Specifically, it will be elaborated that the abstraction from trained particular stimuli, as it is epitomized in the phenomenon of *category learning* (*concept formation*), is a complex but fundamental learning phenomenon for which auditory cortex is a relevant structure harboring the necessary functional organization.

Information Processing by Neuronal Populations, ed. Christian Hölscher and Matthias Munk.
Published by Cambridge University Press. © Cambridge University Press 2009.

Learning-induced plasticity in auditory cortex

A fundamental role in the process of moving away from the view of auditory cortex as a mere bottom–up feature detector was played by the discovery that learning procedures can alter cortical representations of otherwise constant stimuli, both on the level of entire cortical areas activated by such stimuli (Gonzalez-Lima and Scheich, 1986) as well as on the level of single neurons' firing behavior (Kraus and Disterhoft, 1982). Subsequent studies focusing on the frequency *tuning* of single neurons have described two basic types of learning-induced receptive field retuning. The first type is a learning-induced shift of a cortical neuron's best frequency ("BF shift") towards a frequency that has gained behavioral relevance in a classical conditioning task (for overview see Weinberger, 2004). The second type is an increase of local slope on the tuning curve, i.e. spectral sensitivity, in the spectral neighborhood of the conditioned frequency (Ohl and Scheich, 1996, 1997). As a functional interpretation of the BF shift it has been argued that such a retuning facilitates neuronal responses to the tonal conditioned stimulus at the expense of other frequencies (Weinberger, 2007). A similar argument was used (Weinberger, 2007) to interpret the functional relevance of increased map representations that were found in experiments with perceptual discrimination learning (Recanzone *et al.*, 1993). From a theoretical standpoint it has been argued (Ghose, 2004) that at least under the assumption that a decision in a sensory discrimination task is based on the comparison of a sum of neuronal responses such a type of receptive field retuning would in fact *reduce* discrimination performance. This is because sensitivity to changes in frequency, as reflected in the local absolute slope of a tuning curve, is minimal (in fact, zero) at its peak. Also, similar to the findings in auditory cortex (Ohl and Scheich, 1996, 1997), orientation discrimination training in the visual system has led to local steepening of orientation tuning curves in units of V1 (Schoups *et al.*, 2001) and V4 (Raiguel *et al.*, 2006), i.e. to a selective increase of discriminability around the trained orientation. Therefore, somewhat contrary to the above sketched interpretation of the functional role of BF shift, a suppression-based model of perceptual learning was proposed (Ghose, 2004) that can account for both the selective slope increases and the findings of little changes, or reduced response strengths of neurons for trained stimulus parameters that have been reported in visual (e.g. Ghose *et al.*, 2001) and auditory (Ohl and Scheich, 1996, 1997; Beitel *et al.*, 2003; Witte and Kipke, 2005) experiments. Mechanistically it is not clear for any of these forms of retuning how modulated excitatory or inhibitory synaptic mechanisms are responsible for the observed effects, although several studies implicate the modulation of inhibitory systems (Ohl and Scheich, 1996,

1997; Fritz *et al.*, 2003, 2005); for a recent review of this issue see Irvine and Wright (2005).

Implicit in the discussion about physiological correlates of learning in sensory systems are assumptions about the different roles of "tuning peak" and "tuning slope" for "coding" sensory information. Using an information-theoretic approach, i.e. investigating the probability of identifying a stimulus among a set of possible stimuli on the basis of an observed neuronal response, Butts and Goldman could show that the mean specific information of a response (the latter being the entropy of the stimulus set minus the entropy of the stimulus probability conditional on the measured response) is a function that can have its absolute maximum either at a tuning curve's peak or close to the maximum slope (Butts and Goldman, 2006). For a rather broad class of noise models they could show that the former is the case in conditions with high background noise and the latter in case of low noise conditions. In their account noise is the control parameter which affects the relative importance of fine discrimination, which favors stimuli located in high-slope regions on the tuning curve, versus coarse discrimination which favors stimuli located near the tuning curve peak. In our current conceptualization the relative importance of fine discrimination versus coarse discrimination or detection are already determined directly by the experimental task.

The study of category learning as a strategic approach

Apparently, both mechanisms and functional interpretation of different forms of learning-induced retuning in auditory cortex neurons are presently only incompletely understood. An aggravating aspect for the functional interpretation of these results is the fact that the learning paradigms utilized to address receptive field retuning and cortical map reorganization in auditory cortex have traditionally used pure tone classical conditioning, pure tone detection, or pure tone discrimination tasks. For these tasks, however, receptive field retuning and map reorganization in auditory cortex are neither sufficient nor necessary. They are not sufficient, because map reorganization induced by alternative means (electrical microstimulation) does not alter frequency discrimination performance (Talwar and Gerstein, 2001), and they are not necessary, because tonal memory can develop even after bilateral ablation of auditory cortex (Butler *et al.*, 1957; Thompson, 1960; Goldberg and Neff, 1961; Diamond *et al.*, 1962; see Ohl *et al.*, 1999 for a more detailed discussion). The present chapter therefore focuses on experimental paradigms for which the relevance of auditory cortex has been demonstrated in independent experiments, namely discrimination of frequency-modulated tones (Ohl *et al.*, 1999; Rybalko *et al.*, 2006) and the discrimination on tone sequences (Diamond and Neff, 1957; Kelly, 1973).

It should be emphasized that variation of spectral content with time, as the additional dimension introduced by these two stimulus classes, renders these sets of stimuli as sufficiently complex to allow addressing the issue of abstraction from some particular stimulus features while identifying others as relevant for a certain behavioral task. This leads to the opportunity of studying categorization and category learning (for a rigorous definition of these terms see next section). It is a characteristic of category learning that it involves abstraction from once learned particulars and "goes beyond the information given" (Kommatsu, 1992). Category learning therefore transcends the simpler forms of associative learning and generalization typically studied in behavioral neuroscience (for review see Miller *et al.*, 2003). In the research to be described, the occurrence of the transfer of the learned discrimination behaviors to novel stimuli will be used as a marker event for the study of the physiological correlates of category learning. The particular focus is the investigation of those neuronal processes underlying the construction of that "which goes beyond the information given." It will be emphasized that beyond a mere extraction of statistical invariants (which could be done on the basis of classic bottom–up feature detectors retaining the metric of evoked activity reflecting physical stimulus features) this construction involves recruitment of novel patterns of neuronal activity establishing a new metric of evoked activity which reflects the subjective perceptual scaling of stimuli instead.

Category learning

Theoretical underpinnings of the study of category learning

We define category learning as the process by which the ability to categorize stimuli or other perceived situations is acquired. Categorization describes the phenomenon that a multitude of stimuli or situations encountered or imagined is structured by grouping these into separable sets (categories) although members of a given category could, in principle, still be discriminated from each other. The nature of this phenomenon is subject to intellectual debates which originated at least with Aristotle but are continued into our present times. The Aristotelian view (often referred to as the *Classical Theory*) considers having or establishing a set of necessary and/or sufficient criteria to be met by stimuli or situations as being the essence of determining their membership to a category. This view, enriched by an appropriate formal framework, is also maintained by some "artificially intelligent" approaches to the categorization problem. A main objection that was brought up against this view is that for many "natural" categories such sets of criteria cannot be found (e.g. Rosch, 1973): neither seems category membership always to be defined by features which can unequivocally be attributed to all members, nor can features always be evaluated to indicate membership or no membership to a category. Rather, varying degrees of *typicalness* of features and/or category members seem to be the rule. Consequently, a number of accounts have been proposed which can be summarized under the heading of *prototype theories* (e.g. Posner and Keele, 1968). Prototype theories hold that some form of idealized representation of category members exists and actual membership of a given stimulus is determined by some scaling of *similarity* of this member to the prototype. Prototype theories in some instances have faced the problem of providing a convincing theory for establishing the required similarity relations (e.g. Ashby and Perrin, 1988) or could not account for non-prototype members having more pronounced effects on categorization performance than the prototype itself (e.g. Brooks, 1978; Medin and Schaffer, 1978). Both studies on human (Rosch, 1975) and animal (Lea and Ryan, 1990) categorization have argued that some of these problems can be attenuated by lifting the requirement for local comparisons with a singular prototype and instead requiring global comparisons with multiple (in the extreme form with all) category members (Estes, 1986; Hintzman, 1986; Nosofsky, 1986) for which reason such theories are summarized under the heading *exemplar theories*. While historically these (and other) approaches have been initiated and put forward by their authors with very different rationales in mind they can be more objectively compared to each other using a suitably formalized reformulation (Ashby and Maddox, 1993) which can be derived from *general recognition theory* (Ashby and Townsend, 1986). It should be noted, however, that many experimental observables can be accounted for by any of the theoretical viewpoints provided, at least if appropriate modifications to the extreme positions of such theories are allowed (Pearce, 1994). This is notwithstanding the fact that category learning, rather than being a coherently definable mechanism, seems to include several separable types of cognitive processes probably mediated by separable physiological processes in separable anatomical structures of the brain (for review see Kéri, 2003).

Category learning in human and non-human species

While in human studies the above sketched (and other) viewpoints are maintained without questioning the existence of *concepts*, as mental constructs, as a possible basis for our categorizations, this option is not so clear for non-human species, simply because they cannot report on having a concept that might guide their categorization behavior. It has been repeatedly argued that defining a concept would require reference to language so that non-human species could not have concepts at all (e.g. Chater and Heyes, 1994). A less extreme viewpoint suggests that some instances of animal categorization might reflect mere discriminations albeit with very complex stimuli (Wasserman and Astley, 1994). In the spirit of cognitive theories on category learning it has been argued that the demonstration of concepts in non-human species might be possible by suitably designed transfer experiments with appropriately constructed control stimulus sets (Lea, 1984). Transfer of learned discriminations to novel stimuli is a relevant phenomenon because it demonstrates that during an instantiation of category learning more has happened than mere associations of responses to trained particular stimuli (Lea and Ryan, 1990).

It is reasonable to assume that not all possible categorizations can equally be established in different species. For example, while pigeons could not be trained to categorize (and in some cases even discriminate) photographs showing two different individual pigeons, or showing two different individual chickens, bantam cockerels could be trained easily using the same stimulus material (Ryan, 1982). While in particular cases it might be difficult to trace species-specific differences in categorization behavior back to differences in the social behavior of different species – for example it can be excluded that the discrimination of individual conspecifics were simply irrelevant to pigeons (Ryan, 1982) – it seems clear that category learning might serve species-specific needs. In the vervet monkey it has been demonstrated, for example, that maturing individuals are able to develop fine categorization of the production of and response to three different types of alarm calls (each one signaling a different predator category) while young individuals might be wrong, e.g. by uttering an "eagle alarm call" in response to a harmless bird (Seyfarth *et al.*, 1980a, 1980b). More generally, it is reasonable to assume that categorization of stimuli on privately learned exemplar experiences is advantageous to categorization based on a species-specific prototype (Nelson and Marler, 1990).

Categorization and generalization

As a note of caution, it seems appropriate to consider the terms *categorization* and *generalization* at this point in more detail as they are inconsistently used in the literature and have a bearing on the experiments to be described in this chapter.

When a (human or non-human) subject is trained to discriminate a stimulus A from a stimulus B the discrimination performance typically develops gradually over some time as is manifest in the various forms of functions usually referred to as *learning curves*. A typical form is schematized in Fig. 10.1a which displays the temporal evolution of the hit rate and false alarm rate in a GO/(NO-GO) discrimination experiment. Other depictions of the discrimination learning behavior are possible depending on the kind of experiment performed (be these symmetric choice experiments, signal detection approaches, etc.)

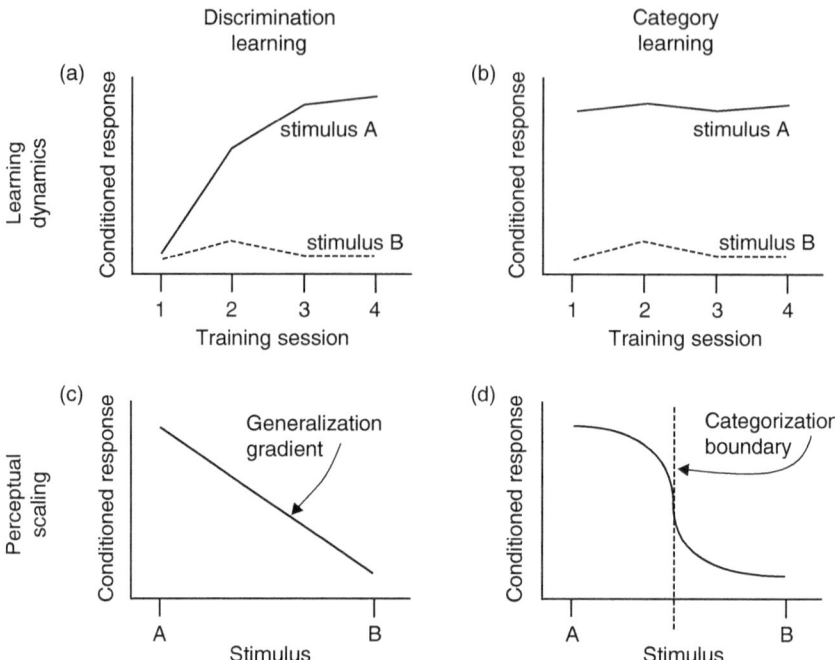

Figure 10.1 Characteristic differences between non-categorical discrimination learning (left panels: (a) and (c)) and category learning (right panels: (b) and (d)) in terms of learning dynamics (upper panels: (a) and (b)) and perceptual scaling (bottom panels: (c) and (d)). Learning curves in a GO/(NO-GO) paradigm demonstrate gradual development of discrimination performance in case of non-categorial discrimination learning (a) or spontaneous transfer of discrimination in case of category learning (b). Psychometric functions are characterized by gradually falling generalization gradients in case of non-categorical discrimination learning (c) or sigmoid curves in case of categorical perception (d).

and the choice of behavioral observables (e.g. the various transformations of hit rate and false alarm rates which are suitable under given conditions). Generally, however, the changing conditional rate of occurrence of some behavior must be assessed at some point in the analysis. The typically asymptotic response rate depends on various parameters (vigilance, internal response biases, strength of the learned association, etc.) but also shows some degree of stimulus specificity. This can be demonstrated by generalization gradients: In a GO/(NO-GO) experiment in which a subject is trained to show the GO reaction in response to stimulus A and the NO-GO reaction in response to stimulus B, a generalization gradient can be assessed by measuring the conditioned response (quantified by its frequency of occurrence across trials, by the strength of its expression, or by some other suited observable) as a function of a physical distance between parameters characterizing stimuli A and B. The latter defines a path through the parameter space connecting stimuli A and B. When physical stimulus parameters are varied along this path a more or less gradual fall-off in the A-specific response amplitude is typically observed and referred to as a generalization gradient (Fig. 10.1c).

Conversely, when a subject has formed categories it can recognize even novel stimuli as representatives of the learned categories. A depiction of behavioral variables analogous to a learning curve would therefore indicate a high discrimination performance even in the first training session (Fig. 10.1b). The experimental demonstration of this behavior is critical for assessing category learning and is sometimes referred to as the criterion of the "transfer of learned behaviors to novel stimuli." Therefore, category learning is distinguished from simple or discriminative conditioning also by the psychometric functions it produces. Instead of gradual generalization gradients we find sigmoid psychometric functions with a more or less sharp boundary at some location in the stimulus parameter space, called the categorization boundary (Fig. 10.1d). Categories develop as cognitive constructs; they epitomize subjective "hypotheses" that are expressible as parcelations of the set of actually perceived or imagined stimuli, conditions or actions, into equivalence classes of meaning in particular contexts. Transfer of learned behaviors to novel stimuli therefore follows the subjective laws of this parcelation, rather than being guided by (physical) similarity relations between stimuli or stimulus features. These are important criteria for cognitive structures that have to be accounted for by physiological models of learning.

In a nutshell, generalization is a general feature of learned stimulus-cued behaviors reflecting the converse of stimulus specificity, while categorization is a cognitive process based on the parcelation of the represented world into equivalence classes of meaning, valid for an individual in a particular context and in a particular time.

Neurodynamics of category formation in rodent auditory cortex

Physiological correlates of category learning

In the search for physiological correlates of category learning is has to be taken into account that the relationship between learning and its physiological correlates, in general, is non-trivial, because the former is a psychological/ethological concept while the latter is a physiological concept. Conceptual difficulties in interrelating these two domains are therefore predictably similar to other situations in science where conceptually different levels have to be linked, like in the case of relationship between Newtonian physics and thermodynamics, or in the case of the "mind–body problem." In the former case, a partial solution to the problem of linking the two levels of has been achieved via the development of statistical physics, which worked by introducing scientific concepts that were not motivated by either Newtonian physics or thermodynamics alone. The solution must be considered only partial because it is sufficiently developed only for the equilibrium state of matter (see Haken, 1983). In this section it will be argued that traditional physiological accounts for learning phenomena are insufficient to characterize the physiological basis of category learning.

Traditional physiological accounts for learning phenomena view the *capacity for rerouting* the flow of excitation through a neuronal system as the key element of a physiological correlate of a learning process. This capacity is thought to underlie a stimulus processing which is altered by the learning experience. This concept is very much encouraged by the study of Pavlovian conditioning, probably the one learning phenomenon that has been most intensively studied by physiologists. Pavlovian conditioning can be described, and has in fact traditionally been defined, as a process by which an initially behaviorally neutral stimulus can elicit a particular behavior, after having been paired with a stimulus, the *unconditional stimulus* (US), which unconditionally triggers this behavior. Figure 10.2a shows a scheme of the flow of information in Pavlovian conditioning. For the case of Pavlovian conditioning it has proved successful to build models of the physiological basis of learning by translating the flow of information within the conditioning scheme into a flow of neuronal excitation through a neuronal substrate (Fig. 10.2b). In the example of the conditioned gill withdrawal reflex in *Aplysia*, for example, this concept is manifest in the feedforward convergence of – in the simplest case – two sensory neurons on an interneuron which projects onto a motor neuron. The straightforwardness of the translation of the flow scheme of information into a flow scheme of neuronal excitation has led to consider the latter as elemental to physiological accounts of learning (see however Schouten and De Long, 1999). This is expressed, for example, in the metaphors of the "cellular alphabet" (Hawkins and Kandel, 1984) or the "molecular alphabet" of learning.

In contrast to Pavlovian conditioning category learning encompasses aspects which go beyond mere stimulus–response associations. In particular, the transfer of learned responses to novel stimuli (Lea, 1984), outside of generalization gradients established by associative training, implies the insufficiency of a simple scheme as in Fig. 10.2b, because novel stimuli (which are per definition not encountered before) cannot be associated with unconditional triggers (Fig. 10.2c).

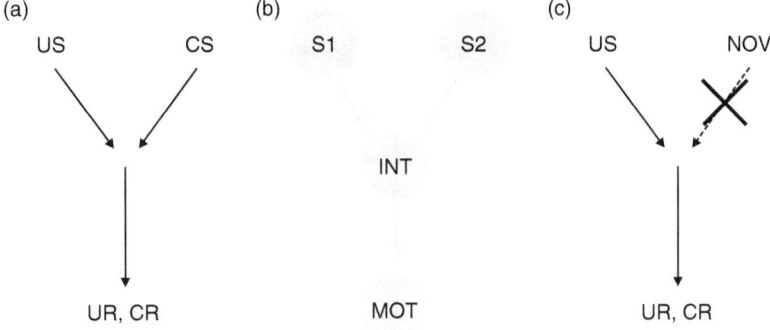

Figure 10.2 (a) Schematic depiction of information flow in Pavlovian conditioning. Before conditioning, an unconditioned stimulus (US) will elicit a particular behavior, then referred to as unconditioned response (UR). After conditioning, a previously behaviorally neutral stimulus can elicit this behavior as a conditioned response (CR) and is then referred to as the conditioned stimulus (CS). (b) A direct translation of the scheme of information flow in Pavlovian conditioning into a flow of neuronal excitation within a simple neuronal substrate involving two sensory neurons (S1 and S2), an interneuron (INT), and a motor neuron (MOT). (c) The architecture in (b) cannot explain responses to novel stimuli (NOV), because novel stimuli have by definition not been previously associated with unconditional triggers and, consequently, cannot be conditioned.

A rodent model of auditory category learning

In the last decade the Mongolian gerbil (*Meriones unguiculatus*) has been introduced as a rodent model of category learning (Wetzel et al., 1998) amenable to detailed physiological treatment (Ohl et al., 2001, 2003a, 2003b). With respect to the background summarized above (section "The study of category learning as a strategic approach"), the focus will be on the transfer of learned behaviors to novel stimuli. To avoid confound of the learning behavior with potential innate biases towards any of the training stimuli care was taken to choose a stimulus set that is demonstratively behaviorally neutral to the naïve gerbil. Moreover, stimuli were so selected that they did not differ in the ease with which they could be associated with the particular behaviors used in the training experiments. Furthermore, the stimulus set had to fulfill the requirement that it should be possible to observe the transition from discrimination behavior to categorization behavior simultaneously in behavioral and neurophysiological data.

These conditions were met by linearly frequency-modulated tones traversing a frequency range of 1 octave in 250 ms in rising or falling fashion. In frequency modulated tones a number of parameters can be (co-)varied, like duration, intensity, frequency ranges covered, and modulation rate, i.e. the rate of change of the tones' instantaneous frequency. Most of the stimuli (all, except those with zero modulation rate, i.e. pure tones) can be categorized as either "rising" or "falling" frequency-modulated tones, depending on whether the instantaneous frequency changes from low to high or from high to low, respectively. For the experiment stimuli were so designed that they fell outside the generalization gradients established by training naïve animals to the two neighboring (in the parameter space) stimuli. This ensures that observed learning curves would resemble the schemes in Fig. 10.1a and 10.1b indicating that the subject has not or has categorized the stimuli, respectively.

Training consisted of a GO/(NO-GO) avoidance paradigm carried out in a two-compartment shuttle box, with the two compartments separated by a little hurdle, ensuring a low rate of spontaneous hurdle crossings. Animals were trained to cross the hurdle in response to a rising frequency-modulated tone and to stay in the current compartment in response to a falling frequency-modulated tone. Training was organized in so-called "training blocks" during which the discrimination of one particular rising frequency-modulated tone from a tone traversing the identical frequency range in falling direction was trained. A training block consisted of a number of training sessions, with one session per day, and was continued until no further changes in conditioned response rates were achieved in three consecutive sessions. Then another training block was initiated in which the discrimination of another pair of a rising and falling frequency-modulated tone was trained. A training session encompassed the randomized presentation of 30 rising and 30 falling frequency-modulated tones with the animals' false responses (misses and false alarms) being negatively reinforced by a mild electrodermal stimulation through a metal grid forming the cage floor. Control groups were run with the opposite contingencies to test for potential biases in this behavior which have been reported for the dog (McConnell, 1990). We showed that for the stimulus parameters tested no such biases existed for the gerbil.

Behavioral results

Animals trained on one or more training blocks never generalized to pure tones of any frequency (e.g. start or stop frequencies of the modulated tone, or frequencies traversed by the modulation or extrapolated from the modulation). This could be demonstrated by direct transfer experiments (Ohl et al., 2001, supplementary material) or by measuring generalization gradients for modulation rate which never encompassed zero modulation rates (Ohl et al., 2001).

Categorization was demonstrated by the transfer of the conditioned response behavior (changing compartment in response to tones of one category and remaining in the current compartment in response to tones of the other) to *novel* stimuli as measured by the response rates in the first session of a new training block. This sequential design allowed the experimenter to determine the moment in which an individual would change its response behavior from a *discrimination phase* (Fig. 10.1a and c) to a *categorization phase* (Fig. 10.1b and d). All animals tested were able to categorize novel frequency-modulated tones but, most notably, different individuals showed the transition from the discrimination phase to the categorization phase at different points in time in their training histories, although all had been trained with the same sequence of training blocks (Ohl et al., 2001). Also, the transition from the discrimination phase to the categorization phase occurred abruptly rather than gradually, i.e. in the first session of a training block either no discrimination performance (discrimination phase) or the full performance was observed (categorization phase). A third property of the transition was that after its occurrence the discrimination performance remained stable for the rest of the training blocks. These three properties of the transition, individuality of the time point of occurrence, abruptness of occurrence, and behavioral stability after its occurrence, make it resemble a state transition in dynamic systems. We used this transition as a marker in the individual learning history of a subject to guide our search for physiological correlates of this behavioral state transition.

Electrophysiological results

A suitable level for studying electrophysiological correlates of perceptual organization is the mesoscopic level of neurodyamics (Freeman, 2000), which defines the spatial scale of phenomena observed and provides focus on electrical phenomena emergent from the mass action of ensembles of some 10^4 to 10^5 neurons. Since discrimination of the direction of frequency-modulated tones is known to require a functional auditory cortex in gerbil (Ohl et al., 1999; Kraus et al., 2002) and rat (Rybalko et al., 2006) the training procedures were combined with the parallel measurement of the neurodynamics in auditory cortex. For cortical structures, this level of description is accessible by measurement of the electrocorticogram. We therefore combined the described category learning paradigm with the measurement of the electrocorticogram using arrays (3–6) of micro-electrodes chronically implanted on the dura over the primary auditory cortex. The spatial configuration and inter-electrode distance (600 μm) of the recording array were designed to cover the tonotopic representation of the frequency-modulated stimuli used and avoid spatial aliasing of electrocorticogram activity (Ohl et al., 2000a). The spatial organization of the thalamic input into the auditory cortex can be studied by averaging electrocorticograms across multiple stimulus presentations, yielding the well-known auditory evoked potential (Barth and Di, 1990, 1991). Our studies of evoked potentials in primary auditory cortex, field AI, induced by pure tones (Ohl et al., 2000a) or frequency-modulated tones (Ohl et al., 2000b) revealed topographically organized early components (P1 and N1) located at positions within the tonotopic gradient of the field that corresponded to the frequency interval traversed by the frequency modulation, while their late components (P2 and N2) were not. On a finer spatial scale, the location of the early components of rising and falling frequency-modulated tones was found to be shifted towards tonotopic representations of the respective end frequencies of the modulations, i.e. towards higher frequencies for rising modulations and towards lower frequencies for falling modulations. These "tonotopic shifts" (Ohl et al., 2000b) could be explained by the finding that single neurons are usually activated more strongly when the frequency modulation is towards the neuron's best frequency than when it is away from it (Phillips et al., 1985). In the former case, the activation of frequency channels in the neighborhood of the best frequency of a single neuron are recruited more synchronously than in the latter case, due to the increasing response latency with increasing spectral distance from the neuron's best frequency. If this asymmetry is transferred onto a tonotopically organized array of neurons the described tonotopic shift is the result. Comparable tonotopic shifts have previously been reported in the cortex analogue of the chick (Heil et al., 1992).

Since physiological correlates of category learning could not be expected to occur time-locked to stimulus presentation we analyzed electrocorticograms recorded during the training with a single trial type of analysis. Instantaneous spatial patterns in the ongoing cortical activity was described by state vectors (Ohl et al., 2001). State vectors were formed from estimates of signal power in 120-ms time windows obtained for each channel. A state vector moved through the state space along a trajectory according to the temporal evolution of the spatial pattern over a 6-s period starting 2s before stimulus delivery. For each trial, the Euclidean distance (parameterized by time) to a reference trajectory was calculated and termed "dissimilarity function." In each case, the reference trajectory was calculated as the centroid over trajectories measured during presentation of stimuli belonging to the respective other category in the same training session. Thus, each trajectory associated with a rising frequency-modulated tone was compared to the centroid over all trajectories associated falling frequency-modulated tones in the same session, and vice versa. Comparison of single trajectories with centroids of trajectories, rather than other single trajectories, ensured that, on a statistical basis, transient increases in the pattern dissimilarity (peaks in the dissimilarity function) were due to pattern changes in the observed trajectory rather than in the centroid. In naïve animals, dissimilarity functions showed a "baseline behavior" with a sharp peak (2–7 standard deviations of baseline amplitude) after stimulus onset. This peak occurred predictably because of the topographically dissimilar patterns (tonotopic shifts) of early evoked responses that rising and falling frequency-modulated tones produce (Ohl et al., 2000b). With learning, additional peaks emerged from the ongoing activity. Those labeled spatial activity patterns in single trials with transiently increased dissimilarity to the reference trajectory indicating a potential relevance for representing category-specific information processing. These patterns were therefore termed "marked states."

To test whether marked states do in fact represent processing of category-specific information we analyzed the similarity and dissimilarity relations among them in the entire course of the training. While animals were in their discrimination phases (prior to the formation of categories), we observed that dissimilarities between marked states within categories were of the same order of magnitude as between categories (Fig. 10.3). After an individual animal had entered its categorization phase dissimilarities within a category were significantly smaller than between categories (Ohl et al., 2001) (Fig. 10.3). This indicated the existence of a metric which reflected the parcelation of stimuli into equivalence classes of meaning. This type of metric reflects subjective aspects of stimulus meaning, namely its belonging to categories formed by previous experience and, therefore, it is different from the known tonotopic one, which reflects similarity relations of physical stimulus parameters, namely spectral composition.

The spatial organization of the emerging marked states (late activity) was analyzed in more detail and compared to that of the early evoked activity (also yielding peaks in the dissimilarity function) by a multivariate discriminant analysis, identifying the regions in the recording area which maximally contributed to the dissimilarity between the observed pattern and the reference pattern (Ohl et al., 2003b), or identifiying the regions that contribute most information about the pattern (Ohl et al., 2003a). This analysis revealed that two separate coding schemes coexist in the same neuronal tissue of auditory cortex: the early activity represents physical stimulus parameters (like frequency content or modulation direction) in the well-known tonotopic framework while the late activity shows a metric that parallels the psychophysical scaling for each individual animal corresponding to its current behavioral state in the process of going from discrimination learning to category learning.

It should be noted that the lack of invariance with the mere physical parameters challenges the interpretation of the marked states as "sensory representations." The observed metastability of the spatiotemporal activity patterns was hypothesized to reflect context aspects of the stimulation as well as the perceptual history of the individual, and it was inferred that such patterns reflect

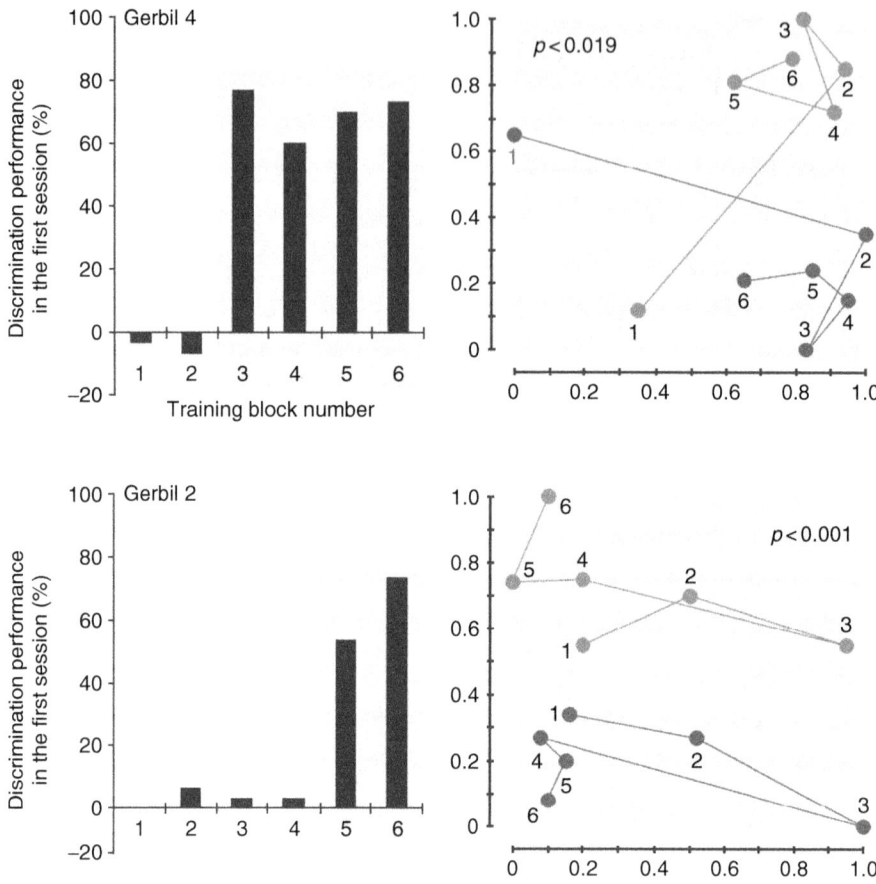

Figure 10.3 Correspondence between the behavioral transition from the discrimination phase to the categorization phase (left column) and the emergence of category-representing mesoscopic activity patterns (right column) in two representative animals (top and bottom row). (Left column) Bars represent discrimination performance for novel stimuli at the beginning of a training block as a function of training block index. Categorization phase is indicated by the abruptly increased discrimination performance (emphasized by gray rectangles) and occurred at different times for different individuals. (Right column) Similarity–dissimilarity relations between activity patterns found for rising and falling frequency modulated tones (top and bottom line, respectively) in the sequence of training blocks (numbers). The relative dissimilarity between any pair of states is represented by the distance of the corresponding state points in this two-dimensional display. Note the emergence of point clustering (indicating high state similarity) within categories when transfer from discrimination phase to categorization phase happened. (Modified from Ohl et al., 2001.)

subjectively relevant cognitive structures (for summary see Freeman (2000) and references therein). The results described above critically confirm this interpretation: the category learning paradigm, first, allows determination of the point in time when a particular cognitive structure (the formation of the categories "rising frequency-modulated tone" and "falling frequency-modulated tone") emerges, and second, predicts that the main source of variance in the stimuli (the spectral interval traversed by the frequency modulation) is no longer a relevant feature after a subject's transition to categorization. Consequently, it was found that the dissimilarity between marked states associated with stimuli belonging to the same category was significantly reduced after the transition to categorization, although the physical dissimilarity of the corresponding stimuli was still high, as also reflected in the topographic organization of the stimulus-locked peaks in the dissimilarity function and the fact the dissimilarities remained high in individuals that had not yet formed categories.

In this sense, the utilized paradigm and analysis strategy provided an objective (for the experimenter) window of a subjective cognitive structure (that of the animal) providing an example for the reconciliation between experimenter-centered and subject-centered views of neural activity relevant for a functional interpretation of dynamic neuronal processes (Eggermont 2007).

Frequency-contour categorization in monkey

The experiments to be described in this section were carried out in the macaque auditory cortex and served to extend the insight into cortical categorization mechanisms (1) to the level of information provided by the firing of neurons, and (2) to categorization that depends on the sequence aspect of stimuli (Brosch *et al.*, 1999, 2004, 2005, 2006; Brosch and Schreiner, 2000; Selezneva *et al.*, 2006). Tone sequences at particular frequency intervals are characteristic of melodies. One of the most important cues for recognizing melodies is the contour of tone sequences with rising or falling frequency steps or constant frequency steps (Dowling, 1978). If melodies are transposed in frequency they are still recognized as the same, a process requiring categorization of such frequency steps. In some sense this categorization in the time-frequency domain is related to the frequency-modulated directional categorization which however does not involve relationships between sound events separated in time.

In these experiments, monkeys were trained to categorize upward and downward frequency steps in tone sequences irrespective of the absolute tone frequencies used. They received a reward upon correct identification of a downward step. The monkeys learned the categorization by the following procedure: upon a light signal in a trial the monkeys could grasp a holding bar and thereby elicited themselves an unpredictable tone sequence. When successive tones remained at the same frequency or formed any upward step they should keep holding the bar but should release it after a downward step. In auditory cortex (fields AI and CM) of the fully trained monkeys auditory neurons not only responded to the tone steps in a categorical fashion (see below). Also the light signal, the grasp and release of the bar, and the reward, elicited firing in a large proportion of neurons that also responded to the tones (Fig. 10.4). The intriguing aspect of these results is not so much that a relationship is established between the tones and the polymodal cues of the behavioral context of the task. This can be assumed for any learning task in which the behavioral meaning of the tones have to be deduced from the context (as explained in the first section) but that this relationship is expressed by responses of auditory neurons in auditory cortex, and not only in relevant other parts of the brain (as may be assumed) points to the existence of states in auditory cortex that include information on the non-auditory, interpretative, side of the sounds. These non-auditory activations of the same auditory neurons were not observed when the monkeys during unit recording were made to switch to a visual discrimination task with the same behavioral contingencies. This provided direct evidence that for auditory cortex this non-auditory information is only relevant when it can serve the interpretation of an auditory input, i.e. the solution of the task. These results demonstrate that, after learning, the concept of sound meaning being related to objects that emit the sounds may be extended to the context and to the listeners' own behavioral role in eliciting the sounds and responding to them.

The recordings in monkey auditory cortex also revealed two other aspects of task-related stimulus representations that shed light on the special state during categorization. One is that neurons in the cortical areas AI and CM which responded phasically to the tones of the sequences showed in a trial-by-trial analysis showed an increase of response only to the rewarded downward steps and no change of response relative to the preceding tone when the frequency changed in upward direction or remained the same. Thus, not the absolute firing level of responses to the tones carried the information about the rewarded downward steps in an ongoing trial, but the response difference relative to the preceding tone. Across the population of neurons this behavior was independent of the best frequency of neurons, i.e. categorical. This directional preference was unlike properties found in naïve animals where in populations of units both directional preferences are found (Brosch *et al.*, 1999; Bartlett and Wang, 2005). This result is similar to the results of the gerbil experiment in as much as selection of a category-relevant property (downward step) is performed irrespective of other properties (absolute frequency). Thus, stimuli with the same meaning were represented similarly.

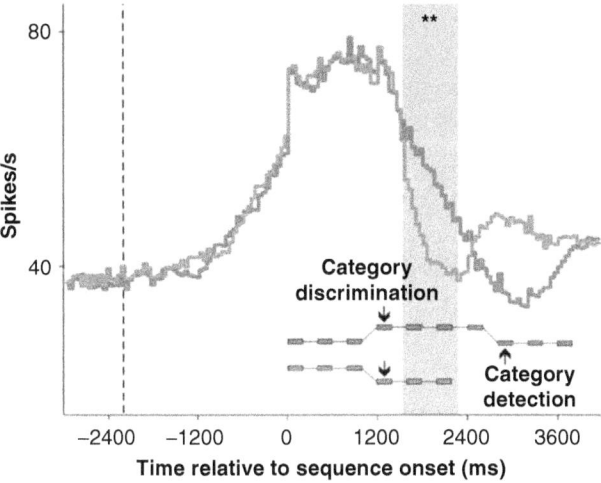

Figure 10.4 Average time courses of 40 tonically firing multi-units in the caudomedial auditory cortex of primates in the two tone sequence conditions (inset with same time scale): sequences with a downward frequency step only are marked in light gray. Sequences with an upward frequency step followed by a downward frequency step are marked in black. Dashed vertical line indicates bar grasping. In the period after the 4th tone indicated by the gray bar (**) the decrease of firing represented by the black and the gray curves was significantly different. The motor reactions (bar release) occurred approximately at the minimum of the curves, i.e. earlier in down sequences (gray). Thus, the slopes of the black and gray curves are predictive of the subsequent behavioral responses. (Modified from Selezneva et al., 2006.)

The second finding in field CM covered neurons that slowly increased their firing as soon as the monkeys grasped the bar (Fig. 10.4). Their firing leveled off during the tone sequence and showed a sharp decline after the first frequency change. If this change was a downward step the decrease of firing was steep, predicting that the monkey would release the bar. If the change was an upward step the decrease of firing was significantly flatter, predicting that the monkey would hold the bar. This prediction of the behavioral response held true also for error trials when the monkeys misinterpreted the direction of steps. Interestingly, the phasically responding units correctly reported the direction of steps in these error trials, i.e. they were better than the monkeys' responses. Thus, the slow units represented the misinterpretations of an upward step by predicting that the monkey would soon release the bar. Together these results in monkey auditory cortex suggest that a categorical state includes category-relevant representation of the stimuli as well as representations of multimodal and cognitive task aspects relevant for the interpretation of stimuli.

Conclusion

The experimental results presented here from gerbil and monkey auditory cortex converge on the conclusion that stimulus representation in auditory cortex as reflected by several correlates of neuronal activity are modified according to the structure of the auditory categorization task. This implies that neuronal activity in auditory cortex in addition to representing physical stimulus parameters (bottom–up processing) is also governed by categorization-task-specific top–down processing (Ohl et al., 2001; Ohl and Scheich, 2005). The top–down processing involves two essential aspects. First, there seems to be a selection of category-relevant properties of stimuli at the expense of currently irrelevant properties which is associated with a reorganization of the stimulus representation in the naïve behavioral state. Second, in the categorizing animal, auditory cortex generates activity states that (1) reflect behavioral contingencies of the categorization task, and (2) predict subsequent behavioral decisions of the animal. Anatomical investigation has demonstrated that auditory cortex is strategically positioned to form an interface between these bottom–up and top–down processes. This is enabled by the quantitative dominance of cortico-cortical connections of auditory cortex and even of the primary auditory field which provide flexible options for task-related interaction with other brain areas (Budinger et al., 2000, 2006; Scheich et al., 2007). As category formation typically involves multiple experiences with different stimuli acquiring the same meaning learning and memory processes are essential mechanistic elements of categorization. The described experiments represent a first step to understand categorization in terms of such mechanisms.

Acknowledgments

This work was supported by BMBF BioFuture grant 0311891, Deutsche Forschungsgemeinschaft SFB-TR31, and EU grant FP6-IST-027787. We thank Kathrin Ohl for technical help during preparation of the manuscript.

References

Ashby, F. G. and Maddox, W. T. (1993). Relations between prototype, exemplar, and decision bound models of categorization. *J Math Psychol* **37**:372–400.

Ashby, F. G. and Perrin, N. A. (1988). Toward a unified theory of similarity and recognition. *Psychol Rev* **95**:124–150.

Ashby, F. G. and Townsend, J. T. (1986). Varieties of perceptual independence. *Psychol Rev* **93**:154–179.

Barth, D. S. and Di, S. (1990). Three-dimensional analysis of auditory-evoked potentials in rat neocortex. *J Neurophysiol.* **64**:1527–1636.

Barth, D. S. and Di, S. (1991). The functional anatomy of middle latency auditory evoked potentials. *Brain Res* **565**:109–115.

Bartlett, E. L. and Wang, X. (2005). Long-lasting modulation by stimulus context in primate auditory cortex. *J Neurophysiol* **94**:83–104.

Beitel, R. E., Schreiner, C. E., Cheung, S. W., Wang, X., and Merzenich, M. M. (2003). Reward-dependent plasticity in the primary auditory cortex of adult monkeys trained to discriminate temporally modulated signals. *Proc Natl Acad Sci USA* **100**:11 070–11 075.

Brooks, L. (1978). Nonanalytic concept formation and memory for instances. In: *Cognition and Categorization*, eds. Rosch, E. and Lloyd, B. B., pp. 169–211. Hillsdale, NJ: Lawrence Erlbaum.

Brosch, M. and Schreiner, C. E. (2000). Sequence sensitivity of neurons in cat primary auditory cortex. *Cereb Cortex* **10**:1155–1167.

Brosch, M., Schulz, A., and Scheich, H. (1999). Processing of sound sequences in macaque auditory cortex: response enhancement. *J Neurophysiol* **82**:1542–1559.

Brosch, M., Selezneva, E., Bucks, C., and Scheich, H. (2004). Macaque monkeys discriminate pitch relationships. *Cognition* **91**:259–272.

Brosch, M., Selezneva, E., and Scheich, H. (2005). Nonauditory events of a behavioural procedure activate auditory cortex of highly trained monkeys. *J Neurosci* **25**:6797–6806.

Brosch, M., Oshurkova, E., Bucks, C., and Scheich, H. (2006). Influence of tone duration and intertone interval on the discrimination of frequency contours in a macaque monkey. *Neurosci Lett* **406**:97–101.

Budinger, E., Heil, P., and Scheich, H. (2000). Functional organization of auditory cortex in the Mongolian gerbil (*Meriones unguiculatus*). III. Anatomical subdivisions and corticocortical connections. *Eur J Neurosci* **12**:2425–2451.

Budinger, E., Heil, P., Hess, A., and Scheich, H. (2006). Multisensory processing via early cortical stages: connections of the primary auditory cortical field with other sensory systems. *Neuroscience* **143**:1065–1083.

Butler, R. A., Diamond, I. T., and Neff, W. D. (1957). Role of auditory cortex in discrimination of changes in frequency. *J Neurophysiol* **20**:108–120.

Butts, D. A. and Goldman, M. S. (2006). Tuning curves, neuronal variability, and sensory coding. *PLoS Biol* **4**:1–8.

Chater, N. and Heyes, C. (1994). Animal concepts: content and discontent. *Mind Lang* **9**:209–247.

Diamond, I. T. and Neff, W. D. (1957). Ablation of temporal cortex and discrimination of auditory patterns. *J Neurophysiol* **20**:300–315.

Diamond, I. T., Goldberg, J. M., and Neff, W. D. (1962). Tonal discrimination after ablation of auditory cortex. *J Neurophysiol* **25**:223–235.

Dowling, W. J. (1978). Scale and contour: two components of a theory of memory for melodies. *Psychol Rev* **85**:341–354.

Eggermont, J. J. (2007). Correlated neural activity as the driving force for functional changes in auditory cortex. *Hear Res* **229**:69–80.

Estes, W. K. (1986). Array models for category learning. *Cogn Psychol* **18**:500–549.

Freeman, W. J. (2000). *Neurodynamics: An Exploration in Mesoscopic Brain Dynamics*. London: Springer.

Fritz, J., Shamma, S., Elhiali, M., et al. (2003). Rapid task-related plasticity of spectrotemporal receptive fields in primary auditory cortex. *Nat Neurosci* **6**:1216–1223.

Fritz, J. B., Elhilali, M., and Shamma, S. A. (2005). Differential dynamic plasticity of A1 receptive fields during multiple spectral tasks. *J Neurosci* **25**:7623–7635.

Ghose, G. M. (2004). Learning in mammalian sensory neocortex. *Curr Opin Neurobiol* **14**:513–518.

Ghose, G. M., Yang, T., and Maunsell, J. H. R. (2001). Physiological correlates of perceptual learning in monkey V1 and V2. *J Neurophysiol* **87**:1867–1888.

Goldberg, J. M. and Neff, W. D. (1961). Frequency discrimination after bilateral ablation of cortical auditory areas. *J Neurophysiol* **24**:119–128.

Gonzalez-Lima, F. and Scheich, H. (1986). Neural substrates for tone-conditioned bradycardia demonstrated with 2-deoxyglucose. II. Auditory cortex plasticity. *Behav Brain Res* **20**:281–293.

Haken, H. (1983). *Synergetics: An Introduction – Non-Equilibrium Phase Transitions and Self-Organization in Physics, Chemistry and Biology*. Berlin: Springer-Verlag.

Hawkins, R. D. and Kandel, E. R. (1984). Is there a cell-biological alphabet for simple forms of learning? *Psychol Rev* **91**:375–391.

Heil, P., Langner, G., and Scheich, H. (1992). Processing of frequency-modulated stimuli in the chick auditory cortex analogue: evidence for topographic representations and possible mechanisms of rate and directional sensitivity. *J Comp Physiol A* **171**:583–600.

Hintzman, D. L. (1986). "Schema abstraction" in a multiple trace memory model. *Psychol Rev* **93**:411–428.

Irvine, D. R. (2007). Auditory cortical plasticity: does it provide evidence for cognitive processing in the auditory cortex? *Hear Res* **229**:158–170.

Irvine, D. R. F. and Wright, B. A. (2005). Plasticity of spectral processing. *Int Rev Neurobiol* **70**:435–472.

Kelly, J. B. (1973). The effects of insular and temporal lesions in cats on two types of auditory pattern discrimination. *Brain Res* **62**:71–87.

Kéri, S. (2003). The cognitive neuroscience of category learning. *Brain Res Rev* **43**:85–109.

Kommatsu, L. K. (1992). Recent views of conceptual structure. *Psychol Bull* **112**:500–526.

Kraus, N. and Disterhoft, J. F. (1982). Response plasticity of single neurons in rabbit auditory association cortex during tone-signalled learning. *Brain Res* **246**:205–215.

Kraus, M., Schicknick, H., Wetzel, W., et al. (2002). Memory consolidation for the discrimination of frequency-modulated tones in Mongolian gerbils is sensitive to protein-synthesis inhibitors applied to auditory cortex. *Learn Mem* **9**:293–303.

Lea, S. E. G. (1984). In what sense do pigeons learn concepts. In: *Animal Cognition*, eds. Terrace, H. S., Bever, T. G., Roitblat, H. L., pp. 263–276. Hillsdale, NJ: Lawrence Erlbaum.

Lea, S. E. G. and Ryan, C. M. E. (1990). Unnatural concepts and the theory of concept discrimination in birds. In: *Quantitative Analyses of Behavior*, vol. 8, *Behavioral*

Approaches to Pattern Recognition and Concept Formation, eds. Commons, M.L., Herrnstein, R.J., Kosslyn, S.M., and Mumford, D.B., pp. 165–185. Hillsdale, NJ: Lawrence Erlbaum.

McConnell, P.B. (1990). Acoustic structure and receiver response in domestic dogs, *Canis familiaris*. *Anim Behav* **39**:897–904.

Medin, D.L. and Schaffer, M.M. (1978). Context theory of classification learning. *Psychol Rev* **85**:207–238.

Miller, E.K., Nieder, A., Freedman, D.J., and Wallis, J.D. (2003). Neural correlates of categories and concepts. *Curr Opin Neurobiol* **13**:198–203.

Nelson, D.A. and Marler, P. (1990). The perception of birdsong and an ecological concept of signal space. In: *Comparative Perception: Complex Signals*, vol. 2, eds. Stebbins, W.C. and Berkeley, M.A., pp. 443–477. New York: John Wiley.

Nosofsky, R.M. (1986). Attention and learning processes in the identification and categorization of integral stimuli. *J Exp Psychol: Learn Mem Cogn* **13**:87–108.

Ohl, F.W. and Scheich, H. (1996). Differential frequency conditioning enhances spectral contrast sensitivity of units in the auditory cortex (field AI) of the alert Mongolian gerbil. *Eur J Neurosci* **8**:1001–1017.

Ohl, F.W. and Scheich, H. (1997). Learning-induced dynamic receptive field changes in primary auditory cortex of the unanaesthetized Mongolian gerbil. *J Comp Physiol A* **181**:685–696.

Ohl, F.W. and Scheich, H. (2005). Learning-induced plasticity in animal and human auditory cortex. *Curr Opin Neurobiol* **15**:470–477.

Ohl, F.W., Wetzel, W., Wagner, T., Rech, A., and Scheich, H. (1999). Bilateral ablation of auditory cortex in Mongolian gerbil affects discrimination of frequency modulated tones but not of pure tones. *Learn Mem* **6**:347–362.

Ohl, F.W., Scheich, H., and Freeman, W.J. (2000a). Topographic analysis of epidural pure-tone-evoked potentials in gerbil auditory cortex. *J Neurophysiol* **83**:3123–3132.

Ohl, F.W., Scheich, H., and Freeman, W.J. (2000b). Spatial representation of frequency-modulated tones in gerbil auditory cortex revealed by epidural electrocorticography. *J Physiol (Paris)* **94**:549–554.

Ohl, F.W., Scheich, H., and Freeman, W.J. (2001). Change in pattern of ongoing cortical activity with auditory category learning. *Nature* **412**:733–736.

Ohl, F.W., Deliano, M., Scheich, H., and Freeman, W.J. (2003a). Early and late patterns of stimulus-related activity in auditory cortex of trained animals. *Biol Cybernet* **88**:374–379.

Ohl, F.W., Deliano, M., Scheich, H., and Freeman, W.J. (2003b). Analysis of evoked and emergent patterns of stimulus-related auditory cortical activity. *Rev Neurosci* **14**:35–42.

Pearce, J.M. (1994). Discrimination and categorization. In: *Animal Learning and Cognition*, ed. Mackintosh, N.J., pp. 109–134. San Diego, CA: Academic Press.

Phillips, D.P., Mendelson, J.R., Cynader, M.S., and Douglas, R.M. (1985). Response of single neurons in the cat auditory cortex to time-varying stimuli: frequency-modulated tones of narrow excursion. *Exp Brain Res* **58**:443–454.

Posner, M.I. and Keele, S.W. (1968). On the genesis of abstract ideas. *J Exp Psychol* **77**:353–363.

Raiguel, S., Vogels, R., Mysore, S.G., et al. (2006). Learning to see the difference specifically alters the most informative V4 neurons. *J Neurosci* **26**:6589–6602.

Recanzone, G.H., Schreiner, C.E., and Merzenich, M.M. (1993). Plasticity in the frequency representation of primary auditory cortex following discrimination training in adult owl monkeys. *J Neurosci* **13**:87–103.

Ryan, C.M.E. (1982). Concept formation and individual recognition in the domestic chicken (*Gallus gallus*). *Behav Anal Lett* **2**:213–220.

Rosch, E. (1973). Natural categories. *Cogn Psychol* **4**:328–350.

Rosch, E. (1975). Cognitive reference points. *Cogn Psychol* **7**:192–238.

Rybalko, N., Suta, D., Nwabueze-Ogbo, F., and Syka, J. (2006). Effect of auditory cortex lesions on the discrimination of frequency-modulated tones in rats. *Eur J Neurosci* **23**:1614–1622.

Scheich, H., Brechmann, A., Brosch, M., Budinger, E., and Ohl, F.W. (2007). The cognitive auditory cortex: task-specificity of stimulus representations. *Hear Res* **229**:213–224.

Schoups, A., Vogels, R.N.Q., and Orban, G. (2001). Practising orientation identification improves orientation coding in V1 neurons. *Nature* **412**:549–553.

Schouten, M.K.D. and De Long, L. (1999). Reduction, elimination, and levels: the case of the LTP-learning link. *Phil Psychol* **12**:237–262.

Selezneva, E., Scheich, H., and Brosch, M. (2006). Dual time scales for categorical decision making in auditory cortex. *Curr Biol* **16**:2428–2433.

Seyfarth, R.M., Cheney, D.L., and Marler, P. (1980a). Monkey responses to three different alarm calls: evidence of predator classification and semantic communication. *Science* **210**:801–803.

Seyfarth, R.M., Cheney, D.L., and Marler, P. (1980b). Vervet monkey alarm calls: semantic communication in a free-ranging primate. *Anim Behav* **28**:1070–1094.

Talwar, S.K. and Gerstein, G.L. (2001). Reorganization in awake rat auditory cortex by local microstimulation and its effect on frequency-discrimination behaviour. *J Neurophysiol* **86**:1555–1572.

Thompson, R.F. (1960). Function of auditory cortex of cat in frequency discrimination. *J Neurophysiol* **23**:321–334.

Wasserman, E.A. and Astley, S.L. (1994). A behavioural analysis of concepts: its application to pigeons and children. *Psychol Learn Motiv* **31**:73–132.

Weinberger, N.M. (2004). Specific long-term memory traces in primary auditory cortex. *Nat Rev Neurosci* **5**:279–290.

Weinberger, N.M. (2007). Auditory associative memory and representational plasticity in the primary auditory cortex. *Hear Res* **229**:54–68.

Wetzel, W., Wagner, T., Ohl, F.W., and Scheich, H. (1998). Categorical discrimination of direction in frequency-modulated tones by Mongolian gerbils. *Behav Brain Res* **91**:29–39.

Witte, R.S. and Kipke, D.R. (2005). Enhanced contrast sensitivity in auditory cortex as cats learn to discriminate sound frequencies. *Cogn Brain Res* **23**:171–184.

11

The construction of olfactory representations

THOMAS A. CLELAND

Introduction

Sensory information progresses centrally from the primary sensors in the periphery to the central neural structures that derive relevant environmental information from these sensory data and determine appropriate physiological and behavioral responses. In this chapter, I present a general theory of early olfactory sensory processing in the primary olfactory epithelium and olfactory bulb (OB). The theory depicts olfactory sensory processing as a cascade of representations, each of which exhibits characteristic physical properties and is sampled by appropriate neural mechanisms in order to construct the subsequent representation. The primary olfactory representation is mediated by the activation pattern across the population of primary olfactory sensory neurons (OSNs) in the sensory epithelium. The secondary olfactory representation is similarly mediated by the activation pattern across the population of principal neurons immediately postsynaptic to the OSNs, known as mitral cells. (Mitral cell axons diverge dramatically, projecting to roughly ten different central structures within the brain; the resulting tertiary and subsequent olfactory representations are constructed outside the olfactory bulb and are not discussed at length herein.) The transformation between the primary and secondary representations is a robust, intricate, two-stage process that corrects for artefacts that can hinder the recognition of odor qualities, regulates stimulus selectivity, and transduces the underlying mechanics from a robust but costly rate-coding scheme on a slow respiratory (theta-band) timescale to a sparse

Information Processing by Neuronal Populations, ed. Christian Hölscher and Matthias Munk. Published by Cambridge University Press. © Cambridge University Press 2009.

dynamical representation operating on the beta- and gamma-band timescales and suitable for integration with other central neural processes. Odor memory mechanisms within the OB, dynamical coordination with central structures, and the regulation of sampling behaviors all modulate the patterned spiking output of OB principal neurons, thereby selectively sampling and filtering afferent sensory information pursuant to the needs of the organism.

Stimulus encoding in olfactory sensory neurons

Olfactory sensory neurons are embedded within a sensory epithelium that lines a portion of the nasal cavity and is covered with a layer of mucus. They extend a protrusion into this mucus layer known as the dendritic knob, from which extend several cilia between 1 and 200 µm in length (ciliary lengths vary both within and between species) (Getchell, 1986; Morrison and Costanzo, 1990, 1992) (Fig. 11.1). Odorous molecules inhaled into the nose dissolve into the mucus layer and associate with the extracellular binding sites of odorant receptor proteins located on these OSN cilia. These associations trigger transduction cascades within OSNs that form the basis of the *primary olfactory representation*.

Odorant receptors (ORs) are G-protein-coupled seven-transmembrane (7TM) receptors related to other 7TM proteins such as rhodopsin and β-adrenergic receptors (Buck and Axel, 1991; Mombaerts, 1999). Their association with certain odorant molecules elicits a conformational change which is propagated through the plasma membrane, releasing intracellular G proteins which, in turn, activate membrane-associated adenylate cyclase enzymes that generate the second messenger cyclic AMP (cAMP). One target of this cAMP is a cyclic nucleotide-gated (CNG) ion channel permeant to sodium and calcium ions, the opening of which depolarizes the OSN. The activation of calcium-dependent chloride channels by this influx of calcium – the third messenger – magnifies the evoked potential by enabling the efflux of chloride ions into the chloride-poor mucus, potentiating the depolarization of the OSN. There are specific subclasses of OSN that utilize a cyclic guanosine monophosphate (cGMP)-based transduction cascade instead (Meyer *et al.*, 2000), and other transduction mechanisms may also play a role (Restrepo *et al.*, 1990; Kaur *et al.*, 2001), but the principle is the same: odorant binding evokes a cascade response in OSNs that can be dramatically amplified by the second- and third-messenger cascades that it activates. The resulting depolarization of the OSN evokes a train of action potentials, the frequency of which depends systematically on the intensity of the odor-evoked depolarization (Getchell and Shepherd, 1978; Rospars *et al.*, 2000). These spike trains are temporally unsophisticated and (as is typical for primary sensory neurons across modalities) are generally considered to be simple *rate codes*

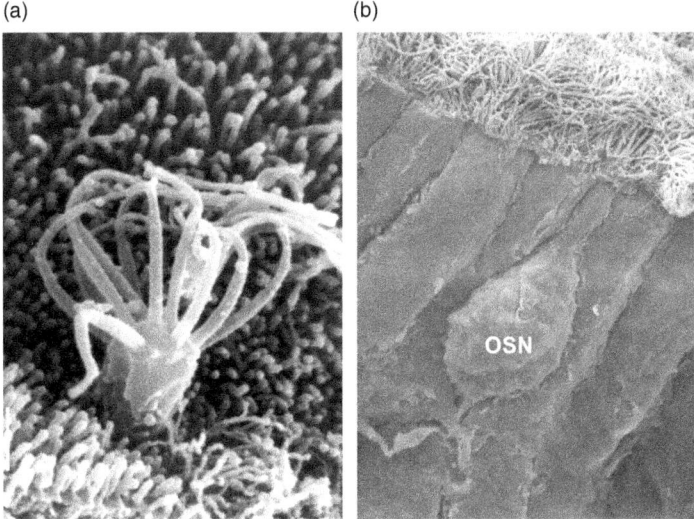

Figure 11.1 Olfactory sensory neuron morphology. (a) Scanning electron micrograph of the sensory surface of the human olfactory epithelium, showing the protruding dendritic knob and sensory cilia of an OSN. The microvilli of sustentacular (supporting) cells surround the OSN. (b) Scanning electron micrograph of a fractured human olfactory epithelium in cross-section, showing an olfactory sensory neuron (OSN) surrounded by sustentacular cells. The columnar sustentacular cells extend the full depth of the epithelium. The OSN axon extends down and to the left, towards the cribriform plate and olfactory bulb. (Micrographs courtesy of Richard M. Costanzo, Virginia Commonwealth University.)

reporting the intensity of activation of the corresponding sensory neuron, modulated to a moderate extent by intracellular adaptation processes that serve to emphasize transient changes in the intensity of activation (Zufall and Leinders-Zufall, 2000).

Elemental odor stimuli

In a formal sense, essential for a quantitative understanding of olfactory processing, the elemental odor stimulus is not an odorant molecule per se, but rather "that aspect of an odorant molecule that associates with a single type of odorant receptor" or even "the net effect of a given odorant molecule on a single type of odorant receptor." This theoretical element has been termed an odor epitope, or odotope (the term *odotope* is preferable as it does not connote a literal correspondence with molecular structural features), and reflects the fact that, just as any given OR can bind to many different odorant molecules, single odorant molecules also bind to (and activate) many different ORs. It is important for theoretical purposes to dissociate the abstract odotope concept from the physical binding sites on odorant molecules from which it is derived; for example, increasing the concentration of an odorant will increase the number of odotopes binding at significant levels to their corresponding ORs, but one would not say that a given odotope comes to associate with more than one OR type. The fact that a given molecular structural feature may associate with multiple ORs and hence contribute physically to multiple odotopes is not relevant at this level of analysis, nor is the fact that the precise balance of ligand–receptor associations that constitute each odotope (i.e. that contribute to the net activation of a given OR by a single given odorant) will change with concentration. In this odotopic parlance, the intricate matrix of associations between odorant structural features and ORs reduces to the simple statement that "odorant molecules are composed of a characteristic combination of odotopes, each of which associates with a specific OR."

The association of odotopes with their receptors is described by standard pharmacological laws; indeed, a number of "puzzling" problems in olfaction are simple corollaries of these laws. Ligand–receptor binding is described by rate constants for association and dissociation, yielding sigmoidal (or sums-of-sigmoidal) binding curves that depict the equilibrium probability (or proportion) of ligand–receptor association as a function of ligand concentration. The ligand concentration at which 50% association is achieved is the dissociation constant (K_d), a measure of ligand–receptor *affinity*. The maximum slope of the binding curve (also at 50% association) determines the Hill coefficient, or cooperativity, of the reaction. (Odotope–receptor associations that exhibit multiple affinities yield more complex, sums-of-sigmoidal binding curves, but the underlying principle is the same.) Of equally critical importance is a separate term, *efficacy*, which describes the effectiveness of the odotope – once bound to the OR – at activating the OSN transduction cascade. Affinity and efficacy are distinct properties; for example, an odotope with high affinity and low efficacy represents a receptor antagonist that will not be directly detected by the olfactory system when presented alone but which can interfere with the ability of other, simultaneously presented odorants to activate that OR. While this phenomenon is predictable from first principles it has only recently been explicitly described in the olfactory system (Araneda *et al.*, 2004; Oka *et al.*, 2004).

The deployment of metabotropic 7TM receptors as ORs enables a further dissociation between the ligand–receptor binding curve per se and the analogous dose–response curve of cellular activation. Specifically, the decoupling of the 7TM receptor from its effector ion channels enables a receptor with a modest binding affinity to a given ligand to mediate a cellular response of arbitrarily high sensitivity to the presence of that ligand (Fig. 11.2). Furthermore, the degree of functional sensitivity can be smoothly regulated by the modulation of

(a)

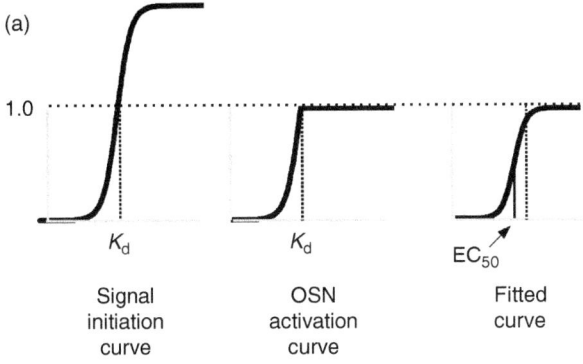

Signal initiation curve   OSN activation curve   Fitted curve

(b)

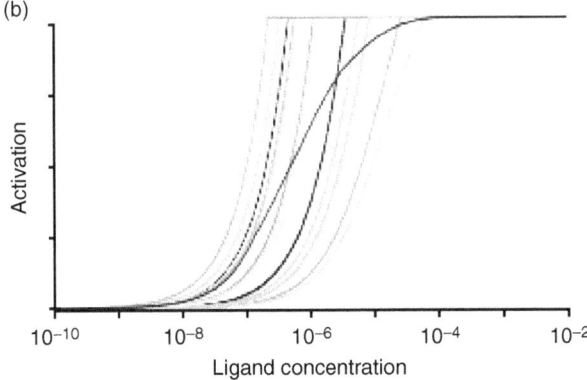

Figure 11.2 Illustration of a mechanism by which neurons expressing metabotropic receptors such as ORs can achieve arbitrarily high sensitivities (EC_{50}) to those receptors' ligands irrespective of their binding affinities (dissociation constant, K_d). Other terminology depicted is that of Cleland and Linster (1999). (a) (Left panel) If twice the number of receptors are expressed as are necessary to maximally activate the existing population of effector channels in a given neuron, the dissociation constant is unaffected and ligand–receptor binding over the level of maximum effector activation (1.0) has no further effect. (Center panel) The dose–response curve then measured in that neuron would be cut off at the level of maximum effector activation. (Right panel) Fitting this cut-off curve to a sigmoid yields an half-activation concentration (EC_{50}) that is lower than the dissociation constant, and this difference increases in proportion to the overexpression of receptors with respect to effectors (or, equivalently, to the effective gain of the intracellular G protein cascade). The abscissa represents ligand concentration as in (b). (b) A convergent population of neurons, such as OSNs, that each express a different (or loosely regulated) degree of receptor overexpression or intracellular gain will generate correspondingly different activation curves (each as in (a), center panel). The summation of their output produces a collective dose–response curve (*sigmoid curve*) that is broader than any of those of the individual converging neurons, as is observed in the collective responses of OSNs visualized in glomerular imaging studies. (Figures adapted from Cleland and Linster, 1999.)

intracellular cascades or the differential deployment of receptor and effector populations, enabling a population of OSNs expressing the same type of odorant receptor (and hence possessing identical receptive fields) to be differently tuned for concentration. A convergent population of such neurons will collectively express an arbitrarily broad dose–response function, much like those observed in OB glomeruli (see below), without distorting the receptive field. This model (Cleland and Linster, 1999) both explains the extreme sensitivity exhibited by some animals' olfactory systems without relying on implausibly strong ligand–receptor affinities and resolves the conundrum of how the collective dose–response curves measured by glomerular imaging (and in mitral cell responses) can be rendered substantially broader than those measured in individual OSN recordings. These principles illustrate the sophistication of odor representation at even the most peripheral levels, suggesting how OSN properties may be organized so as to take advantage of physical laws to increase their collective coding capacity and reduce the computational burden on subsequent processing stages.

Defining the odotope as the elemental odor stimulus offers several critical advantages to quantitative analysis. The association of each odotope with its receptor exhibits a characteristic set of ligand–receptor binding constants. Indeed, as noted above, it is likely that a given odotope–receptor association will have multiple sets of binding constants, each with a characteristic efficacy, although it suffices for most purposes to model these complex odotope–receptor interactions as a single sigmoidal binding function with a single efficacy. The primary sensory representation of an odorant molecule presented at a given concentration, then, is simply the sum of the primary representations of its constituent odotopes at that concentration. The problem of representing odor mixtures is also conceptually simplified. From an odotopic perspective, single odorant molecules *are* mixtures (of odotopes); combinations of multiple odorant molecules are conceptually no different. However, the concentrations of odotopes associated with different odorant molecules may now vary with respect to one another, and multiple odotopes (associated with different odorant molecules) may now compete for the same receptor. As odotopes by definition associate with a single OR type, computing these competitive interactions is elementary. The primary representation of complex mixtures, rendered as an intractable, recursive problem from the perspective of odorant molecules, is thereby reduced to a conceptually simple list of independent competitive interactions. The problem of "subtractive" interactions, for example, in which particular OSNs respond less to a mixture of A + B than they do to A alone (at the same concentration), is explained as a simple matter of competition between a high-efficacy odotope for that receptor contained in A and a low-efficacy odotope of comparable affinity contained in B.

An odotopic basis for quantifying odor representations also renders tractable the problem of the dimensionality of odor space. Whereas primary auditory stimuli can be mapped along the single dimension of frequency, and visual stimuli can be retinotopically mapped in two dimensions, the similarity spaces in which olfactory stimuli can be mapped are irreducibly high-dimensional. Furthermore, unlike frequency or distance, the axes of variation among chemical stimuli themselves cannot be unambiguously identified, much less rendered orthogonal to one another. As with the definition of the odotope, which utilizes the population of expressed ORs to organize the stimulus properties of odorant molecules, the solution lies in the inclusion of sensor properties within the definition of the sensory space. Specifically, each axis of variation is defined as the activation level of a single OR type; hence, the dimensionality of odor space is identical to the number of different OR types expressed (roughly 350 in humans, 1000 in mice: Mombaerts, 1999, 2001). These axes should be construed as orthogonal, in that each OR type can in principle be independently activated. This orthogonality principle is made clear by studies employing "artificial odor" stimuli via the direct electrical stimulation of glomeruli (Mouly et al., 1990, 1993), in which any combination of OR-specific elemental stimuli can be administered, and holds despite the fact that, in any finite odor environment, consistent correlations will be measurable between the activation levels of particular ORs. A capacity for orthogonality is also important to ensure that the olfactory system is appropriately responsive in diverse odor environments, as the statistics of odorant prevalence have a strong impact on measures of elemental similarity.

An odotopic basis function for odor representation significantly disambiguates and simplifies the representation and quantification of odor stimuli. It is explicitly species-specific in that it depends quantitatively on the complement of ORs expressed by a given subject; hence, phenomena such as the overlap between two odor representations cannot be directly compared between two species that may express different OR complements. While this is clearly appropriate and does not constitute a weakness in the odotopic approach for its intended purpose, odotopic depictions of odorant similarity consequently are of limited value outside of a biological context.

Olfactory sensory neuron receptive fields

Stimulus selectivity in the olfactory system is primarily generated by the diversity of ORs that are expressed among the OSN population. Different populations of OSNs express different ORs that are responsive to different odotopes, such that the representation of any given odor arises from the degree to which its constituent odotopes bind to their cognate ORs and activate particular subpopulations of OSNs. Hence, a given odor, presented at a given concentration and in the absence of background odors or other disruptive stimuli, evokes a characteristic, replicable profile of activation across the set of all such OSN subpopulations. This stimulus specificity at the primary receptor level in olfaction offers a potent advantage (over, for example, vision) to computational studies of sensory processing, because the basis for the perception of stimulus quality is derived directly from the *receptive fields* of primary OSNs. The quantification of olfactory receptive fields (also known as *molecular receptive ranges*: Mori and Shepherd, 1994), however, exhibits challenges similar to those encountered in the definition of elemental odor stimuli, which can be met by adherence to the following principles:

(1) The receptive field of a neuron must be defined in terms of the output activity that different odor stimuli elicit in that neuron, rather than ligand–receptor binding per se. For present purposes, receptive fields will be defined as that range of stimuli that evoke a net excitatory response in a neuron. For principal neurons such as OSNs and mitral cells, "output" implies the evocation of action potentials. Hence, a stimulus evoking an excitatory synaptic input to a neuron that is overcome by simultaneous heterosynaptic inhibition such that the neuron produces no action potentials will not be considered part of that neuron's receptive field. Similarly, an odor that does not activate a given neuron is not part of that neuron's receptive field even if some of its component molecules activate that neuron when presented separately.

(2) The receptive fields of neurons must be considered as distinct from the receptive fields of the receptors that they express. Hence, an OSN expressing two types of odorant receptor would have a single receptive field combining those of both receptors. Similarly, the receptive fields of central neurons must be considered independently of the receptive fields of neurons presynaptic to them, as the transfer function between the two is highly stimulus-dependent and can be arbitrarily complex.

(3) Receptive fields are dynamic and contextual properties that may vary with stimulus intensity (odorant concentration), centrifugal neuromodulation, mixture properties, or a number of other factors. Experimental measurements of receptive fields also depend, of course, on the range of test stimuli used to probe them, and hence will nearly always yield underestimates of their scope.

This framework for defining olfactory receptive fields clarifies some idiosyncratic olfactory phenomena for theoretical purposes, two examples of which follow. First, canonically, in mammals, each OSN expresses exactly one

OR – indeed, in mice, allelic inactivation ensures that only one of the two alleles of each OR is expressed (Chess et al., 1994; Strotmann et al., 2000). All other OR genes in that cell are silenced. However, this strict pattern of one OR per OSN is not uniformly exhibited in all species, some of which appear to express multiple OR types in a single OSN; even in mammals it is not certain that this pattern is strictly maintained (Mombaerts, 2004). This indicates that the one-OR-per-OSN property is not a critical feature of the olfactory system, but a variant with (presumably) some species-specific utility. Indeed, while the expression of only one OR per OSN (and just one allele thereof) renders the receptive fields of individual OSNs as narrow as can be achieved solely by regulating OR expression, it does not qualitatively affect the nature of the neural computation. OSNs expressing any complement of receptors will exhibit a single, overall receptive field. The formation of narrower receptive fields through highly selective receptor expression and/or allelic inactivation is simply one strategy with characteristic costs and benefits; specifically, sensors with narrower receptive fields can improve the capacity of the olfactory system to discriminate among highly similar odorants, but require deployment of a greater number of differently tuned sensors in order to remain comparably sensitive to the same range of different odors. Indeed, the one-receptor-type-per-OSN strategy was identified in mice, which are macrosomatic mammals expressing large numbers of different OR species. In the nematode *Caenorhabditis elegans* and the fruit fly *Drosophila melanogaster*, in contrast, recent work indicates that this principle is not strictly respected (Troemel et al., 1995; Goldman et al., 2005).

Second, it often has been suggested that ORs exhibit uniquely broad receptive fields among 7TM receptors, based on the well-known sensitivity of OSNs and ORs to a reasonably broad range of odor ligands. There is as yet no valid reason to believe that this is true. All receptors in general, and 7TM receptors in particular, are sensitive to any ligand able to induce their reconformation into the active state. Multiple agonists, weak agonists, and weak and strong antagonists have been identified for all well-studied receptor species (although typically only strong agonists and strong antagonists become widely known because of their experimental utility). The primary functional difference between the receptive fields of odorant receptors and other 7TM receptors is not a property of the receptor per se, but of the statistical structure of their normal stimulus environments. Most receptors are embedded in a highly regulated environment in which only a single effective agonist is normally present, hence establishing their nominal status as "glutamate receptors" or "acetylcholine receptors" largely on the basis of that limited environment. The actual receptive field of β_1-adrenergic receptors, for example, is considerably broader, including epinephrine, norepinephrine, isoproterenol, xamoterol, and denopamine as full agonists, whereas the inhibitory surround (antagonists) includes alprenolol, propranolol, pindolol, betaxolol, and atenolol. Odorant receptors, in contrast, are deployed directly into the unregulated chemical environment of the external world, wherein the scope and complexity of their receptive fields cannot be functionally overlooked, and indeed is essential to their function in that context.

Olfactory sensory neuron convergence

The axons of OSNs project via the olfactory nerve to synapse with mitral cells and multiple classes of interneurons within the OB, a telencephalic cortex devoted to the processing of olfactory sensory information. Critically, the axons of OSNs expressing the same OR converge together, intertwining their axonal arbors within the surface layer of the OB to form *glomeruli*. Glomeruli are surrounded by glia (and hence visible as spheroid structures under the light microscope, particularly in mammals) and contain no cell bodies. Mitral cells (the principal neurons of the OB), as well as multiple classes of periglomerular and tufted cells, extend dendrites into one or a few glomeruli, wherein they receive synaptic inputs from OSN axons and from one another. In particular, mitral cells in mammals tend to sample strictly from a single glomerulus, although – like the one-OR-per-OSN hypothesis – this principle is not strictly adhered to across species, illustrating another advantage of treating mitral cell receptive fields independently from those of their presynaptic OSNs. Interestingly, most classes of bulbar interneurons – at least in mammals – also confine the majority of their synaptic interactions to single glomeruli and their associated mitral cells, such that the olfactory bulb exhibits a *columnar* chemotopic structure. There are several exceptions to this, of course, notably the short-axon cells of the deep glomerular layer (Aungst et al., 2003) and the sparse axons of periglomerular cells. Interestingly, even granule–mitral interactions, previously an obvious counter-example to this generalization about OB columns, now appear to adhere to this columnar organization (Willhite et al., 2006).

The selective clustering into glomeruli of the axon terminals of OSNs expressing the same OR, irrespective of the locations of their somata within the nasal epithelium, is the single most important feature of OSN convergence. The convergence of OSNs with similar receptive fields enables the promulgation of chemoselectivity to subsequent OB processing layers, as coordinated neural computations can be performed across populations of similarly tuned neurons, in principle improving both coding capacity and odor sensitivity (van Drongelen et al., 1978; Duchamp-Viret et al., 1989; Cleland and Linster, 2005) as well as extending the range of concentrations across which consistent representations can be maintained (Cleland and Linster, 1999). Moreover, the clustering of thousands of similarly tuned axonal arbors into discrete regions (glomeruli) transforms the surface of the OB into a visualizable map of the activation levels of each OR-specific OSN type expressed. This phenomenon has been an experimental windfall; visualization of glomerular activity across the OB surface using any of a number of techniques enables direct measurement of the primary olfactory representation – that is, of the population-average activation levels of each OR-specific population of OSNs.

Degenerate feature maps of odor stimulus qualities

The patterns of glomerular activity recorded by such imaging techniques constitute degenerate odotopic feature maps. That is, for a single odorant presented at a single concentration they are specific to and diagnostic for the presence of that odorant, being composed of a map of the relative efficacies of all the odotopes comprising that odorant. Increasing the concentration of an odorant, however, damages this characteristic relational activity pattern by recruiting new glomeruli into the active ensemble and increasing the activity of existing glomerular signals (generally but not necessarily monotonically) until a maximum is reached, as is predictable by the law of mass action for increasing ligand concentrations. Furthermore, the simultaneous presence of multiple odorants typically results in competition between their odotopes for access to certain ORs, the result of which may be anywhere from fully additive (summation) to subtractive (blockade), introducing significant ambiguity into the combined representation. As most common odors – such as the smell of apples or green grass – are composed of dozens to hundreds of odorant species in characteristic ratios, and will nearly always be perceived in the presence of simultaneous unrelated but comparably complex odor sources (cow manure, baking bread), the degeneracy of complex odor representations imposes a critical limitation on sensory processing. Indeed, such limitations substantially define the perceptual problems that downstream sensory processing must solve and constrain the computational mechanisms that it can effectively use.

A second important constraint on chemosensory processing mechanisms is the lack of an ordered topography of stimulus quality across the OB surface, which limits the mechanisms that can be deployed to perform *similarity-dependent computations* on the primary representation – that is, processes such as contrast enhancement that rely upon a representation of the similarities among stimuli in their inputs. Olfactory contrast enhancement clearly occurs at the level of mitral cells (Yokoi *et al.*, 1995), despite the fact that, as discussed above, the similarities among odorant stimuli are distributed high-dimensionally. While the choice of basis function for this distribution is essentially arbitrary, the minimum dimensionality of an odor representation in an unrestricted or unpredictable chemosensory environment is equal to the number of differentiable sensors (i.e. the number of different OR types); in any species this number is substantially greater than two. This latter point is critical, as a dimensionality greater than two precludes the representation of stimulus similarity by physical proximity within cortical or other layered neural structures (Kohonen, 1982; Kohonen and Hari, 1999; Cleland and Sethupathy, 2006). Hence, there can be no olfactory analogue to the frequency maps observed along the cochlea or inferior colliculus, nor to the two-dimensional retinotopic map of visual space. The physical location of glomeruli can not and does not connote the chemoselective properties of their associated ORs. This does not imply that all glomeruli are distributed randomly with respect to their chemoselectivity – data regarding putatively activity-dependent glomerular segregation (Ishii *et al.*, 2001; Tozaki *et al.*, 2004) and sensitivity to whole-molecule properties (Schoenfeld and Cleland, 2005, 2006) provide two counter-examples – but it does mean that neighborhood relationships among glomeruli are not reliable bases for the neural representation of stimulus similarity. Because of this, nearest-neighbor synaptic projections will not reliably target neurons with correspondingly similar receptive fields; consequently, despite efforts to argue the contrary, similarity-dependent computations such as olfactory contrast enhancement cannot be mediated by proximity-dependent mechanisms such as nearest-neighbor lateral inhibition in the olfactory system (Cleland and Sethupathy, 2006). This property places additional major constraints on bulbar processing mechanisms, as discussed below.

Stimulus processing in the olfactory bulb

Despite the many computational operations deployed in OSNs and in the architecture of their convergence, these are only the beginnings of the sensory processing cascade. Highly redundant rate coding on a slow timescale, as exhibited by the OSN population, is advantageous to primary olfactory sampling but also extraordinarily inefficient and metabolically costly (Attwell and Laughlin, 2001). (Timescale in this context refers to *precision*, i.e. the degree of variance in spike timing tolerated in a neuron before the meaning of its signal is altered from the perspective of a given follower.) Central cortical representations and processing mechanisms, in contrast, are sparse and timing-sensitive on considerably faster timescales than populations of isolated OSNs can support. Hence, the information contained in the primary olfactory representation must be transformed to become physically compatible with the processing rules and architecture of the cerebral cortex. Furthermore, OSN receptive fields are unregulated beyond that which is inherent in OR structure. Features of odor stimulus quality are not fully disambiguated from artefacts of concentration, nor are the responses to odotopes associated with different odorants (or odors) disambiguated from one another so that they can be appropriately identified or learned. The centrifugal regulatory influences of selective attention, perceptual learning, and the like have had little or no capacity to influence stimulus processing. Hence, in addition to the physical transformation, the afferent representation must be integrated with descending and neuromodulatory influences as well as locally situated memory effects. This process begins in the OB.

The OB is a multilayered telencephalic cortical structure. While it contains a substantial diversity of cell types interconnected within an intricate synaptic architecture (Fig. 11.3a), and much remains to be learned about its capacities, it can be rendered relatively tractable for study by emphasizing two of its architectural properties. First, the same mitral (and middle/deep tufted) principal neurons that are directly postsynaptic to OSNs constitute the only output of the OB network. Consequently, all bulbar processing converges onto one target effector: the pattern of spikes generated by these principal neurons. Second, synaptic inputs onto mitral cells within the OB are located in one of two discrete regions: either within the glomerular layer (OSNs, external tufted cells, GABAergic and dopaminergic periglomerular cells) or within the external plexiform layer (EPL; granule cells). Centrifugal descending and neuromodulatory inputs affecting mitral cells also generally conform to these two regions, or else exert their effects indirectly by influencing interneurons. These principles frame the following presentation of OB function.

Glomerular computations

The first processing stage for OSN spike trains entering the OB is within the glomerular layer itself. OSNs form synapses not only onto mitral cell dendrites, but also onto two classes of interneuron: periglomerular cells and external tufted cells. The former are GABAergic and dopaminergic inhibitory interneurons that inhibit mitral cells via a $GABA_A$-ergic mechanism as well as presynaptically inhibiting OSN terminals with $GABA_B$ or dopamine D2 pharmacology. The latter are glutamatergic excitatory interneurons that synapse onto periglomerular cells and one another as well as with another class of glomerular-layer interneurons known as short-axon cells. Critically, these two cell types mediate the only two feedforward sources of afferent information that can modify mitral cell activity. In contrast, feedback circuits such as mitral–granule lateral interactions in the EPL depend on mitral cell activation as the sole source of afferent input by which to modify mitral cell firing properties. Consequently, the strengths and capabilities of the glomerular and EPL networks differ substantively, as illustrated below.

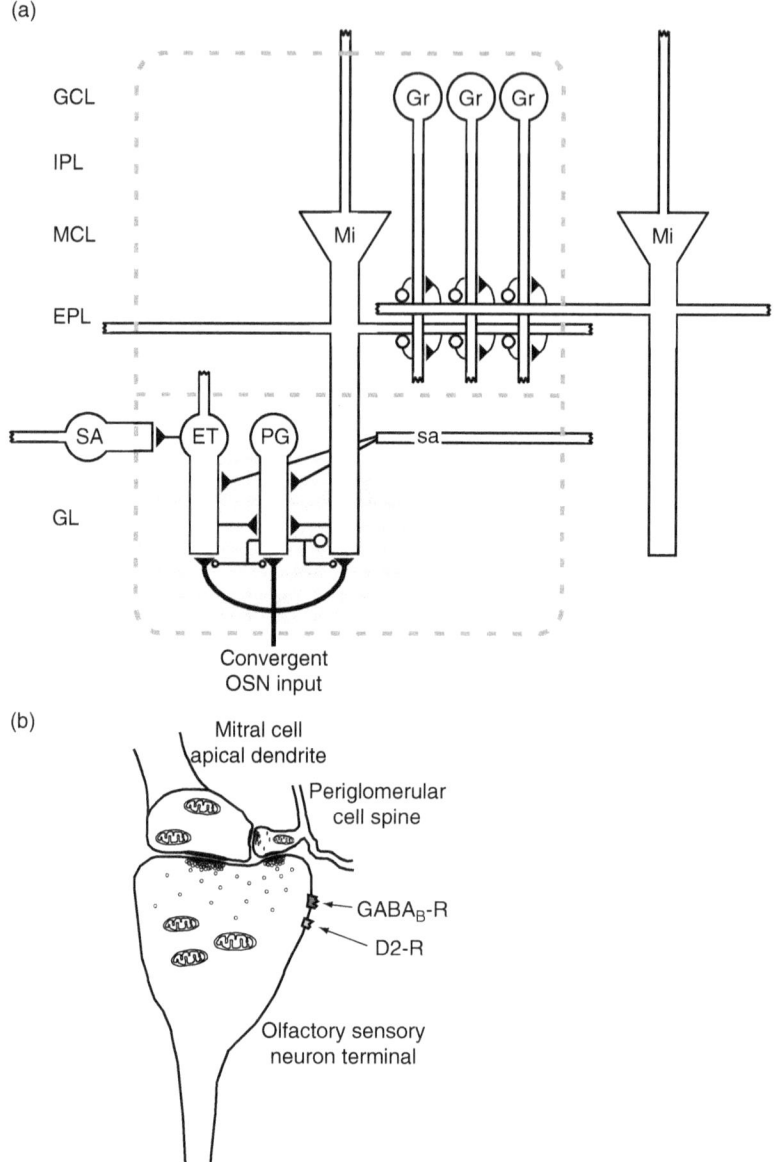

Figure 11.3 Olfactory bulb circuit architecture. (a) Illustration of major OB circuit elements. OSN input arises from the bottom and forms excitatory synapses (filled triangles) onto mitral (Mi), periglomerular (PG), and external tufted (ET) cells. External tufted cells in turn excite PG cells, short-axon (SA) cells, and each other. Periglomerular cells inhibit mitral cell apical dendrites via $GABA_A$-mediated shunt inhibition (open circles) and OSN axon terminals via $GABA_B$ and dopamine D2 presynaptic receptors (small open circles). Mitral cell lateral dendrites extend along the external plexiform layer (EPL) and form reciprocal synapses with the dendritic spines of inhibitory granule cells (Gr), delivering recurrent inhibition onto

Despite their direct innervation by OSNs, the most common response of mitral cells to odor stimuli is inhibition. While odor-inhibited mitral cells are often excited by direct OSN inputs, their concomitant inhibition by periglomerular cells co-activated by the same odor stimuli shunts this excitation and prevents spike generation. Theoretical models demonstrate that glomerular circuitry is well suited for this function (Cleland and Sethupathy, 2006). Inputs from OSNs to periglomerular cells occur in the same tiny spines that are presynaptic to mitral cell dendrites (Pinching and Powell, 1971) (Fig. 11.3b); the electrotonic compactness and high input resistance of these spines performs a much more potent current-to-voltage transformation than does the considerably larger mitral cell dendrite, enabling this di-synaptic pathway to exert its inhibitory effect upon the depolarizing mitral cell dendrite comfortably before the latter can initiate spiking (as demonstrated by the prevalence of inhibitory odor responses in mitral cells and the observation of early transient hyperpolarizations even in excited mitral cells: Hamilton and Kauer, 1989; Kauer et al., 1990; Wellis and Scott, 1990). The question then becomes: under what circumstances does the direct excitation of mitral cells overpower this inhibitory influence so as to initiate spiking? The most parsimonious hypothesis is that it is a simple matter of input strength. That is, while weak and moderate inputs result in the inhibition of mitral cells, the capacity of this inhibitory circuit saturates, such that stronger odor stimuli generate excitatory inputs that

Caption for Figure 11.3 (cont.)

themselves and lateral inhibition onto the lateral dendrites of other mitral cells. SA cells are not affiliated with any given glomerulus, but extend between them, forming a lateral excitatory network in the deep glomerular layer (Aungst et al., 2003; Cleland et al., 2007). Lower-case labels denote incoming processes originating in other glomeruli. The shaded area connotes the approximate physical boundaries of the visible glomerulus. The dotted box connotes the column of neurons associated with a particular glomerulus; the lower, glomerular layer (GL) section of the box contains the circuitry associated with the first, slow-timescale stage of OB processing, whereas the upper section contains the circuitry associated with the second, dynamical stage of processing. Middle/deep tufted cells and (deep) Blanes cells have been omitted for clarity. Olfactory bulb layers, surface to deep: GL, glomerular layer; EPL, external plexiform layer; MCL, mitral cell layer; IPL, internal plexiform layer; GCL, granule cell layer. (b) Close-up rendition of the OSN–PG–mitral cell synaptic triad in the OB glomerulus. OSN axon terminals concomitantly excite mitral cell apical dendrites and periglomerular cell spines (also known as gemmules); the excited periglomerular spines then deliver inhibition onto those same mitral cell dendrites (Pinching and Powell, 1971). Periglomerular cells also deliver $GABA_B$-ergic and dopamine D2 presynaptic inhibition onto OSN terminals.

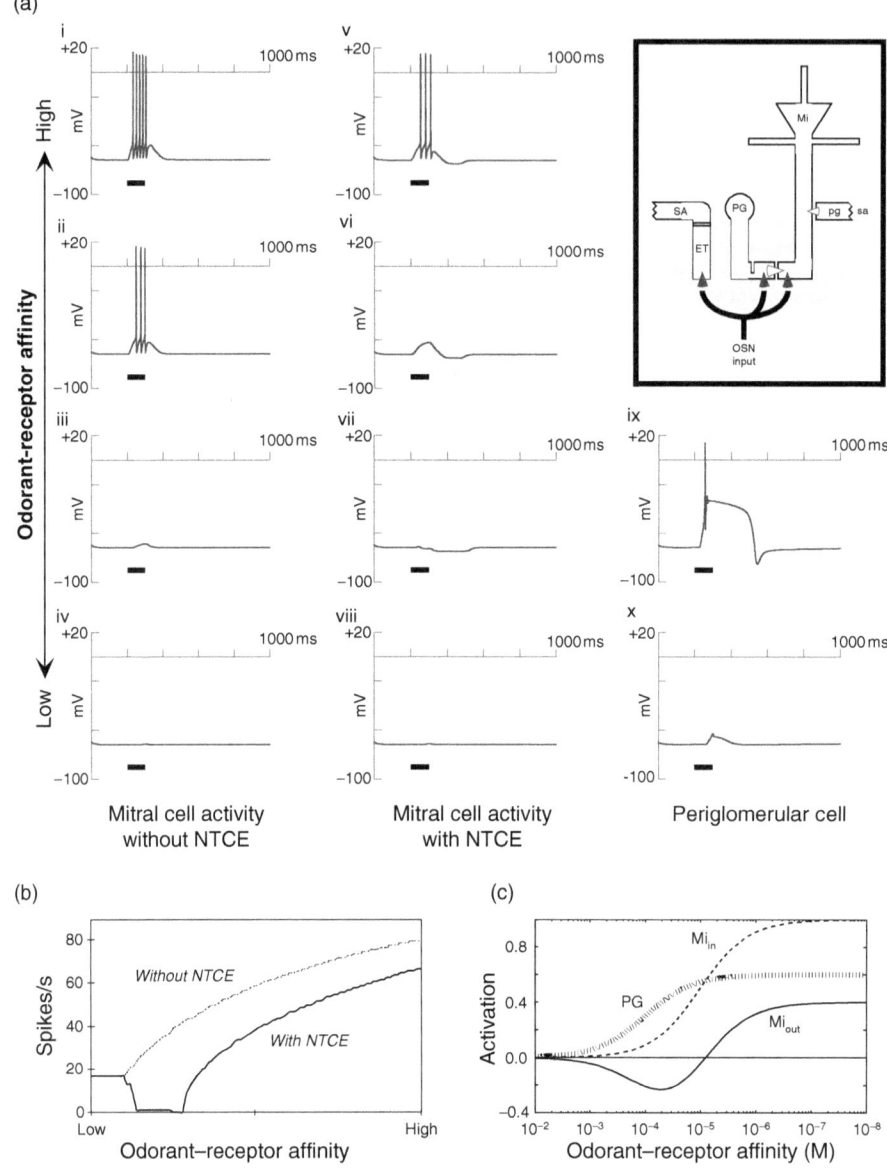

Figure 11.4 Illustration of non-topographical contrast enhancement (NTCE) using a computational model based on the synaptic triad circuit in the OB glomerular layer (Fig. 11.3b). (a)(i–iv) Odor-evoked activity in model mitral cells as a function of odor ligand–receptor affinity, in the absence of periglomerular inhibition and neglecting stimulus concentration. Increasing odor ligand–receptor affinity generates a monotonic increase in mitral cell activation. (v–viii) The addition of periglomerular inhibition upon local mitral cells creates a contrast enhancement generator element by first inhibiting (panel vii), and then exciting (panels vi, v), mitral cells as odor ligand–receptor affinity increases. Inhibition was held constant, and panels v–viii

overpower the concomitant inhibition, depolarizing the mitral cell and evoking action potentials (Fig. 11.4a). This mechanism has two great strengths. First, it results in an improved relational representation, as the mitral cells sampling from the strongest-activated glomeruli (those with receptive fields best tuned

Caption for Figure 11.4 (cont.)
depict the same four odor ligand–receptor affinities as are shown in panels i–iv. (ix–x) Periglomerular cell activation by the two lower-affinity odorant stimuli. While the depolarizing current input to periglomerular and mitral cells is identical, the greater input resistance and smaller volume of PG spines compared to mitral cell dendrites result in a greater voltage deflection in and hence a greater activation of PG cells (compare panels iv and x). Additionally, low-threshold T-type calcium current (McQuiston and Katz, 2001) evokes a near-maximal burst response from PG cells even at low input levels (panel ix), which mediates the mitral cell inhibition shown in panel vii. (Inset) Model architecture. OSN synaptic input activates mitral cell apical dendrite, periglomerular dendritic spine, and a combined ET–SA–PG cascade that projects inhibition onto other mitral cells. Filled triangles: excitatory synapses; open triangles: inhibitory synapses; lower-case labels denote incoming processes originating in other glomeruli. (b) Spike count in a model mitral cell over a 1-s stimulus in the absence and presence of PG cell-mediated NTCE, illustrating the NTCE half-hat function (Cleland and Sethupathy, 2006). In order to illustrate the effects of mitral cell inhibition, a 150-pA depolarizing current was continuously injected into the model mitral cell soma to elicit a baseline spike rate. Mitral cell spiking was employed solely as an index of activation as the model used did not include complex spike patterning mechanisms. With intact NTCE, the mitral cell activation level reflects a half-hat function as odor ligand–receptor affinity increases. (c) Simplified theoretical model of the NTCE half-hat function depicted as the difference of two sigmoids. A principal neuron (mitral cell; Mi_{in}, dashed line) and a local inhibitory interneuron (periglomerular cell; PG, dotted line) are both directly, sigmoidally activated by increasing input levels (abscissa; here depicted as odorant–receptor affinity and neglecting odorant concentration). The local interneuron exhibits greater sensitivity to this input (i.e. it is half-activated by a weaker degree of odorant–receptor affinity owing, for example, to its electrotonic compactness) while the larger principal neuron has a correspondingly greater maximum output amplitude. While input levels in a chemical binding context can conflate ligand–receptor affinity and ligand concentration, this ambiguity can be resolved by global feedback mechanisms (Cleland and Sethupathy, 2006; Cleland et al., 2007). When the two neurons are driven by the same input and the local interneuron inhibits the principal neuron, the net output activity of the principal neuron can become non-monotonic with respect to input level, exhibiting a half-hat function capable of mediating contrast enhancement (mitral cell; Mi_{out}, solid line). That is, with respect to the receptive field of any glomerulus, the mitral cell output profile after NTCE (Mi_{out}) will exhibit a narrower selectivity for odorants than do its associated ORs (Mi_{in}). (Figures adapted from Cleland and Sethupathy, 2006.)

to the stimulus) will be consistently activated while those sampling from less specifically activated glomeruli will be inhibited out of the active ensemble. With the inclusion of global feedback circuitry, odor quality representations can be constructed in mitral cells that are largely independent of concentration (i.e. normalized: Cleland and Sethupathy, 2006; Cleland et al., 2007), at least on the slower timescales associated with rate coding (about tens to hundreds of milliseconds: Chalansonnet and Chaput, 1998). Second, to the extent that concentration effects can be mitigated by normalization, this circuit generates a unidirectional on-center/inhibitory surround function (*half-hat function*, or half of a "Mexican hat" function) (Fig. 11.4b, c) with ligand–receptor affinity as the independent variable. By regulating the competitive efficacies of the excitatory and inhibitory influences on mitral cell activation, this critical circuit property enables regulation of the stringency of odor selectivity via contrast enhancement with respect to the high-dimensional similarity space in which odotope qualities are distributed, irrespective of the particular topology of that space (Cleland and Sethupathy, 2006). This independence from topology is an important feature, as topologies of odor similarity depend on the statistics and features of the current odor environment and hence are unpredictable. Moreover, this hypothesized mechanism of contrast enhancement requires no targeted lateral inhibitory projections (unlike the analogous operations in the visual and auditory modalities), and hence is entirely independent of the physical location of glomeruli on the surface of the OB. In contrast to the visual and auditory modalities, location is not used as a surrogate for stimulus quality, and physical proximity does not connote similarity. As the physical distribution of chemoselective glomeruli on the bulbar surface consequently is irrelevant to the representation or processing of stimulus quality, neither normal variations in glomerular positioning (Strotmann et al., 2000) nor the experimental generation of novel glomeruli (Serizawa *et al.*, 2000; Ishii *et al.*, 2001; Tozaki *et al.*, 2004) would be expected to affect the integrity of olfactory perception. This also demonstrates the critical importance of broad receptive fields among OSNs, as the ordering of different ORs' receptive fields in a high-dimensional similarity space – and hence the perception of similarity – relies upon measurable degrees of overlap.

External plexiform layer computations

The second layer of processing in the OB takes place in the external plexiform layer (EPL) between the extensive lateral dendrites of mitral cells and the dendrites of inhibitory granule cells. The architecture of this interaction differs substantially from that of the glomerular synaptic network. Mitral cells are the only afferent input to the system and also comprise the only output of interest; consequently, granule cell effects on mitral cell activity are entirely dependent on feedback. Reciprocal synapses between mitral cells (excitatory) and granule cells (inhibitory) mediate lateral inhibitory interactions between mitral cells and also support synchronized field oscillations across the OB as discussed below. However, these lateral inhibitory interactions do not play the classical contrast enhancement role in olfaction that they do in other modalities. First, the granule cell synapses mediating this inhibition are electrotonically very distant from the excitatory afferent driver currents in the distal apical dendrite, within the glomerulus, from which mitral cell spikes are initiated. As the efficacy of shunt inhibition is strongly dependent on location (Koch *et al.*, 1983; Liu, 2004; Mel and Schiller, 2004), lateral inhibitory interactions in the EPL are poorly situated to prevent mitral cell spike initiation. (In contrast, as discussed above, intraglomerular feedforward shunt inhibition, delivered directly to the location of OSN excitatory inputs by periglomerular spines, is ideally situated to regulate spike initiation.) Furthermore, of course, mitral-granule lateral inhibition is dependent on the prior initiation of spiking activity in mitral cells, and hence cannot be the source of the initial inhibitory response of mitral cells to odor presentation that precedes even the most rapid excitatory responses (Hamilton and Kauer, 1989; Kauer *et al.*, 1990; Wellis and Scott, 1990). These data indicate that mitral-granule interactions are not the source of the classical contrast enhancement processes that have been observed in mitral cell odor responses (Yokoi *et al.*, 1995); in this sense, the OB does not resemble the retina. Rather, as discussed above, this function appears to be mediated by glomerular circuitry.

Lateral inhibition among mitral cells in the EPL does, however, appear capable of influencing the timing of mitral cell spikes (Lagier *et al.*, 2004; Bathellier *et al.*, 2006; Lledo and Lagier, 2006; David *et al.*, 2008), a function essential to its role in the dynamical synchronization properties of the OB as discussed below. However, lateral inhibition in the OB does not imply a retinal, on-center/inhibitory surround architecture; odor stimulus qualities are still mapped high-dimensionally, and lateral inhibition that is localized within a two-dimensional neighborhood cannot effectively process higher-dimensional maps. Furthermore, the absence of parallel, feedforward inhibitory afferent inputs in the EPL precludes the inheritance of a topology of similarity from the external environment such as occurs in the glomerular layer, because there is no physical basis by which to map the inhibitory surround. In the glomerular layer, high-dimensional processing relies upon the feedforward interactions of one excitatory and one inhibitory process with the latter exhibiting greater sensitivity (or, equivalently, a broader receptive field at any given concentration) than the former. This information is not available to the EPL network. Hence, whereas the EPL processing layer is architecturally capable of mediating lateral inhibition in arbitrarily high dimensions – as required for useful olfactory representation and processing – it lacks a means to map this capability onto an externally defined topology of similarity. The clear implication is that these considerable computational resources instead mediate secondary transformations on olfactory representations based not on physical stimulus attributes but rather on internally defined combinatorial and psychological features not directly related to chemical structure, such as odor binding (Roskies, 1999; Treisman, 1999) and olfactory learning (reviewed in Wilson and Stevenson, 2006). Indeed, odor learning has profound effects on odor processing properties and even OB cellular architecture. Associative learning increases the prevalence of inhibitory responses by mitral cells to reinforced odor stimuli (Wilson and Leon, 1988), improves odor detection performance, and affects animals' capacities to discriminate related odorants (Fletcher and Wilson, 2002). Even non-motivated experience with odorants reduces and retunes mitral cell odor responses (Buonviso et al., 1998; Buonviso and Chaput, 2000; Fletcher and Wilson, 2003), a process that may be mediated by enhanced granule cell activity coupled with effects on N-methyl-D-aspartate (NMDA)-dependent synaptic connectivity (Lincoln et al., 1988; Brennan et al., 1990; Garcia et al., 1995). On a longer timescale, the survival of newly generated granule and periglomerular cells maturing within the OB also depends on olfactory experience (Rochefort et al., 2002; Mandairon et al., 2003), and the newly integrated neurons are distributed in an odor and task-specific manner reflecting olfactory learning (Alonso et al., 2006; Lledo et al., 2006; Mandairon et al., 2006a). Learned associations between disparate elements of odor representations may be essential for binding the structurally unrelated components of odors into unitary phenomena; if this is indeed one of the functions of the mitral–granule network in the EPL, then it is clearly beneficial that its computational topology is not constrained by the physically defined structural similarity of chemical odotopes that define that of the glomerular layer.

A separate problem from that of topology is, of course, mechanism. How are odor representations physically transformed by EPL computations? The emerging answer is that this second stage of olfactory stimulus processing operates in a dynamical regime, with modifications to the secondary olfactory representation mediated not by the generation or prevention of spiking but by the subtler regulation of spike timing. As with similarity mapping in the glomerular layer, physical proximity again appears to play little or no role in the topology of stimulus processing. Active membrane properties within mitral cell lateral dendrites appear to propagate excitation laterally without appreciable distance-dependent loss (Xiong and Chen, 2002; Debarbieux et al., 2003).

In contrast, the effects of shunt inhibition do not propagate and consequently remain distance-dependent (Lowe, 2002; David et al., 2008). Indeed, it has been hypothesized that lateral signal propagation in the EPL may depend solely on non-decrementing excitation along mitral cell lateral dendrites, whereas only those granule cell inputs nearly adjacent to the soma of any given mitral cell would deliver shunt inhibition with appreciable efficacy (Lowe, 2002; Willhite et al., 2006), so as, for example, to influence the timing of its spikes. This hypothesis is theoretically attractive for several reasons: not only does it remove the disruptive effects of physical proximity imposed by differentially located sources of shunt inhibition, it also resolves most of the combinatorial problem arising from bidirectional information transfer in mitral cell lateral dendrites, in which accumulating shunt inhibition interferes with the outgoing active propagation of excitation (Lowe, 2002). Finally, and most importantly, it suggests a columnar architecture for the organization of granule cells reflecting that of glomeruli and their associated mitral cells; specifically, those granule cells physically adjacent to a given mitral cell would be associated with delivering effects onto that mitral cell, whereas mitral cell output onto granule cells would be independent of proximity. Precisely this neighborhood-independent columnar architecture has been proposed based on recent imaging work, contingent on the hypothesis that the efficiency of retrograde transmission of pseudo-rabies virus tracer is activity dependent or otherwise related to synaptic efficacy (Willhite et al., 2006).

If this hypothesis is correct, then what might be the function of electrotonically distant inhibitory synapses on mitral cell lateral dendrites? Reciprocal synapses between mitral and granule cells are distributed along the full length of the mitral cell lateral dendrite. One possibility is that these reciprocal synapses contribute to maintaining the synchronicity of field oscillations across the OB, constituting a background coordinating process that is not directly relevant to odor stimulus processing. A more provocative hypothesis is that these inhibitory inputs may selectively attenuate the lateral propagation of excitation so as to regulate the pattern of associations among activated mitral cells (Lowe, 2002). This possibility exemplifies the capacity of the EPL network to process high-dimensional representations without reference to the physical similarities among stimuli on which glomerular layer processing depends. Rather, as noted above, EPL processing is likely to reflect bulbar learning, as suggested by the dependence of these synaptic interactions on NMDA receptors (Schoppa et al., 1998; Chen et al., 2000). Indeed, both of these hypotheses may have value; calcium imaging studies have demonstrated that relatively weak mitral cell activation produces activity in granule cells that is confined to the spine, presumably minimizing the degree of lateral inhibition compared to recurrent (self-) inhibition, whereas stronger activation excites broader regions of the granule cell dendritic tree (Egger et al., 2003, 2005). The dependence of the latter on T-type calcium currents further suggests a robust, all-or-none effect.

The genesis of spike timing and synchronization

Glomerular computations yield, at first approximation, an initial pattern of activation across the array of mitral cells that is sparser in space than the primary (glomerular) representation on which it depends, owing to periglomerular cell-mediated inhibition, but similarly structured in time. Mitral cells' activity is modulated by breathing (Yokoi et al., 1995; Buonviso et al., 2006; Roux et al., 2006) and they respond with a substantial latency (hundreds of milliseconds) to the presentation of odor stimuli. The roughly theta-band (in rats and mice) timescale of olfactory inputs is physically limited by the low-pass filtering properties of the inhalation cycle, the fluid mechanics of the nasal cavity, the time required for odorants to adsorb to and diffuse through the nasal mucus layer, the transduction and integration time within OSNs, and the inhibition delay imposed by the OSN–periglomerular–mitral synaptic triad. There is no known basis for afferent olfactory information to be encoded by OSNs on any faster timescale than this.

Cortical processing, in contrast, operates at a substantially faster timescale. Cortical spiking, while sparse, is believed to be tightly regulated in time. This temporal precision is essential for the coordination of spike timing among convergent inputs to enable critical computations such as heterosynaptic facilitation, long-term potentiation, and spike-timing dependent plasticity (Song et al., 2000; Cleland and Linster, 2002). In order to integrate into the cortical signaling network, primary olfactory representations must be transformed appropriately in timescale and sparseness, and in such a way that the perceptual information of interest is retained within the resulting secondary representation (mitral cell spike patterning). This transformation occurs largely within the OB, as mitral cell spiking is regulated on cortical timescales (Lagier et al., 2004; Lledo and Lagier, 2006) and their axons diverge to at least ten cortical and subcortical destinations (Cleland and Linster, 2003). The mechanisms underlying the transformation of the slow-timescale primary representation to the fast-timescale secondary representation appear to arise from a combination of endogenous OB properties and descending inputs, according to the following theoretical model.

Mitral cells are intrinsically resonant in the beta band, exhibiting intrinsic subthreshold oscillations at a frequency dependent upon the membrane potential (from 10 Hz at −67 mV to as high as 40 Hz at −59 mV). Inhibitory inputs to mitral cells reset the phase of these oscillations, whereas the effects of excitatory inputs are modulated by the oscillatory phase such that mitral cell spiking output is phase-constrained (Desmaisons et al., 1999; Rubin and Cleland, 2006). These properties are remarkably potent when framed in the general model of the OB synaptic triad presented above. Olfactory input, presented to odor receptors synchronously at the slower (theta-band) timescale of inhalation, evokes a potent inhibition in all mitral cells postsynaptic to activated OSNs, thereby synchronizing the phases of their intrinsic oscillations. A subset of these mitral cells – those sampling from the most strongly activated glomeruli – then overcome this inhibition and fire action potentials. The timing of these action potentials is phase-constrained by the intrinsic membrane properties of the mitral cells, all of which are transiently phase-locked by the preceding inhibitory pulse. The net effect is twofold: first, the initial ensemble of mitral cell spiking is regulated on a roughly beta-band timescale rather than the theta (or slower) band of the inhalation–exhalation cycle. Second, the distribution of spikes within the first resulting phase window is likely to be ordered by the intensity of activation of the mitral cells that generate them, essentially because stronger inputs yield shorter-latency spikes. Hence, at first approximation, the intensity of activation of each glomerulus is likely to be reflected not in the intensity of mitral cell activation, but in the timing of the first spikes evoked by the corresponding mitral cells (White et al., 1992; Hopfield, 1995; Cleland and Linster, 2002). Indeed, just this pattern is observable in mitral cell responses to increasing odorant concentrations: higher concentrations tend to evoke shorter spike latencies in cells that respond to a given odorant presentation with excitation, although this effect becomes conflated with other factors in extended concentration series and hence is not a universal rule. Furthermore, using timing-dependent synaptic facilitation and learning rules, many common neural computations such as contrast enhancement can be performed in this timing-dependent, dynamical regime (Linster and Cleland, 2001; Cleland and Linster, 2002) (Fig. 11.5). In sum, according to this working model, the initial stimulus-dependent response across the mitral cell population consists of a relatively sparse active ensemble that is dependent on odor quality but substantially independent of concentration, with the relative intensities of activation of neurons within this phase-constrained ensemble encoded by spike precedence.

Mitral–granule oscillogenesis

Sustained gamma-band oscillations within the OB are credited with constraining mitral cell spike times and maintaining a coordinated clock across the mitral cell ensemble (Kashiwadani et al., 1999). The origin and mechanisms

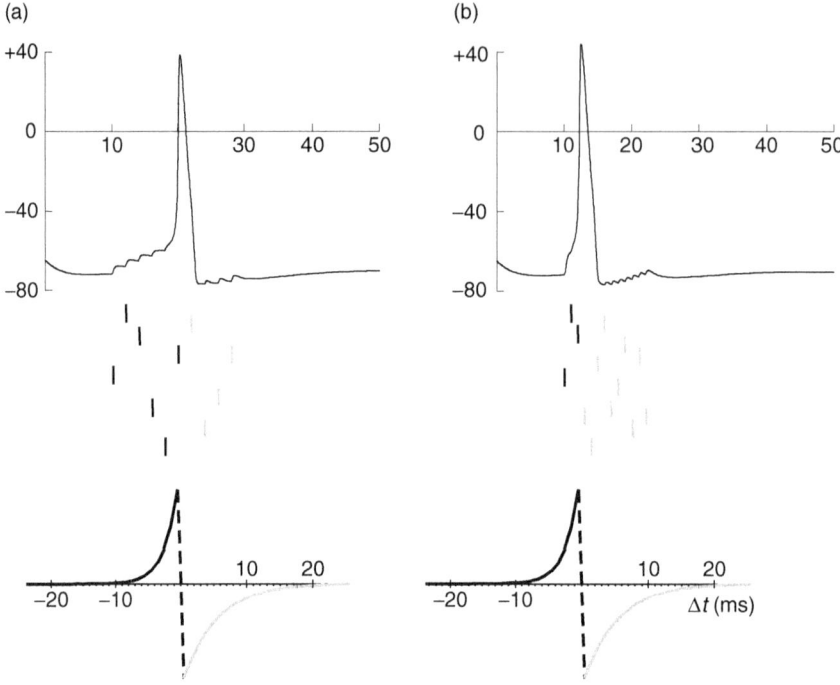

Figure 11.5 Illustration of a dynamical mechanism for contrast enhancement mediated by spike-timing-dependent plasticity (STDP) (Song et al., 2000). The upper panels are aligned in time with their respective lower panels, which depict the effect of the STDP learning rule: presynaptic spikes that precede the postsynaptic spike that they help evoke are strengthened (with the most strongly potentiated synapse being that delivering the presynaptic spike most immediately preceding the postsynaptic spike; black curve), whereas presynaptic spikes that follow the evoked postsynaptic spike are weakened (gray curve). (a) Response of a model cell activated by a weak stimulus. Six input spikes are accumulated before a spike is evoked in the model cell, such that six of the ten input synapses are strengthened to varying degrees (black raster marks) and four weakened (gray raster marks). (b) Response of a model cell activated by a stronger stimulus (in terms either of binding affinity or of concentration). Strong odor stimuli evoke oscillations of similar frequency but higher power (Stopfer et al., 2003), which has the effect of more tightly phase-constraining evoked spikes (Linster and Cleland, 2001; Cleland and Linster, 2002). The increased spike density (within the active phase of each cycle) evokes a postsynaptic spike after only three input spikes have accumulated, reducing the number of input synapses that are strengthened (black) and increasing the number that are weakened (gray). As the earlier spikes represent the responses of the neurons best tuned to the stimulus presented (see text), this STDP-mediated learning process enhances the best tuned input synapses and specifically targets the moderately tuned inputs for weakening, generating a dynamical variant of the half-hat function. The implication for olfactory representation is that mitral cell spikes may occur that do not contribute to the determination of downstream neural activity, hence differentiating *effective activity* from *ineffective activity* in the mitral cell ensemble. In this dynamical regime of effective activity, tighter synchrony mediates narrower receptive fields.

underlying these oscillations are complex and under investigation, but they certainly depend on reciprocal interactions between the lateral dendrites of mitral cells and the dendrites of inhibitory granule cells within the EPL (Schoppa, 2006a), and are closely coordinated with oscillatory processes in central olfactory structures such as the piriform cortex (Bressler et al., 1993; Kay and Freeman, 1998). (During certain phases of olfactory learning, these intrinsic gamma oscillations give way to centrally coordinated beta-band oscillations, reviewed by Lledo and Lagier (2006), which will not be discussed in detail here.) Several network models have been proposed to underlie OB intrinsic oscillatory properties. Generally, these models are based on an excitatory-inhibitory oscillator mechanism by virtue of the prominent mitral–granule reciprocal synapses present in the OB (Li and Hopfield, 1989; Eeckman and Freeman, 1990), although alternative models have been presented that emphasize different characteristic properties of the OB network. In particular, models based on mutual synaptic inhibition may more robustly exhibit the independence of oscillatory frequency from input intensity that is observed in the OB (Linster and Gervais, 1996), and are lent credence by evidence that disruption of GABA receptors specifically located on granule cells affects the oscillatory properties of the OB (Nusser et al., 2001) and by the recent description of a source for GABAergic inputs to granule cells (granule–granule synapses are experimentally contraindicated): a population of OB interneurons known as Blanes cells (Pressler and Strowbridge, 2006; Schoppa, 2006b). Alternatively, or additionally, GABAergic projections from the basal forebrain (Zaborszky et al., 1986) may play a role.

Despite the ongoing debate regarding dynamical mechanisms, it is clear that OB field oscillations are dependent on coordinated mitral–granule synaptic interactions, and that disruption of these interactions influences perception (e.g. Nusser et al., 2001) as first shown in the analogous insect system (Stopfer et al., 1997). However, coordinated network oscillations are slower to arise from inputs that are disorganized in time than from inputs that are all relatively synchronous. The phase resetting and spike-time constraining properties of the glomerular synaptic triad coupled with mitral cell intrinsic resonance properties appear to initiate OB oscillations efficiently and replicably, bypassing an indeterminate period of poorly coordinated activity in the OB network preceding its settlement into a synchronized state, and hence improving reaction times and preventing the loss of potentially critical sensory information. Just as the theta-band coordination among OSNs due to respiration is essential for priming the beta-band synchronization of the onset of mitral cell spiking, so is this transient beta-band coordination essential for the rapid transition to gamma-band spike synchronization among mitral cells.

Active sampling, neuromodulation, and natural scenes

Deeper aspects of neural representations also can feed back onto sensory processing. Centrifugal neuromodulation of the OB – presumably mediating attention, motivation, and other aspects of behavioral state – is a prominent regulator of olfactory perception (Linster and Cleland, 2002; Yuan et al., 2003; Mandairon et al., 2006b), and OB gamma-band oscillations are coordinated with those of other cortices and also depend on behavioral state (Kay and Freeman, 1998; Kay and Laurent, 1999; Kay, 2003; Ravel et al., 2003; Martin et al., 2004, 2006). Furthermore, odor-sampling behavior itself (sniffing) is behaviorally modulated; some effects of sniffing behavior on odor representations may optimize sampling based on anatomical specializations (Schoenfeld and Cleland, 2006), while others appear more temporally sophisticated, suggesting that a more detailed exploration of the interactions among timescales in this complex system is warranted. In particular, repeated active sniffing generates a non-stationary series of odor representations (Spors et al., 2006), potentially providing additional mechanisms for odor discrimination or identification.

The final frontier in olfactory stimulus processing is odor segmentation, or the identification of a relevant odor despite its conflation with other environmental odorants. There is no a priori reason why a particular combination of odotopes should be treated as a single stimulus, much less when the characteristic odotope ratios that might identify an odorant are degraded by the simultaneous presence of other odorants incorporating odotopes that compete for the same ORs. While efforts have been made to create feedforward models of odor segmentation, these have by and large been unenlightening. It is likely that the solution to mixture processing and odor segmentation relies critically on olfactory memory to identify and segregate learned representations from the olfactory milieu, though the mechanisms that may underlie such a process are far from clear. Study of the properties of tertiary olfactory representations in the piriform cortex (Illig and Haberly, 2003; Kadohisa and Wilson, 2006; Linster et al., 2007) and other postbulbar structures (Haberly, 2001; Lei et al., 2006), as well as of OB interactions with putative associative memory networks (Haberly and Bower, 1989; Barkai et al., 1994), will be necessary in order to shed light on the neural mechanisms underlying segmentation and the putative OB contribution to the process.

Summary: the two-stage cascade hypothesis of olfactory bulb stimulus processing

Sensory processing can be understood as a cascade of sequential representations, each with characteristic properties and mechanisms that contribute to the construction of the subsequent representation. As described herein, two sequential odor representations and two stages of processing are contained within the OB; subsequently, the great divergence of mitral/tufted cell axons implies that on the order of ten separate – and potentially quite different – tertiary representations will be constructed in different regions of the brain (Cleland and Linster, 2003), although among these by far the greatest attention has been paid to the piriform cortex. A consistent model of OB processing and information content is hence an important prerequisite to the considerably larger problem of higher-order sensory processing across the brain. The following model is proposed as a framework for understanding and quantifying OB computational processing.

The primary olfactory representation is constructed across the OSN population – the primary olfactory sensory neurons – and reflected in the glomerular maps across the OB input layer that are dominated by OSN axonal arbors. Several cellular and network specializations facilitate the consistent sampling of the odor environment – mitigating intensity differences, emphasizing environmental changes – and set the stage for subsequent processing. The resulting primary representation is based on a highly redundant rate code (in other words, a slow timescale) and hence is energetically costly. As with primary sensory neurons in general, this is probably a necessary adaptation to its function, in which unpredictable time-varying stimulus properties are likely to dominate any attempt at fast-timescale temporal control, but it also introduces the need to transform the representation into a more energetically efficient representation compatible with other cortical signaling rules.

The secondary olfactory representation – the patterned spiking output of the OB – is a remarkable transformation of the primary representation. It is sparser in its use of spikes, temporally precise on a markedly faster timescale, robust to the disruptive effects of concentration variance, and integrates afferent sensory processing with centrifugal descending and neuromodulatory influences. The construction of the secondary representation is mediated by intricate neural computations that can be grouped into two stages with markedly different features. The first stage takes place in the glomerular layer. It operates directly on OSN output activity and hence is characterized by effects on a slower timescale (theta-band or slower) consistent with OSN rate coding. Effective inhibition implies the prevention of spiking in mitral/tufted cells, and effective activity is judged largely by *whether* spikes are evoked (equivalently, *when* on a slow timescale). This first stage of processing is feedforward, operates in a topology of similarity defined by the external chemical environment and mediated by the overlap in OR receptive fields, and specifically includes computations that rely upon the grouping or differentiation of structurally similar stimulus features. The second stage is mediated in the EPL and is based on dynamical interactions between mitral and granule cells. It operates on mitral cell activation profiles and acts to influence activity in those same mitral cells, and hence is based on feedback. Owing to mitral cell intrinsic resonance properties that elevate the timescale to beta-band and mitral–granule synaptic dynamical properties that further elevate it to gamma-band, the timescale of the secondary representation is considerably faster than that of the primary representation, while spiking is rendered concomitantly sparser, in part due to the normalization of concentration effects. Effective inhibition implies the delay of mitral cell spiking, and effective mitral cell activity is judged by *when* – on a millisecond timescale – spikes are evoked. EPL processing is potentially high-dimensional, as is glomerular processing, but owing to its feedback architecture the EPL topology is isolated from the externally defined similarity space that defines glomerular processing. Consequently, it remains free to support arbitrary associations among OB columns, which appear to reflect odor experience and olfactory learning and may contribute to complex processing such as the binding of multiple features into unitary odor percepts. In sum, the OB is a sophisticated sensory parsing engine, transforming sensory representations in form, bias, and context so that they can be usefully integrated into the global operations of the central nervous system and thereby serve the needs of the organism.

References

Alonso, M., Viollet, C., Gabellec, M. M., et al. (2006). Olfactory discrimination learning increases the survival of adult-born neurons in the olfactory bulb. *J Neurosci* **26**:10508–10513.

Araneda, R. C., Peterlin, Z., Zhang, X., Chesler, A., and Firestein, S. (2004). A pharmacological profile of the aldehyde receptor repertoire in rat olfactory epithelium. *J Physiol* **555**:743–756.

Attwell, D. and Laughlin, S. B. (2001). An energy budget for signaling in the grey matter of the brain. *J Cereb Blood Flow Metab* **21**:1133–1145.

Aungst, J. L., Heyward, P. M., Puche, A. C., et al. (2003). Centre-surround inhibition among olfactory bulb glomeruli. *Nature* **426**:623–629.

Barkai, E., Bergman, R. E., Horwitz, G., and Hasselmo, M. E. (1994). Modulation of associative memory function in a biophysical simulation of rat piriform cortex. *J Neurophysiol* **72**:659–677.

Bathellier, B., Lagier, S., Faure, P., and Lledo, P. M. (2006). Circuit properties generating gamma oscillations in a network model of the olfactory bulb. *J Neurophysiol* **95**:2678–2691.

Brennan, P., Kaba, H., and Keverne, E. B. (1990). Olfactory recognition: a simple memory system. *Science* **250**:1223–1226.

Bressler, S. L., Coppola, R., and Nakamura, R. (1993). Episodic multiregional cortical coherence at multiple frequencies during visual task performance. *Nature* **366**:153–156.

Buck, L. and Axel, R. (1991). A novel multigene family may encode odorant receptors: a molecular basis for odor recognition. *Cell* **65**:175–187.

Buonviso, N. and Chaput, M. (2000). Olfactory experience decreases responsiveness of the olfactory bulb in the adult rat. *Neuroscience* **95**:325–332.

Buonviso, N., Gervais, R., Chalansonnet, M., and Chaput, M. (1998). Short-lasting exposure to one odour decreases general reactivity in the olfactory bulb of adult rats. *Eur J Neurosci* **10**:2472–2475.

Buonviso, N., Amat, C., and Litaudon, P. (2006). Respiratory modulation of olfactory neurons in the rodent brain. *Chem Senses* **31**:145–154.

Chalansonnet, M. and Chaput, M. A. (1998). Olfactory bulb output cell temporal response patterns to increasing odor concentrations in freely breathing rats. *Chem Senses* **23**:1–9.

Chen, W. R., Xiong, W., and Shepherd, G. M. (2000). Analysis of relations between NMDA receptors and GABA release at olfactory bulb reciprocal synapses. *Neuron* **25**:625–633.

Chess, A., Simon, I., Cedar, H., and Axel, R. (1994). Allelic inactivation regulates olfactory receptor gene expression. *Cell* **78**:823–834.

Cleland, T. A. and Linster, C. (1999). Concentration tuning mediated by spare receptor capacity in olfactory sensory neurons: a theoretical study. *Neur Comput* **11**:1673–1690.

Cleland, T. A. and Linster, C. (2002). How synchronization properties among second-order sensory neurons can mediate stimulus salience. *Behav Neurosci* **116**:212–221.

Cleland, T. A. and Linster, C. (2003). Central olfactory processing. In: *Handbook of Olfaction and Gustation*, 2nd edn, ed. Doty R. L., pp. 165–180. New York: Marcel Dekker.

Cleland, T. A. and Linster, C. (2005). Computation in the olfactory system. *Chem Senses* **30**:801–813.

Cleland, T. A. and Sethupathy, P. (2006). Non-topographical contrast enhancement in the olfactory bulb. *BMC Neurosci* **7**:7.

Cleland, T. A., Johnson, B. A., Leon, M., and Linster, C. (2007). Relational representation in the olfactory system. *Proc Natl Acad Sci USA* **104**:1953–1958.

David, F., Linster, C. and Cleland, T. A. (2008). Lateral dendritic shunt inhibition can regularize mitral cell spike patterning. *J Comput Neurosci* PMID:18060489.

Debarbieux, F., Audinat, E., and Charpak, S. (2003). Action potential propagation in dendrites of rat mitral cells in vivo. *J Neurosci* **23**:5553–5560.

Desmaisons, D., Vincent, J. D., and Lledo, P. M. (1999). Control of action potential timing by intrinsic subthreshold oscillations in olfactory bulb output neurons. *J Neurosci* **19**:10727–10737.

Duchamp-Viret, P., Duchamp, A., and Vigouroux, M. (1989). Amplifying role of convergence in olfactory system a comparative study of receptor cell and second-order neuron sensitivities. *J Neurophysiol* **61**:1085–1094.

Eeckman, F. H. and Freeman, W. J. (1990). Correlations between unit firing and EEG in the rat olfactory system. *Brain Res* **528**:238–244.

Egger, V., Svoboda, K., and Mainen, Z. F. (2003). Mechanisms of lateral inhibition in the olfactory bulb: efficiency and modulation of spike-evoked calcium influx into granule cells. *J Neurosci* **23**:7551–7558.

Egger, V., Svoboda, K., and Mainen, Z. F. (2005). Dendrodendritic synaptic signals in olfactory bulb granule cells: local spine boost and global low-threshold spike. *J Neurosci* **25**:3521–3530.

Fletcher, M. L. and Wilson, D. A. (2002). Experience modifies olfactory acuity: acetylcholine-dependent learning decreases behavioral generalization between similar odorants. *J Neurosci* **22**:RC201.

Fletcher, M. L. and Wilson, D. A. (2003). Olfactory bulb mitral-tufted cell plasticity: odorant-specific tuning reflects previous odorant exposure. *J Neurosci* **23**:6946–6955.

Garcia, Y., Ibarra, C., and Jaffe, E. H. (1995). NMDA and non-NMDA receptor-mediated release of [^{3}H]GABA from granule cell dendrites of rat olfactory bulb. *J Neurochem* **64**:662–669.

Getchell, T. V. (1986). Functional properties of vertebrate olfactory receptor neurons. *Physiol Rev* **66**:772–818.

Getchell, T. V. and Shepherd, G. M. (1978). Responses of olfactory receptor cells to step pulses of odour at different concentrations in the salamander. *J Physiol* **282**:521–540.

Goldman, A. L., Van der Goes van Naters, W., Lessing, D., Warr, C. G., and Carlson, J. R. (2005). Coexpression of two functional odor receptors in one neuron. *Neuron* **45**:661–666.

Haberly, L. B. (2001). Parallel-distributed processing in olfactory cortex: new insights from morphological and physiological analysis of neuronal circuitry. *Chem Senses* **26**:551–576.

Haberly, L. B. and Bower, J. M. (1989). Olfactory cortex: model circuit for study of associative memory? *Trends Neurosci* **12**:258–264.

Hamilton, K. A. and Kauer, J. S. (1989). Patterns of intracellular potentials in salamander mitral/tufted cells in response to odor stimulation. *J Neurophysiol* **62**:609–625.

Hopfield, J. J. (1995). Pattern recognition computation using action potential timing for stimulus representation. *Nature* **376**:33–36.

Illig, K. R. and Haberly, L. B. (2003). Odor-evoked activity is spatially distributed in piriform cortex. *J Comp Neurol* **457**:361–373.

Ishii, T., Serizawa, S., Kohda, A., et al. (2001). Monoallelic expression of the odourant receptor gene and axonal projection of olfactory sensory neurones. *Genes Cells* **6**:71–78.

Kadohisa, M. and Wilson, D. A. (2006). Olfactory cortical adaptation facilitates detection of odors against background. *J Neurophysiol* **95**:1888–1896.

Kashiwadani, H., Sasaki, Y. F., Uchida, N., and Mori, K. (1999). Synchronized oscillatory discharges of mitral/tufted cells with different molecular receptive ranges in the rabbit olfactory bulb. *J Neurophysiol* **82**:1786–1792.

Kauer, J.S., Hamilton, K.A., Neff, S.R., and Cinelli, A.R. (1990). Temporal patterns of membrane potential in the olfactory bulb observed with intracellular recording and voltage-sensitive dye imaging: early hyperpolarization. In: *Chemosensory Information Processing*, ed. Schild, D., pp. 305–314. Berlin: Springer-Verlag.

Kaur, R., Zhu, X.O., Moorhouse, A.J., and Barry, P.H. (2001). IP3-gated channels and their occurrence relative to CNG channels in the soma and dendritic knob of rat olfactory receptor neurons. *J Membr Biol* **181**:91–105.

Kay, L.M. (2003). Two species of gamma oscillations in the olfactory bulb: dependence on behavioral state and synaptic interactions. *J Integr Neurosci* **2**:31–44.

Kay, L.M. and Freeman, W.J. (1998). Bidirectional processing in the olfactory–limbic axis during olfactory behavior. *Behav Neurosci* **112**:541–553.

Kay, L.M. and Laurent, G. (1999). Odor- and context-dependent modulation of mitral cell activity in behaving rats. *Nat Neurosci* **2**:1003–1009.

Koch, C., Poggio, T., and Torre, V. (1983). Nonlinear interactions in a dendritic tree: localization, timing, and role in information processing. *Proc Natl Acad Sci USA* **80**:2799–2802.

Kohonen, T. (1982). Self-organized formation of topology correct feature maps. *Biol Cybernet* **43**:59–69.

Kohonen, T. and Hari, R. (1999). Where the abstract feature maps of the brain might come from. *Trends Neurosci* **22**:135–139.

Lagier, S., Carleton, A., and Lledo, P.M. (2004). Interplay between local GABAergic interneurons and relay neurons generates gamma oscillations in the rat olfactory bulb. *J Neurosci* **24**:4382–4392.

Lei, H., Mooney, R., and Katz, L.C. (2006). Synaptic integration of olfactory information in mouse anterior olfactory nucleus. *J Neurosci* **26**:12023–12032.

Li, Z. and Hopfield, J.J. (1989). Modeling the olfactory bulb and its neural oscillatory processings. *Biol Cybernet* **61**:379–392.

Lincoln, J., Coopersmith, R., Harris, E.W., Cotman, C.W., and Leon, M. (1988). NMDA receptor activation and early olfactory learning. *Brain Res* **467**:309–312.

Linster, C. and Cleland, T.A. (2001). How spike synchronization among olfactory neurons can contribute to sensory discrimination. *J Comput Neurosci* **10**:187–193.

Linster, C. and Cleland, T.A. (2002). Cholinergic modulation of sensory representations in the olfactory bulb. *Neur Networks* **15**:709–717.

Linster, C. and Gervais, R. (1996). Investigation of the role of interneurons and their modulation by centrifugal fibers in a neural model of the olfactory bulb. *J Comput Neurosci* **3**:225–246.

Linster, C., Henry, L., Kadohisa, M., and Wilson, D.A. (2007). Synaptic adaptation and odor-background segmentation. *Neurobiol Learn Mem* **87**:352–360.

Liu, G. (2004). Local structural balance and functional interaction of excitatory and inhibitory synapses in hippocampal dendrites. *Nat Neurosci* **7**:373–379.

Lledo, P.M. and Lagier, S. (2006). Adjusting neurophysiological computations in the adult olfactory bulb. *Semin Cell Dev Biol* **17**:443–453.

Lledo, P.M., Alonso, M., and Grubb, M.S. (2006). Adult neurogenesis and functional plasticity in neuronal circuits. *Nat Rev Neurosci* **7**:179–193.

Lowe, G. (2002). Inhibition of backpropagating action potentials in mitral cell secondary dendrites. *J Neurophysiol* **88**:64–85.

Mandairon, N., Jourdan, F., and Didier, A. (2003). Deprivation of sensory inputs to the olfactory bulb up-regulates cell death and proliferation in the subventricular zone of adult mice. *Neuroscience* **119**:507–516.

Mandairon, N., Sacquet, J., Jourdan, F., and Didier, A. (2006a). Long-term fate and distribution of newborn cells in the adult mouse olfactory bulb: influences of olfactory deprivation. *Neuroscience* **141**:443–451.

Mandairon, N., Ferretti, C. J., Stack, C. M., *et al.* (2006b). Cholinergic modulation in the olfactory bulb influences spontaneous olfactory discrimination in adult rats. *Eur J Neurosci* **24**:3234–3244.

Martin, C., Gervais, R., Hugues, E., Messaoudi, B., and Ravel, N. (2004). Learning modulation of odor-induced oscillatory responses in the rat olfactory bulb: a correlate of odor recognition? *J Neurosci* **24**:389–397.

Martin, C., Gervais, R., Messaoudi, B., and Ravel, N. (2006). Learning-induced oscillatory activities correlated to odour recognition: a network activity. *Eur J Neurosci* **23**:1801–1810.

McQuiston, A. R. and Katz, L. C. (2001). Electrophysiology of interneurons in the glomerular layer of the rat olfactory bulb. *J Neurophysiol* **86**:1899–1907.

Mel, B. W. and Schiller, J. (2004). On the fight between excitation and inhibition: location is everything. *Sci STKE* 2004:PE44.

Meyer, M. R., Angele, A., Kremmer, E., Kaupp, U. B., and Muller, F. (2000). A cGMP-signaling pathway in a subset of olfactory sensory neurons. *Proc Natl Acad Sci USA* **97**:10595–10600.

Mombaerts, P. (1999). Seven-transmembrane proteins as odorant and chemosensory receptors. *Science* **286**:707–711.

Mombaerts, P. (2001). The human repertoire of odorant receptor genes and pseudogenes. *Annu Rev Genomics Hum Genet* **2**:493–510.

Mombaerts, P. (2004). Odorant receptor gene choice in olfactory sensory neurons: the one receptor-one neuron hypothesis revisited. *Curr Opin Neurobiol* **14**:31–36.

Mori, K. and Shepherd, G. M. (1994). Emerging principles of molecular signal processing by mitral/tufted cells in the olfactory bulb. *Semin Cell Biol* **5**:65–74.

Morrison, E. E. and Costanzo, R. M. (1990). Morphology of the human olfactory epithelium. *J Comp Neurol* **297**:1–13.

Morrison, E. E. and Costanzo, R. M. (1992). Morphology of olfactory epithelium in humans and other vertebrates. *Microsc Res Tech* **23**:49–61.

Mouly, A. M., Gervais, R., and Holley, A. (1990). Evidence for the involvement of rat olfactory bulb in processes supporting long-term olfactory memory. *Eur J Neurosci* **2**:978–984.

Mouly, A. M., Kindermann, U., Gervais, R., and Holley, A. (1993). Involvement of the olfactory bulb in consolidation processes associated with long-term memory in rats. *Behav Neurosci* **107**:451–457.

Nusser, Z., Kay, L. M., Laurent, G., Homanics, G. E., and Mody, I. (2001). Disruption of GABA(A) receptors on GABAergic interneurons leads to increased oscillatory power in the olfactory bulb network. *J Neurophysiol* **86**:2823–2833.

Oka, Y., Omura, M., Kataoka, H., and Touhara, K. (2004). Olfactory receptor antagonism between odorants. *EMBO J* **23**:120–126.

Pinching, A. J. and Powell, T. P. (1971). The neuropil of the glomeruli of the olfactory bulb. *J Cell Sci* **9**:347–377.

Pressler, R. T. and Strowbridge, B. W. (2006). Blanes cells mediate persistent feedforward inhibition onto granule cells in the olfactory bulb. *Neuron* **49**:889–904.

Ravel, N., Chabaud, P., Martin, C., *et al.* (2003). Olfactory learning modifies the expression of odour-induced oscillatory responses in the gamma (60–90 Hz) and beta (15–40 Hz) bands in the rat olfactory bulb. *Eur J Neurosci* **17**:350–358.

Restrepo, D., Miyamoto, T., Bryant, B. P., and Teeter, J. H. (1990). Odor stimuli trigger influx of calcium into olfactory neurons of the channel catfish. *Science* **249**:1166–1168.

Rochefort, C., Gheusi, G., Vincent, J. D., and Lledo, P. M. (2002). Enriched odor exposure increases the number of newborn neurons in the adult olfactory bulb and improves odor memory. *J Neurosci* **22**:2679–2689.

Roskies, A. L. (1999). The binding problem. *Neuron* **24**:7–9.

Rospars, J. P., Lansky, P., Duchamp-Viret, P., and Duchamp, A. (2000). Spiking frequency versus odorant concentration in olfactory receptor neurons. *Biosystems* **58**:133–141.

Roux, S. G., Garcia, S., Bertrand, B., *et al.* (2006). Respiratory cycle as time basis: an improved method for averaging olfactory neural events. *J Neurosci Methods* **152**:173–178.

Rubin, D. B. and Cleland, T. A. (2006). Dynamical mechanisms of odor processing in olfactory bulb mitral cells. *J Neurophysiol* **96**:555–568.

Schoenfeld, T. A. and Cleland, T. A. (2005). The anatomical logic of smell. *Trends Neurosci* **28**:620–627.

Schoenfeld, T. A. and Cleland, T. A. (2006). Anatomical contributions to odorant sampling and representation in rodents: zoning in on sniffing behavior. *Chem Senses* **31**:131–144.

Schoppa, N. E. (2006a). Synchronization of olfactory bulb mitral cells by precisely timed inhibitory inputs. *Neuron* **49**:271–283.

Schoppa, N. E. (2006b). A novel local circuit in the olfactory bulb involving an old short-axon cell. *Neuron* **49**:783–784.

Schoppa, N. E., Kinzie, J. M., Sahara, Y., Segerson, T. P., and Westbrook, G. L. (1998). Dendrodendritic inhibition in the olfactory bulb is driven by NMDA receptors. *J Neurosci* **18**:6790–6802.

Serizawa, S., Ishii, T., Nakatani, H., *et al.* (2000). Mutually exclusive expression of odorant receptor transgenes. *Nat Neurosci* **3**:687–693.

Song, S., Miller, K. D., and Abbott, L. F. (2000). Competitive Hebbian learning through spike-timing-dependent synaptic plasticity. *Nat Neurosci* **3**:919–926.

Spors, H., Wachowiak, M., Cohen, L. B., and Friedrich, R. W. (2006). Temporal dynamics and latency patterns of receptor neuron input to the olfactory bulb. *J Neurosci* **26**:1247–1259.

Stopfer, M., Bhagavan, S., Smith, B. H., and Laurent, G. (1997). Impaired odour discrimination on desynchronization of odour-encoding neural assemblies. *Nature* **390**:70–74.

Stopfer, M., Jayaraman, V., and Laurent, G. (2003). Intensity versus identity coding in an olfactory system. *Neuron* **39**:991–1004.

Strotmann, J., Conzelmann, S., Beck, A., *et al.* (2000). Local permutations in the glomerular array of the mouse olfactory bulb. *J Neurosci* **20**:6927–6938.

Tozaki, H., Tanaka, S., and Hirata, T. (2004). Theoretical consideration of olfactory axon projection with an activity-dependent neural network model. *Mol Cell Neurosci* **26**:503–517.

Treisman, A. (1999). Solutions to the binding problem: progress through controversy and convergence. *Neuron* **24**:105–110.

Troemel, E. R., Chou, J. H., Dwyer, N. D., Colbert, H. A. and Bargmann, C. I. (1995). Divergent seven transmembrane receptors are candidate chemosensory receptors in *C. elegans*. *Cell* **83**:207–218.

van Drongelen, W., Holley, A., and Doving, K. B. (1978). Convergence in the olfactory system: quantitative aspects of odour sensitivity. *J Theor Biol* **71**:39–48.

Wellis, D. P. and Scott, J. W. (1990). Intracellular responses of identified rat olfactory bulb interneurons to electrical and odor stimulation. *J Neurophysiol* **64**:932–947.

White, J., Hamilton, K. A., Neff, S. R., and Kauer, J. S. (1992). Emergent properties of odor information coding in a representational model of the salamander olfactory bulb. *J Neurosci* **12**:1772–1780.

Willhite, D. C., Nguyen, K. T., Masurkar, A. V., *et al.* (2006). Viral tracing identifies distributed columnar organization in the olfactory bulb. *Proc Natl Acad Sci USA* **103**:12592–12597.

Wilson, D. A. and Leon, M. (1988). Spatial patterns of olfactory bulb single-unit responses to learned olfactory cues in young rats. *J Neurophysiol* **59**:1770–1782.

Wilson, D. A. and Stevenson, R. J. (2006). *Learning to Smell: Olfactory Perception from Neurobiology to Behavior*. Baltimore, MD: Johns Hopkins University Press.

Xiong, W. and Chen, W. R. (2002). Dynamic gating of spike propagation in the mitral cell lateral dendrites. *Neuron* **34**:115–126.

Yokoi, M., Mori, K., and Nakanishi, S. (1995). Refinement of odor molecule tuning by dendrodendritic synaptic inhibition in the olfactory bulb. *Proc Natl Acad Sci USA* **92**:3371–3375.

Yuan, Q., Harley, C. W., and McLean, J. H. (2003). Mitral cell beta1 and 5-HT2A receptor colocalization and cAMP coregulation: a new model of norepinephrine-induced learning in the olfactory bulb. *Learn Mem* **10**:5–15.

Zaborszky, L., Carlsen, J., Brashear, H. R., and Heimer, L. (1986). Cholinergic and GABAergic afferents to the olfactory bulb in the rat with special emphasis on the projection neurons in the nucleus of the horizontal limb of the diagonal band. *J Comp Neurol* **243**:488–509.

Zufall, F. and Leinders-Zufall, T. (2000). The cellular and molecular basis of odor adaptation. *Chem Senses* **25**:473–481.

Part IV FUNCTIONAL INTEGRATION OF DIFFERENT BRAIN AREAS IN INFORMATION PROCESSING AND PLASTICITY

12

Anatomical, physiological, and pharmacological properties underlying hippocampal sensorimotor integration

BRIAN H. BLAND

The cellular basis of theta-band oscillation and synchrony

The limbic cortex represents multiple synchronizing systems (Bland and Colom, 1993). Populations of cells in these structures display membrane potential oscillations as a result of intrinsic properties of membrane currents. These cells also receive inputs from other cells in the same structure and inputs from cells extrinsic to the structure, many of the latter from nuclei contributing to the ascending brainstem hippocampal synchronizing pathways. Theta-band oscillation and synchrony in the hippocampal formation (HPC) and related limbic structures is recorded as an extracellular field potential consisting of a sinusoidal-like waveform with an amplitude up to 2 mV and a narrow band frequency range of 3–12 Hz in mammals. The asynchronous activity termed large-amplitude irregular activity (LIA) is an irregular waveform with a broadband frequency range of 0.5–25 Hz (Leung et al., 1982). Kramis et al. (1975) were the first formally to propose the existence of two types of hippocampal theta activity, in both the rabbit and the rat (see review by Bland, 1986). One type was termed atropine-sensitive theta, since it were abolished by the administration of atropine sulfate. Atropine-sensitive theta occurred during immobility in rabbits in the normal state and occurred in both rabbits and rats during immobility produced by ethyl ether or urethane treatment. The other type of theta was termed atropine-resistant, since it was not sensitive to treatment with

Information Processing by Neuronal Populations, ed. Christian Hölscher and Matthias Munk. Published by Cambridge University Press. © Cambridge University Press 2009.

atropine sulfate but was abolished by anesthetics. Atropine-resistant theta was the movement-related theta originally described by Vanderwolf (1969) and is thought to be sensitive to serotonin antagonists (Vanderwolf and Baker, 1986). Since atropine-sensitive theta became operationally defined as theta occurring during one example of a Type 2 behavior, i.e. immobility, it became known as Type 2 (immobility-related) theta. Atropine-resistant theta became known as Type 1 theta, since it occurred during Type 1 (voluntary) motor behaviors, such as walking, rearing, and postural adjustments (Vanderwolf, 1975). Type 2 theta thus originally became operationally defined as theta that occurred in the complete absence of movement. The sensorimotor integration model assumes that Type 2 theta in the HPC is the electrical sign of processing sensory stimuli that are relevant to the initiation and maintenance of voluntary motor behaviors and that it is always coincidently active whenever the Type 1 theta subsystem is active. Thus, an animal may be immobile and generate Type 2 theta in isolation, but whenever Type 1 movements occur, Type 2 theta is always occurring coincidently. As discussed below, there is evidence for a septohippocampal glutamatergic system being involved in theta generation and behavior.

Many populations of cells in the HPC and related structures exhibit discharge properties that are precisely related to hippocampal theta field activity. Such theta-related cells comprise two distinct populations termed theta-ON and theta-OFF, first described in acute preparations by Colom et al. (1987), followed by a detailed cell classification paper by Colom and Bland (1987) and subsequently used to classify theta-related cells in the HPC in a number of studies (McNaughton et al., 1983; Bland and Colom, 1988, 1989; Mizimori et al., 1990; Colom et al., 1991; Smythe et al., 1991; Konopacki et al., 1992, 2006; Bland et al., 1996). Theta-ON and theta-OFF cells have also been recorded in the medial septal nucleus and nucleus of the diagonal band of Broca (MS/vDBB) (Ford et al., 1989; Bland et al., 1990, 1994; Colom and Bland, 1991), the entorhinal cortex (Dickson et al., 1994, 1995), cingulate cortex (Colom et al., 1988), caudal diencephalon (Bland et al., 1995; Kirk et al., 1996), median raphe nucleus (Viana Di Prisco et al., 2002), rostral pontine region (Hanada et al., 1999), the superior colliculus (Natsume et al., 1999), the basal ganglia (Hallworth and Bland, 2004), the red nucleus (Dypvik and Bland, 2004), and the neocortex (Lukatch and MacIver, 1997). Theta-ON cells increase their activity during theta field activity as reflected by an overall mean increase in discharge rate or as a linear positive increase in discharge rate in relation to increasing frequencies of simultaneously recorded theta field activity. Theta-OFF cells decrease their activity during theta field activity as reflected by an overall mean decrease in discharge rate (to zero in many cases) or by a linear negative increase in discharge rate as theta field frequency declines. A further criterion relates to the pattern of cell discharges.

A given theta-related cell discharges in one of two characteristic patterns during theta field activity. The first is a rhythmic bursting pattern we have termed phasic, since each cell burst occurs with a consistent phase relation to each cycle of theta field activity. The second pattern is either regular or irregular discharges we have termed tonic, since they consist of a non-bursting discharge pattern with no consistent phase relation to theta field activity. Both theta-ON and theta-OFF cells have phasic and tonic subtypes. The rhythmic discharges of all phasic theta-ON cells during Type 2 theta was abolished by the administration of atropine sulfate, and are therefore mediated by acetylcholine (ACh). The inhibition of all phasic theta-OFF cells during Type 2 theta was resistant to the administration of atropine sulfate. Theta-ON and theta-OFF cells represent a general organization of the cellular mechanisms underlying "theta-band" oscillation and synchrony in the HPC and related structures.

Theta-band oscillations may also be recorded intracellularly in some populations of cells in the HPC, during the simultaneous occurrence of the extracellular theta field oscillations. In agreement with previous work (Leung and Yim, 1991) we adopted the term membrane potential oscillations (MPOs) to designate the slow intracellular oscillations that occur at theta frequencies in subsets of hippocampal cells. The occurrence of MPOs in HPC pyramidal cells, dentate granule cells, and interneurons has been well documented (Fujita and Sato, 1964; Fox et al., 1983; Leung and Yim, 1986, 1988, 1991; Nunez et al., 1987, 1990a, 1990b, 1990c; Bland et al., 1988, 2002, 2005; Fox, 1989; MacVicar and Tse, 1989; Munoz et al., 1990; Konopacki et al., 1992; Ylinen et al., 1995; Leung and Yu, 1998; Chapman and LaCaille, 1999). The morphological identity of theta-ON and theta-OFF cells is crucial to the understanding of the cellular interactions involved in the generation of theta field activity. Earlier studies by Fox and Ranck (1975, 1981) provided indirect evidence that theta cells (theta-ON cells in our scheme) in the HPC were interneurons, a view supported by some recent work on identified cells (see review by Buzsáki, 2002). On the other hand, Bland and Colom (reviewed in Bland and Colom, 1993) proposed, also based on indirect evidence, that a subpopulation of HPC projection cells (pyramidal and granule cells) were theta-ON cells and a subpopulation of HPC interneurons were theta-OFF cells. More recent work by Bland et al. (2002) provided evidence supporting the following conclusions concerning cells classified as phasic theta-ON cells: (1) morphologically identified hippocampal CA1 (see Fig. 12.1) and CA3 pyramidal cells represent a subset of cells meeting the criteria for classification as phasic theta-ON cells, supporting the findings of a number of previous studies (Fujita and Sato, 1964; Nunez et al., 1987, 1990a; Leung and Yim, 1988, 1991); (2) MPOs occurred only during theta field activity, their onset signaled by a 5–10-mV depolarizing shift in membrane potential;

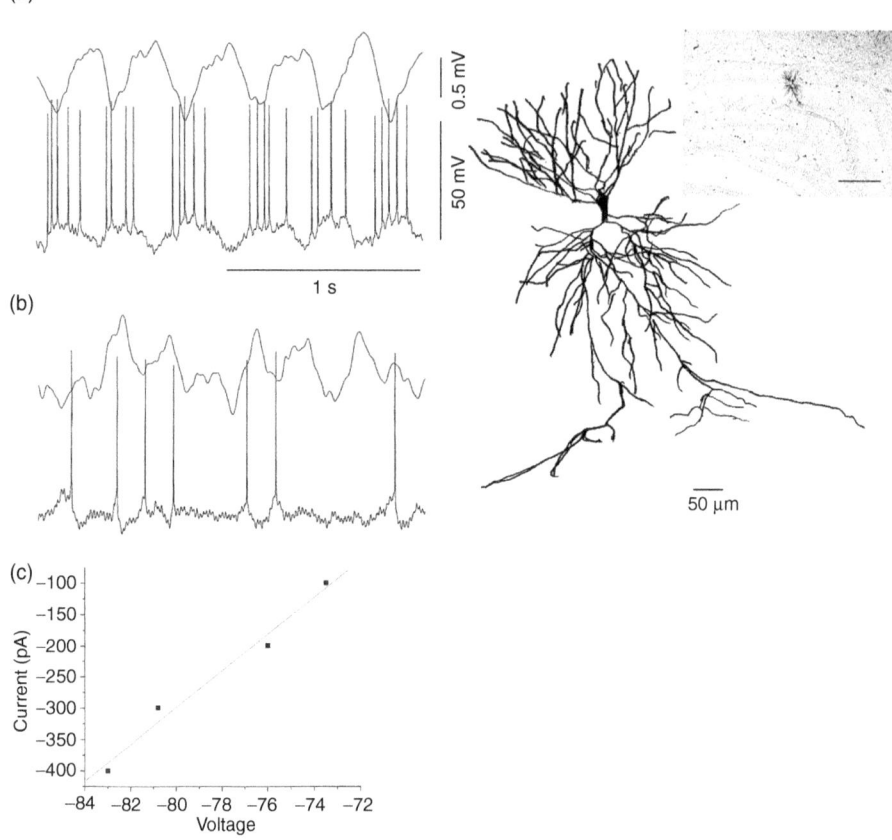

Figure 12.1 (a) Relationship between spontaneously occurring hippocampal theta field activity and discharges of a CA1 pyramidal cell. (Top) The hippocampal field activity recorded dorsal to the CA1 pyramidal cell layer in stratum oriens; (bottom) the discharge pattern of a CA1 pyramidal cell (256) classified as a phasic theta-ON cell (positivity up in all traces). The cell discharged in a rhythmic pattern with 4–5 spikes riding on the positive phase of the MPOs and the negative phase of the locally recorded extracellular theta field. (b) Relationship of the same cell to spontaneously occurring hippocampal LIA field activity. Note the overall lower discharge rate, irregular discharge pattern, and lack of MPOs. (c) Current–voltage plot of cell 256, input resistance = 29 MΩ. Low-power photomicrograph insert shows the location of the cell in the CA1 pyramidal layer while the higher-power magnification insert confirmed its identity as a CA1 pyramidal cell. (Reprinted with permission from Elsevier Science Publishers BV.)

(3) the amplitude of membrane potential oscillations in CA1 pyramidal phasic theta-ON cells was voltage-dependent and frequency was voltage independent; (4) there were no phase changes observed during current injections; however, amplitude analysis of MPOs revealed an inverted U-shaped curve asymmetrically

distributed around the average value of the membrane potential occurring during the spontaneous theta (no current) control condition; (5) the rate of rhythmic spike discharges in the CA1 pyramidal phasic theta-ON cells during the theta condition was precisely controlled within a critical range of membrane potential values from approximately −57 to −68 mV, corresponding to a range of MPO amplitudes of approximately 4 to 7 mV. Outside the critical range, rhythmical discharges were abolished. Also in that study, 22 cells met the criteria for classification as phasic theta-OFF cells and provided evidence to support the following conclusions: (1) morphologically identified CA1 pyramidal layer basket cells (see Fig. 12.2), mossy hilar cells, and granule cells formed a subset of cells meeting the criteria for classification as phasic theta-OFF cells; (2) MPOs occurred only during theta field activity, their onset signaled by a hyperpolarizing shift of 5–10 mV in membrane potential; (3) the amplitude of MPOs in CA1 pyramidal layer basket cells was voltage dependent and frequency was voltage independent; (4) the phase of MPOs in CA1 pyramidal layer basket cells underwent an approximately 180° phase reversal when the membrane potential was depolarized to around −65 mV; (5) the occurrence and rate of rhythmic spike discharges in the CA1 pyramidal layer basket cell phasic theta-OFF cells during the theta condition was precisely controlled within a critical range of membrane potential values from approximately −62 to −60 mV, corresponding to a range of MPO amplitudes of approximately 7 to 7.5 mV. Outside the critical range, spikes were absent or occurred singly. In a subsequent intracellular recording and labeling study, Bland *et al.* (2005) demonstrated that hippocampal pyramidal cells were functionally heterogeneous in relation to the generation of theta-band oscillation and synchrony. In field CA1 pyramidal cells formed theta-related subsets of phasic theta-ON cells and tonic theta-ON cells and non-theta-related subsets of simple spike discharging cells, complex spike discharging cells, and "silent" cells. Similar findings were evident for CA3 pyramidal cells. Recently, in an amazing technical breakthrough, Lee *et al.* (2006) were able to carry out whole-cell recording of a CA1 pyramidal cell in a freely moving rat. During a head movement the membrane potential displayed oscillatory (MPO) activity in the theta frequency range, thus verifying our results in hippocampal slices and acute anesthetized preparations.

The ascending brainstem hippocampal synchronizing pathways

The rostral pontine region

Experiments utilizing electrical stimulation techniques revealed that the origins of the ascending brainstem hippocampal synchronizing pathways were the nucleus reticularis pontis oralis (RPO) and the pedunculopontine

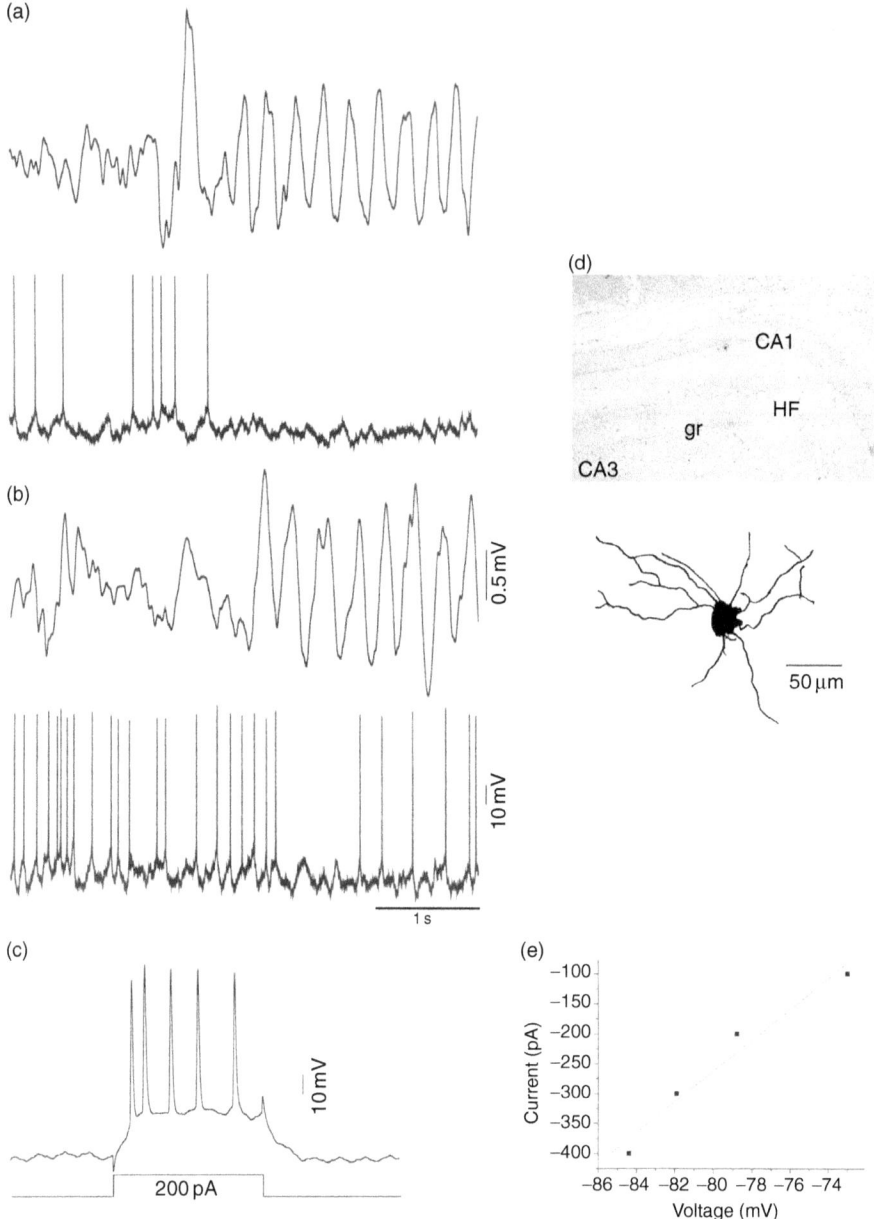

Figure 12.2 (a) and (b) Relationships between spontaneously occurring HPC field activity and the spike discharges of a CA1 layer basket cell (#229) classified as a phasic theta-OFF cell, in the no current control condition. The upper trace in each panel is the HPC field activity recorded from the molecular layer of the dentate region and the lower trace is the discharge pattern of the cell. The first half of the panels in (a) and (b) show the irregular cell discharge pattern occurring during HPC LIA. Note the absence of MPOs. The second half of the panel in (a) shows the complete cessation

tegmental nucleus (PPT) (Macadar et al., 1974; Vertes, 1981; Vertes et al., 1993; Oddie et al., 1994; Vertes and Kocsis, 1997; Bland and Oddie, 1998). Vertes et al. (1993) also demonstrated that microinjections of the cholinergic agonist carbachol into the RPO and PPT were very effective in eliciting theta field activity in the HPC. Subsequently, Bland et al. (1994) showed that either electrical stimulation or the microinfusion of carbachol administered to the RPO resulted in the intense activation of phasic and tonic theta-ON cells in the medial septal nuclei and the nucleus of the vertical limb of the diagonal band of Broca (MS/vDDB). Kirk et al. (1996) investigated the effects of electrical stimulation of the RPO on hippocampal field activity and the activity of cells in the posterior diencephalic region. They showed that electrical stimulation of the RPO produced theta field activity in the hippocampus and a regular non-bursting (tonic) increase in the discharge rate of the posterior hypothalamic (PH) cells, as well as a rhythmic bursting pattern of cells in the supramammillary (SUM) and medial mammillary (MM) nucleus. All studies to date investigating cellular activity in the nuclei of the RPO and PPT in relation to hippocampal theta generation have revealed only irregular (tonic) discharge patterns (Siegal et al., 1977; Vertes, 1977, 1979; Nunez et al., 1991; Hanada et al., 1999). Nunez et al. (1991) also reported similar discharge patterns in 18 RPO cells recorded during hippocampal field activity evoked both by sensory stimulation and by microinjections of carbachol into the pontine region. Subsequent injection of atropine abolished the carbachol effects.

The caudal diencephalic region

Fibers from the RPO and PPT ascend and synapse with caudal diencephalic nuclei, primarily the posterior hypothalamic (PH) nucleus and the supramammillary (SUM) nucleus (Vertes, 1992; Vertes et al., 1995). Physiological and

Caption for Figure 12.2 (cont.)
of spike discharges during theta field activity at a higher frequency (4.3 Hz) and the occurrence of MPOs. The second half of the panel in (b) shows that as theta field frequency slowed to 3.4 Hz, phase-locked spike discharges began to occur. Again, MPOs were recorded during theta field activity. (c) Intracellular depolarizing current pulse (200 pA, 100 ms duration) applied to the cell during spontaneously occurring theta. (d) The intracellular injection of Neurobiotin™ into cell #229 resulted in the labeling of a cell identified as a CA1 pyramidal layer basket cell. The upper panel is a low-power magnification showing the location of the cell in the CA1 cell pyramidal layer. The lower panel is a higher-power magnification showing the details of cell morphology. HF, hippocampal formation; gr, granule layer. (e) Current–voltage plot of the cell shown in 12.1a, input resistance = 29.4 MΩ. (Reprinted with permission from the American Physiological Society.)

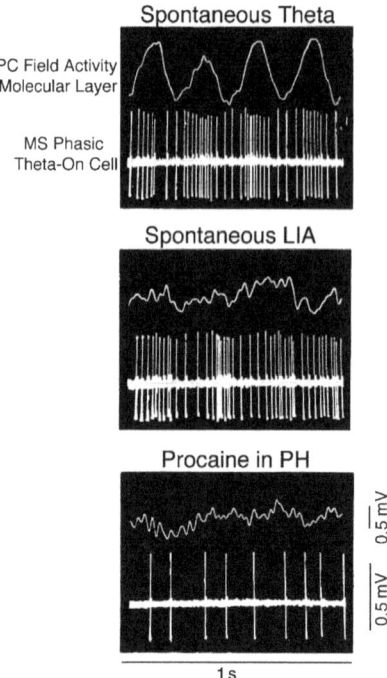

Figure 12.3 Sample recordings of an MS/vDBB phasic theta-ON cell that was rhythmic during HPC LIA, during the conditions of HPC theta (upper panels), spontaneous LIA (middle panels), and microinfusion of procaine hydrochloride into the PH (lower panels). Post-procaine data collected between 1 and 10 min after infusion of procaine. (Reprinted with permission from John Wiley and Sons, Inc.)

pharmacological studies support the view that the PH and SUM nuclei of the caudal diencephalon are a critical part of the ascending synchronizing pathways linking the rostral pontine region with the septohippocampal pathways. Kirk and McNaughton (1993) demonstrated that procaine infused into the SUM abolished HPC theta field activity produced by electrical stimulation of the reticular formation. Figure 12.3 from a study by Oddie *et al.* (1994) illustrates that the microinfusion of procaine into the PH abolished the spontaneous occurrence of HPC theta field activity and the theta field activity produced in response to electrical stimulation of the RPO. The microinfusion of procaine into the PH also abolished the rhythmicity of all phasic theta-ON cells in the MS/vDBB that was previously induced by electrical stimulation of the RPO.

Oddie *et al.* (1994) also showed that the microinfusion of carbachol into the PH and SUM region resulted in the continuous generation of HPC theta field activity that was higher in frequency than the mean frequency of spontaneously occurring theta and significantly larger in amplitude.

Electrical stimulation of the PH produces hippocampal theta activity, the frequency of which is linearly related to the intensity of hypothalamic stimulation (Bland and Vanderwolf, 1972a). Electrical stimulation of the PH also results in the intense activation of theta-ON cells in the MS/vDBB (Bland et al., 1990, 1994), the hippocampal formation (Colom et al., 1987; Smythe et al., 1991), and the entorhinal cortex (Dickson et al., 1995), as well as theta field activity in the entorhinal cortex (Dickson et al., 1994). In the Bland et al. (1994) study, five MS/vDBB phasic theta-ON cells were tested consecutively with electrical stimulation of the RPO and the PH and were shown to be activated in a similar manner in either condition. The intensity of activation of MS/vDBB cells by electrical stimulation of the RPO and PH was relayed on to theta-related cells in the HPC in a very precise manner. Colom et al. (1987) demonstrated by linear regression analysis that phasic theta-ON cells in the HPC translated the level of activation of the ascending synchronizing pathways through their discharge rates (shown in Fig. 12.4). Figure 12.5 shows a phasic theta-OFF cell during the transition from higher-frequency theta to lower-frequency theta to LIA. Phasic theta-OFF cells were inhibited by PH stimulation and the inhibition was not abolished by the administration of atropine sulfate (shown in Fig. 12.6).

Studies investigating the discharge properties of theta-related cells in the caudal diencephalon have revealed both rhythmic and non-rhythmic patterns. Kirk and McNaughton (1991) were the first to demonstrate, based on multi-unit recordings, that cells in the SUM discharged in a rhythmically bursting pattern in phase with HPC theta field activity. Kocsis and Vertes (1994) verified this finding with single-unit recordings in the SUM and, in addition, demonstrated that cells in the MM nuclei also discharged in a rhythmically bursting pattern in phase with HPC theta field activity. Bland et al. (1995) characterized the discharge patterns of cells in specific nuclei of the caudal diencephalon in relation to simultaneously recorded field activity from the stratum moleculare of the dentate gyrus, according to the criteria of Colom and Bland (1987). Of 54 cells recorded in the PH 43 (80%) were classified as tonic theta-ON and 11 (20%) as non-related. Tonic theta-ON cells in the PH discharged at significantly higher rates during theta, either occurring spontaneously or elicited with a tail pinch, than during LIA. Nine thalamic centromedial (CM) cells were recorded, seven of which were classified as tonic theta-ON cells and two of which were non-related (see Fig. 12.7).

Twenty cells were recorded in the border region of the PH/SUM. Of these, 15 (75%) were classified as tonic theta-OFF cells discharging at significantly higher rates during LIA than during either spontaneously occurring theta or tail-pinch-induced theta (see Fig. 12.8). Five cells recorded in the PH/SUM border region were non-related.

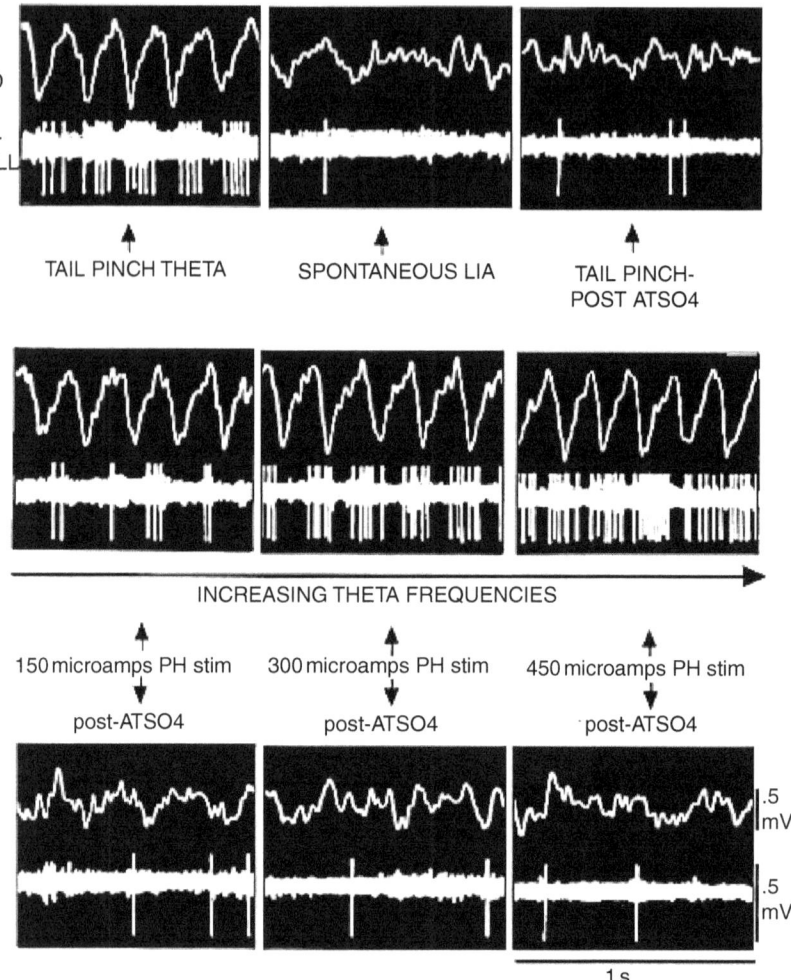

Figure 12.4 Responses of a dentate area theta-ON cell to a tail pinch, spontaneous LIA, and a tail pinch post-atropine (upper panels), increasing levels of PH stimulation (middle panels), and increasing levels of PH stimulation post-atropine (lower panels). (Reprinted with permission from Elsevier Science Publishers BV.)

All 16 cells (100%) recorded from the SUM were classified as phasic theta-ON cells. Unlike phasic theta-ON cells recorded in other brain regions, cells in the SUM did not significantly increase their discharge rate during the transition from LIA to theta field activity. Of the 23 cells recorded from the MM, 19 (83%) were also classified as phasic theta-ON cells and the remaining four cells were non-related (see Fig. 12.9).

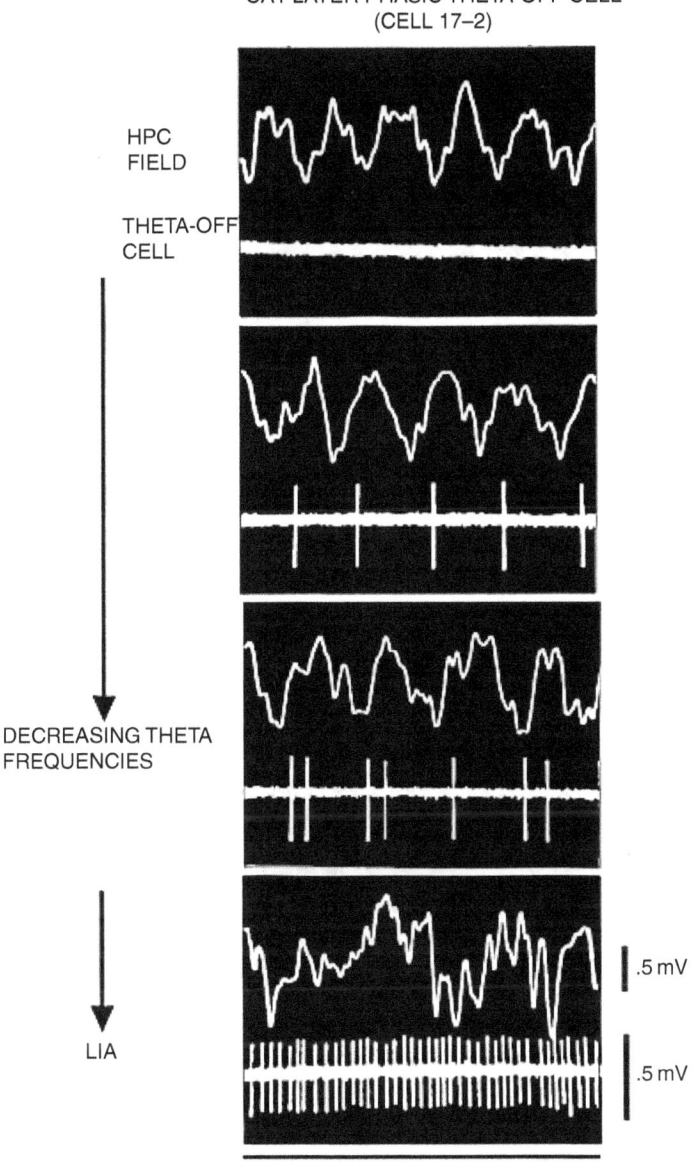

Figure 12.5 Responses of a CA1 area theta-OFF cell following PH stimulation. Note this type of cell was silent during slow wave frequencies above 5 Hz (first second post-stimulation). (Reprinted with permission from Elsevier Science Publishers BV.)

While theta-related cells in the PH and SUM nuclei are activated by the ascending brainstem hippocampal synchronizing pathways, the production of rhythmic cell bursting in the MM appeared to be dependent on descending activation from the septohippocampal pathways (Kirk et al., 1996).

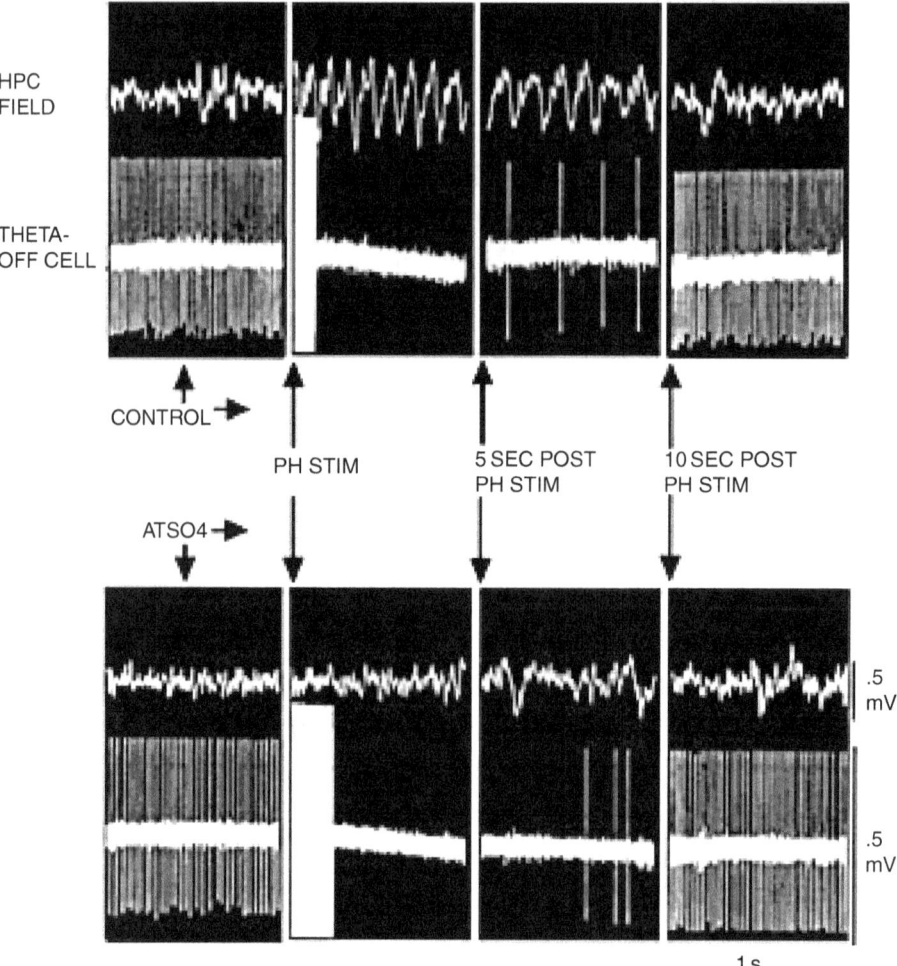

Figure 12.6 The relation between the discharge patterns of a phasic linear theta-OFF cell in the dentate granule layer and the simultaneously recorded field activity from the dentate stratum moleculare. (Upper panels) Spontaneous LIA and hypothalamic stimulation (500 microamps) during control conditions, and (lower panels) spontaneous activity and hypothalamic stimulation (500 microamps) following the administration of atropine sulfate. (Reprinted with permission from Elsevier Science Publishers BV.)

The medial septal region

The medial septal region (MS/vDBB) functions as the node of the ascending synchronizing pathways, distributing inputs to the posterior cingulate cortex, entorhinal cortex, and the hippocampal formation (Bland, 2000).

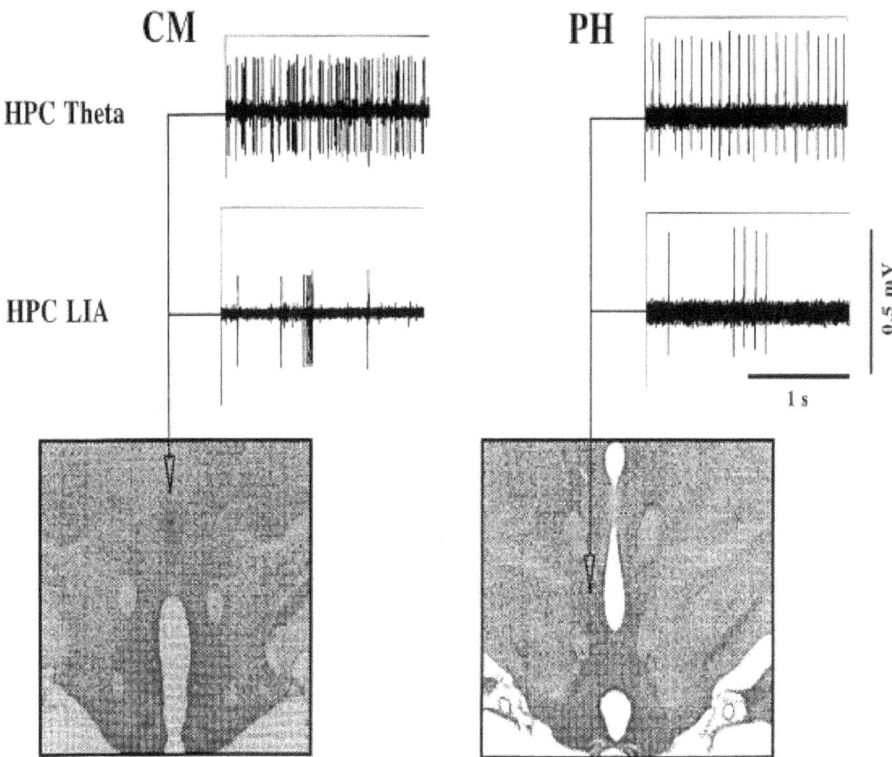

Figure 12.7 (Top) Discharge patterns of representative CM (left) and PH (right) tonic theta-ON cells during spontaneously occurring hippocampal theta and large-amplitude irregular activity (LIA). (Bottom) Digitized coronal sections through the caudal diencephalon. Arrows: locations of the pontamine sky blue dot deposited from the tips of the glass microelectrodes used to record the cells shown above. Tonic theta-ON cells were located predominately in the CM and PH. (Reprinted with permission from the American Physiological Society.)

The cholinergic septohippocampal projection was the first pathway to be documented, using a variety of experimental techniques, showing that HPC pyramidal and granule cells were the main recipients of these cholinergic inputs (Frotscher and Leranth, 1985). The activation of cholinergic receptors on cells in the MS/vDBB through the microinfusion of carbachol resulted in the production of theta field activity in the HPC in both acute (Oddie et al., 1994) and freely moving rats (Monmaur and Breton, 1991; Lawson and Bland, 1993). Smythe et al. (1992) proposed that cholinergic and GABAergic projections originating in the MS/vDBB act synergistically to modulate the synchronizing activity from the ascending brainstem pathways. Cholinergic projections provide a steady tonic excitatory afferent drive for HPC theta-ON cells, and GABAergic projections act

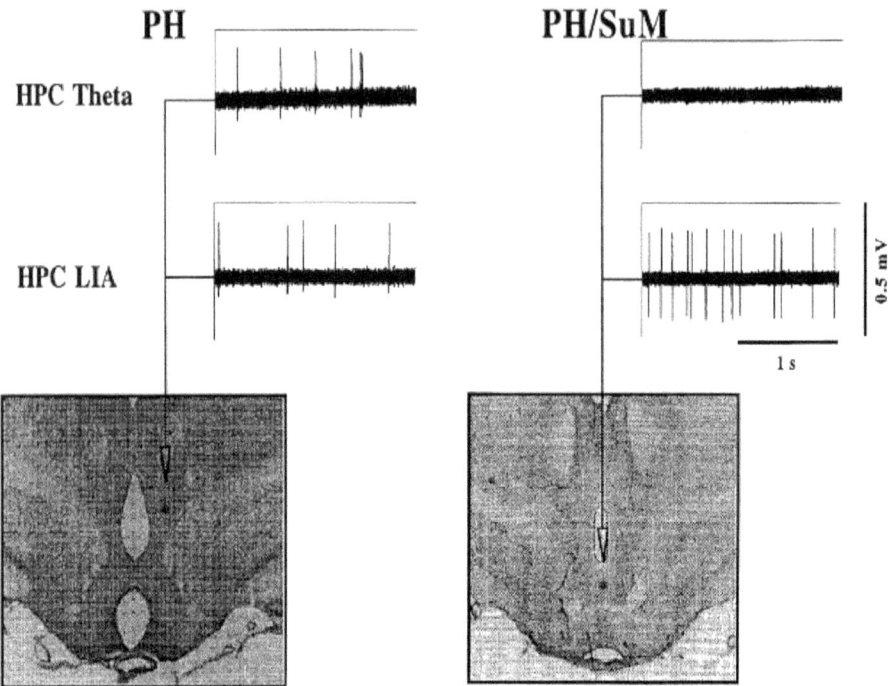

Figure 12.8 (Top) Discharge patterns of a non-related cell in the PH (left) and a tonic theta-cell in the PH/SUM border region (right) during spontaneously occurring hippocampal theta and LIA. (Bottom) Digitized coronal sections through the caudal diencephalon. Arrows: locations of the pontamine sky blue dot deposited from the tips of the glass microelectrodes used to record the cells shown above. Non-related cells were found throughout the caudal diencephalon, whereas tonic theta-OFF cells were found predominately in the PH/SUM border region. (Reprinted with permission from the American Physiological Society.)

to reduce the overall level of inhibition by inhibiting HPC GABAergic interneurons (theta-OFF cells). Both activities must be present for the generation of HPC theta field and cellular activities. The balance between the cholinergic and GABAergic systems determines whether hippocampal synchrony (theta) or asynchrony (LIA) occurs. Destruction of the medial septal region completely abolished theta field activity in the HPC (Bland, 1986). Bland and Bland (1986) demonstrated that both Type 1 and Type 2 HPC theta field activity was abolished in freely moving rabbits, along with theta-ON cell rhythmicity during both these conditions.

Subsequent research led to the description of a GABAergic septohippocampal projection (Misgeld and Frotscher, 1986) and these projection form synaptic contacts on all identified HPC interneurons and other "non-pyramidal" neurons

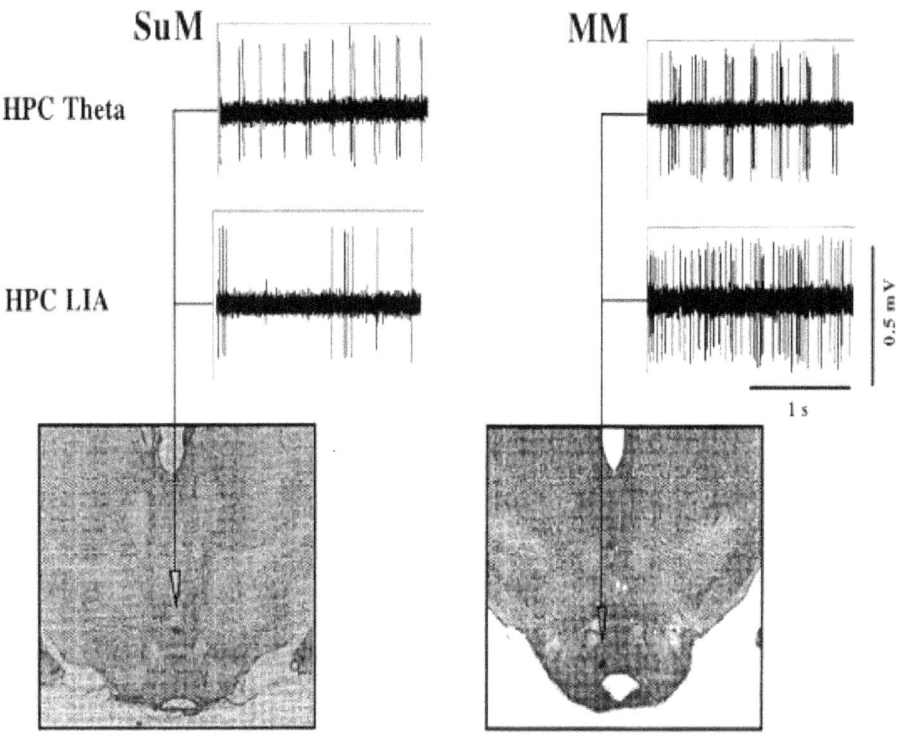

Figure 12.9 (Top) Discharge patterns of representative SUM (left) and MM (right) phasic theta-ON cells during spontaneously occurring hippocampal theta and LIA. (Bottom) Digitized coronal sections through the caudal diencephalon. Arrows: locations of the pontamine sky blue dot deposited from the tips of the glass microelectrodes used to record the cells shown above. Phasic theta-ON cells were located predominately in the SUM and the MM. (Reprinted with permission from the American Physiological Society.)

(Acsady et al., 1993; Freund, 1989; Freund and Antal, 1988; Gulyas et al., 1990; Meittinen and Freund, 1992). Bland et al. (1996) microinfused the $GABA_A$ agonist muscimol into the MS/vDBB nuclei while recording hippocampal field and theta-ON cell activity occurring both spontaneously and in response to electrical stimulation of the PH, in urethane-anesthetized rats. The microinfusion of 5.0–12.5 nmol of muscimol resulted in the progressive reduction in the power (amplitude) and finally the total loss of hippocampal theta field activity. In contrast, the frequency of HPC theta remained unaffected during the entire post-infusion period that theta field activity was present. Overall, the effects of intraseptal microinfusions of muscimol on HPC theta field activity produced by electrical stimulation of the PH were the same as those reported for spontaneously occurring HPC theta field activity. In the time immediately following the first 1 min infusion of 5 nmol muscimol (before changes in theta amplitude occurred), a brief period of increased theta-ON cell excitability was observed, manifested as an increase in the number of discharges per burst. Associated with the progressive reduction of HPC theta amplitude, phasic theta-ON cell discharge rates progressively decreased. Just prior to the disappearance of theta, phasic theta-ON cells ceased discharging. During the period when HPC field activity was replaced with low-amplitude asynchronous activity, however, phasic theta-ON cells discharged in bursts correlated with every occurrence of hippocampal sharp wave field activity. The loss of rhythmic bursting of hippocampal phasic theta-ON cells following the microinfusion of muscimol into the MS/vDBB paralleled the loss of theta amplitude. Bland et al. (1996) interpreted the results of the intraseptal infusion of muscimol on HPC theta field and cellular activity in the following manner. The brief excitatory effect on hippocampal theta-ON cell discharges may be correlated pharmacologically with an initial brief increase in the turnover of ACh in the hippocampus. The subsequent reduction of phasic theta-ON cell discharges and theta field activity may be correlated with a longer-lasting reduction of turnover of ACh in the hippocampus, that was controlled by MS/vDBB $GABA_A$ inputs to MS/vDBB cholinergic septohippocampal neurons, possibly along with a direct inhibition of the GABAergic septohippocampal projection; the contribution of the MS/vDBB nuclei, as the node of the ascending brainstem hippocampal synchronizing pathways, was in relaying the ascending theta frequency code, the modulation of theta amplitude, and the correlated discharges of hippocampal formation theta-ON cells. Furthermore, the HPC participation of phasic theta-ON cells in the generation of hippocampal theta field activity and sharp waves was mediated by separate inputs.

Recent evidence supporting the existence of a third septohippocampal projection has accumulated, beginning with the knowledge that a population of MS/vDBB neurons could not be identified with either cholinergic or GABAergic markers and could possibly be glutamatergic (Gritti et al., 1997; Kiss et al., 1997). Next, analysis using phosphate-activated glutaminase and by the retrograde transport of [^{3}H]aspartate in septal neurons provided indirect evidence for MS/vDBB glutamate neurons (Gonzalo-Ruiz and Morte, 2000; Manns et al., 2001; Kiss et al., 2002). Two types of vesicular glutamate transporters have now been well documented in excitatory synapses (VGLUT1 and VGLUT2) (Fremeau et al., 2001). Strong evidence for the presence of glutamate neurons in the MS/vDBB has been provided by single-cell multiplex RT-PCR that has identified a subpopulation of septohippocampal neurons expressing mRNAs for VGLUT1 and VGLUT2 (Danik et al., 2003, 2005; Sotty et al., 2003). Colom et al. (2005) used an antiglutamate antibody and a retrograde tracing technique to identify a subpopulation of septohippocampal glutamatergic neurons in the medial septal region. Using stereological probes these authors concluded that in the rat the septohippocampal glutamatergic population comprised approximately 16000 neurons. In addition, based on triple immunostaining, they provided evidence that most glutamatergic neurons did not immunoreact with cholinergic or GABAergic neuronal markers in the medial septum. Previous evidence in the literature suggested glutamate might play a role in HPC theta-band oscillation and synchrony. Carre and Harley (2000) demonstrated that the injection of glutamate into the medial septum resulted in the generation of hippocampal theta. Bonansco and Buno (2003) showed that theta-like rhythmic oscillations of CA1 cell MPOs and rhythmic spike discharges could be induced by the microiontophoresis of N-methyl-D-aspartate (NMDA) at the apical dendrites of CA1 pyramidal cells in the in vitro hippocampal slice preparation. Bland *et al.* (2007a) recently demonstrated that the microinfusion of NMDA into apical dendrites of hippocampal CA1 pyramidal cells of urethane-anesthetized rats resulted in long-lasting (20–30mins) induction of hippocampal synchrony at the field and cellular level (see Fig. 12.10).

Power of NMDA-induced theta was significantly greater than tail-pinch-induced theta activity but frequency did not differ. This effect was antagonized by intrahippocampal infusion of AP5, but unaffected by intravenous ATSO4. During AP5 blockade tail-pinch theta frequency and power were significantly reduced. Microinfusion of NMDA into the medial septum also resulted in the induction of HPC theta field activity significantly higher in frequency than tail-pinch-induced theta activity, while power did not differ. Microinfusion of AP5 into the medial septum significantly lowered the power of tail-pinch-induced theta but did not affect frequency. These findings supported the conclusion that the glutamatergic septohippocampal projection represents a third pathway capable of generating hippocampal field and cellular synchrony, independent of that generated by the septohippocampal cholinergic and GABAergic projections. The possible functional significance of this system will be discussed later.

One of the widely accepted beliefs concerning the role of the medial septal region in hippocampal theta generation is that it serves a "pacemaker" function. Although this is subject to various interpretations, in the strictest sense this implies that the generation of rhythmic activity in the hippocampus is produced as a result of rhythmic inputs from the septum. The acceptance of this notion is easy to understand, given that many medial septal cells discharge in a rhythmic pattern and lesions of the region abolish theta field activity in the HPC. However, Bland and Colom (1993) reviewed evidence supporting their belief that the septal "pacemaker" hypothesis in its strictest interpretation was not tenable. Chief among this evidence was the demonstration that carbachol applied to

(a) NMDA + ATSO4 into the HPC

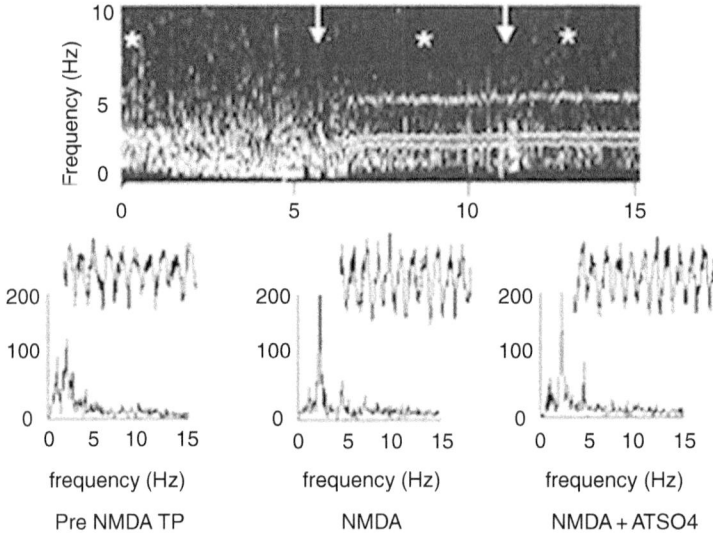

Pre NMDA TP   NMDA   NMDA + ATSO4

(b) NMDA + AP5 into the HPC

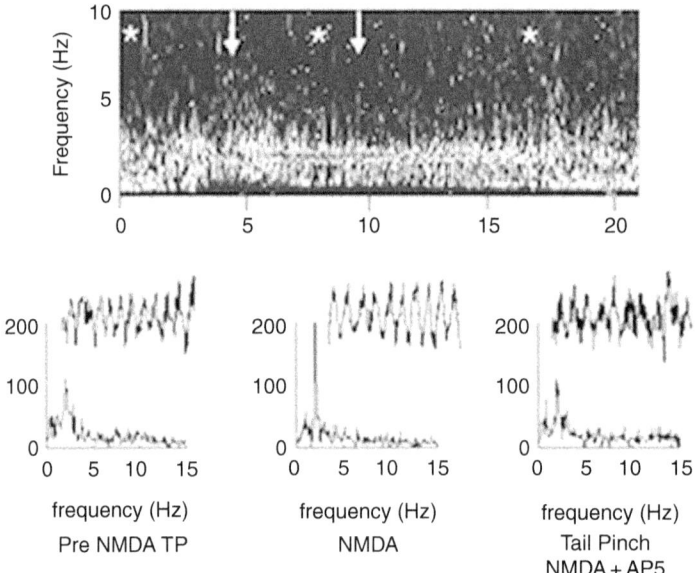

Pre NMDA TP   NMDA   Tail Pinch NMDA + AP5

Figure 12.10 (a) Time-frequency contour plot and fast Fourier transform (FFT) analyses of hippocampal field activity during various experimental conditions. Time-frequency analysis in upper panel shows the sequence of a control tail-pinch, followed by NMDA microinfusion into the HPC (first arrow), and then the administration of atropine sulfate (ATSO4) (second arrow). Lower left FFT is a control tail-pinch (first asterisk), middle FFT is a sample of NMDA-induced theta

isolated hippocampal slices resulted in the generation of theta field activity (Konopacki *et al.*, 1987a, 1987b, 1987c, 1987d, 1988a, 1988b, 1988c; Bland *et al.*, 1988). Rowntree and Bland (1986) had previously demonstrated that the intrahippocampal microinfusion of carbachol in septally intact urethane-anesthetized rats was capable of generating hippocampal theta field activity. Despite these findings, Colom *et al.* (1991) were unable to induce hippocampal theta in urethane-anesthetized rats following intrahippocampal microinfusions of carbachol, with the medial septum temporarily blocked with procaine hydrochloride. These authors reasoned that this failure could be due to the fact the removal of septal GABAergic inputs in the anesthetized preparations may leave the hippocampus "hyperinhibited." This did not happen in transverse hippocampal slices since many dendritic arbors of GABAergic interneurons ran parallel to the main hippocampal axis and were severed in transverse cuts. Slices thus contained a reduced population of inhibitory interneurons and therefore reduced inhibition. Colom *et al.* (1991) demonstrated that the combination of carbachol and bicuculline microinfusions into the HPC of septally deafferented rats produced theta-like field oscillations and rhythmic discharges of phasic theta-ON cells, both of which were antagonized by atropine sulfate (see Fig. 12.11).

If the medial septal region does not serve as a pacemaker what role does it play in hippocampal theta generation? Recent findings support the view that hippocampal theta field frequency is primarily determined by brainstem nuclei below the medial septum (the pontine and diencephalic nuclei) while the medial septum relays frequency information and contributes primarily to theta amplitude (Kirk and MacNaughton, 1993; Bland and Oddie, 2001; Pan and MacNaughton, 2004; Woodnorth and MacNaughton, 2005; Bland *et al.*, 2006a;

Caption for Figure 12.10 (cont.)
(second asterisk), and right side FFT is a sample of theta field activity at the peak frequency following the intravenous administration of ATSO4. Note the lack of effect of ATSO4 on the power of NMDA-induced theta field oscillations. (B) Time-frequency contour plot and FFT analyses of hippocampal field activity during various experimental conditions. Time-spectral analysis in upper panel shows the sequence of a control tail-pinch, followed by NMDA microinfusion into the HPC (first arrow), and then the microinfusion of AP5 into the HPC (second arrow). Lower left FFT is a control tail-pinch (first asterisk), middle FFT is a sample of NMDA-induced theta (second asterisk), and right side FFT is a sample of theta field activity at the peak frequency following the microinfusion of AP5 into the HPC. Note that the NMDA-induced theta field oscillations were abolished by AP5 and that the power of the primarily cholinergically mediated tail-pinch theta field activity was reduced compared to that produced before the AP5 treatment. (Reprinted with permission from John Wiley and Sons, Inc.)

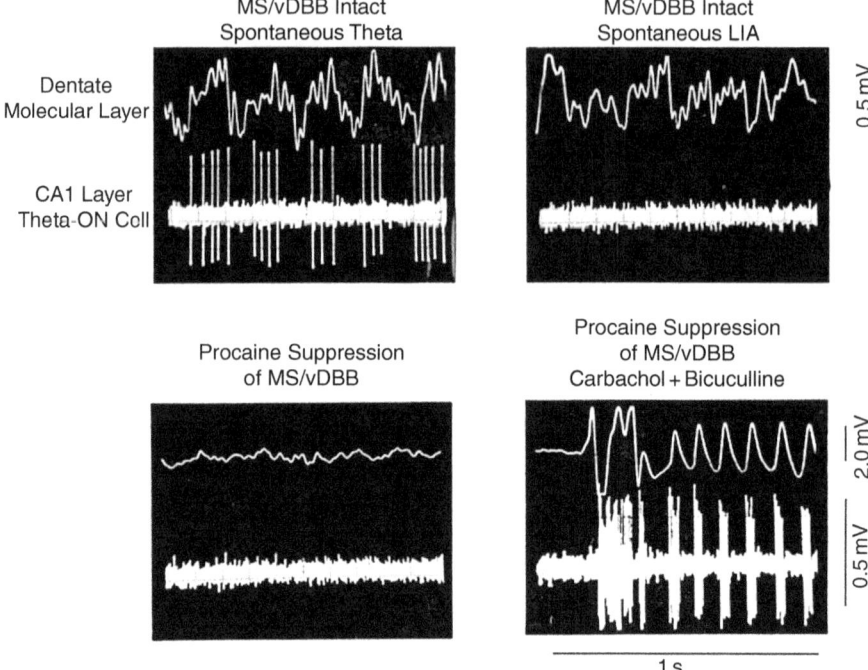

Figure 12.11 The effects of intrahippocampal microinfusions of carbachol plus bicuculline on field activity and theta-ON cell discharges. Analogues are the simultaneously recorded field and cellular activity from the dentate molecular layer (upper trace) and a phasic theta-ON cell in the CA1 layer (lower trace) during various pre- and post-procaine conditions. (Reprinted with permission from John Wiley and Sons, Inc.)

Jackson and Bland, 2006). Jackson and Bland (2006) used independent and combined electrical stimulation pairings of the pontine nucleus, posterior hypothalamus and medial septum in urethane-anesthetized rats to determine the contribution of the septum to the frequency and amplitude parameters of HPC theta generation. Their findings supported several conclusions: (1) the major theta generating activity of the ascending brainstem hippocampal synchronizing pathways involved projections from the pontine region to the posterior diencephalic region, relayed through the medial septal region to the hippocampus; (2) the medial septal region directly controlled theta amplitude and secondarily translated the level of ascending brainstem activity into the appropriate frequency of HPC theta.

Another long-standing paradox in the theta literature concerned the effects of electrical stimulation of the medial septal region. Electrical stimulation of the medial septum of urethane-anesthetized rats at appropriate parameters produced hippocampal theta field activity (Brucke et al., 1959; Stumpf, 1965; Gray and Ball, 1970). Similar stimulation of the medial septum in freely moving rats also produced theta, but it was dissociated from the normal behavior correlates (Kramis and Routenberg, 1977), in direct contrast to the effects of posterior hypothalamic stimulation (Bland and Vanderwolf, 1972a; Oddie et al., 1996). A paper by Scarlett et al. (2004) provided a possible explanation for the dissociation between the induced hippocampal theta field activity and its behavioral correlates. These authors showed that theta induced by electrical stimulation of the septum indeed had the same depth profile as spontaneously occurring theta. However, the responses of theta-related cells to medial septal stimulation were very different from the discharge properties of these cells in relation to spontaneously occurring theta, and to those accompanying electrical stimulation of the pons and posterior hypothalamic region. Thus, on the basis of cellular evidence, electrical stimulation of the medial septum activated hippocampal neural circuitry involved in the generation of theta field activity in a non-physiological manner that is, in addition, devoid of behaviorally relevant sensorimotor inputs from the posterior hypothalamic region.

The fact that lesions of the medial septal region abolished theta field activity was discussed above. Smythe et al. (1991), also showed that reversible blockade of the MS/vDBB nuclei with procaine abolished spontaneously occurring HPC theta and HPC theta produced by electrical stimulation of the PH. In addition, they demonstrated that the blockade abolished the rhythmic discharges of HPC phasic theta-ON cells and caused the release (disinhibition) of the discharges of phasic theta-OFF cells. During the period of the blockade when the discharges of phasic theta-OFF cells were released, electrical stimulation of the PH was no longer effective in inhibiting these discharges. The study thus confirmed that the medial septal region mediated the synchronizing influences of PH stimulation on HPC field and phasic theta-ON cell activity, as well as the inhibition of phasic theta-OFF cells.

Simultaneous recordings of septohippocampal cells have provided more information concerning the critical role of the medial septum in the control of oscillation and synchrony in the HPC. Macadar et al. (1970) and Alonso et al. (1987) were the first to do such studies, utilizing time-averaged cross-correlation techniques. Bland et al. (1997, 1999) subsequently carried out experiments utilizing urethane-anesthetized rats in which 18 simultaneously recorded septohippocampal cell pairs (36 individual cells), each classified as theta-related according to the criteria of Colom and Bland (1987), were studied during four spontaneously occurring HPC field conditions: (1) large-amplitude irregular activity (LIA) only; (2) the transition from LIA to theta; (3) theta only; and (4) the transition from theta to LIA. The main objective was to examine the temporal relationships and degree of neural synchrony between the discharges of cell pairs during the four conditions, utilizing both time-averaged and time-dependent (joint per stimulus time histogram analysis: JPSTH) cross-correlation techniques, in order to determine their contribution to the control of oscillation and synchrony (theta) in the HPC. The JPSTH analysis, which assessed the time-dependent correlation between the spontaneous discharges of two cells, has generally been interpreted as indicative of their interconnectivity and/or the sharing of a common input (Gerstein, 1970). The findings of Bland et al. (1999) demonstrated that the transition from the LIA state to the theta field state in the HPC required a temporal sequence of changes in theta-related cellular activity occurring an average of 500 ms preceding the transition, which were suggested to be: (1) the medial septum inhibited HPC theta-OFF cells; (2) tonic MS/vDBB theta-ON cells provided tonic depolarizing inputs to initiate MPOs in HPC phasic theta-ON cells, whereas phasic MS/vDBB theta-ON cells synchronized the MPOs of phasic HPC theta-ON cells and the discharges of tonic HPC theta-ON cells. Much of the time preceding the LIA to theta field transition was accounted for by recruitment of these theta-related cell populations. On the other hand, the "turning off" of the theta state occurred abruptly and involved the medial septal activation of hippocampal theta-OFF cells.

The "turning off" of the theta state, that is, the initiation of hippocampal desynchronization, is mediated by another group of ascending brainstem pathways that originate in the nucleus of the median raphe. The ascending brainstem hippocampal synchronizing and desynchronizing systems are likely to interact at a number of levels, with the medial septal region as the "node" for both systems. Bland and Vanderwolf (1972b) may have activated the terminal portions of the desynchronizing system in their early study. They demonstrated that electrical stimulation of sites in the dentate-CA4 region of the HPC produced short latency evoked potentials bilaterally in the HPC that supplanted the normal theta field activity and resulted in the immediate behavioral arrest of Type 1 movements. This same stimulation did not interfere with ongoing Type 2 behaviors such as conditioned immobility, shivering, or licking. In this same series of experiments, Type 1 movement related theta was recorded from the posterior hypothalamus and electrical stimulation of the dentate-CA4 region supplanted this activity as well. The raphe system will be discussed in the following section.

The ascending brainstem hippocampal desynchronizing pathways

The median raphe

The median raphe nucleus (MR) is a serotonin-containing cell group located in the midbrain, sending projections to many forebrain regions (see review by Vertes *et al.*, 2004). Among these are very strong projections to the medial septal region as well as direct connections to the HPC and SUM regions (Vertes *et al.*, 1999; Aznar *et al.*, 2004). One of the earliest documented findings was the demonstration that electrical stimulation of the MR resulted in desynchronization of hippocampal field activity (Macadar *et al.*, 1974; Assaf and Miller, 1978; Vertes, 1981) while lesions of the MR resulted in the continuous release of theta (Maru *et al.*, 1979; Yamamoto *et al.*, 1979). These effects are likely to be mediated by serotonergic cells in the MR since injections of drugs that either suppressed serotonergic neurons in the MR of anesthetized rats (5-hydroxytryptamine (5-HT$_{1A}$) autoreceptor agonists or GABA agonists) or reduced excitatory drive to them (excitatory amino acid antagonists) produced long-lasting theta field activity (Kinney *et al.*, 1994, 1995, 1996; Vertes *et al.*, 1994). Varga *et al.* (2002) reported that GABA$_B$ receptors are found on serotonergic MR cells and that their activation by the agonist baclofen also resulted in long-lasting theta generation. The MR also contains a population of GABAergic cells that likely inhibit serotonergic MR cells (see review by Vertes *et al.*, 2004). Viani Di Prisco *et al.* (2002) demonstrated that approximately 80% of MR cells could be categorized as theta-ON or theta-OFF cells and further related their discharge properties to putative serotonergic or GABAergic cells.

Electrical stimulation of the MR in anesthetized rats also has a disruptive effect on rhythmically discharging medial septal neurons (Assaf and Miller, 1978) while in freely moving rabbits such stimulation also disrupted the rhythmic discharges of both medial septal cells and hippocampal theta (Kitchigina *et al.*, 1999; Vinogradova *et al.*, 1999). These authors also reported that the suppression of the MR by lidocaine increased the frequency and regularity of rhythmically discharging cells in the medial septum and hippocampus as well as producing continuous hippocampal theta field activity. A study by Segal (1975) demonstrated that electrical stimulation of the MR resulted in the inhibition of 48% of recorded hippocampal pyramidal cells, although in this study the cells were not rigorously characterized according to theta-related properties. Recent work in our laboratory has shown that electrical stimulation of the MR in urethane-anesthetized rats that suppressed ongoing theta also resulted in a short latency inhibition of 100% of all phasic theta-ON cells tested (Jackson *et al.*, 2006). The effects of electrical stimulation of the MR on hippocampal field and cellular activity may be mediated by a number of different pathways, either direct or indirect. Given the data discussed above that such stimulation affects medial septal cell activity, a route through the septum is suggestive but not proven. A seemingly obvious test in anesthetized rats would be to inject procaine into the septum to see if the effects of MR stimulation (abolishing theta) would be disrupted. However, this manipulation itself abolishes theta. Crooks *et al.*

(R. Crooks, J. Jackson, and B.H. Bland, unpublished data) solved this problem by looking at the effects of procaine suppression of the medial septum on the release of theta produced by either procaine injections or injections of 8-hydroxy-2-(di-*n*-propylamino)-tetraline (8-OH-DPAT) in the MR. The experiments revealed that procaine injections into the medial septum abolished the released theta, thus supporting the view that the effect of electrical stimulation of the MR on the hippocampus was modulated by pathways through the septum.

The sensorimotor integration model would predict that MR stimulation in the freely moving rat should result in the inhibition of Type 1 theta-related behaviors. Several earlier studies could be interpreted as support for this prediction (although the authors had different interpretations). Graeff and Silveira Filho (1978) and Graeff *et al.* (1980) reported seeing behavioral inhibition or behavioral "freezing" as a result of MR stimulation in freely moving rats. Robinson and Vanderwolf (1978) demonstrated that MR stimulation at a frequency of 100 Hz resulted in the generation of slow 4–6 Hz theta and transient behavioral arrest. Peck and Vanderwolf (1991) showed that stimulation of some sites in the MR of freely moving rats induced theta and locomotion while stimulation of most sites in the MR resulted in behavioral freezing and theta. Treatment with scopolamine abolished the theta, replacing it with hippocampal suppression, while not affecting the behavioral suppression. Jackson *et al.* (2006) carried out a study investigating the effects of MR stimulation on wheel-running and hippocampal theta field activity induced by posterior hypothalamic stimulation. Electrical stimulation of the MR resulted in the total inhibition of wheel-running, accompanied by slow-frequency theta, even during the continued application of posterior hypothalamic stimulation. When the MR stimulation was turned off, the rat immediately resumed intense wheel-running accompanied by higher-frequency theta. These data thus support the hypothesis that the functional significance of the MR HPC desynchronizing system is to terminate Type 1 theta-related movements.

The motor feedback role of the posterior hypothalamic region

As discussed earlier, significant additions to the updated sensorimotor model included the description of the ascending brainstem hippocampal synchronizing pathways as the anatomical basis of the model, with the posterior hypothalamic region assigned the role of providing motor feedback to the hippocampal formation. In the updated version, Type 2 theta inputs ascended from the pontine region to the midline diencephalic region, through to the medial septum and then input to the hippocampus. In the case where Type 1 movements were not yet been initiated, the hippocampus sent only Type 2 inputs to motor systems. Initiation of Type 1 movements by motor systems sent Type 2 and Type 1 inputs to the PH that again ascended through their respective pathways to the medial septum and to the hippocampus. Experiments investigating the effects of electrical stimulation of nuclei contributing to the ascending brainstem hippocampal synchronizing pathways on hippocampal field activity and behavior of freely moving rats provided strong support for these ideas. Robinson and Vanderwolf (1978) showed that stimulation of many brainstem sites (nucleus cuneiformis; subnucleus compactus; nucleus reticularis oralis, caudalis, and medial gigantocellularis; the pontine central gray; nuclei adjacent to the midbrain central gray; centralis superior of the raphe; locus coeruleus) at lower levels of intensity produced Type 2 (atropine-sensitive) theta during behavioral immobility. At slightly higher levels of stimulation of most sites walking or circling behavior occurred, accompanied by Type 1 (atropine-resistant) theta. Similar studies of the PH region have revealed that this area differs from other brainstem sites in that the Type 1 movements elicited were under environmental control. Bland and Vanderwolf (1972a) demonstrated that rats receiving electrical stimulation of the PH could turn to avoid obstacles and reversed direction when necessary and speed of locomotion was increased as stimulation intensity increased. Type 2 behaviors never occurred during the periods of electrical stimulation. Theta frequency was related systematically to the speed of initiation of Type 1 movements. In a running-wheel experiment, the speed of initiation of running was directly related to the level of PH stimulation and the onset frequency of HPC theta. These experiments did not prove that the theta and movements induced by electrical stimulation of the PH were both dependent on activity ascending to the hippocampus. Later experiments did provide support for this being the case. Oddie et al. (1996) replicated and extended Bland and Vanderwolf's (1972a) wheel-running results by adding in the manipulation of septohippocampal pathways. Following the baseline condition the medial septal region was reversibly inactivated by the microinfusion of procaine hydrochloride. The stimulation induced wheel-running and HPC theta were both abolished. Subsequent multiple regression analysis revealed that the recovery of wheel-running speeds more closely paralleled the recovery of HPC theta frequency rather than HPC theta amplitude.

Experiments carried out in Sinnamon's laboratory have also provided important data supporting the relationship between hypothalamic sites and locomotor behavior. Utilizing a paradigm involving lightly anesthetized rats suspended in a sling, Sinnamon demonstrated that electrical stimulation of hypothalamic nuclei induced locomotor stepping (Sinnamon, 1993). He has also demonstrated that the locomotor stepping induced by electrical stimulation of these hypothalamic nuclei was accompanied by HPC theta, albeit at lower frequencies than in the awake rat (Sinnamon et al., 2000). A subsequent study (Sinnamon, 2000) demonstrated that the association between HPC theta activity in the 3–6 Hz range and the excitability of locomotor initiation was sufficiently specific to allow prediction of the magnitude of stepping by the prior power levels of HPC 3–6 Hz theta.

The experiments discussed above demonstrated that activation of the PH induced Type 1 motor behaviors correlated with Type 1 theta. Destruction of this area should profoundly affect motor behavior and theta and indeed this was the case. Robinson and Whishaw (1974) showed that large lesions of the PH produced profound akinesia together with the loss of Type 1 movement related theta.

Data supporting the sensorimotor integration model of hippocampal function

The sensorimotor integration model was based on the assumption there were at least two separate theta inputs to the HPC, one for generating Type 1 movement related theta and one for generating Type 2 sensory processing theta, and that the Type 2 theta subsystem was always coincidently active whenever the Type 1 theta subsystem was active. Early support for the "two thetas" concept came from the observations that immobility-related Type 2 theta was atropine sensitive while movement related Type 1 theta was atropine resistant (Bland, 1986) and the demonstration by Leung (1984a) that the gradual phase-shift of theta observed in depth profiles made through the CA1 stratum radiatum of the HPC of freely moving rats was related to the relative participation of the Type 2 and Type 1 theta generating pathways. Activation of the Type 2 pathway by itself resulted in the rapid 180° phase-shift observed in urethane-anesthetized rats. Addition of Type 1 theta during the occurrence of Type 2 theta resulted in a gradual phase shift. In a subsequent paper Leung (1984b) provided a model of the CA1 pyramidal region that successfully explained the type of theta profiles one would expect to see under various experimental conditions, including providing support for the idea that Type 1 and Type 2 theta inputs were both active during Type 1 movements. Sinclair *et al.* (1982) demonstrated that phasic theta-ON cells (simply termed theta cells at that time) discharged in rhythmic bursts during both Type 2 immobility related theta and Type 1 movement theta (see Fig. 12.12).

Interestingly, the number of discharges per rhythmic burst was always lower during Type 2 theta compared to Type 1 theta, even when the theta field frequencies accompanying the two behaviors were identical. This provided physiological evidence there was a difference in inputs to the same cell during sensory processing and movement. Furthermore, although not investigated

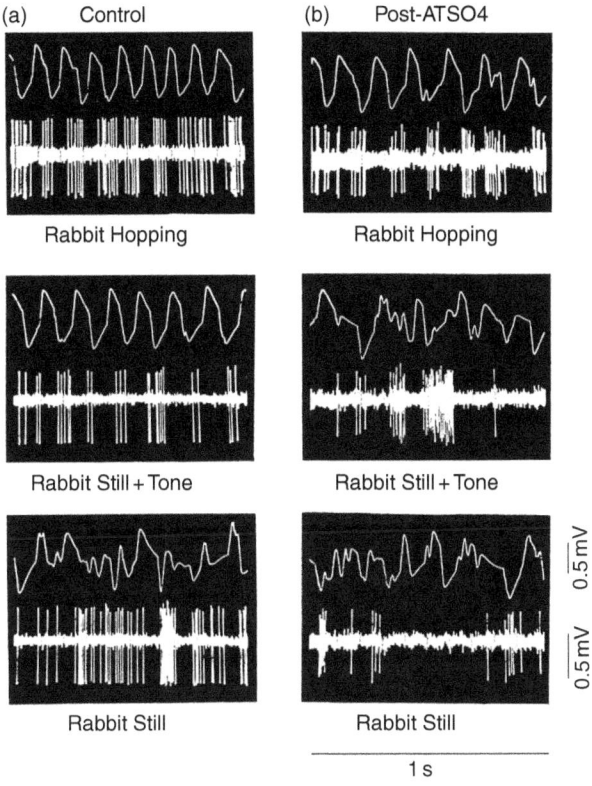

Figure 12.12 Relationships between the discharge patterns of a dentate area theta-ON cell to hippocampal field activity recorded from the stratum moleculare recorded from a freely moving rabbit. (a) Pre-drug condition. Recordings taken during hopping (top panel), immobility during a tone presentation (middle panel), and during immobility (bottom panel). (b) Post-atropine administration. Recordings taken during hopping (top panel), immobility during a tone presentation (middle panel), and during immobility (bottom panel). (Reprinted with permission from Elsevier Science Publishers BV.)

systematically, the study provided evidence that the higher the theta field frequency, the greater the number of discharges per burst. This relationship was investigated in detail in a subsequent study (Bland et al., 1983), and shown to hold for both Type 2 and Type 1 theta conditions. These observations supported the idea that individual theta-ON cells coded for increasing levels of activation of sensory inputs and increasing levels of activation (speed of initiation) of movements. In a subsequent study Bland et al. (1984) administered atropine sulfate to rabbits during the presentation of sensory stimuli while immobile. As predicted, both Type 2 theta and the theta-ON cell rhythmic discharges were abolished by the administration of atropine sulfate but continued to discharge in a rhythmically bursting pattern during the Type 1 theta accompanying movement. However, the number of discharges per rhythmic burst was reduced compared to the pre-atropine condition (see Fig. 12.12).

Recent work by Shin et al. (2005) has provided additional strong support for the presence of both atropine-sensitive and atropine-resistant theta using a genetic knockout model. The PLC-β^{-1-} isoenzyme is the most critical of the four identified PLC-β isoenzymes in studies of hippocampal theta rhythms as it is coupled to muscarinic receptors (Shin et al., 2005) and Group 1 metabotropic glutamate receptors in the hippocampus (Chuang et al., 2001). Knockout mice for the PLC-β^{-1-} gene were shown to lack atropine-sensitive (Type 2) theta while atropine-resistant (Type 1) theta was intact. Carbachol-induced theta oscillations were abolished in hippocampal slices from PLC-β^{-1-} knockout mice, and in urethane-anesthetized PLC-β^{-1-} knockout mice Type 2 theta was absent. In freely moving PLC-β^{-1-} knockout mice, theta was preserved during Type 1 theta-related behaviors such as walking and running.

As discussed earlier, there is strong evidence for a septohippocampal glutamatergic pathway and support for it being involved in theta generation. Leung and Desborough (1988) previously reported that the infusion of AP5 into the lateral ventricles of freely moving rats resulted in the attenuation of HPC theta rhythm and the theta phase-shift at the apical dendrites in the CA1 region. These authors suggested that this effect was a selective suppression of Type 2 (atropine-sensitive) theta. Leung and Shen (2004) carried out a study to test whether the effects of the intraventricular AP5 administration they reported earlier were due to its action on the septum or HPC. They showed that AP5 infusions directly into either the HPC or the medial septum reduced the power but not the frequency of HPC theta. What might be the functional significance of this third theta-generating pathway? Bland et al. (2007a) showed that wheel-running behavior of rats induced by low levels of electrical stimulation of the posterior hypothalamic nucleus was completely abolished by microinfusion of the NMDA antagonist AP5 into the medial septum, accompanied by a significant reduction in theta amplitude (power). Wheel-running and theta were maintained at control levels in a high level PH stimulation condition. Bland et al. (2007a) hypothesized that the glutamatergic septohippocampal projection provided the excitatory drive for the rapid initiation of movement, superimposed on the tonic cholinergic/GABAergic drive occurring during the sensory processing period prior to movement. A study by Bland et al. (2006b) provided some insight as to how this might work. In these experiments rats were trained in an avoidance task to jump out of a box, the distance of which could be varied at three different heights. Amplitude and frequency of HPC Type 2 theta was measured in the immobility period just prior to jump initiation. The same parameters were measured for the Type 1 theta "jump" wave as well as measurements of the phase of the jump wave in relation to the moment of jump initiation. The results demonstrated that the immobility period prior to the execution of the jump could be divided into two components: a sensory processing period and a movement preparation period (see Fig. 12.13).

Comparing these two periods, average amplitudes were higher while frequency remained relatively constant during the sensory processing period. During the movement preparation period there was a negative correlation between amplitude and frequency: amplitude declined rapidly and frequency increased rapidly. During the execution of the jump, theta (Type 1) amplitude and frequency were positively correlated, both reaching peak values. Both the amplitude and the frequency of the Type 1 theta jump wave increased as jump height increased. These results supported the hypotheses that Type 1 theta amplitude was associated with the magnitude of the movement and Type 1 theta frequency was associated with the speed of initiation of the movement. A significant phase preference was demonstrated for the highest jump height, with movement initiation occurring around the trough of theta recorded from the stratum moleculare of the dentate region. These data suggested that Type 2 theta amplitude (power) during the sensory processing period might be associated with the determination of movement initiation while the frequency of Type 2 theta during the movement preparation period was associated with the intensity of initiation of that movement. In the wheel-running experiment discussed above, the microinfusion of AP5 into the medial septum resulted in the reduced amplitude of theta and blocking of PH-induced wheel-running at lower levels of PH stimulation. Increasing the level of PH stimulation resulted in an increase in theta amplitude, and running occurred.

Bland *et al.* (2007b) recently carried out another set of experiments designed to determine whether there was a relationship between Type 2 theta and subsequent movement. One group of rats were trained to escape shock in a runway avoidance task while another group received the same number and intensity of shocks but could not escape the start box. Twenty-four hours later both groups received a single shock probe in an open field test box. Most rats responded to the shock probe with "freezing" behavior (immobility). Rats trained to avoid generated Type 2 theta during immobility, while rats that could not escape produced LIA during the immobility period. Similar results were first reported by Balleine and Curthoys (1991). Our interpretation of these results was that rats receiving inescapable shock learned that the shock was not associated with the possibility of movement, and thus in the probe test produced only LIA, contrary to the avoidance-trained rats. To test this hypothesis the rats

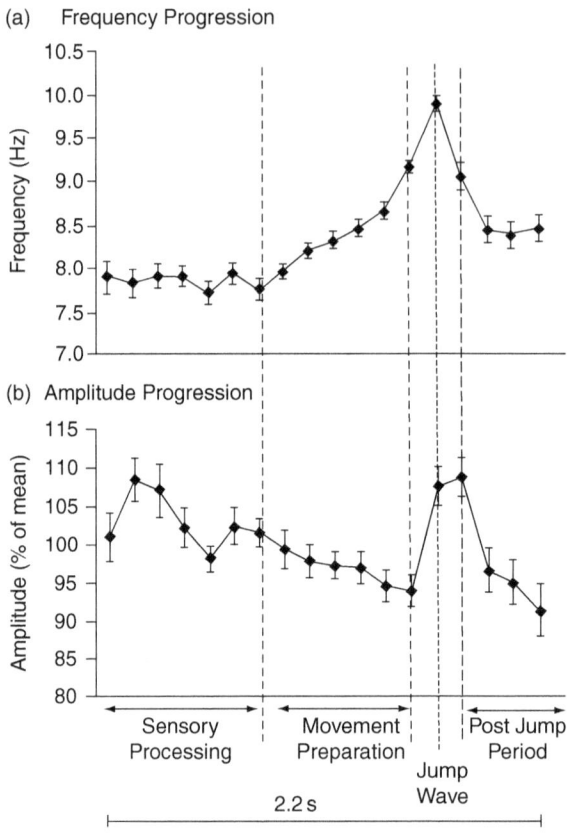

Figure 12.13 (a) Graph of group data collapsed across the three jump heights showing the progression of theta frequency before, during, and after the jumps. Each data point is the grand mean ± standard error derived from the averages of 30 measurements for each of the seven animals. Note the steep increase in frequency during the movement preparation period before the jump, and including the jump wave. Frequency declined beginning with the first post jump wave. (b) Graph of group data collapsed across the three jump heights showing the progression of theta amplitude before, during, and after the jumps. Each data point is the grand mean ± standard error derived from the averages of 30 measurements for each of the seven animals. Amplitudes for each animal were normalized before averaging. Note the increase in amplitude during the sensory processing period and for the jump wave. Amplitude continued to increase up to the first post jump wave and subsequently declined. (Reprinted with permission from John Wiley and Sons, Inc.)

received reversal training: the avoidance group received inescapable shock while the inescapable group was trained to avoid. The avoidance group retrained with inescapable shock now produced LIA in the shock probe test while the inescapable group retrained in avoidance produced Type 2 theta, thus supporting the hypothesis that previous experience determined whether movement preparation would occur.

Several other studies have carefully examined changes in hippocampal theta related to sensorimotor behavior (see review by Oddie and Bland, 1998). Oddie et al. (1997) tested the idea that Type 2 theta played a role in sensory integration and the neural processing required for the initiation of Type 1 movements, using a behavioral paradigm called "ducking and robbing." One of a pair of hungry rats, the victim, was given food that the other rat, the robber, would attempt to steal. Because the victim dodged from the robber with a latency, distance, and velocity dependent on the size of the food, elapsed eating time, and proximity to the robber, the movement required sensory integration and planning of subsequent movements. The study showed that although eating behavior continued and Type 1 theta was still recorded during movement, the intraseptal microinfusion of atropine sulfate abolished Type 2 theta and dodging behavior was severely disrupted. Wyble et al. (2004) reported a decrease in theta power in a runway task when rats were to receive reward following voluntary movement, and no decrease when the task went unrewarded. Both amplitude and frequency of theta have been shown to decrease as rats are completing a locomotor approach sequence (Sinnamon, 2005a). Sinnamon (2005b) also identified changes in the time course of maximal frequency and amplitude of hippocampal theta. This study identified a dissociation between frequency and amplitude as rats prepared to initiate or inhibit locomotion toward the possible presentation of a food pellet. Specifically, amplitude increased before the food pellet was presented, indicating that the rat was preparing to process the upcoming sensory information, and frequency increased in the moments before movement during certain trials. During consummatory behaviors (such as milk-lapping and consumption of food rewards) theta amplitude is reported to decrease in power, likely due to the automatic nature of the behavior (Wyble et al., 2004; Sinnamon, 2005b). Similarly, van Lier et al. (2003) demonstrated that during behavioral transitions, amplitude increased following a Type 1 behavior (voluntary movement) and was reduced when following an automatic, Type 2 behavior.

Much of the work discussed in the context of the sensorimotor integration model has dealt with how ascending sensory information is processed by the hippocampus. The model predicts that there would be relationships between the neural activity underlying theta-band oscillation and synchrony in the

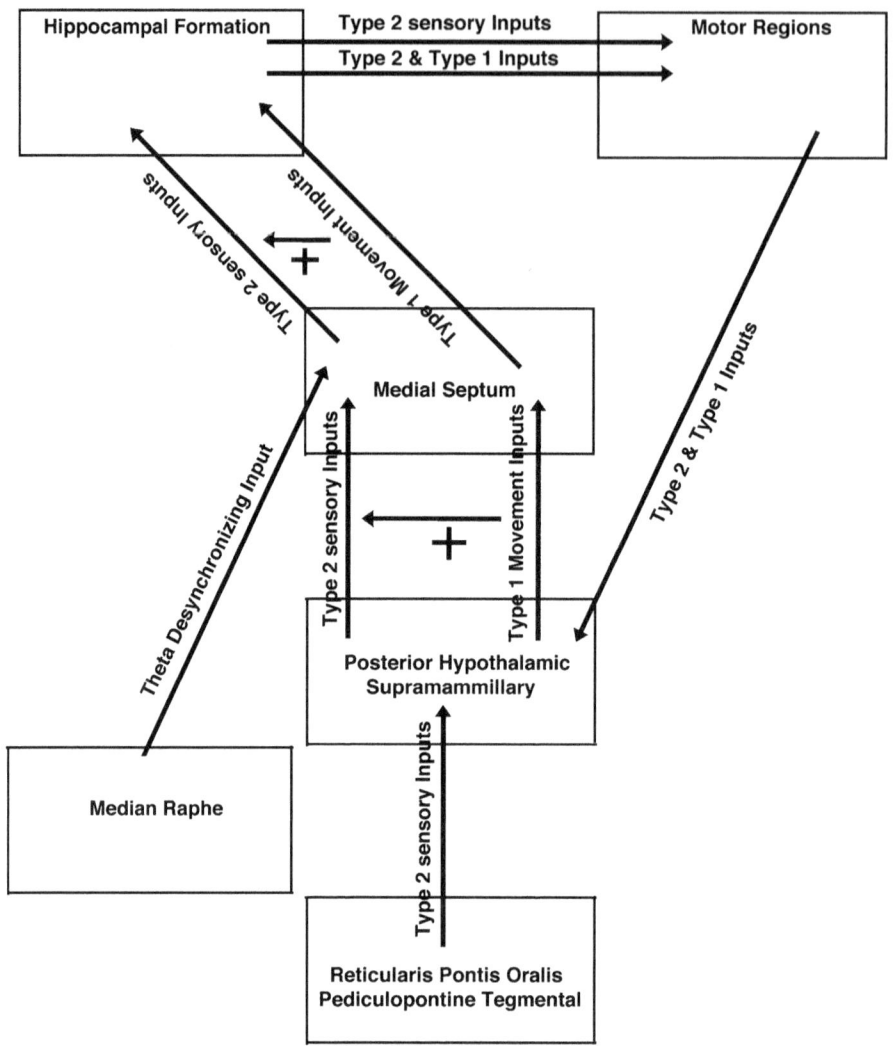

Figure 12.14 An updated diagrammatic representation of the sensorimotor model for the hippocampal formation theta subsystems. Ascending hippocampal synchronizing Type 2 sensory processing inputs illustrate that Type 2 theta can occur in isolation. Once a Type 1 (voluntary) movement is initiated, there is always a co-activation of Type 1 and Type 2 inputs, as illustrated by the addition Type 1 movement inputs from motor regions. The main median raphe ascending hippocampal desynchronizing input is also mediated by the medial septum. (Reprinted with permission from Elsevier Science Publishers BV.)

hippocampal formation (and related structures) and the neural activity in motor structures. Examinations of these possible relationships between the hippocampus, deep layers of the superior colliculus, the basal ganglia, and the red nucleus have indeed provided support for this hypothesis (Natsume *et al.*, 1999; Dypvik and Bland, 2004; Hallworth and Bland, 2004).

Since the data discussed above are based on animal research a fair question to ask is whether the sensorimotor integration model of hippocampal function relevant to humans? A long-standing controversy in the literature revolved around the question of whether theta-band oscillation and synchrony could even be recorded from the hippocampus of primates and humans. The answer is yes it can and it appears to be related to sensorimotor integration (Caplan *et al.*, 2003, Ekstrom *et al.*, 2005).

Figure 12.14 presents a diagrammatic representation of the anatomical basis of the model, including the ascending brainstem hippocampal synchronizing pathways as the basis of initiating sensorimotor integration and the ascending brainstem desynchronizing pathways as the basis of terminating sensorimotor integration. The connectivity is oversimplified for clarity.

Acknowledgments

This work was supported by Natural Sciences and Engineering Research Council (NSERC) Grant A9935 to BHB. I dedicate this chapter to my wife Cheryl E. Bland in recognition of her many years of love, support, and encouragement. I am indebted to M. Bruce MacIver for valuable suggestions on content and substantial improvements of the model in Fig. 12.14.

References

Acsady, L., Halasy, K. and Freund, T.F. (1993). Calretinin is present in non-pyramidal cells of the rat hippocampus: Their inputs from the median raphe and medial septal nuclei. *Neuroscience* **52**:829–841.

Alonso A., Gaztelu, J.M. and Buno, W., Jr., and Garcia-Austt, E. (1987). Cross-correlation analysis of septohippocampal neurons during θ-rhythm. *Brain Res* **413**:135–146.

Assaf, S.Y. and Miller, J.J. (1978). Role of a raphe serotonin system in control of septal unit activity and hippocampal desynchronization. *Neuroscience* **3**:539–550.

Aznar, S., Qian, Z.X., and Knudsen, G.M. (2004). Non-serotonergic dorsal and median raphe projection onto parvalbumin- and calbindin-containing neurons in the hippocampus and septum. *Neuroscience* **124**:573–581.

Balleine, B.W. and Curthoys, I.S. (1991). Differential effects of escapable and inescapable foot shock on hippocampal theta activity. *Behav Neurosci* **105**:202–209.

Bland, B.H. (1986). The physiology and pharmacology of hippocampal formation theta rhythms. *Progr Neurobiol* **26**:1–54.

Bland, B. H. (2000). The medial septum: node of the ascending brainstem hippocampal synchronizing pathways. In: *The Behavioral Neuroscience of the Septal Region*, ed. Numan, R., vol. 6, pp. 115–145. New York: Springer.

Bland, S. K. and Bland, B. H. (1986). Medial septal modulation of the hippocampal theta cell discharges. *Brain Res* **375**:1102–1111.

Bland, B. H. and Colom, L. V. (1988). Responses of phasic and tonic hippocampal theta-on cells to cholinergics: differential effects of muscarinic and nicotinic activation. *Brain Res* **440**:167–171.

Bland, B. H. and Colom, L. V. (1989). Preliminary observations on the physiology and pharmacology of hippocampal theta-off cells. *Brain Res* **505**:333–336.

Bland, B. H. and Colom, L. V. (1993). Extrinsic and intrinsic properties underlying oscillation and synchrony in limbic cortex. *Prog Neurobiol* **41**:157–208.

Bland, B. H. and Oddie, S. D. (1998). Anatomical, electrophysiological and pharmacological studies of ascending brainstem hippocampal synchronizing pathways. *Neurosci Biobehav Rev* **22**:259–273.

Bland, B. H. and Oddie, S. D. (2001). Theta band oscillation and synchrony in the hippocampal formation and related structures: the case for its role in sensorimotor integration. *Behav Brain Res* **127**:119–136.

Bland, B. H. and Vanderwolf, C. H. (1972a). Diencephalic and hippocampal mechanisms of motor activity in the rat: effects of posterior hypothalamic stimulation on behavior and hippocampal slow wave activity. *Brain Res* **43**:67–88.

Bland, B. H. and Vanderwolf, C. H. (1972b). Electrical stimulation of the hippocampal formation: behavioral and bioelectrical effects. *Brain Res* **43**:89–106.

Bland, B. H., Seto, M., Sinclair, B. R., and Fraser, S. M. (1983). The pharmacology of hippocampal theta cells: evidence the sensory processing correlate is cholinergic. *Brain Res* **299**:121–131.

Bland, B. H., Seto, M., and Rowntree, C. J. (1984). The relation of multiple hippocampal theta cell discharge rates to slow wave theta frequency. *Physiol Behav* **31**:111–117.

Bland, B. H., Colom, L. V., Konopacki, J., and Roth, S. B. (1988). Intracellular records of carbachol-induced theta rhythm in hippocampal slices. *Brain Res* **447**:364–368.

Bland, B. H., Colom, L. V., and Ford, R. D. (1990). Responses of septal theta-on and theta-off cells to activation of the dorsomedial-posterior hypothalamic region. *Brain Res Bull* **24**:71–79.

Bland, B. H., Oddie, S. D., Colom, L. V., and Vertes, R. P. (1994). Extrinsic modulation of medial septal cell discharges by the ascending brain stem hippocampal synchronizing pathway. *Hippocampus* **4**:649–660.

Bland, B. H., Konopacki, J., Kirk, I. J., Oddie, S. D., and Dickson, C. T. (1995). Discharge patterns of hippocampal theta-related cells in the caudal diencephalon of the urethane-anesthetized rat. *J Neurophysiol* **74**:322–333.

Bland, B. H., Trepel, C., Oddie, S. D., and Kirk, I. J. (1996). Intraseptal microinfusion of muscimol: effects on hippocampal formation field activity and phasic theta discharges. *Exp Neurol* **138**:286–297.

Bland, B. H., Colom, L. V., Oddie, D., Kirk, I. J., and Scarlett, D. (1997). Mechanisms of hippocampal theta generation: evidence from simultaneous recordings of medial septal and hippocampal cells. *Soc Neurosci Abst* **23**:486.

Bland, B. H., Oddie, S. D., and Colom, L. V. (1999). Mechanisms of neural synchrony in the septohippocampal pathways underlying hippocampal theta generation. *J Neurosci* **19**:3223–3237.

Bland, B. H., Konopacki, J., and Dyck, R. H. (2002). Relationship between membrane potential oscillation and rhythmic discharges in identified hippocampal theta-related cells. *J Neurophysiol* **88**:3046–3066.

Bland, B. H., Konopacki, J., and Dyck, R. H. (2005). Heterogeneity among hippocampal pyramidal neurons revealed by their relation to theta-band oscillation and synchrony. *Exp Neurol* **195**:458–474.

Bland, B. H., Bird, J., Jackson, J., and Natsume, K. (2006a). Medial septal modulation of the ascending brainstem hippocampal synchronizing pathways in the freely moving rat. *Hippocampus* **16**:11–19.

Bland, B. H., Jackson, J., Derie-Gillespie, D., et al. (2006b). Amplitude, frequency, and phase analysis of hippocampal theta during sensorimotor processing in a jump avoidance task. *Hippocampus* **16**:673–681.

Bland, B. H., DeClerck, S., Jackson, J., Glasgow, S., and Oddie, S. D. (2007a). Septohippocampal properties of N-methyl-D-aspartate-induced theta band oscillation and synchrony. *Synapse* **61**:185–197.

Bland, B. H., Mestek, P., Jackson, J., Crooks, R., and Cormican, A. (2007b). To move or not: previous experience in a runway avoidance task determines the appearance of hippocampal type 2 sensory processing. *Behav Brain Res* **179**:299–304.

Bonansco, C. and Buno, W. (2003). Cellular mechanisms underlying the rhythmic bursts induced by NMDA microiontophoresis at the apical dendrites of CA1 pyramidal neurons. *Hippocampus* **13**:150–163.

Brucke, F., Petsche, H., Pillat, B., and Deisenhammer, E. (1959). Die beeinflussung der "Hippocampusarousal-reaktion" beim Kaninchen durch elektrische Reizung im Septum. *Pflugers Arch Ges Physiol* **269**:319–338.

Buzsáki, G. (2002). Theta oscillations in the hippocampus. *Neuron* **33**:325–340.

Caplan, J. B., Madsen, J. R., Schulze-Bonhage, A., et al. (2003). Human oscillations related to sensorimotor integration and spatial learning. *J Neurosci* **23**:4726–4736.

Carre, G. P. and Harley, C. W. (2000). Glutamatergic activation of the medial septum complex: an enhancement of the dentate gyrus population spike and accompanying EEG and unit changes. *Brain Res* **861**:16–25.

Chapman, A. C. and LaCaille, J. C. (1999). Intrinsic theta-frequency membrane potential oscillations in CA1 interneurons of stratum moleculare. *J Neurophysiol* **81**:1296–1307.

Chuang, S. C., Bianchi, R., Kim, D., Shin, H. S. and Wong, R. K. (2001). Group 1 metabotropic glutamate receptors elicit epileptiform discharges in the hippocampus through PLC! Signaling. *J Neurosci* **21**:6387–6394.

Colom, L. V. and Bland, B. H. (1987). State-dependent spike train dynamics of hippocampal formation neurons: evidence for theta-on and theta-off cells. *Brain Res* **422**:277–286.

Colom, L. V. and Bland, B. H. (1991). Medial septal cell interactions in relation to hippocampal field activity and the effects of atropine. *Hippocampus* **1**:5-30.

Colom, L. V., Ford, R. D., and Bland, B. H. (1987). Hippocampal formation neurons code the level of activation of the cholinergic septohippocampal pathways. *Brain Res* **410**:12-20.

Colom, L. V., Christie, B. R., and Bland, B. H. (1988). Cingulate cell discharges related to hippocampal EEG and their modulation by muscarinic and nicotinic agents. *Brain Res* **460**:329-338.

Colom, L. V., Nassif-Caudarella, S., Dickson, C. T., Smythe, J. W. and Bland, B. H. (1991). *In vivo* intrahippocampal microinfusion of carbachol and bicuculline induces theta-like oscillations in the septally deafferented hippocampus. *Hippocampus* **1**:381-390.

Colom, L. V., Castaneda, M. T., Reyna, T., Hernandez, S., and Garrido-Sanabria, E. (2005). Characterization of medial septal glutamatergic neurons and their projection to the hippocampus. *Synapse* **58**:151-164.

Danik, M., Puma, C., Quirion, R., and Williams, S. (2003). Widely expressed transcripts for chemokine receptor CXCR1 in identified glutamatergic, gamma-aminobutyric acidergic, and cholinergic neurons and astrocytes of the rat brain: a single-cell reverse transcription-multiplex polymerase chain reaction study. *J Neurosci Res* **74**:286-295.

Danik, M., Cassoly, E., Manseau, F., et al. (2005). Frequent coexpression of the vesicular glutamate transporter 1 and 2 genes, as well as coexpression with genes for choline acetyltransferase or glutamic acid decarboxylase in neurons of rat brain. *J Neurosci Res* **81**:506-521.

Dickson, C. T., Trepel, C., and Bland, B. H. (1994). Extrinsic modulation of theta field activity in the entorhinal cortex of the anesthetized rat. *Hippocampus* **4**:37-52.

Dickson, C. T., Kirk, I. J., Oddie, S. D., and Bland, B. H. (1995). Classification of theta-related cells in the entorhinal cortex: cell discharges are controlled by the ascending brainstem synchronizing pathway in parallel with hippocampal theta cells. *Hippocampus* **5**:306-319.

Dypvik, A. and Bland, B. H. (2004). Functional connectivity between the red nucleus and the hippocampus supports the role of hippocampal formation in sensorimotor integration. *J Neurophysiol* **92**:2040-2050.

Ekstrom, A. D., Caplan, J. B., Ho, H., et al. (2005). Human hippocampal theta activity during virtual navigation. *Hippocampus* **15**:881-889.

Ford, R. D., Colom, L. V., and Bland, B. H. (1989). The classification of medial septum-diagonal band cells as theta-on or theta-off in relation to hippocampal EEG states. *Brain Res* **493**:269-282.

Fox, S. E. (1989). Membrane potential and impedance changes in hippocampal pyramidal cells during theta rhythm. *Exp Brain Res* **77**:283-294.

Fox, S. E. and Ranck, J. B. (1975). Localization and anatomical identification of theta and complex spike cells in dorsal hippocampal formation of rats. *Exp Neurol* **49**:229-313.

Fox, S. E. and Ranck, J. B. (1981). Electrophysiological characteristics of hippocampal complex-spike cells and theta cells. *Exp Brain Res* **41**:399–410.

Fox, S. E., Wolfson, S., and Ranck, J. B. (1983). Investigating the mechanisms of hippocampal theta rhythm: approaches and progress. In: *Neurobiology of the Hippocampus*, ed. Seifert, W., pp. 303–319. New York: Academic Press.

Fremeau, R. T., Jr., Troyer, M. D., Pahner, I., *et al.* (2001). The expression of vesicular glutamate transporters defines two classes of excitatory synapse. *Neuron* **31**:247–260.

Freund, T. F. (1989). GABA-ergic septohippocampal neurons contain parvalbumin. *Brain Res* **478**:375–381.

Freund, T. F. and Antal, M. (1988). GABA-containing neurons in the septum control inhibitory interneurons in the hippocampus. *Nature* **336**:170–173.

Frotscher, M. and Leranth, C. (1985). Cholinergic innervation of the rat hippocampus as revealed by choline-acetyltransferase immunocytochemistry: a combined light and electron-microscope study. *J Comp Neurol* **239**:237–246.

Fujita, Y. and Sato, T. (1964). Intracellular records from hippocampal pyramidal cells in rabbits during theta rhythm activity. *J Neurophysiol* **27**:1011–1025.

Gerstein, G. L. (1970). Functional associations of neurons: detection and interpretation. In: *The Neurosciences: Second Study Program*, ed. Schmitt, F. O., pp. 648–661. New York: Rockefeller University Press.

Gonzalo-Ruiz, A. and Morte, L. (2000). Localization of amino acids, neuropeptides, cholinergic markers in neurons of the septum-diagonal band complex projecting to the retrosplenial granular cortex of the rat. *Brain Res Bull* **52**:499–510.

Graeff, F. G. and Silveira Filho, N. G. (1978). Behavioral inhibition produced by electrical stimulation of the median raphe nucleus of the rat. *Physiol Behav* **21**:477–484.

Graeff, F. G., Quintero, S., and Gray, J. S. (1980). Median raphe stimulation, hippocampal theta and threat-induced behavioural inhibition. *Physiol Behav* **25**:253–261.

Gray, J. A. and Ball, C. G. (1970). Frequency-specific relation between hippocampal theta rhythm. *Science* **168**:1246–1248.

Gritti, I., Mainville, L., Mancia, M., and Jones, B. E. (1997). GABAergic and other noncholinergic basal forebrain neurons, together with cholinergic neurons, project to the mesocortex and isocortex in the rat. *J Comp Neurol* **383**:163–177.

Gulyas, A. I., Gorcs, T. J., and Freund, T. F. (1990). Innervation of different peptide-containing neurons in the hippocampus by GABAergic septal afferents. *Neuroscience* **37**:31–44.

Hallworth, N. E. and Bland, B. H. (2004). Basal ganglia–hippocampal interactions support the role of the hippocampal formation in sensorimotor integration. *Exp Neurol* **188**:430–443.

Hanada, Y., Hallworth, N. E., Szgatti, T. L., and Bland, B. H. (1999). The distribution and analysis of hippocampal theta-related cells in the pontine region of the urethane-anestheized rat. *Hippocampus* **9**:288–302.

Jackson, J. C. and Bland, B. H. (2006). Medial septal modulation of the ascending brainstem hippocampal synchronizing pathways in the acute rat. *Hippocampus* **16**:1–10.

Jackson, J. C., Cormican, A. M., and Bland, B. H. (2006). Median raphe stimulation inhibits hippocampal theta-ON cells and posterior hypothalamic induced running behavior. *Soc Neurosci Abstr* **210**:2.

Kinney, G. G., Kocsis, B., and Vertes, R. P. (1994). Injections of excitatory amino acid antagonists into the median raphe nucleus produce hippocampal theta rhythm in the urethane anesthetized rat. *Brain Res* **654**:96–104.

Kinney, G. G., Kocsis, B., and Vertes, R. P. (1995). Injections of muscimol into the median raphe nucleus produce hippocampal theta rhythm in the urethane anesthetized rat. *Psychopharmacology* **120**:244–248.

Kinney, G. G., Kocsis, B., and Vertes, R. P. (1996). Medial septal unit firing characteristics following injections of 8-OH-DPAT into the median raphe nucleus. *Brain Res* **708**:116–122.

Kirk, I. J. and McNaughton, N. (1991). Supramammillary cell firing and hippocampal rhythmical slow activity. *Neuroreport* **2**:723–725.

Kirk, I. J. and McNaughton, N. (1993). Mapping the differential effects of procaine on the frequency and amplitude of reticularly elicited rhythmical slow activity. *Hippocampus* **3**:517–526.

Kirk, I. J., Oddie, S. D., Konopacki, J., and Bland, B. H. (1996). Evidence for differential control of posterior hypothalamic, supramammillary, and medial mammillary theta-related cellular discharge by ascending and descending pathways. *J Neurosci* **16**:5547–5554.

Kiss, J., Magloczky, Z., Somogyi, J., and Freund, T. F. (1997). Distribution of calretinin-containing neurons relative to other neurochemically identified cell types in the medial septum of the rat. *Neuroscience* **78**:399–410.

Kiss, J., Csaki, A., Bokor, H., Kocsis, K., and Kocsis, B. (2002). Possible glutamatergic/aspartatergic projections to the supramammillary nucleus and their origins in the rat studied by selective [(3)H] D-aspartate labeling and immunocytochemistry. *Neuroscience* **111**:671–691.

Kitchigina, V. F., Kudina, T. A., Kutyreva, E. V., and Vinogradova, O. S. (1999). Neuronal activity of the septal pacemaker of theta rhythm under the influence of stimulation and blockade of the median raphe nucleus in the awake rabbit. *Neuroscience* **94**:453–463.

Kocsis, B., and Vertes, R. P. (1994). Characterization of neurons in the supramammillary nucleus and mammillary body that discharge rhythmically with the hippocampal theta rhythm in the rat. *J Neurosci* **14**:7040–7052.

Konopacki, J., MacIver, M. B., Bland, B. H., and Roth, S. H. (1987a). Theta in hippocampal slices: relation to synaptic responses of dentate neurons. *Brain Res Bull* **18**:25–27.

Konopacki, J., MacIver, M. B., Roth, S. H., and Bland, B. H. (1987b). Carbachol-induced EEG "theta" activity in hippocampal brain slices. *Brain Res* **405**:196–198.

Konopacki, J., Bland, B. H., and Roth, S. H. (1987c). Phase shifting of CA1 and dentate EEG "theta" in hippocampal formation slices. *Brain Res* **417**:399–402.

Konopacki, J., Bland, B. H., MacIver, M. B., and Roth, S. H. (1987d). Cholinergic theta rhythm in transected hippocampal slices: independent CA1 and dentate generators. *Brain Res* **436**:217–222.

Konopacki, J., Bland, B. H., and Roth, S. H. (1988a). Carbachol-induced EEG theta in hippocampal formation slices: evidence for a third generator of theta in CA3C area. *Brain Res* **451**:33–42.

Konopacki, J., Bland, B. H., and Roth, S. H. (1988b). Evidence that activation of in vitro hippocampal theta rhythm involves only muscarinic receptors. *Brain Res* **455**:110–114.

Konopacki, J., Bland, B. H., and Roth, S. H. (1988c). The development of carbachol-induced EEG theta examined in hippocampal formation slices. *Dev Brain Res* **38**:229–232.

Konopacki, J., Bland, B. H., Colom, L. V., and Oddie, S. D. (1992). In vivo intracellular correlates of hippocampal formation theta-on and theta-off cells. *Brain Res* **586**:247–255.

Konopacki, J., Eckersdorf, B., Kowalczyk, T., and Golebiewski, H. (2006). Firing cell repertoire during carbachol-induced theta rhythm in rat hippocampal formation slices. *Eur J Neurosci* **23**:1811–1818.

Kramis, R. C. and Routenberg, A. (1977). Dissociation of hippocampal EEG from its behavioral correlates by septal and hippocampal electrical stimulation. *Brain Res* **125**:37–49.

Kramis, R. C., Vanderwolf, C. H., and Bland, B. H. (1975). Two types of hippocampal rhythmical slow activity in both the rabbit and the rat: relations to behavior and effects of atropine, diethyl, ether, urethane, and pentobarbital. *Exp Neurol* **49**:58–85.

Lawson, V. H. and Bland, B. H. (1993). The role of the septohippocampal pathways in the regulation of hippocampal field activity and behavior: analysis by the intraseptal microinfusion of carbachol, atropine, and procaine. *Exp Neurol* **129**:132–144.

Lee, A. K., Manns, I. D., Sakmann, B., and Brecht, M. (2006). Whole cell recordings in freely moving rats. *Neuron* **51**:399–407.

Leung, L. S. (1984a). Pharmacology of theta phase shift in the hippocampal CA1 region of freely moving rats. *Electroencephalogr Clin Neurophysiol* **60**:457–466.

Leung, L. S. (1984b). Theta rhythm during REM and waking: correlation between power, phase and frequency. *Electroencephalogr Clin Neurophysiol* **58**:553–564.

Leung, L. S. and Shen, B. (2004). Glutamatergic synaptic transmission participates in generating the hippocampal EEG. *Hippocampus* **14**:510–525.

Leung, L. S. and Yim, C. Y. (1986). Intracellular records of theta rhythm in hippocampal CA1 cells of the rat. *Brain Res* **367**:323–327.

Leung, L. S. and Yim, C. Y. (1988). Membrane potential oscillations in hippocampal neurons in vitro induced by carbachol or depolarizing currents. *Neurosci Res Commun* **2**:159–167.

Leung, L. S. and Yim, C. Y. (1991). Intrinsic membrane potential oscillations in hippocampal neurons in vitro. *Brain Res* **553**:261–274.

Leung, L. S. and Yu, H. W. (1998). Theta frequency resonance in hippocampal CA1 neurons in vivo demonstrated by sinusoidal current injection. *J Neurophysiol* **79**:1592–1596.

Leung, L. S., Lopes Da Silva, F. H., and Wadman, W. J. (1982). Spectral characteristics of the hippocampal EEG in freely moving rat. *Electroencephalogr Clin Neurophysiol* **54**:203–219.

Leung, L. W. and Desborough, K. A. (1988). APV, an N-methyl-D-aspartate receptor antagonist, blocks the hippocampal theta rhythm in behaving rats. *Brain Res* **463**:148–152.

Lukatch, H. S. and MacIver, M. B. (1997). Physiology, pharmacology and topography of cholinergic neocortical oscillations in vitro. *J Neurophysiol* **77**:2427–2445.

Macadar, S. W., Roig, A., Monti, M., and Budelli, R. (1970). The functional relationship between septal and hippocampal unit activity and hippocampal theta rhythm. *Physiol Behav* **5**:1443–1449.

Macadar, S. W., Chalupa, L. M., and Lindsley, D. B. (1974). Differentiation of brain stem loci which affect hippocampal and neocortical activity. *Exp Neurol* **43**:499–514.

MacVicar, B. A. and Tse, F. W. Y. (1989). Local neuronal circuitry underlying cholinergic rhythmical slow activity in CA3 area of rat hippocampal slices. *J Physiol* **417**:197–212.

Manns, I. D., Mainville, L., and Jones, B. E. (2001). Evidence for glutamate, in addition to acetylcholine and GABA, neurotransmitter synthesis in basal forebrain neurons projecting to the entorhinal cortex. *Neuroscience* **107**:249–263.

Maru, E., Takahashi, K., and Iwahara, S. (1979). Effects of median raphe nucleus lesions on hippocampal EEG in the freely moving rat. *Brain Res* **163**:223–234.

McNaughton, B. L., Barnes, C. A., and O'Keefe, J. (1983). The contributions of position, direction and velocity to single unit activity in the hippocampus of freely-moving rats. *Exp Brain Res* **52**:41–49.

Meittinen, R. and Freund, T. F. (1992). Neuropeptide Y-containing interneurons in the hippocampus receive synaptic input from median raphe and GABAergic septal afferents. *Neuropeptides* **22**:185–193.

Misgeld, U. and Frotscher, M. (1986). Postsynaptic-GABA-ergic inhibition of non-pyramidal neurons in the guinea pig hippocampus. *Neuroscience* **19**:185–193.

Mizumori, S. J. Y., Barnes, C. A., and McNaughton, B. L. (1990). Behavioral correlates of theta-on and theta-off cells recorded from hippocampal formation of mature young and aged rats. *Exp Brain Res* **80**:365–373.

Monmaur, P. and Breton, P. (1991). Elicitation of hippocampal theta by intraseptal carbachol injection in freely moving rats. *Brain Res* **544**:150–155.

Munoz, M. D., Nunez, A., and Garcia-Austt, E. (1990). *In vivo* intracellular analysis of rat dentate granule cells. *Brain Res* **509**:91–98.

Natsume, K., Hallworth, N. E., Szagatti, T. L., and Bland, B. H. (1999). Hippocampal theta related cellular activity in the superior colliculus of the urethane-anesthetized rat. *Hippocampus* **9**:500–509.

Nunez, A., Garcia-Austt, E., and Buno, W., Jr. (1987). Intracellular theta rhythm generation in identified hippocampal pyramids. *Brain Res* **416**:289–300.

Nunez, A., Garcia-Ausst, E., and Buno, W., Jr. (1990a). Synaptic contributions to theta rhythm genesis in rat CA1–CA3 hippocampal pyramidal neurons *in vivo*. *Brain Res* **533**:176–179.

Nunez, A., Garcia-Austt, E., and Buno, W., Jr. (1990b). In vivo electrophysiological analysis of Lucifer yellow-coupled hippocampal pyramids. *Exp Neurol* **108**:76–82.

Nunez, A., Garcia-Austt, E., and Buno, W., Jr., (1990c). Slow intrinsic spikes recorded in vivo in rat CA1–CA3 hippocampal pyramidal neurons. *Exp Neurol* **109**:294–299.

Nunez, A., de Andres, I., and Garcia-Austt, E. (1991). Relationship of nucleus reticularis pontis caudalis neuronal discharge with sensory and carbachol evoked hippocampal theta rhythm. *Exp Brain Res* **87**:303–308.

Oddie, S. D. and Bland, B. H. (1998). Hippocampal formation theta activity and movement selection. *Neurosci Biobehav Rev* **22**:21–231.

Oddie, S. D., Bland, B. H., Colom, L. V., and Vertes, R. P. (1994). The midline posterior hypothalamic region comprises a critical part of the ascending brainstem hippocampal synchronizing pathway. *Hippocampus* **4**:454–473.

Oddie, S. D., Stefanek, W., Kirk, I. J., and Bland, B. H. (1996). Intraseptal procaine abolishes hypothalamic stimulation-induced wheel-running and hippocampal theta field activity in rats. *J Neurosci* **16**:1948–1956.

Oddie, S. D., Kirk, I. J., Whishaw, I. Q., and Bland, B. H. (1997). Hippocampal formation is involved in movement selection: evidence from medial septal cholinergic modulation and concurrent slow-wave (theta rhythm) recording. *Behav Brain Res* **88**:169–180.

Pan, W. X. and MacNaughton, N. (2004). The supramammillary area: its organization, functions and relationship to the hippocampus. *Prog Neurobiol* **74**:127–166.

Peck, B. K. and Vanderwolf, C. H. (1991). Effects of raphe stimulation on hippocampal and neocortical activity and behavior. *Brain Res* **568**:244–252.

Robinson, T. E. and Vanderwolf, C. H. (1978). Electrical stimulation of the brainstem in freely moving rats. II. Effects on hippocampal and neocortical electrical activity, and relations to behavior. *Exp Neurol* **61**:485–515.

Robinson, T. E. and Whishaw, I. Q. (1974). Effects of posterior hypothalamic lesions on voluntary behavior and hippocampal electroencephalograms in the rat. *J Comp Physiol Psychol* **86**:768–786.

Rowntree, C. I. and Bland, B. H. (1986). An analysis of cholinoceptive neurons in the hippocampal formation by direct microinfusion. *Brain Res* **362**:98–113.

Scarlett, D., Dypvik, A. T., and Bland, B. H. (2004). Comparison of spontaneous and septally driven hippocampal theta field and theta-related cellular activity. *Hippocampus* **14**:99–106.

Segal, M. (1975). Physiological and pharmacological evidence for a serotonergic projection to the hippocampus. *Brain Res* **94**:115–131.

Shin, J., Kim, D., Bianchi, R., Wong, R. K. S. and Shin, H. (2005). Genetic dissection of theta rhythm heterogeneity in mice. *Proc Natl Acad Sci USA* **102**:18165–18170.

Siegal, J. M., McGinty, D. J., and Breedlove, S. M. (1977). Sleep and waking activity of pontine gigantocellular field neurons. *Exp Neurol* **56**:553–573.

Sinclair, B. R., Seto, G. S., and Bland, B. H. (1982). Cells in CA1 and dentate layers of hippocampal formation: relations to slow-wave activity and motor behavior in the freely moving rabbit. *J Neurophysiol* **48**:1214–1222.

Sinnamon, H. M. (1993). Preoptic and hypothalamic neurons and the initiation of locomotion the anesthetized rat. *Prog Neurobiol* **41**:323–344.

Sinnamon, H. M. (2000). Priming pattern determines the correlation between hippocampal theta activity and locomotor stepping elicited by stimulation in anesthetized rats. *Neuroscience* **98**:459–470.

Sinnamon, H. M. (2005a). Hippocampal theta activity related to elicitation and inhibition of approach locomotion. *Behav Brain Res* **160**:236–249.

Sinnamon H. M. (2005b). Hippocampal theta activity and behavioral sequences in a reward-directed approach task. *Hippocampus* **15**:518–534.

Sinnamon, H. M., Jassen, A. K., and Ilch, C. P. (2000). Hippocampal theta activity and facilitated locomotor stepping produced by GABA injections in the midbrain raphe region. *Behav Brain Res* **107**:93–1103.

Smythe, J. W., Christie, B. R., Colom, L. V., Lawson, V. H., and Bland, B. H. (1991). Hippocampal theta field activity and theta-on/theta-off cell discharges are controlled by an ascending hypothalamo-septal pathway. *J Neurosci* **11**:2241–2248.

Smythe, J. W., Colom, L. V., and Bland, B. H. (1992). The extrinsic modulation of hippocampal theta depends on the coactivation of cholinergic and GABA-ergic medial septal inputs. *Neurosci Biobehav Rev* **16**:289–308.

Sotty, F., Danik, M., Manseau, F., et al. (2003). Distinct electrophysiological properties of glutamatergic, cholinergic and GABAergic rat septohippocampal neurons: novel implications for hippocampal rhythmicity. *J Physiol* **551**:927–943.

Stumpf, C. (1965). The fast component in the electrical activity of rabbit's hippocampus. *Electroencephalogr Clin Neurophysiol* **18**:477–486.

Vanderwolf, C. H. (1969). Hippocampal electrical activity and voluntary movement in the rat. *Electroencephalogr Clin Neurophysiol* **26**:407–418.

Vanderwolf, C. H. (1975). Neocortical and hippocampal activation in relation to behavior: effects of atropine, eserine, phenothiazines, and amphetamine. *J Comp Physiol Psychol* **88**:306–323.

Vanderwolf, C. H. and Baker, G. B. (1986). Evidence that serotonin mediates non-cholinergic neocortical low voltage fast activity, non-cholinergic hippocampal rhythmical slow activity and cognitive abilities. *Brain Res* **374**:342–356.

Van Lier, H., Coenen, A. M. L., and Drinkenberg, W. H. (2003). Behavioral transitions modulate hippocampal electroencephalogram correlates of open field behavior in the rat: support for the sensorimotor function of hippocampal rhythmical synchronous activity. *J Neurosci* **23**:2459–2465.

Varga, V., Sik, A., Freund, T. F., and Kocsis, B. (2002). GABA (B) receptors in the median raphe nucleus: distribution and role in the serotonergic control of hippocampal activity. *Neuroscience* **109**:119–132.

Vertes, R. P. (1977). Selective firing of rat pontine gigantocellular neurons during movement and REM sleep. *Brain Res* **128**:146–152.

Vertes, R. P. (1979). Brain stem gigantocellular neurons: patterns of activity during behavior and sleep in the freely moving rat. *J Neurophysiol* **42**:214–228.

Vertes, R. P. (1981). An analysis of ascending brain stem systems involved in hippocampal synchronization and desynchronization. *J Neurophysiol* **46**:1140–1159.

Vertes, R. P. (1992). PHA-L analysis of projections from the supramammillary nucleus in the rat. *J Comp Neurol* **326**:595–622.

Vertes, R. P. and Kocsis, B. (1997). Brainstem-diencephalo-septohippocampal systems controlling the theta rhythm of the hippocampus. *Neuroscience* **81**:893–926.

Vertes, R. P., Colom, L. V., Fortin, W. J., and Bland, B. H. (1993). Brainstem sites for the carbachol elicitation of the hippocampal theta rhythm in the rat. *Exp Brain Res* **96**:419–429.

Vertes, R. P., Kinney, G. G., Kocsis, B., and Fortin, W. J. (1994). Pharmacological suppression of the median raphe nucleus with serotonin agonists, 8-OH-DPAT and busiperone, produces hippocampal theta rhythm in the rat. *Neuroscience* **60**:441–451.

Vertes, R. P., Crane, A. M., Colom, L. V., and Bland, B. H. (1995). Ascending projections of the posterior nucleus of the hypothalamus: a PHA-L analysis in the rat. *J Comp Neurol* **359**:90–116.

Vertes, R. P., Fortin, W., and Crane, A. M. (1999). Projections of the median raphe nucleus in the rat. *J Comp Neurol* **407**:555–582.

Vertes, R. P., Hoover, W., and Di Prisco, G. V. (2004). Theta rhythm of the hippocampus: subcortical control and functional significance. *Behav Cogn Neurosci Rev* **3**:173–200.

Viana Di Prisco, G., Albo, Z., Vertes, R. P., and Kocsis, B. (2002). Discharge properties of neurons of the median raphe nucleus during hippocampal theta rhythm in the rat. *Exp Brain Res* **145**:383–394.

Vinogradova, O. S., Kitchigina, V. F., Kudina, T. A., and Zenchenko, K. I. (1999). Spontaneous activity and sensory responses of hippocampal neurons during persistant theta-rhythm evoked by median raphe nucleus blockade in rabbit. *Neuroscience* **94**:745–753.

Woodnorth, M. A. and MacNaughton, N. (2005). Different systems in the posterior hypothalamic nucleus of rats control theta frequency and trigger movement. *Behav Brain Res* **163**:107–114.

Wyble, B. P., Hyman, J. M., Rossa, C. A., and Hasselmo, M. E. (2004). Analysis of theta power in hippocampal EEG during bar pressing and running in rats during distinct behavioral contexts. *Hippocampus* **14**:662–674.

Yamamoto, T., Watanabe, S., Oshi, R., and Ueki, S. (1979). Effects of midbrain raphe stimulation and lesion on EEG activity in rats. *Brain Res Bull* **4**:491–495.

Ylinen, A., Solstez, I., Bragin, A., *et al.* (1995). Intracellular correlates of hippocampal theta rhythm in identified pyramidal cells, granule cells, and basket cells. *Hippocampus* **5**:78–90.

13

A face in the crowd: which groups of neurons process face stimuli, and how do they interact?

KARI L. HOFFMANN

Introduction

Neural responses to face stimuli may seem like an unwieldy subject for investigating population activity: neurons with face-selective responses are many synapses removed from sensory input, the coding for faces appears to be very sparse, and the stimuli are complex making "proper" control stimuli difficult to come by. So why bother? To the extent that population coding underlies certain cognitive abilities, then those activities that are biological imperatives for the animal should be given "neural priority." In the rat, foraging and spatial localization relative to "home" points is one critical natural behavior. In primates, social cognition is essential. With the face at the heart of social communication and identification of social status, it should not come as a surprise that neurons appear to "care" about face stimuli in a way not seen for many non-face objects. But the nature of perceiving and learning about facial signals, in terms of population dynamics, is very under-explored territory. Surprisingly, in regions most often associated with face-selective responses, the conclusion of some researchers has been that population activity may add little to nothing to the perception of faces. The current state of knowledge regarding neural bases of face perception will be discussed. The role, if any, of population dynamics, will then be explored. Specifically, the population interactions of face-processing systems across space (e.g. circuits), and time (e.g. oscillations) will be discussed.

Information Processing by Neuronal Populations, ed. Christian Hölscher and Matthias Munk. Published by Cambridge University Press. © Cambridge University Press 2009.

The many facets of faces: what can be extracted from a face

The face stands as the single most reliable and versatile indicator of social signals. Where studied, non-human primates whose social structures include dominance hierarchies, vocal communication, and communication through facial expressions appear to share many of the face-processing abilities demonstrated in humans. In the literature on the neural basis of face processing, nearly all of the non-human primate studies come from macaques. It is, therefore, worth considering what signals macaques can extract from a face.

Like humans, monkeys are able to identify conspecifics based on their faces (Rosenfeld and Van Hoesen, 1979), they recognize kin relationships and rank in the dominance hierarchy (Haude et al., 1976; Dasser, 1988), discriminate facial expressions (Redican et al., 1971; Haude and Detwiler, 1976) and gender (Koba and Izumi, 2006), and are sensitive to changes in gaze direction (Emery et al., 1997; Sato and Nakamura, 2001; Deaner and Platt, 2003), while maintaining an ability to generalize across viewpoints when recognizing an individual (Rosenfeld and Van Hoesen, 1979; Heywood and Cowey, 1992). Physiological markers of arousal also indicate that macaques differentiate directed and averted gaze (Hoffman et al., 2007a).

Beyond general sensitivity to gaze, monkeys share with humans the tendency to look or attend in the direction of seen gaze (Emery et al., 1997; Deaner and Platt, 2003). This "gaze following" produces faster responses to stimuli presented in locations consistent with seen gaze, even when the gaze is not predictive of object location in the context of the task. The effect, though rapid and considered "automatic," has also been shown to depend on the dominance rank of the cuing and subject monkeys (Shepherd et al., 2006).

Some species- or genus-specific signals must also be considered: reddening of the face occurs during the mating season in male macaques, and such reddened male faces were preferentially viewed by female rhesus monkeys, suggesting that sexual receptivity or fitness may also be signaled through the face (Waitt et al., 2003).

Perhaps as a consequence of the plethora and importance of signals generated by faces, the perception of faces differs from the perception of other homogeneous object classes. For example, rhesus macaques naturally differentiate conspecifics better than other animals (Humphrey, 1974). The perceived difference between pairs of unfamiliar rhesus faces is greater than that between other unfamiliar animals' faces, including other species of monkeys (Dahl et al., 2007). Without explicit training or reinforcement, macaques show holistic face-processing effects expressed as increased sensitivity to changes in aligned versus misaligned face parts, and in some cases, they show sensitivity to configural manipulations of eye spacing (Dahl et al., 2007). Macaques have also shown, albeit less consistently, a face inversion effect (Swartz, 1983; Perrett et al., 1988; Vermeire and Hamilton, 1998) (but see Rosenfeld and Van Hoesen, 1979; Bruce, 1982; Parr et al., 1999; Gothard et al., 2004).

In conclusion, there is evidence that monkeys, like humans, are attuned to the faces of their conspecifics. Information provided by the face includes the gender and identity of the monkey, and his or her associated position in the social hierarchy. Monkeys are sensitive to communicative gestures such as silent facial expressions and vocalizations, as well as to their interactions with head and gaze direction, indicating the intended target of facial expressions. More generally, another's gaze direction appears to be used as an indication of salient objects or locations in the environment.

Common neural specializations for processing facial signals may explain the various face-processing abilities that appear to be conserved across species. The next section considers some of the key nodes in a face-processing network that may be conserved across social primate species.

Where in the brain does face processing occur?

Human resources

In humans, the recognition of individuals, their facial expressions, and their direction of gaze involves a broad network of brain structures. Imaging studies have shown activation in several key regions including the *lateral fusiform gyrus* (LFG, also called the fusiform "face" area, or FFA) (Sergent et al., 1992; Haxby et al., 1994; Clark et al., 1996; Kanwisher et al., 1997; McCarthy et al., 1997), *inferior occipital gyrus (and sulcus)*, sometimes referred to as the lateral occipital complex (LOC) (Malach et al., 1995; Halgren et al., 1999; Gauthier et al., 2000; Hoffman and Haxby, 2000; Grill Spector et al., 2001), *superior temporal sulcus* (Puce et al., 1998; Chao et al., 1999; Halgren et al., 1999; Hoffman and Haxby, 2000; Winston et al., 2004; Ishai et al., 2005), and the *amygdala* (Morris et al., 1996; Phillips et al., 1997; Vuilleumier et al., 2001; Blonder et al., 2004; Killgore and Yurgelun-Todd, 2004; Ishai et al., 2005). Lesions (or stimulation: e.g. Allison et al., 1994) of these regions have revealed the importance of these structures in processing some characteristics of faces (Meadows, 1974; Benton, 1980; Damasio et al., 1982; Sergent and Signoret, 1992; Adolphs et al., 1994; Young et al., 1995) as well as an interdependence of function across structures (Vuilleumier et al., 2004). Although the above regions have received the most attention in the literature, reliable activation is also seen in areas such as the temporal pole, intraparietal sulcus, anterior cingulate, hippocampus, parahippocampal gyrus, superior colliculus, putamen, inferior frontal and orbitofrontal cortex, insula and hypothalamus for various aspects of face processing (Sergent *et al.*, 1992; Phillips *et al.*, 1997; Hoffman and Haxby, 2000; Vuilleumier *et al.*, 2001; Ishai *et al.*, 2004; Killgore and Yurgelun-Todd, 2004).

Direct verification that some of the key structures generated the expected neural responses comes from intracranial electrophysiology in humans (Allison *et al.*, 1994, 1999; McCarthy *et al.*, 1999; Puce *et al.*, 1999; Privman *et al.*, 2007). Some studies have demonstrated that the entorhinal cortex, amygdala, and hippocampus contain neurons that respond selectively to faces (Fried *et al.*, 1997; Kreiman *et al.*, 2000) and sometimes even to specific individuals (Heit *et al.*, 1988; Quiroga *et al.*, 2005).

Monkey homologues

The advent of imaging studies in monkeys has revealed regions of activation for faces that are not unlike those seen in human brains (Fig. 13.1), with clusters of activation in numerous regions of the temporal lobe including the *superior temporal sulcus* (STS), *middle temporal gyrus* (MTG), and along the ventral surface including the *occipitotemporal sulcus* (OTS) or *anterior middle temporal sulcus* (AMTS) (Logothetis *et al.*, 1999; Tsao *et al.*, 2003; Pinsk *et al.*, 2005; Hoffman *et al.*, 2007a), the *amygdala* (Logothetis *et al.*, 1999; Hoffman *et al.*, 2007a), as well as parietal and frontal areas including the *intraparietal sulcus*, *arcuate sulcus*, and *prefrontal cortex* (Pinsk *et al.*, 2006). These activations reflect hemodynamic changes that are greater for faces than for objects, grid-scrambled faces, or Fourier-phase scrambled faces.

Inactivation of temporal lobe regions including the STS, MTG, or ITG produces equal face discrimination deficits, despite preserved discrimination of simple visual stimuli (Horel *et al.*, 1987). In some cases impairments are small and include impairments for objects, as well (Heywood and Cowey, 1992); nevertheless, monkeys show deficits for discriminating faces even when the discrimination is confined to configural changes of internal features (Horel, 1993). This was true regardless of whether anterior or posterior regions of MTG and the adjacent, lateral lower bank of STS were inactivated. Deficits were even greater when *both* anterior and posterior regions were inactivated – demonstrating independent contributions from anterior and posterior areas. When both banks of the STS are lesioned, monkeys have difficulty making fine discriminations in gaze direction (Heywood and Cowey, 1992). More extensive lesions of IT cortex that disrupt face processing are non-selective, producing deficits for other objects as well (Gross, 1978). Finally, early postnatal lesions of the MTG and lower-bank STS (area TE) or of the amygdala and hippocampus lead to deficits in socioemotional behavior in infants, but only the amygdalo-hippocampal damage leads to deficits that persist into adulthood (Malkova *et al.*, 1997;

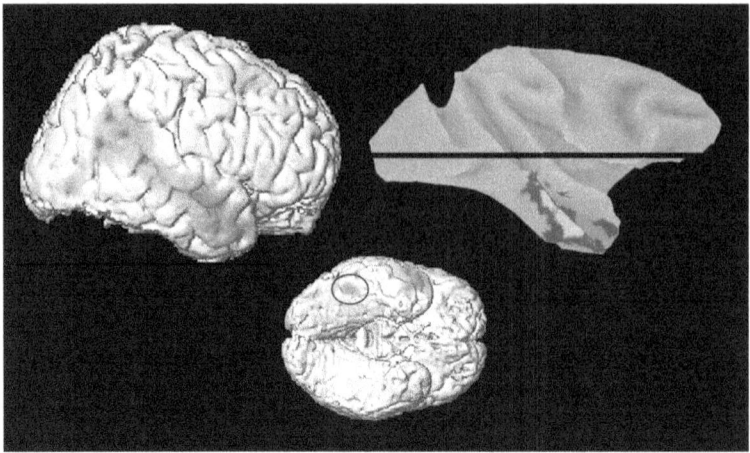

Figure 13.1 Activation for faces in monkeys and humans as visualized with functional magnetic resonance imaging (fMRI). The top left and bottom panels show sagittal and ventral views of the human brain, respectively, and the superimposed BOLD activation maps in response to faces that were present in movie clips. The contributions of color, language, and human bodies in these movie clips have already been factored out of the activation maps. The darker gray shading indicates the level of significance of the statistical map, group-averaged and thresholded at $p < 0.05$ corrected. Numerous regions are active, including a broad extent of STS. The small but significant activation in the lateral fusiform gyrus has been circled. The right panel shows a sagittal view of a gray-matter-eroded macaque brain, with superimposed BOLD activation for faces versus Fourier phase scrambled versions of those faces, thresholded at $p < 0.001$, uncorrected. As in the human activation map, a broad extent of STS was active, extending to include ventral temporal lobe areas. With no clustering used, the activation nevertheless extends over a contiguous range of approximately 200 voxels in each hemisphere. (Modified with permission from Bartels and Zeki (2004) and Hoffman et al. (2007a).)

Bachevalier et al., 2001). The face-processing abilities in these monkeys were not tested, so it is not clear what role these structures play in face processing per se, but the suggestion would be that early socioemotional behaviors normally mediated by TE can be reallocated to other structures over the course of development (Webster et al., 1991). Taken together, these results indicate that numerous locations within the temporal lobe neocortex support face processing – both along dorsoventral and anteroposterior axes. It's difficult to reconcile these findings with the notion of a unitary (monolithic) locus for face processing in the temporal lobe, akin to the role that some have ascribed to the lateral fusiform gyrus in humans.

Consistent with the spatial extent of the lesions implicated in face processing, neural responses to faces have been commonly observed throughout a large territory of temporal lobe, including the upper and lower banks of the STS, extending along the lateral convexity of IT cortex to AMTS, including areas TPO, TEa, TEm, TE1, TE2, and TE3 (Gross et al., 1972; Bruce et al., 1981; Perrett et al., 1982; Desimone et al., 1984; see Perrett et al., 1992, Figure 4 for summary). One study mapped the responses to faces, objects, simple static images, movies, and auditory and tactile stimuli for each of ten subdivisions of the temporal lobe neocortex (Baylis et al., 1987), revealing many face-selective regions, but only a subset of multimodal and motion-sensitive regions. In addition to the widespread regions along the STS/MTG, face-selective cells have also been reported in the inferior convexity and orbitofrontal cortex, ventral premotor cortex, the hippocampus, and the amygdala (Thorpe et al., 1983; Leonard et al., 1985; Nakamura et al., 1992; Rolls, 1996; O'Scalaidhe et al., 1997; Ferrari et al., 2003).

Spatial clustering

Even in the initial studies to record from temporal lobe neurons in the awake monkey, neurons with similar responses were observed to cluster together (Harries and Perrett, 1991; Perrett et al., 1992) a result that has been replicated and extended (Kreiman et al., 2006), and supported by anatomical (Harries and Perrett, 1991) and optical imaging methods (Wang et al., 1996). The clustering of neurons with similar response preferences has been offered as evidence for a modular organization of the STS/IT regions (Tanaka, 2003). These studies suggest clustering of similar responses extends somewhere between 0.5 and 2 mm (or 3–5 mm: Perrett et al., 1992) spread laterally across the cortical sheet. More recently, sampling a 1–2-mm section within a $\sim$16-mm^2 face-selective patch identified by functional magnetic resonance imaging (fMRI) in posterior STS revealed that almost all neurons sampled were selective for faces as a class compared to various non-face stimuli as a class (Tsao et al., 2006). In addition to replicating some of the findings of common responses within an electrode track and in tracks within 2 mm of each other (e.g. Perrett et al., 1992), this study has been offered as evidence that fMRI can, indeed, detect regions containing neurons with common response preferences, i.e. clusters.

But for all their merits, fMRI and electrophysiological methods are not well suited to deciphering the spatial distribution of neurons with similar response selectivity. Electrodes detect a small, biased sample of cells, often after additional biases in the electrode-positioning procedures used by the experimenter. Electrodes can only record individual neurons within approximately a 100-μm region around the electrode tip (Buzsáki, 2004); and sampling from additional electrode tracks is extremely sparse relative to the area of neocortex in question. In contrast, MRI can be used to sample the whole brain, but only via a sluggish hemodynamic response that, at best, samples from a pool of thousands to millions of neurons, based on a poorly understood transfer function. Fortunately, an emerging imaging method can help to fill in these gaps, providing both the single-neuron spatial resolution of electrophysiology and the whole-brain spatial coverage of fMRI.

Immediate early gene (IEG) imaging has been used to identify neural responses to faces and objects throughout the temporal lobe in vervet monkeys (Zangenehpour and Chaudhuri, 2005). By carefully timing the presentation of blocks of face or object stimuli, neural responses during the presentation period would lead to *zif268* the activation of the immediate early gene (1) in the cytoplasm (mRNA labeled) for the most recently presented block, (2) within the neuron's nucleus (protein labeled) for the remotely presented block, (3) in both the nucleus and cytoplasm for responses during both blocks. In this way, the selectivity for faces, objects or both was visible for *each* neuron in the STS/IT cortex (see Chaudhuri *et al.*, 1997; Guzowski *et al.*, 1999 for descriptions of the general technique). Interestingly, a neuron with one response pattern tended to be surrounded by similar neurons, creating dozens of patches of neurons responding exclusively to faces but not objects, to objects but not faces, or to both stimulus classes. Occasionally, activation became much more randomly distributed among the three response types, but this was more the exception than the rule.

Another study suggests that the putative "face cells" found in such clusters may have functional significance: when microstimulation was delivered to the middle of a patch of face-selective cells in the macaque, categorization performance was biased towards faces (Afraz *et al.*, 2006). Here, face-selective regions were interspersed throughout a ~6-mm anteroposterior extent of lower bank STS and TEa, in areas that generally showed patchy *zif268* activation profiles in vervets.

In light of both the inactivation and recent microstimulation studies, it would appear that the numerous IEG-based "face" patches contain neurons that are important for face processing, though this has not been verified directly. These potentially important face patches were nonetheless in the minority: the most common IEG response was for both faces and objects, constituting almost half of the activated cells. Evidence from categorization studies suggests that more information about category membership comes from cells with responses that are biased towards – but not exclusively responsive to – items in one category than from cells that are only responsive to items within one category (Thomas *et al.*, 2001). This could have serious consequences for some methods used to localize face-processing regions in the brain.

Roughly speaking, fMRI activation for faces versus objects would likely only result from those labeled neurons that were selective for faces (~25% in the IEG study) *and* that had sufficient numbers of similar neighbors to activate a voxel. Thus, more than half of the *active* cells, those whose selectivity is distributed and those whose spatial distribution is too fine-grained for fMRI, would go unaccounted for. Importantly, this underscores how fMRI, using the standard "cognitive subtraction" paradigm, is designed to detect spatially clustered rather than distributed activity. As such, the standard application of this method is unsuitable for comparing the extent to which cells are clustered or distributed.

What about when more lenient contrasts are used, such as faces > scrambled images? The resultant fMRI activation is consistent with the activation for the "face only" and "both" categories of IEG responses. Namely, broad expanses of STS and IT are active, and this corresponds well with the broad STS activation for faces seen in humans (Fig. 13.1). In fact, in order to come close to the four to five discrete patches per hemisphere reported in some monkey fMRI studies, the cellular-resolution zif268 signal was smoothed with a Gaussian filter, producing roughly ten clusters. That's double the number reported from fMRI, even with considerable smoothing. Working back to the neural origins of this signal, these results indicate that the underlying neural pool of the fMRI activation is probably far more "spotty" than would be consistent with the two or three modules reported in humans and some monkey studies. This doesn't mean that the spots or regions reported from fMRI are false; rather, they may reflect the tip of the iceberg when it comes to the neurons involved in face processing.

What are the roles played by "face cells" found within the well-delineated clusters compared to those distributed among cells with other response preferences? Are cells responding exclusively to faces the rule or exception, and what response distributions are optimal for various aspects of face processing? For the "hows" and "whys" of the neural processing of face signals, we turn to recordings of individual neurons within our broadly defined regions of interest in the temporal lobe.

Face encoding I: single-unit information

In the beginning came the discovery of a cell in the MTG and STS whose firing rate clearly increased to the sight of a hand, and this appeared to be the most effective stimulus for driving that cell (Gross *et al.*, 1972). Shortly thereafter, preferred stimuli that were found for other cells included a flaming Q-tip and, not infrequently, faces. In contrast, simple visual stimuli (such as gratings) that are so effective at driving other areas of visual cortex were rarely effective for cells in the STS/MTG (Desimone *et al.*, 1984). Depending on the area, 4–20% of visually responsive cells responded at least twice as strongly to the most effective face stimulus as to the most effective non-face stimulus (Rolls, 1984; Baylis *et al.*, 1987). Typically, ANOVAs followed by post-hoc tests are then used to determine whether the response to the most effective face stimulus is significantly greater than the response to the most effective non-face stimulus. These are the most standard criteria used to dub a "face cell," though other methods have been used which, by design, downplay within-category response differences, focusing on across-category response differences (Tsao et al., 2006).

Face-selective responses in STS/MTG show the following general characeristics (see Rolls, 2007 for a recent review). Responses relative to baseline (or to non-face stimuli) are nearly always excitatory. Response onset latencies are typically between 80 and 110ms, with sustained responses for the duration of the images. Coding is considered to be through firing rate in most studies (Rolls et al., 2003). Correlated activity in IT cells in response to simple stimuli was nearly identical when timing information was added into that of response magnitude (Gawne and Richmond, 1993). Likewise, there was little difference in stimulus decoding accuracy when considering simultaneously recorded versus trial-shuffled responses in IT (Gochin et al., 1994) suggesting a minor role, at best, for spike timing within a population in stimulus encoding. However, more recent evidence suggests that spike correlations among cells in IT is greater for "preferred" feature configurations (Hirabayashi and Miyashita, 2005), and correlated state transitions have been observed in pairs of IT neurons (Uchida et al., 2006), suggesting there could, in principle, be a role for spike timing or oscillatory modulation of spikes in the processing face stimuli.

In addition, latency differences have provided information about stimulus type (Richmond et al., 1987; Tovee et al., 1993; Eifuku et al., 2004; Kiani et al., 2005). Closer examination of the seminal work of Baylis and Rolls (1987) reveals that the latency histograms for some of the STS regions did, indeed, have a bimodality that included responses from 60 to 100ms. It is possible, therefore, that the response windows of 100–500ms commonly used in studies of STS/MTG may be omitting some of the shortest-onset responses. Finally, although there does not appear to be the sufficient temporal precision between stimulus and response onset necessary for a temporal coding of that stimulus, the firing rates of a population of IT cells within a remarkably small temporal window were both consistent and informative (Tovee et al., 1993; Hung et al., 2005). Starting at 100ms after stimulus onset, a mere 12.5-ms response window was sufficient for decoding categories or individual examples of visual stimuli (Hung et al., 2005). Such a short time window reduced the contributions of each neuron to nearly a binary code – either a cell fired during the window or not. This suggests that a rate code that is highly distributed across the population of IT cells is a meaningful currency with which to evaluate seen objects.

The issue of what aspect of faces is coded in STS/MTG has received a great deal of attention. Elaboration of the canonical response selectivity of "face cells" revealed several subclasses of cells: (1) "holistic" responses, occurring only when facial features are joined in the proper "whole-face" configuration, or "part" responses that occurred to a subset of facial features (Perrett *et al.*, 1982); (2) head direction, with preferred head orientations (Perrett *et al.*, 1985; Eifuku *et al.*, 2004); (3) gaze (Perrett *et al.*, 1985); (4) expression, or more specifically, closed or open-mouth expressions (Hasselmo *et al.*, 1989); (5) motion (Oram *et al.*, 1993), in particular biological motion (Oram and Perrett, 1994; see also Puce and Perrett, 2003), with some cells selective for the end points of particular actions (Jellema and Perrett, 2003); (6) identity – many face cells respond very strongly to the faces of one or a few individuals, less for various other faces, and very little for most (but not all) non-face stimuli (Hasselmo *et al.*, 1989; Young and Yamane, 1992; Eifuku *et al.*, 2004). This allows substantial information about identity to be extracted from populations of these cells (Young and Yamane, 1992; Rolls *et al.*, 1997). The "face space" that could underlie coding of identity was recently described by Leopold and colleagues (2006). Thus, many aspects of the face as a stimulus can be read out from fluctuations in the spiking activity in STS and MTG.

In general, neural coding about faces is sparse and distributed (Young and Yamane, 1992). Here, sparse is meant in the sense that a given neuron will fire strongly to only a few stimuli. Distributed is meant in the sense that a given stimulus will elicit activity in many neurons. Sparsity leads to greater capacity – neural population activity won't "saturate" as quickly. Distributed responses enable generalization and pattern completion from incomplete or degraded stimuli, or from biological noise. Errors or lack of responses in a few cells won't throw off the whole representation of the stimulus, since that load is distributed over many neurons. Moreover, distributed representations are decoded more quickly than local, "grandmother cell" representations.

From groups of neurons within focal areas to the presence of multiple modules, all such groups are in need of some mode of concerted information processing. How is this accomplished? What are the neocortical rules of engagement?

Face encoding II: how is information integrated across single units?

One of the ways that faces (or the significance of a face) might be represented in a distributed code across neurons and even brain regions is through the synchronized activity of relevant neurons. As described in other chapters in this volume (see Chapters 4, 7, 15, and 17), Hebb postulated that the neural representation of an event could be formed by cell assemblies: co-active groups of neurons, whose activity is self-reinforcing. The cell assembly is one attractive means of integrating and representing information such as that contained in faces.

Note that, in principle, any relatively fixed pattern of spikes from a set of neurons could constitute an assembly, though according to Hebb's postulate, spiking within the same temporal window has the effect of reinforcing the assembly. Thus, some degree of synchrony across constituent neurons is thought to be critical to the maintenance of an assembly, though the concept can be extended in time based on synchronized cascades of assemblies, such as synfire chains (Abeles *et al.*, 1993). But before jumping into the relationship between face representation and cell assemblies, let's take a step back and ask *whether* activity among these populations would need to be coordinated in the first place.

If responses to faces are exclusively stimulus-driven, face-processing regions could build up representations for the features unique to an individual or at least diagnostic for individuation on the whole. The same could be said for regions processing facial expressions, gender, dominance status, etc. If properly "tuned up," presentation of an individual could elicit reliable, decodable activity in a subset of neurons across several face-processing regions. Because these cells are "driven," there is no additional need for them to act as an assembly – their evoked responses would constitute the signature for that particular stimulus.

But what happens when much of the face is occluded, or only a vocalization heard? Here, the visual stimulus would likely (or certainly) be insufficient to drive the same set of neurons. One solution leading to correct recognition would be pattern completion, which could result from learned associations, but this then becomes part of an association network, and not something strictly stimulus-driven. Cell assemblies, as conceived here, are one example of such an associative network.

In most situations, facial expressions are not generated into thin air – they are a means of signaling to other individuals. If discrete regions process facial expressions and identity, what happens during a social encounter, when multiple faces and expressions are seen? Do we frequently confuse which individual delivered which expression? This "cluttered" scenario is just the social version of the old binding problem. But humans and monkeys have no trouble allocating various signals generated by a face to the appropriate face. There is no known illusion where the emotion of one is mistaken for another. This is difficult to reconcile with the characterization of face representations as complex passive filters of external stimuli. As with the binding problem, if the activity of certain groups of neurons could be reliably and simultaneously elicited, as with cell assemblies, then the temporal grouping of responses could enable proper attribution of which expression belonged to which individual, for example. The foveation of only one face at a time may not be sufficient to disambiguate the stimuli, given the large receptive fields (position invariance) of face cells (Desimone and Gross, 1979).

Taking the idea one step further, we are able to imagine faces and even complete social scenarios. Imaging studies suggest that brain regions responding to actual face stimuli are also selectively active during imagined faces (Ishai *et al.*, 2002). Here, clearly, activity is not stimulus-driven. How, then, does the brain accomplish a coherent reflection of particular individuals and their role in specific scenarios, including likely expressions?

The answers to theses problems are not straightforward; numerous solutions can be imagined. But the cell assembly offers a convenient first-pass mechanism allowing proper grouping of relevant distributed groups of neurons in time, and their elicitation as a whole from incomplete or even absent sensory inputs. What evidence exists for cell assembly representation of faces, particularly in the necessary temporal lobe regions? Unfortunately, precious little, but after considering some of the evidence for synchronized grouping of cells (i.e. signs of cell assemblies) in other systems, I will come back to what circumstantial evidence exists within the field of face processing.

Multi-site synchronization and evidence for cell assemblies

Several variations on the theme of Hebb's cell assemblies have emerged over the last several decades (Freeman, 1975; Damasio, 1989; McNaughton and Barnes, 1990; Varela *et al.*, 2001). In addition to early work addressing global synchronization (Bressler *et al.*, 1993) and the well-known observations of binding through gamma synchronization within striate cortex and with EEG in humans (see Roskies, 1999) a few recent studies are revealing synchronous interactions directly recorded from distant brain regions, as described in the next section.

In one study, the lead–lag relationship between two brain regions was reversed when comparing stimulus contexts that required bottom–up versus top–down processing (Buschman and Miller, 2007). Here the two tasks involved the pop-out of a stimulus based on its physical difference to other, distractor stimuli. Under "pop-out" conditions, the response appeared in brain region LIP prior to the response in PFC; the latency order was reversed during the more difficult search involving the same target stimulus, but with more similar distractors. Moreover, these two regions showed synchronized, or coherent, oscillations, though the frequencies with greatest coherence differed across conditions. Optimal frequencies for the top–down condition were slower (22–34 Hz) than those of the bottom–up condition (35–55 Hz). One implication is that assembly recruitment may occur asymmetrically, depending on the regions responding initially.

Another example of inter-area synchronization during task performance was reported between areas MT and parietal cortex (Saalmann et al., 2007). In a search task, similar to the top–down search task mentioned above, monkeys were required to identify matches of both location and orientation or sets of lines. Here, MT and a major projection area of MT – the parietal cortex – shared coherent oscillations between 25 and 45Hz, with the parietal neurons firing slightly in advance of MT neurons. This temporal offset may indicate feedback not unlike that described by Buschman and Miller (2007) between PFC and parietal areas.

These experiments demonstrate that, across vast neocortical territory and diverse behavioral demands, brain regions associated with task processing tend to oscillate together, and the preferred frequencies may depend on the regions involved and the task demands. Spiking activity can be restricted to the relevant oscillatory windows, making the induction or creation of spatially distributed cell assemblies a possibility. These experiments do not address whether the relevant subset of neurons are brought together as an assembly – only that the brain regions known to be important for solving a task demonstrate coherent oscillations. These experiments also leave open the possibility that the oscillations are epiphenomenal, and not directly or causally influencing behavior.

The former issue – that of binding the relevant subsets of cells – is addressed elsewhere in this book (see Chapters 1, 3, 4, 6, and 7), but, in brief, the work on theta phase precession and cells contributing to sharp wave-associated ripples, both in the rodent hippocampus, are but two examples of selective ensemble activity occurring within oscillations. Rodent neocortex also shows specific patterns of ensemble activity confined by or grouped within oscillatory windows (MacLean et al., 2005; Ji and Wilson, 2007; Luczak et al., 2007). A clear benefit for shaping the timing of responses and even for plasticity was revealed between projection neurons and the activity of Kenyon cells in the locust mushroom body (Cassenaer and Laurent, 2007). Although a very different model system, it is one that is clearly "paced" through 20-Hz oscillatory activity. Critically, the conduction timing delays between areas, and the time constant of observed spike-timing dependent plasticity (STDP) in synapses connecting the two regions, led to a most remarkable observation: spikes falling eccentric to the "preferred" phase of the oscillation would undergo facilitation or depression, whichever had the effect of pulling in subsequent spikes towards the preferred phase. That is, the STDP effectively "focused" spikes within a tight phase of the oscillation. Although nothing quite so elegant has been demonstrated in the mammalian neocortex, there is evidence for a relationship between optimal phase and the oscillation strength that follows (Womelsdorf et al., 2007); thus, the principles of how oscillations may help maintain a high-fidelity signal and how this might enable longer-term pattern "memory" are not a matter of thought experiment; these principles can be tested by direct experimentation.

More critically, if oscillations support cell assembly formation, this should lead to behavioral benefits. Some evidence for behavioral relevance of oscillations comes from correlations with behavior: the coherence of the theta oscillations between the olfactory bulb and hippocampus of rodents is associated with enhanced performance, and this correlation is specific to a discrimination task (Kay, 2005). In humans, increased coherence of gamma oscillations between motor cortex and spinal cord neurons predicts shorter reaction times (Schoffelen et al., 2005). Within the medial temporal lobe, depth electrode recordings in humans showed that a time-limited gamma coherence between hippocampus and rhinal cortex at the time of memory encoding was predictive of subsequent memory performance (Fell et al., 2001), and the power of gamma and/or theta oscillations in temporal and frontal lobes is also associated with strength of encoding (Sederberg et al., 2003, 2007). And in macaques, during an object shape match-to-sample task, one pair of electrodes from an array implanted over V4/TEO/caudal TE demonstrated greater beta coherence on correct than on incorrect trials (Tallon-Baudry et al., 2004). Finally, a causal relationship between one type of oscillation and behavior has been described (Marshall et al., 2006). Induction of slow waves in sleeping humans led to: (1) increased oscillations at another frequency band and (2) better subsequent memory recall. The memory enhancement was frequency- and task-specific, demonstrating a causal role of slow waves in memory consolidation during sleep. Although memory consolidation may not involve cell assembly formation, it is interesting to note that the oscillation in question (and the concomitant behavioral state) are the ones during which the most compelling observations of cell assemblies have been made (e.g. during memory trace reactivation: see Hoffman et al., 2007b).

In sum, there is growing evidence that oscillations can coordinate activity among spatially distributed neurons, as in cell assemblies, and that this results in behavioral benefits; yet the evidence within the temporal lobe and for face processing is much more limited. In a rare example of synchronization involving face perception, one study has looked for synchronization across regions during face processing (Rodriguez et al., 1999). Compared to EEG fields when faces were made difficult to perceive, easy-to-perceive faces produced synchrony across a select but broad extent of cortex. This synchrony was followed by a period of phase scattering before synchrony associated with motor output

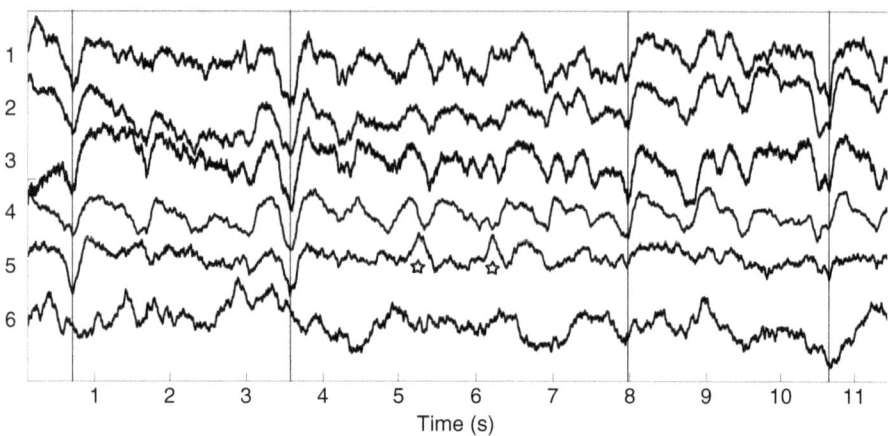

Figure 13.2 Field potentials plotted as a function of time from simultaneously recorded electrodes in the monkey temporal lobe obtained during slow-wave sleep. Slow oscillations, indicated by vertical lines, can be seen as synchronized deflections that are largely conserved across electrodes. These deflections appear to have local components: not all electrodes show the oscillations at any given time, e.g. electrode 6. Moreover, the relative amplitudes over multiple slow oscillations can vary by electrode, such as electrode 5, which shows synchronized deflections with electrodes 1–4, but only during the first two indicated deflections and not the third and fourth. In addition, electrode 5 shows two positive waves indicated by stars, when electrodes 2–4 show negative waves. The temporal lobe appears to contain multi-site synchronized oscillations, but the spatiotemporal patterns and their behavioral significance are not yet understood.

was observed. Although a very important step in the right direction, probes indicated frontal and occipitotemporal regions, without mentioning possible activity in temporal lobe structures. This absence may relate to the size of relevant neural pools and their location relative to the scalp EEG probes.

The scant tests of assembly activity in the temporal lobe neocortex have produced mixed results. It was reported that simultaneously recorded electrodes showed no effects of trial-by-trial variance (Gochin *et al.*, 1994), and it has been stated that, unlike occipital regions, neural activity in temporal lobe neocortex does not oscillate in response to static face stimuli (Tovee and Rolls, 1992). It is interesting, then, that there were changes in firing rate correlations to manipulations of images with face-like configurations (Hirabayashi and Miyashita, 2005), and low-frequency oscillations were observed in IT, provided multiple stimuli were presented (Rollenhagen and Olson, 2005). Large-amplitude oscillations in the temporal lobe are also easily observed during sleep (Fig. 13.2). It would appear that coordinated neural activity can occur in the temporal lobe, as oscillations, and for face stimuli, though it is not entirely clear why none were seen in the earliest investigations. Face stimuli elicit coordinated activity over a distributed territory, yet we have little basis for suggesting what the key oscillations are, what their relations to behavior or behavioral state might be, and, critically, whether grouping of neural ensembles into assemblies is an underlying process.

Where to go from here

From the research described above, one conclusion stands out above the rest: the available information-rich signals in the brain have been underexplored. Nevertheless, existing studies suggest that the temporal lobe is "concerned" with faces, as part of a network of face-processing regions. Further, stimulus coding (or "tuning") in neurons appears to be selective for a small number of stimuli (i.e. "sparse") but with populations of cells responding (i.e. "distributed"). The spatial distribution of face-selective cells in the temporal lobe tends to be clustered, albeit in clusters of varying sizes and of varying breadths of selectivity. The debate of oscillations in the temporal lobe is only beginning; the key frequencies and causal relationship with behavior effectively unexplored. And, aside from the absence of evidence for trial-to-trial fluctuations in population activity, there is very little to inform us about whether cell assemblies are created and used for face processing in the temporal lobe. Yet evidence from other systems gives ample justification for a concerted effort in the future.

The population-grouping ability of oscillations and their behavioral relevance has been described in invertebrate circuits and in primate neocortex alike. Likewise, understanding the nature of coding in the temporal lobe for behaviorally relevant stimuli is an attainable goal, and one that could serve as a "Rosetta Stone" for understanding some of the links between brain and behavior.

References

Abeles, M., Vaadia, E., Bergman, H., *et al.* (1993). Dynamics of neuronal interactions in the frontal cortex of behaving monkeys. *Concepts Neurosci* **4**:131–158.

Adolphs, R., Tranel, D., Damasio, H., and Damasio, A. (1994). Impaired recognition of emotion in facial expressions following bilateral damage to the human amygdala. *Nature* **372**:669–672.

Afraz, S.-R., Kiani, R., and Esteky, H. (2006). Microstimulation of inferotemporal cortex influences face categorization. *Nature* **442**:692–695.

Allison, T., Ginter, H., McCarthy, G., *et al.* (1994). Face recognition in human extrastriate cortex. *J Neurophysiol* **71**:821–825.

Allison, T., Puce, A., Spencer, D.D., and McCarthy, G. (1999). Electrophysiological studies of human face perception. I. Potentials generated in occipitotemporal cortex by face and non-face stimuli. *Cereb Cortex* **9**:415–430.

Bachevalier, J., Malkova, L., and Mishkin, M. (2001). Effects of selective neonatal temporal lobe lesions on socioemotional behavior in infant rhesus monkeys (*Macaca mulatta*). *Behavi Neurosci* **115**:545–559.

Bartels, A. and Zeki, S. (2004). Functional brain mapping during free viewing of natural scenes. *Hum Brain Map* **21**:75–85.

Baylis, G.C., Rolls, E.T., and Leonard, C.M. (1987). Functional subdivisions of the temporal lobe neocortex. *J Neurosci* **7**:330–342.

Benton, A.L. (1980). The neuropsychology of facial recognition. *Am Psychol* **35**:176–186.

Blonder, L.X., Smith, C.D., Davis, C.E., et al. (2004). Regional brain response to faces of humans and dogs. *Cogn Brain Res* **20**:384–394.

Bressler, S.L., Coppola, R., and Nakamura, R. (1993). Episodic multiregional cortical coherence at multiple frequencies during visual task performance. *Nature* **366**:153–156.

Bruce, C. (1982). Face recognition by monkeys: absence of an inversion effect. *Neuropsychologia* **20**:515–521.

Bruce, C., Desimone, R., and Gross, C.G. (1981). Visual properties of neurons in a polysensory area in superior temporal sulcus of the macaque. *J Neurophysiol* **46**:369–384.

Buschman, T.J. and Miller, E.K. (2007). Top–down versus bottom–up control of attention in the prefrontal and posterior parietal cortices. *Science* **315**:1860–1862.

Buzsáki, G. (2004). Large-scale recording of neuronal ensembles. *Nat Neurosci* **7**:446–451.

Cassenaer, S. and Laurent, G. (2007). Hebbian STDP in mushroom bodies facilitates the synchronous flow of olfactory information in locusts. *Nature* **448**:709–713.

Chao, L.L., Martin, A., and Haxby, J.V. (1999). Are face-responsive regions selective only for faces? *Neuroreport* **10**:2945–2950.

Chaudhuri, A., Nissanov, J., Larocque, S., and Rioux, L. (1997). Dual activity maps in primate visual cortex produced by different temporal patterns of zif268 mRNA and protein expression. *Proc Natl Acad Sci USA* **94**:2671–2675.

Clark, V.P., Keil, K., Maisog, J.M., et al. (1996). Functional magnetic resonance imaging of human visual cortex during face matching: a comparison with positron emission tomography. *Neuroimage* **4**:1–15.

Dahl, C.D., Logothetis, N.K., and Hoffman, K.L. (2007). Individuation and holistic processing of faces in rhesus monkeys. *Proc R Soc Lond B* **274**:2069–2076.

Damasio, A.R. (1989). Time-locked multiregional retroactivation: a systems-level proposal for the neural substrates of recall and recognition. *Cognition* **33**:25–62.

Damasio, A.R., Damasio, H., and Van Hoesen, G.W. (1982). Prosopagnosia: anatomical basis and behavioral mechanisms. *Neurology* **32**:331–341.

Dasser, V. (1988). A social concept in Java monkeys. *Anim Behav* **36**:225–230.

Deaner, R.O. and Platt, M.L. (2003). Reflexive social attention in monkeys and humans. *Curr Biol* **13**:1609–1613.

Desimone, R. and Gross, C. G. (1979). Visual areas in the temporal cortex of the macaque. *Brain Res* **178**:363–380.

Desimone, R., Albright, T. D., Gross, C. G., and Bruce, C. (1984). Stimulus-selective properties of inferior temporal neurons in the macaque. *J Neurosci* **4**:2051–2062.

Eifuku, S., De Souza, W. C., Tamura, R., Nishijo, H., and Ono, T. (2004). Neuronal correlates of face identification in the monkey anterior temporal cortical areas. *J Neurophysiol* **91**:358–371.

Emery, N. J., Lorincz, E. N., Perrett, D. I., Oram, M. W., and Baker, C. I. (1997). Gaze following and joint attention in rhesus monkeys (*Macaca mulatta*). *J Comp Psychol* **111**:286–293.

Fell, J., Klaver, P., Lehnertz, K., et al. (2001). Human memory formation is accompanied by rhinal-hippocampal coupling and decoupling. *Nat Neurosci* **4**:1259–1264.

Ferrari, P. F., Gallese, V., Rizzolatti, G., and Fogassi, L. (2003). Mirror neurons responding to the observation of ingestive and communicative mouth actions in the monkey ventral premotor cortex. *Eur J Neurosci* **17**:1703–1714.

Freeman, W. J. (1975). *Mass Action in the Nervous System*. New York: Academic Press.

Fried, I., MacDonald, K. A., and Wilson, C. L. (1997). Single neuron activity in human hippocampus and amygdala during recognition of faces and objects. *Neuron* **18**:753–765.

Gauthier, I., Tarr, M. J., Moylan, J., et al. (2000). The fusiform "face area" is part of a network that processes faces at the individual level. *J Cogn Neurosci* **12**:495–504.

Gawne, T. J. and Richmond, B. J. (1993). How independent are the messages carried by adjacent inferior temporal cortical neurons? *J Neurosci* **13**:2758–2771.

Gochin, P. M., Colombo, M., Dorfman, G. A., Gerstein, G. L., and Gross, C. G. (1994). Neural ensemble coding in inferior temporal cortex. *J Neurophysiol* **71**:2325–2337.

Gothard, K. M., Erickson, C. A., and Amaral, D. G. (2004). How do rhesus monkeys (*Macaca mulatta*) scan faces in a visual paired comparison task? *Anim Cogn* **7**:25–36.

Grill Spector, K., Kourtzi, Z., and Kanwisher, N. (2001). The lateral occipital complex and its role in object recognition. *Vision Res* **41**:1409–1422.

Gross, C. G. (1978). Inferior temporal lesions do not impair discrimination of rotated patterns in monkeys. *J Comp Physiol Psychol* **92**:1095–1109.

Gross, C. G., Rocha-Miranda, C. E., and Bender, D. B. (1972). Visual properties of neurons in inferotemporal cortex of the macaque. *J Neurophysiol* **35**:96–111.

Guzowski, J. F., McNaughton, B. L., Barnes, C. A., and Worley, P. F. (1999). Environment-specific expression of the immediate-early gene *Arc* in hippocampal neuronal ensembles. *Nat Neurosci* **2**:1120–1124.

Halgren, E., Dale, A. M., Sereno, M. I., et al. (1999). Location of human face-selective cortex with respect to retinotopic areas. *Hum Brain Map* **7**:29–37.

Harries, M. H. and Perrett, D. I. (1991). Visual processing of faces in temporal cortex: physiological evidence for a modular organization and possible anatomical correlates. *J Cogn Neurosci* **3**:9–24.

Hasselmo, M. E., Rolls, E. T., and Baylis, G. C. (1989). The role of expression and identity in the face-selective responses of neurons in the temporal visual cortex of the monkey. *Behav Brain Res* **32**:203–218.

Haude, R. H. and Detwiler, D. H. (1976). Visual observing by rhesus monkeys: influence of potentially threatening stimuli. *Percept Motor Skills* **43**:231–237.

Haude, R. H., Graber, J. G., and Farres, A. G. (1976). Visual observing by rhesus monkeys: some relationships with social dominance ranks. *Anim Learn Behav* **4**:163–166.

Haxby, J. V., Horwitz, B., Ungerleider, L. G., et al. (1994). The functional organization of human extrastriate cortex: a PET-rCBF study of selective attention to faces and locations. *J Neurosci* **14**:6336–6353.

Heit, G., Smith, M. E., and Halgren, E. (1988). Neural encoding of individual words and faces by the human hippocampus and amygdala. *Nature* **333**:773–775.

Heywood, C. A. and Cowey, A. (1992). The role of the "face-cell" area in the discrimination and recognition of faces by monkeys. *Phil Trans R Soc Lond B* **335**:31–38.

Hirabayashi, T. and Miyashita, Y. (2005). Dynamically modulated spike correlation in monkey inferior temporal cortex depending on the feature configuration within a whole object. *J Neurosci* **25**:10299–10307.

Hoffman, E. A. and Haxby, J. V. (2000). Distinct representations of eye gaze and identity in the distributed human neural system for face perception. *Nat Neurosci* **3**:80–84.

Hoffman, K. L., Gothard, K. M., Schmid, M. C., and Logothetis, N. K. (2007a). Facial-expression and gaze-selective responses in the monkey amygdala. *Curr Biol* **17**:766–772.

Hoffman, K. L., Battaglia, F. P., Harris, K. D., et al. (2007b). The upshot of up states in neocortex: from slow oscillations to memory formation. *J Neurosci* **27**:11838–11841.

Horel, J. A. (1993). Retrieval of a face discrimination during suppression of monkey temporal cortex with cold. *Neuropsychologia* **31**:1067–1077.

Horel, J. A., Pytko-Joiner, D. E., Voytko, M. L., and Salsbury, K. (1987). The performance of visual tasks while segments of the inferotemporal cortex are suppressed by cold. *Behav Brain Res* **23**:29–42.

Humphrey, N. K. (1974). Species and individuals in the perceptual world of monkeys. *Perception* **3**:105–114.

Hung, C. P., Kreiman, G., Poggio, T., and DiCarlo, J. J. (2005). Fast readout of object identity from macaque inferior temporal cortex. *Science* **310**:863–866.

Ishai, A., Haxby, J. V., and Ungerleider, L. G. (2002). Visual imagery of famous faces: effects of memory and attention revealed by fMRI. *Neuroimage* **17**:1729–1741.

Ishai, A., Pessoa, L., Bikle, P. C., and Ungerleider, L. G. (2004). Repetition suppression of faces is modulated by emotion. *Proc Nat Acad Sci USA* **101**:9827–9832.

Ishai, A., Schmidt, C. F., and Boesiger, P. (2005). Face perception is mediated by a distributed cortical network. *Brain Res Bull* **67**:87–93.

Jellema, T. and Perrett, D. I. (2003). Perceptual history influences neural responses to face and body postures. *J Cogn Neurosci* **15**:961–971.

Ji, D. and Wilson, M. A. (2007). Coordinated memory replay in the visual cortex and hippocampus during sleep. *Nat Neurosci* **10**:100–107.

Kanwisher, N., McDermott, J., and Chun, M.M. (1997). The fusiform face area: a module in human extrastriate cortex specialized for face perception. *J Neurosci* **17**:4302–4311.

Kay, L.M. (2005). Theta oscillations and sensorimotor performance. *Proc Nat Acad Sci USA* **102**:3863–3868.

Kiani, R., Esteky, H., and Tanaka, K. (2005). Differences in onset latency of macaque inferotemporal neural responses to primate and non-primate faces. *J Neurophysiol* **94**:1587–1596.

Killgore, W.D. and Yurgelun-Todd, D.A. (2004). Activation of the amygdala and anterior cingulate during nonconscious processing of sad versus happy faces. *Neuroimage* **21**:1215–1223.

Koba, R. and Izumi, A. (2006). Sex categorization of conspecific pictures in Japanese monkeys (*Macaca fuscata*). *Anim Cogn* **9**:183–191.

Kreiman, G., Koch, C., and Fried, I. (2000). Category-specific visual responses of single neurons in the human medial temporal lobe. *Nat Neurosci* **3**:946–953.

Kreiman, G., Hung, C.P., Kraskov, A., *et al.* (2006). Object selectivity of local field potentials and spikes in the macaque inferior temporal cortex. *Neuron* **49**:433–445.

Leonard, C.M., Rolls, E.T., Wilson, F.A.W., and Baylis, G.C. (1985). Neurons in the amygdala of the monkey with responses selective for faces. *Behav Brain Res* **15**:159–176.

Leopold, D.A., Bondar, I.V., and Giese, M.A. (2006). Norm-based face encoding by single neurons in the monkey inferotemporal cortex. *Nature* **442**:572–575.

Logothetis, N.K., Guggenberger, H., Peled, S., and Pauls, J. (1999). Functional imaging of the monkey brain. *Nat Neurosci* **2**:555–562.

Luczak, A., Bartho, P., Marguet, S.L., Buzsáki, G., and Harris, K.D. (2007). Sequential structure of neocortical spontaneous activity in vivo. *Proc Nat Acad Sci USA* **104**:347–352.

MacLean, J.N., Watson, B.O., Aaron, G.B., and Yuste, R. (2005). Internal dynamics determine the cortical response to thalamic stimulation. *Neuron* **48**:811–823.

Malach, R., Reppas, J.B., Benson, R.R., *et al.* (1995). Object-related activity revealed by functional magnetic resonance imaging in human occipital cortex. *Proc Natl Acad Sci USA* **92**:8135–8139.

Malkova, L., Mishkin, M., Suomi, S.J., and Bachevalier, J. (1997). Socioemotional behavior in adult rhesus monkeys after early versus late lesions of the medial temporal lobe. *Ann N Y Acad Sci* **807**:538–540.

Marshall, L., Helgadóttir, H., Mölle, M., and Born, J. (2006). Boosting slow oscillations during sleep potentiates memory. *Nature* **444**:610–613.

McCarthy, G., Puce, A., Gore, J.C., and Allison, T. (1997). Face-specific processing in the human fusiform gyrus. *J Cogn Neurosci* **9**:605–610.

McCarthy, G., Puce, A., Belger, A., and Allison, T. (1999). Electrophysiological studies of human face perception. II. Response properties of face-specific potentials generated in occipitotemporal cortex. *Cereb Cortex* **5**:431–444.

McNaughton, B.L. and Barnes, C.A. (1990). From cooperative synaptic enhancement to associative memory: bridging the abyss. *Sem Neurosci* **2**:403–416.

Meadows, J. C. (1974). The anatomical basis of prosopagnosia. *J Neurol Neurosurg Psychiat* **37**:489–501.

Morris, J. S., Frith, C. D., Perrett, D. I., et al. (1996). A differential neural response in the human amygdala to fearful and happy facial expressions. *Nature* **383**:812–815.

Nakamura, K., Mikami, A., and Kubota, K. (1992). Activity of single neurons in the monkey amygdala during performance of a visual discrimination task. *J Neurophysiol* **67**:1447–1463.

Oram, M. W. and Perrett, D. I. (1994). Responses of anterior superior temporal polysensory (STPa) neurons to "biological motion" stimuli. *J Cogn Neurosci* **6**:99–116.

Oram, M. W., Perrett, D. I., and Hietanen, J. K. (1993). Directional tuning of motion-sensitive cells in the anterior superior temporal polysensory area of the macaque. *Exp Brain Res* **97**:274–294.

O'Scalaidhe, S. P., Wilson, F. A., and Goldman-Rakic, P. S. (1997). Areal segregation of face-processing neurons in prefrontal cortex. *Science* **278**:1135–1138.

Parr, L., Winslow, J. T., and Hopkins, W. (1999). Is the inversion effect in rhesus monkeys face-specific? *Anim Cogn* **2**:123–129.

Perrett, D. I., Rolls, E. T., and Caan, W. (1982). Visual neurones responsive to faces in the monkey temporal cortex. *Exp Brain Res* **47**:329–342.

Perrett, D. I., Smith, P. A. J., Potter, D. D., et al. (1985). Visual cells in the temporal cortex sensitive to face view and gaze direction. *Proc R Soc Lond B* **223**:293–317.

Perrett, D. I., Mistlin, A. J., Chitty, A. J., et al. (1988). Specialized face processing and hemispheric asymmetry in man and monkey: evidence from single unit and reaction time studies. *Behav Brain Res* **29**:245–258.

Perrett, D. I., Hietanen, J. K., Oram, M. W., and Benson, P. J. (1992). Organization and functions of cells responsive to faces in the temporal cortex. *Phil Trans R Soc Lond B* **335**:23–30.

Phillips, M. L., Young, A. W., Senior, C., et al. (1997). A specific neural substrate for perceiving facial expressions of disgust. *Nature* **389**:495–498.

Pinsk, M. A., DeSimone, K., Moore, T., Gross, C. G., and Kastner, S. (2005). Representations of faces and body parts in macaque temporal cortex: a functional MRI study. *Proc Natl Acad Sci USA* **102**:6996–7001.

Pinsk, M. A., Weiner, K. S., Gross, C. G., Ghazanfar, A. A., and Kastner, S. (2006). Activation of a face processing network in the macaque using dynamic stimuli. *Soc Neurosci Annu Mtg* **438**.413.

Privman, E., Nir, Y., Kramer, U., et al. (2007). Enhanced category tuning revealed by intracranial electroencephalograms in high-order human visual areas. *J Neurosci* **27**:6234–6242.

Puce, A. and Perrett, D. (2003). Electrophysiology and brain imaging of biological motion. *Phil Trans R Soc Lond B* **358**:435–445.

Puce, A., Allison, T., Bentin, S., Gore, J. C., and McCarthy, G. (1998). Temporal cortex activation in humans viewing eye and mouth movements. *J Neurosci* **18**:2188–2199.

Puce, A., Allison, T., and McCarthy, G. (1999). Electrophysiological studies of human face perception. III. Effects of top–down processing on face-specific potentials. *Cereb Cortex* **9**:445–458.

Quiroga, R. Q., Reddy, L., Kreiman, G., Koch, C., and Fried, I. (2005). Invariant visual representation by single neurons in the human brain. *Nature* **435**:1102–1107.

Redican, W. K., Kellicut, M. H., and Mitchell, G. (1971). Preferences for facial expressions in juvenile rhesus monkeys. *Dev Psychol* **5**:539–642.

Richmond, B. J., Optican, L. M., Podell, M., and Spitzer, H. (1987). Temporal encoding of two-dimensional patterns by single units in primate inferior temporal cortex. I. Response characteristics. *J Neurophysiol* **57**:132–146.

Rodriguez, E., George, N., Lachaux, J.-P., et al. (1999). Perception's shadow: long-distance synchronization of human brain activity. *Nature* **397**:430.

Rollenhagen, J. E. and Olson, C. R. (2005). Low-frequency oscillations arising from competitive interactions between visual stimuli in macaque inferotemporal cortex. *J Neurophysiol* **94**:3368–3387.

Rolls, E. T. (1984). Neurons in the cortex of the temporal lobe and in the amygdala of the monkey with responses selective for faces. *Hum Neurobiol* **3**:209–222.

Rolls, E. T. (1996). The orbitofrontal cortex. *Phil Trans R Soc Lond B* **351**:1433–1443; discussion 1443–1434.

Rolls, E. T. (2007). The representation of information about faces in the temporal and frontal lobes. *Neuropsychologia* **45**:124–143.

Rolls, E. T., Treves, A., and Tovee, M. J. (1997). The representational capacity of the distributed encoding of information provided by populations of neurons in primate temporal visual cortex. *Exp Brain Res* **114**:149–162.

Rolls, E. T., Franco, L., Aggelopoulos, N. C., and Reece, S. (2003). An information theoretic approach to the contributions of the firing rates and the correlations between the firing of neurons. *J Neurophysiol* **89**:2810–2822.

Rosenfeld, S. A. and Van Hoesen, G. W. (1979). Face recognition in the rhesus monkey. *Neuropsychologia* **17**:503–509.

Roskies, A. L. (1999). The binding problem. *Neuron (Special Issue)* **24**:7–125.

Saalmann, Y. B., Pigarev, I. N., and Vidyasagar, T. R. (2007). Neural mechanisms of visual attention: how top–down feedback highlights relevant locations. *Science* **316**:1612–1614.

Sato, N. and Nakamura, K. (2001). Detection of directed gaze in rhesus monkeys (*Macaca mulatta*). *J Comp Psychol* **115**:115–121.

Schoffelen, J. M., Oostenveld, R., and Fries, P. (2005). Neuronal coherence as a mechanism of effective corticospinal interaction. *Science* **308**:111–113.

Sederberg, P. B., Kahana, M. J., Howard, M. W., Donner, E. J., and Madsen, J. R. (2003). Theta and gamma oscillations during encoding predict subsequent recall. *J Neurosci* **23**:10809–10814.

Sederberg, P. B., Schulze-Bonhage, A., Madsen, J. R., et al. (2007). Hippocampal and neocortical gamma oscillations predict memory formation in humans. *Cereb Cortex* **17**:1190–1196.

Sergent, J. and Signoret, J. L. (1992). Varieties of functional deficits in prosopagnosia. *Cereb Cortex* **2**:375–388.

Sergent, J., Ohta, S., and MacDonald, B. (1992). Functional neuroanatomy of face and object processing: a positron emission tomography study. *Brain* **115**:15–36.

Shepherd, S. V., Deaner, R. O., and Platt, M. L. (2006). Social status gates social attention in monkeys. *Curr Biol* **16**:R119–R120.

Swartz, K. B. (1983). Species discrimination in infant pigtail macaques with pictorial stimuli. *Dev Psychobiol* **16**:219–231.

Tallon-Baudry, C., Mandon, S., Freiwald, W. A., and Kreiter, A. K. (2004). Oscillatory synchrony in the monkey temporal lobe correlates with performance in a visual short-term memory task. *Cereb Cortex* **14**:713–720.

Tanaka, K. (2003). Columns for complex visual object features in the inferotemporal cortex: clustering of cells with similar but slightly different stimulus selectivities. *Cereb Cortex* **13**:90–99.

Thomas, E., Van Hulle, M. M., and Vogel, R. (2001). Encoding of categories by noncategory-specific neurons in the inferior temporal cortex. *J Cogn Neurosci* **13**:190–200.

Thorpe, S. J., Rolls, E. T., and Maddison, S. (1983). The orbitofrontal cortex: neuronal activity in the behaving monkey. *Exp Brain Res* **49**:93–115.

Tovee, M. J. and Rolls, E. T. (1992). Oscillatory activity is not evident in the primate temporal visual cortex with static stimuli. *Neuroreport* **3**:369–372.

Tovee, M. J., Rolls, E. T., Treves, A., and Bellis, R. P. (1993). Information encoding and the responses of single neurons in the primate temporal visual cortex. *J Neurophysiol* **70**:640–654.

Tsao, D. Y., Freiwald, W. A., Knutsen, T. A., Mandeville, J. B., and Tootell, R. B. (2003). Faces and objects in macaque cerebral cortex. *Nat Neurosci* **6**:989–995.

Tsao, D. Y., Freiwald, W. A., Tootell, R. B., and Livingstone, M. S. (2006). A cortical region consisting entirely of face-selective cells. *Science* **311**:670–674.

Uchida, G., Fukuda, M., and Tanifuji, M. (2006). Correlated transition between two activity states of neurons. *Phys Rev E* **73**:031910.

Varela, F., Lachaux, J. P., Rodriguez, E., and Martinerie, J. (2001). The brainweb: phase synchronization and large-scale integration. *Nat Rev Neurosci* **2**:229–239.

Vermeire, B. A. and Hamilton, C. R. (1998). Inversion effect for faces in split-brain monkeys. *Neuropsychologia* **36**:1003–1014.

Vuilleumier, P., Armony, J. L., Driver, J., and Dolan, R. J. (2001). Effects of attention and emotion on face processing in the human brain: an event-related fMRI study. *Neuron* **30**:829–841.

Vuilleumier, P., Richardson, M. P., Armony, J. L., Driver, J., and Dolan, R. J. (2004). Distant influences of amygdala lesion on visual cortical activation during emotional face processing. *Nat Neurosci* **7**:1271–1278.

Waitt, C., Little, A. C., Wolfensohn, S., *et al.* (2003). Evidence from rhesus macaques suggests that male coloration plays a role in female primate mate choice. *Proc R Soc Lond B* **270**:S144–S146.

Wang, G., Tanaka, K., and Tanifuji, M. (1996). Optical imaging of functional organization in the monkey infereotemporal cortex. *Science* **272**:1665–1668.

Webster, M.J., Ungerleider, L.G., and Bachevalier, J. (1991). Lesions of inferior temporal area TE in infant monkeys alter cortico-amygdalar projections. *Neuroreport* **2**:769–772.

Winston, J.S., Henson, R.N., Fine-Goulden, M.R., and Dolan, R.J. (2004). fMRI-adaptation reveals dissociable neural representations of identity and expression in face perception. *J Neurophysiol* **92**:1830–1839.

Womelsdorf, T., Schoffelen, J.-M., Oostenveld, R., *et al.* (2007). Modulation of neuronal interactions through neuronal synchronization. *Science* **316**:1609–1612.

Young, A.W., Aggleton, J.P., Hellawell, D.J., *et al.* (1995). Face processing impairments after amygdalotomy. *Brain* **118**:15–24.

Young, M.P. and Yamane, S. (1992). Sparse population coding of faces in the inferotemporal cortex. *Science* **256**:1327–1331.

Zangenehpour, S. and Chaudhuri, A. (2005). Patchy organization and asymmetric distribution of the neural correlates of face processing in monkey inferotemporal cortex. *Curr Biol* **15**:993–1005.

14

Using spikes and local field potentials to reveal computational networks in monkey cortex

KRISTINA J. NIELSEN AND GREGOR RAINER

Introduction

Traditionally, neurophysiological investigations in awake non-human primates have largely focused on the study of single-unit activity (SUA), recorded extracellularly in behaving animals using microelectrodes. The general aim of these studies has been to uncover the neural basis of cognition and action by elucidating the relation between brain activity and behavior. This is true for studies in sensory systems such as the visual system, where investigators are interested in how SUA covaries with aspects of visually presented stimuli, as well as for studies in the motor system where SUA covariation with movement targets and dynamics are investigated. In addition to these SUA studies, there has been increasing interest in the local field potential (LFP), a signal that reflects aggregate activity across populations of neurons near the tip of the microelectrode. In this chapter, we will describe recent progress in our understanding of brain function in awake behaving monkeys using LFP recordings. We will show that the combination of recording the activity of single neurons and local populations simultaneously offers a particularly promising way to gain insight into cortical brain mechanisms underlying cognition and memory.

Information Processing by Neuronal Populations, ed. Christian Hölscher and Matthias Munk.
Published by Cambridge University Press. © Cambridge University Press 2009.

Measures of neural activity at the level of neurons and networks

The activity of single neurons (SUA) is estimated by amplifying and collecting the comprehensive broadband electrical signal, which can be detected in the brain by using microelectrodes. This signal is digitized at rates of 20 kHz or higher, and high-pass filtering to remove its low-frequency components at a typical cut-off frequency of 300 Hz. Clustering methods are then used to extract the times of action potentials generated by one or more neurons near the electrode tip, by identifying and gathering the occurrence of waveforms with a particular, predefined shape corresponding to the signature of action potentials emitted by the neuron being studied. The part of the comprehensive signal that has often been filtered away to identify and sort action potentials has received comparatively little attention, but recent accumulating evidence suggests that it carries potentially very useful information. When the unfiltered broadband signal picked up at the microelectrodes is considered, it becomes clear that it actually consists of several components (for a review, see Logothetis, 2002; Logothetis and Wandell, 2004). To extract these components, the broadband signal is usually split into two different frequency bands by high- and low-pass filtering, respectively. High-pass filtering is used to extract the multi-unit activity (MUA), while low-pass filtering isolates the LFP. The MUA, obtained by bandpass filtering the comprehensive signal in a frequency range of 400 Hz to about 3 kHz, represents the weighted average of the spiking activity within a sphere of about 200–300 μm around the electrode tip. This MUA bears close resemblance to SUA, but it differs in that it represents average action potential activity generated by neurons close to the electrode tip rather than those of an individual single neuron. Another difference between the signals is that there tends to be an overrepresentation of large excitatory pyramidal cells in SUA estimates (Henze et al., 2000) due to sampling biases related to the recording technique, whereas MUA is less susceptible to this limitation. According to current estimates, over 70% of excitatory synapses in the cortex remain local, and only about 30% target distant brain regions (Braitenberg and Schüz, 1998; Binzegger et al., 2004). Since both MUA and SUA capture spiking activity, they thus represent local processing within a cortical column as well as the long-range output that targets distant brain regions. The LFP, on the other hand, is extracted from the broadband signal picked up at an electrode by low-pass filtering below about 300 Hz. It measures extracellular fields generated by membrane currents originating from axons, somata, and dendrites surrounding the electrode tip (Mitzdorf, 1985; Logothetis, 2002; Logothetis and Wandell, 2004). Synchronized dendritic activity is thought to have the largest contribution to the LFP (Mitzdorf, 1985), making the LFP a measure of the local processing in a brain region, as well as of the inputs that the brain region receives. The LFP is a mass signal, representing the weighted average of the synaptic signals of a neuronal population within 0.5–3 mm of the electrode tip (Logothetis, 2002). Accordingly, LFPs depend on temporal synchronization between dendritic events in the sampled population, as well as on the spatial alignment of the constituting neurons. Furthermore, the LFP may not only reflect the activity of the large pyramidal neurons, but also the interneuronal activity within a cortical volume. In summary, MUA and SUA capture different aspects of neural processing than the LFP. On the one hand, MUA and SUA contain signals related to the spiking output of a brain region, along with signatures of local processing carried out in that region. On the other hand, the LFP reflects dendritic input to the region near the electrode tip, as well as local processing in that region. The spiking output of a single neuron is related to the synaptic input by a non-linear transformation, and the same is true at the level of neuronal populations. Thus systematically comparing the input to a brain region to its output represents valuable information. Using appropriate experimental designs, one can directly compare neural activity at the level of LFP to MUA/SUA, and the differences between these measures can be thought of as representing the local operations carried out by a particular brain structure.

When comparing LFP and SUA, the different nature of the two signals has to be kept in mind. SUA consists of the occurrence of action potentials from a particular neuron, and is thus a discrete signal. In most cases, SUA is summarized by computing the mean firing rate of a neuron in a selected time interval. In contrast, the LFP is a continuous signal. The LFP is commonly analyzed either in the spectral or the temporal domain. Spectral analysis consists of computing the power spectrum of the LFP, possibly in a time-resolved manner by computation of the spectrogram. The power in selected frequency bands, or the overall shape of the spectrum, is then analyzed further. Analysis in the temporal domain is usually performed by computing the averages of the LFP across multiple trials of the same condition. Before averaging, trials are aligned to a selected time point, such as the onset of a stimulus or a movement. The resulting evoked potentials usually show a series of positive and negative peaks, whose amplitudes can then be analyzed further. MUA is a continuous signal like the LFP, which is often analyzed by computing its average amplitude in a selected time interval.

Previous work on LFP and SUA

Throughout the last years, a number of studies have performed detailed comparisons between the LFP and SUA or MUA in the macaque monkey.

Comparisons were carried out across a number of brain regions, as well as a number of tasks. Many of these studies have largely focused on describing similarities between these signals during various cognitive tasks. Such similarities emerged for example when studying the LFP responses in area V1, the initial cortical stage of visual processing. In agreement with the behavior of single neurons, LFPs recorded in area V1 display sensitivity to the orientation of a grating pattern, as well as the grating's contrast (Frien et al., 2000; Henrie and Shapley, 2005). In both cases, it seems that LFP components with frequencies in the gamma band (i.e. above about 30Hz) are most sensitive to stimulus parameters. Similar results have been obtained for area MT in the visual cortex, which is strongly implicated in motion perception based on the response properties of single MT neurons (for a review, see Born and Bradley, 2005). The LFP recorded in MT, specifically its frequency components above 40Hz, also carries information about the direction and speed of a moving stimulus (Liu and Newsome, 2006). Agreement between LFP and MUA could be observed at individual recording sites, as the preferences for particular stimulus speeds and motion directions were highly correlated between LFP and MUA signals recorded from the same electrode. Furthermore, variability in both LFP and MUA correlated well with the monkey's trial-to-trial performance in a speed discrimination task. LFPs were also recorded in the inferotemporal (IT) cortex, which represents the final stage of the ventral visual processing stream. Neurons in this area show a strong preference for complex objects including faces (for reviews, see Logothetis and Sheinberg, 1996; Tanaka, 1996). Again, the LFP recorded in area IT is in agreement with the behavior established for individual neurons, and LFPs recorded from sites in IT show selectivity for complex objects (Kreiman et al., 2006). In further agreement with the behavior of single neurons, the LFP recorded in IT also shows tolerance to changes in an object's position in space, as well as the object's size. Object selectivity can be observed by analyzing the LFP either in the temporal or in the spectral domain. In the temporal domain, the range of the LFP signal (the difference between the maximum and minimum LFP amplitude) differs between objects, which correlates with a modulation of LFP broadband power in the spectral domain. Finally, a number of studies have assessed the LFP signals related to the planning and execution of hand and eye movements. The LFP recorded in the arm regions of the primary motor area M1, as well as of the supplementary motor area (SMA) has been found to convey information about arm movements. The evoked potentials in these areas contain components that are influenced by the direction of an arm movement. Evoked potentials are furthermore sensitive to which hand is used for a task, and whether the monkey moves only one or both hands (Donchin et al., 2001; Mehring et al., 2003). The same response properties are displayed by single neurons recorded in M1 and SMA (Donchin et al., 1998). Further similarities between LFP and SUA have been documented for the parietal reach region (PRR) and the lateral intraparietal area (LIP) in the posterior parietal cortex. Analyses of single-neuron responses have established that these areas contain maps for the direction of either arm or eye movements that the monkey is intending to perform (for a review, see Andersen and Buneo, 2002). The LFP in areas PRR and LIP also encodes the direction of planned arm and eye movements (Pesaran et al., 2002; Scherberger et al., 2005). In area LIP, tuning widths for movement directions are similar for the LFP and SUA. Consistent with the results obtained for the visual cortex, planning either an eye or an arm movement is most strongly reflected in LFP components with frequencies in the gamma range.

In conclusion, many of the studies performed so far have revealed that the LFP in general shows response properties similar to that of the neurons recorded in the same brain region. However, experiments have also documented interesting differences between LFP and SUA or MUA behavior. A number of these differences could be due to the fact that the LFP pools signals over a larger neuronal population than the other two signals. This means that the neurons contributing to the LFP signals have more diverse response properties than the ones contributing to either SUA or MUA. For example, it has been demonstrated that the LFP recorded on an electrode is a poor predictor of the behavior of single neurons recorded from the same electrode. Instead, the LFP correlates better with the average signal of the neuronal population within a 3-mm radius around the electrode tip (Kreiman et al., 2006). An example of such a discrepancy between LFP and SUA is shown in Fig. 14.1, which illustrates the lack of agreement between LFP and SUA recorded from the same electrode using data collected during an experiment performed by our group. LFP and SUA were recorded in IT while monkeys were presented with four different natural scenes, which were matched in overall contrast and luminance (see Nielsen et al. (2006a) for details on the stimuli). LFP responses were characterized by analyzing the amplitude of a positive peak occurring at about 140 ms after stimulus onset in the visual evoked potentials (VEP). For individual neurons, the mean firing rate in a 300-ms interval during stimulus presentation was used to compute responses to the natural scenes. To test for similarities between the LFP and SUA recorded at the same electrode, we determined the agreement in the stimulus preferences of the two signals. For this purpose, the natural scene evoking the largest response from a single neuron was determined. The scenes were also rank ordered according to their peak amplitudes determined from the VEP. The rank of the best single neuron stimulus was then determined for the LFP signal recorded at the same electrode. Figure 14.1 plots the distribution of

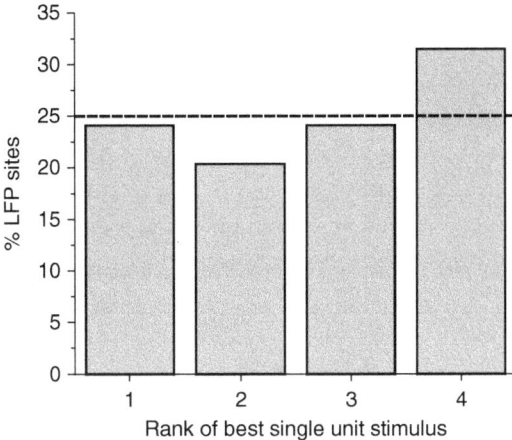

Figure 14.1 Agreement in stimulus selectivity between single inferotemporal (IT) neurons and the LFP signals recorded at the same electrode. Only visually responsive neurons and LFP sites were considered for this analysis. We first identified the best stimulus (out of a set of four natural scenes) for each single neuron. It was then determined how strong the LFP responses to this stimulus were in comparison to the other stimuli. For this purpose, the natural scenes were rank ordered according to the LFP responses that they evoked, and the rank of the best single neuron stimulus was determined for the LFP. Rank 1 cases are cases in which the best single neuron stimulus evokes the largest responses from the LFP. Rank 4 cases imply that the best single neuron stimulus was the worst stimulus for the LFP. The histogram shows the percentage of LFP sites for each rank; the dashed line indicates the chance level of 25%.

ranks across the population of 54 IT neurons, which responded excitatorily to at least one of the full scenes, and their corresponding LFP sites. LFP–SUA pairs were only considered if the LFP was also responsive to at least one stimulus. If stimulus preferences were similar for SUA and LFP, then the best SUA stimulus should rank first for the LFP signal. However, we observed an almost equal distribution of all ranks. There was no indication of a similar scene preference for SUA and LFP. Most importantly, the number of rank 1 cases (i.e. the number of LFP–SUA pairs with the same scene preference) was not significantly larger than chance (χ^2 test, $p = 0.9$). The stimulus preference of the LFP can thus in general not be inferred from the stimulus preference of a locally recorded single neuron. It is possible that this is a consequence of the fact that the LFP pools across substantial brain regions, and thus individual stimulus preferences of local neurons are lost in this averaging process.

Pooling signal from neurons with various orientation preferences may similarly explain why the LFP recorded in V1 is less orientation-selective than the

MUA (Frien et al., 2000). It may also be the reason for the fact that fewer LFP than MUA sites are tuned for stimulus speed (17.3% vs. 2.2%) and motion direction (22.1% vs. 3.4%) in MT (Liu and Newsome, 2006), and that in IT fewer LFP sites than MUA sites are object-selective (44% vs. 72%) (Kreiman et al., 2006). It becomes more difficult to explain how averaging over a diverse population of neurons can be linked to the finding that about 20% of the LFP sites in MT are not visually responsive, as almost all MUA sites respond visually (Liu and Newsome, 2006). It is likely that the local three-dimensional structure of cortex is responsible for this, since dendritic dipoles can cancel each other if oriented opposite one another. Yet, there are a number of additional results that cannot be easily explained by assuming that the LFP is simply an average over a larger neuronal population than MUA or SUA. These findings hint at the different sources generating LFP and SUA or MUA, and highlight the importance of performing combined analyses of these signals. First, both SUA and MUA recorded in area V1 show strong adaptation effects, with responses to a continuously presented stimulus ceasing after about 3s. The LFP response, on the other hand, remains elevated above baseline level throughout the presentation duration (Logothetis et al., 2001). Another discrepancy between the SUA and LFP recorded in V1 pertains to their dependency on stimulus contrast. V1 neurons initially increase their responses to a grating whose contrast is increasing. After a threshold contrast has been reached, the responses of most V1 cells saturate. The LFP shows a similar dependency on stimulus contrast, with increases in responses with increasing contrast. However, the LFP keeps increasing at contrast levels at which the single cell responses have already saturated (Henrie and Shapley, 2005). Furthermore, the comparison of LFP and SUA in M1 and SMA shows that correlations between LFP and SUA recorded from the same electrode may be absent in one brain area, but present in another (Donchin et al., 2001).

Differences between LFP and SUA and MUA have also turned up when these signals were used to predict monkey behavior. In the motor cortex, the LFP can be used to successfully decode a movement direction about 50ms after this is possible based on SUA and MUA. Furthermore, combining LFP and SUA or LFP and MUA results in higher decoding accuracy than possible based on any signal alone, suggesting independent information in these signals (Mehring et al., 2003). In LIP, both SUA and LFP components with frequencies above 30Hz can be used to predict the direction of an eye movement. However, only the LFP can be used to decode the transition from planning an eye movement to executing it. Interestingly, a different frequency band in the LFP carries this information, as decoding performance for the behavioral state is best when based on LFP components with frequencies below 20Hz (Pesaran et al., 2002). Differences between LFP and SUA were also observed in the PRR. Here, SUA could predict better than the LFP which direction a planned eye or arm movement was going to have. In contrast, the LFP was better at distinguishing between eye and arm movements (Scherberger *et al.*, 2005).

Most recently, a study has described interesting differences between LFP and MUA in V1 (Wilke *et al.*, 2006), confirming earlier findings (Gail *et al.*, 2004). Monkeys were trained to indicate the presence of a large circle by pulling a lever. By adding small dots in the periphery around the large circle, the circle could be made to disappear perceptually, despite its continuous physical presence on the screen. At the level of the MUA, effects of the perceptual stimulus disappearance were first observed in V4. MUA in V1 and V2 did not reflect the perceptual disappearance. A largely similar result was obtained when the gamma frequency components of the LFP were analyzed. However, the LFP power in the alpha-band range (9–14 Hz) was modulated by perception in V1, V2, and V4. Interestingly, the modulation in LFP power was observed later than the first influence of perceptual disappearance on the MUA in V4.

All these findings highlight the fact that LFP and SUA or MUA indeed reflect different brain processes. As mentioned at the beginning of the chapter, the LFP reflects input and local processing to a brain region, whereas SUA and MUA represent the output of that region. The similarities between LFP and SUA or MUA therefore suggest a high degree of similarity between the inputs to a brain region and its outputs. The listed discrepancies between LFP and SUA or MUA in contrast are instances where input and output are not this closely related. These are cases that may allow us to identify the unique contributions of individual brain regions. The data from Scherberger *et al.* (2005) for example indicate that while information about the movement type is already present in the input to PRR, the direction of the movement becomes more precisely defined in this area. On the other hand, the results reported by Wilke *et al.* (2006) represent a case in which information is initially absent in the output from a brain region, but is later relayed to the brain region from a different source – likely activated after additional processing. In conclusion, a combined analysis of LFP and SUA or MUA has the power to reveal how different brain regions interact with each other to process information.

Combining LFP to SUA to reveal computational networks across the brain

Recent results from our group provide a compelling example of how a comparison between LFP and SUA can be used to identify how different brain regions participate in the extraction of information from a visual scene (Nielsen *et al.*, 2006a). In this study, two monkeys were trained to discriminate between natural scenes. We then determined the regions of each natural scene on which the monkeys relied to perform the discrimination task. These diagnostic regions were determined by using a behavioral paradigm called "Bubbles" (Gosselin and Schyns, 2001). During each trial of the Bubbles paradigm, the scenes appeared behind randomly constructed occluders, which consisted of a non-transparent surface punctured by randomly placed, round windows. The monkeys continued to perform the discrimination task on the occluded scenes. Depending on which scene regions were occluded on a trial, the monkeys could or could not identify a scene correctly. Occlusion of diagnostic scene regions rendered the monkeys unable to perform the task correctly, while occlusion of non-diagnostic scene regions did not. By collecting performance data for a large number of different occluders, and determining the scene regions that were systematically occluded during incorrect responses, we could determine how relevant each scene region was for the monkeys' behavior (Nielsen et al., 2006b).

Based on the behavioral results, we then constructed unique stimulus sets for each monkey. Stimulus sets consisted of four natural scenes, presented either as their original version, or as one of six modifications. Three modifications showed diagnostic scene regions (diagnostic conditions), and three modifications showed non-diagnostic scene regions (non-diagnostic conditions). All scene regions except the selected ones were covered by an occluder. The three diagnostic conditions varied in how much of the original scene remained visible (10%, 30%, or 50%); the same was the case for the non-diagnostic conditions. All stimuli were matched in luminance and overall contrast. We used this stimulus set to probe the influence of diagnosticity on the responses of single neurons and the LFP in the IT cortex. Monkeys passively viewed the stimuli during the recording sessions. Figure 14.2 shows the responses of a selected neuron, as well as a selected LFP site to the different versions of one natural scene. In this figure, the LFP responses are shown in form of the trial-averaged VEP. As can be seen, the VEP contains a prominent positive peak at about 140 ms after stimulus onset (the P140), whose amplitude differs between diagnostic and non-diagnostic conditions. We quantified single-neuron responses to different stimuli by computing the mean firing rate in a 300-ms interval beginning 100 ms after stimulus onset. For the LFP, stimulus responses were determined as the mean LFP amplitude in a 20-ms bin centered on the P140 maximum.

We could establish that stimulus diagnosticity is represented at the level of SUA in IT as the average response to the diagnostic conditions was significantly larger than the average response to the non-diagnostic condition in the tested single-neuron population (paired t-tests between diagnostic and non-diagnostic conditions showing the same amount of the original scene, $p \leq 0.02$ in all three cases). We then mapped the influences of diagnosticity across IT. For each

Computational networks in monkey cortex 359

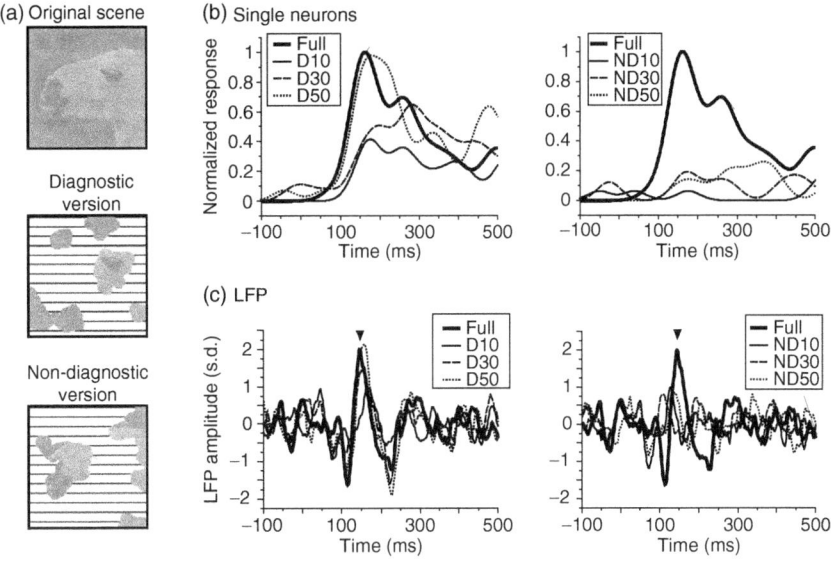

Figure 14.2 Responses of a sample single neuron and sample LFP site. (a) Sample stimuli. The three images show one of the natural scenes in its original version, the diagnostic version of this image as constructed for one of the monkeys, and the matching non-diagnostic scene version. In the latter two images, occluded image parts are indicated by hatched regions. The diagnostic and non-diagnostic stimulus show 30% of the original scene. (b) Responses of a sample neuron. The two plots summarize the responses to the seven versions of one natural scene by plotting spike density functions (spike trains were smoothed with a Gaussian kernel with a standard deviation of 30 ms). The stimulus appeared at time 0 ms and stayed on the screen for 500 ms. Spike density functions were normalized so that the maximum across all conditions equaled 1. The plot on the left shows the responses to the original (full) scene and the three diagnostic versions. The right side plots the responses to the non-diagnostic stimulus versions, the response to the full stimulus is repeated as a reference. In the legend, D indicates diagnostic, ND non-diagnostic conditions; the numbers correspond to the amount of the original image visible in a condition. (c) Responses of a sample LFP site. These plots show the VEP for each condition. VEPs are plotted in units of standard deviation. These units are computed by subtracting the mean LFP amplitude in a 100-ms window preceding stimulus onset from the LFP of each trial, and dividing this signal by the standard deviation calculated from the same baseline period. The arrow points at the P140. The layout of these two plots as is in (b).

single neuron, we quantified the influence of diagnosticity on its responses by computing the amount of variance in the firing rate that could be attributed to differences in responses to diagnostic and non-diagnostic conditions (diagnostic variance). We performed the same analysis on the LFP responses. Using this

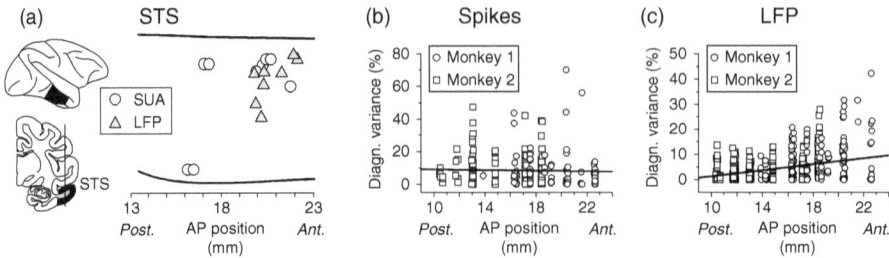

Figure 14.3 Influence of recording position on the properties of single neurons and the LFP. (a), Location of high diagnosticity cases in one monkey, shown on a sagittal view of parts of the temporal lobe. The two small brain pictures on the left indicate the location of the selected brain region. This region is indicated in black in the upper image; it is generated by slicing along the line depicted in the lower image. The right side shows the location of the single neurons (circles) and LFP sites (triangles) strongly influenced by diagnosticity. To allow a better separation of different cases, the anterior–posterior (AP) position of each case was randomly jittered by a small amount for display purposes only. Thick black lines indicate the location of the superior temporal sulcus (STS) and the ventral end of the brain. The position of these landmarks is plotted as estimated during recordings. (b), (c) Diagnostic variance as a function of recording location. In these plots, the diagnostic variance of each case is plotted as a function of its AP position. Symbols indicate the monkey in which a case was recorded; the dashed line plots the regression computed between diagnostic variance and AP position. (b) Single-unit data. (c) LFP data. In all three plots, "Post" and "Ant" label the posterior and anterior end of the recording region, respectively.

measure, we could locate the single neurons that were most strongly influenced by diagnosticity. Figure 14.3a plots their location within the recording region for one of the monkeys. We similarly located the LFP sites most strongly influenced by diagnosticity. The location of these sites for the same monkey is also plotted in Fig. 14.3a. As can be seen, diagnosticity strongly influences single neurons throughout the whole recording region. In contrast, the influence of diagnosticity on the LFP increases from posterior to anterior recording locations, as all LFP sites with strong diagnosticity influences cluster in the anterior half of the recording region. We confirmed these conclusions by plotting the diagnostic variance of all single neurons as a function of their anterior–posterior location (see Fig. 14.3b). The diagnostic variance for the LFP sites was plotted in the same way (Fig. 14.3c). While the diagnostic variance did not depend on the recording location for the single neurons, it increased from posterior to anterior locations for the LFP. Both findings were obtained by analyzing the data from each monkey individually, as well as combining their results (SUA: Pearson correlation coefficients r not significantly different from 0 for each monkey individually and the combined data, $p \geq 0.1$ for the three tests; LFP: $r \geq 0.16$, $p \leq 0.04$ for the three tests).

In summary, these data suggest that diagnosticity is encoded in the output of single neurons throughout IT. However, the encoding of diagnosticity at the input level – as indicated by the LFP – increases from posterior to anterior IT. Our findings highlight a novel way to combine LFP and SUA recordings to reveal computational networks underlying a particular cognitive function, in the present case the neural encoding of diagnostic elements in complex visual displays. Diagnosticity is first encoded by neurons in posterior IT, and then transmitted to more anterior regions. The extraction of diagnosticity is accomplished by neural networks in posterior IT, since its signatures can be seen in the spiking output (SUA) from that region, but not in their input (LFP). Thus even though anterior IT contains signals of diagnosticity at both LFP and SUA levels, extraction of diagnosticity is largely not accomplished in that region itself but is already present in its input signals. The assignment of computational functions to connected brain networks is possible only by having LFP and SUA signals available during a suitable experimental paradigm, which is designed in such a way that critical parametric task condition variations are detectable at the level of LFPs. The insights given by these joint LFP/SUA analyses go a step beyond traditional brain–behavior correlations based on each of the signals considered alone. Our approach can generalize to other behavioral tasks, and promises to allow a delineation of functional networks with far greater accuracy than has previously been possible.

References

Andersen, R.A. and Buneo, C.A. (2002). Intentional maps in posterior parietal cortex. *Annu Rev Neurosci* **25**:189–220.

Binzegger, T., Douglas, R.J., and Martin, K.A.C. (2004). A quantitative map of the circuit of cat primary visual cortex. *J Neurosci* **24**:8441–8453.

Born, R.T. and Bradley, D.C. (2005). Structure and function of visual area MT. *Annu Rev Neurosci* **28**:157–189.

Braitenberg, V. and Schüz, A. (1998). *Cortex: Statistics and Geometry of Neuronal Connectivity*. Heidelberg, Germany: Springer-Verlag.

Donchin, O., Gribova, A., Steinberg, O., Bergman, H., and Vaadia, E. (1998). Primary motor cortex is involved in bimanual coordination. *Nature* **395**:274–278.

Donchin, O., Gribova, A., Steinberg, O., *et al.* (2001). Local field potentials related to bimanual movements in the primary and supplementary motor cortices. *Exp Brain Res* **140**:46–55.

Frien, A., Eckhorn, R., Bauer, R., Woelbern, T., and Gabriel, A. (2000). Fast oscillations display sharper orientation tuning than slower components of the same recordings in striate cortex of the awake monkey. *Eur J Neurosci* **12**:1453–1465.

Gail, A., Brinksmeyer, H. J., and Eckhorn, R. (2004). Perception-related modulations of local field potential power and coherence in primary visual cortex of awake monkey during binocular rivalry. *Cereb Cortex* **14**:300–313.

Gosselin, F. and Schyns, P. G. (2001). Bubbles: a technique to reveal the use of information in recognition tasks. *Vision Res* **41**:2261–2271.

Henrie, J. A. and Shapley, R. (2005). LFP power spectra in V1 cortex: the graded effect of stimulus contrast. *J Neurophysiol* **94**:479–490.

Henze, D. A., Borhegyi, Z., Csicsvari, J., et al. (2000). Intracellular features predicted by extracellular recordings in the hippocampus in vivo. *J Neurophysiol* **84**:390–400.

Kreiman, G., Hung, C. P., Kraskov, A., et al. (2006). Object selectivity of local field potentials and spikes in the macaque inferior temporal cortex. *Neuron* **49**:433–445.

Liu, J. and Newsome, W. T. (2006). Local field potential in cortical area MT: stimulus tuning and behavioral correlations. *J Neurosci* **26**:7779–7790.

Logothetis, N. K. (2002). The neural basis of the blood-oxygen-level-dependent functional magnetic resonance imaging signal. *Phil Trans R Soc Lond B* **357**:1003–1037.

Logothetis, N. K. and Sheinberg, D. L. (1996). Visual object recognition. *Annu Rev Neurosci* **19**:577–621.

Logothetis, N. K. and Wandell, B. A. (2004). Interpreting the BOLD signal. *Annu Rev Physiol* **66**:735–769.

Logothetis, N. K., Pauls, J., Augath, M., Trinath, T., and Oelterman, A. (2001). Neurophysiological investigation of the basis of the fMRI signal. *Nature* **412**:150–157.

Mehring, C., Rickert, J., Vaadia, E., et al. (2003). Inference of hand movements from local field potentials in monkey motor cortex. *Nat Neurosci* **6**:1253–1254.

Mitzdorf, U. (1985). Current source–density method and application in cat cerebral cortex: investigation of evoked potentials and EEG phenomena. *Physiol Rev* **65**:37–100.

Nielsen, K. J., Logothetis, N. K., and Rainer, G. (2006a). Dissociation between local field potentials and spiking activity in macaque inferior temporal cortex reveals diagnosticity-based encoding of complex objects. *J Neurosci* **26**:9639–9645.

Nielsen, K. J., Logothetis, N. K., and Rainer, G. (2006b). Discrimination strategies of humans and rhesus monkeys for complex visual displays. *Curr Biol* **16**:814–820.

Pesaran, B., Pezaris, J. S., Sahani, M., Mitra, P. P., and Andersen, R. A. (2002). Temporal structure in neuronal activity during working memory in macaque parietal cortex. *Nat Neurosci* **5**:805–811.

Scherberger, H., Jarvis, M. R., and Andersen, R. A. (2005). Cortical local field potential encodes movement intentions in the posterior parietal cortex. *Neuron* **46**:347–354.

Tanaka, K. (1996). Inferotemporal cortex and object vision. *Annu Rev Neurosci* **19**:109–139.

Wilke, M., Logothetis, N. K., and Leopold, D. A. (2006). Local field potential reflects perceptual suppression in monkey visual cortex. *Proc Natl Acad Sci USA* **103**:17507–17512.

15

Cortical gamma-band activity during auditory processing: evidence from human magnetoencephalography studies

JOCHEN KAISER AND WERNER LUTZENBERGER

Introduction

Oscillatory synchronization in the gamma-band range (~30–100 Hz) has been proposed as a possible solution to the "binding problem," i.e. the question of how the brain integrates perceptual features that are processed in distant cortical regions to generate a coherent object representation. Intracortical recordings in animals have demonstrated stimulus-specific synchronous oscillations of spatially distributed, feature-selective neurons (Eckhorn et al., 1988; Gray et al., 1989) that may provide a general mechanism for the temporal coordination of activity patterns in spatially separate regions of the cortex (Gray and Singer, 1989; Singer et al., 1997). In addition to visual feature binding, fast oscillations have been found to reflect modulations of arousal (Munk et al., 1996), perceptual integration (Fries et al., 1997), and attentional selection processes (Fries et al., 2001), and have even been proposed as a potential neural correlate of consciousness (Engel and Singer, 2001; Singer, 2001). In the middle of the last decade, the first studies of gamma-band activity (GBA) in human electroencephalogram (EEG) have relied on paradigms analogous to the early animal work (Lutzenberger et al., 1995; Müller et al., 1996). Since then, investigations using scalp EEG, magnetoencephalography (MEG), and intracranial recordings have supported the functional significance of fast oscillatory activity for a wide range of human cognitive functions. The present chapter will first provide a brief overview of the current

Information Processing by Neuronal Populations, ed. Christian Hölscher and Matthias Munk.
Published by Cambridge University Press. © Cambridge University Press 2009.

state of human GBA research related to visual perception, selective attention, and memory. We will then give a more detailed account of MEG investigations of different aspects of auditory processing with a focus on differences between the representation of spatial versus non-spatial acoustic information. In these studies, enhanced gamma-band amplitude and coherence was observed with a high temporal and topographical specificity during acoustic change detection, audiovisual integration, and short-term memory, probably reflecting both the activation of task-specific local networks and interactions between networks underlying different task components. At the end we will draw general conclusions and point out some of the open questions in research on human cortical oscillatory activity.

Human gamma-band activity and visual perception

Early research on fast oscillations in humans was driven by the hypothesis that GBA would reflect the activation of cortical neural assemblies underlying the representation of gestalt-like or meaningful visual stimuli. In line with this assumption, coherently moving light bars elicited GBA enhancements at occipital EEG recording sites (Lutzenberger et al., 1995; Müller et al., 1996, 1997). Increased GBA was also found for unambiguous positions of rotating bistable figures (Keil et al., 1999), for canonical presentations of a point-light walker compared with an inverted walker or a scrambled display (Pavlova et al., 2004), or for the processing of words compared with non-words (Lutzenberger et al., 1994; Pulvermüller et al., 1996, 1997; Krause et al., 1998). Oscillatory synchronization as a neurophysiological mechanism underlying visual binding has been demonstrated by Tallon-Baudry et al. (1996) who found that induced EEG activity at 30–45 Hz distinguished both illusory Kanizsa figures and real triangles from a no-triangle stimulus. Whereas this effect was topographically widespread, a replication study in MEG revealed local GBA enhancements both over midline and lateral occipital areas for the comparison of illusory with no-triangle stimuli, and over posterior parietal cortex when illusory figures were contrasted against real triangles, suggesting an involvement of both ventral and dorsal visual pathways in the processing of Kanizsa figures (Kaiser et al., 2004). More recently, interhemispheric neural coupling in the gamma band (30–40 Hz) was found to be increased over visual cortex when information from the right and left visual hemifields had to be integrated to perceive horizontal motion (Rose and Büchel, 2005), or when local information had to be integrated across hemifields into the percept of a single global shape (Rose et al., 2006). Most of the effects summarized here were observed for induced, non-phase-locked GBA with typical peak latencies of ~200–300 ms post-stimulus onset (Tallon-Baudry and Bertrand, 1999). In contrast, evoked, phase-locked gamma responses have mainly been related to early sensory processing (Karakas & Basar, 1998; Busch et al., 2004). However, evoked gamma responses are also modulated by attention (Tiitinen et al., 1993; Herrmann et al., 1999) (see section on selective attention below), and have been proposed to reflect fast matches between sensory input and existing representations in long-term memory (Herrmann et al., 2004a). An overview of the different types of gamma responses has been provided by Basar-Eroglu et al. (1996).

Human gamma-band activity and selective attention

Induced GBA is not only related to perceptual processes but may also reflect the modulation of cortical network synchronization by top–down processes like selective attention. Attending to a specific stimulus was accompanied by increased EEG gamma power compared to when the same stimulus was ignored (Müller et al., 2000). Cues directing spatial attention to a visual hemifield led to a shift of the gamma response from a broad posterior distribution to electrodes contralateral to the attended side (Gruber et al., 1999). Similarly, directing attention to a non-spatial visual feature like color gave rise to GBA increases (Müller and Keil, 2004). Enhancements of GBA during visual and auditory target processing have also been interpreted in support of top–down influences on oscillatory synchrony. During a visual detection task, evoked 40-Hz responses were higher for Kanizsa figures than for non-Kanizsa figures and within the Kanizsa figures for targets than for non-targets (Herrmann et al., 1999). Similarly, targets in auditory oddball paradigms have been found to elicit increased phase-locked gamma responses in EEG (Debener et al., 2003) and induced GBA in MEG (Kaiser and Lutzenberger, 2004), and spatial selective attention for tactile stimuli led to enhanced induced gamma power over somatosensory cortex (Bauer et al., 2006). Shifts of attention between the visual and auditory modalities were accompanied by marked increases of induced gamma activity in MEG sensors over the modality-specific cortices (Sokolov et al., 2004). Evidence for attentional increases of GBA was also obtained in intracranial recordings during a visual delayed matching-to-sample task, with considerable differences between regions (lateral occipital cortex versus fusiform gyrus) (Tallon-Baudry et al., 2005). Concerning the mechanism underlying attentional GBA enhancements, it has been speculated that selective attention may enhance the saliency of sensory representations in early sensory areas and thereby strengthen their impact on subsequent areas (Bauer et al., 2006), and that neural synchronization may reflect reciprocal information transfer between attentional control structures to stimulus representation networks (Fell et al., 2003).

Human gamma-band activity and visual memory

Perception of an object activates the corresponding memory trace which may be reflected by enhanced fast oscillatory activity in EEG. Initial presentations of line drawings of common objects elicited topographically broad increases of induced GBA and gamma phase synchronization between distant electrode sites (Gruber and Müller, 2002). These effects were reduced to repeated presentations of the same stimuli, possibly reflecting network "sharpening" as a basis of repetition priming. In contrast, novel stimuli were not accompanied by GBA increases upon their first presentation but showed an augmentation of gamma power and an enhancement of inter-electrode phase-locking to repeated presentations, possibly reflecting the formation of a new cortical object representation (Gruber et al., 2004a; Gruber and Müller, 2005). These differential gamma-band responses to familiar versus novel stimuli were clearly separable from changes in evoked potentials but were highly comparable to hemodynamic responses (Fiebach et al., 2005). Increases in GBA during perception have been shown to predict subsequent object recognition in declarative memory tasks both in EEG (Gruber et al., 2004b), MEG (Osipova et al., 2006), and intracranial recordings (Fell et al., 2001; Sederberg et al., 2003). Increased gamma synchronization in sensory areas may either result in a stronger communication with subsequent areas, thus enhancing memory encoding, or it may reflect the top–down-driven reinforcement of object representations (Osipova et al., 2006).

Research on working memory has shown that human GBA reflects the cognitively controlled activation of cortical object representations. When visual shapes were maintained in short-term memory to be compared with a subsequently presented probe stimulus, the memorization phase was characterized by enhanced gamma spectral amplitudes at occipital and bilateral temporal electrode sites (Tallon-Baudry et al., 1998). The time course of this activity was dependent on the duration of the delay phase, suggesting that it reflected the rehearsal of object representations in visual short-term memory (Tallon-Baudry et al., 1999). Furthermore, enhancements of induced GBA were found during the internal activation of an object representation during a visual search task (Tallon-Baudry et al., 1997).

In summary, there is ample evidence for a functional role of human GBA during a variety of cognitive processes that involve the activation and maintenance of existing object representations or the formation of new memory traces. In EEG, these effects have usually been observed with a broad topography, making an attribution to specific cortical areas difficult. In contrast, MEG studies have shown more local effects which may be attributable to the fact that more widespread activations including radial sources may be less visible in MEG.

In the following sections, a series of MEG investigations will be described that have yielded stimulus- and task-dependent local oscillatory activations during different types of auditory and audiovisual processing.

Cortical oscillatory activity in magnetoencephalogram

Methodological considerations

Magnetoencephalogram (MEG) is a non-invasive tool for the measurement of magnetic field changes generated by neuroelectric activity in the human brain. Recordings with an excellent temporal resolution are performed by superconductive quantum interference devices that are located in a helmet-shaped device over the subject's head. Modern MEG systems cover the whole head with a dense sensor array and thus also provide a good spatial resolution. One of the main differences between MEG and EEG is the fact that MEG is unable to detect cortical dipole sources with a radial orientation. In addition, opposing sources in the opposite banks of a sulcus may cancel each other out, resulting in a reduced sensitivity of MEG to widespread activation patterns. On the other hand, unlike electric potentials, magnetic fields are not attenuated by cortical tissue, possibly enabling the detection of weaker sources than EEG. This may be particularly relevant to the investigation of high-frequency oscillatory activity. Furthermore the reference-free recording of neuromagnetic activity facilitates the computation of measures of cortico-cortical functional connectivity.

As will be described in the following sections, task-related oscillatory activity was typically observed in higher frequencies (>50 Hz) and with a more local topography than in most EEG investigations. While the analysis of GBA in EEG has often relied on the visual identification of electrodes and time-frequency windows with clear amplitude enhancements compared to baseline, and effects in some paradigms have even been so widespread that signals could be averaged across groups of recording sites, the topographical heterogeneity of GBA changes in MEG and the larger number of sensors have required a more conservative statistical approach (for detailed descriptions, see e.g. Lutzenberger *et al.*, 2002; Kaiser *et al.*, 2005b). Starting point was a fast Fourier transform-based spectral analysis of single 600-ms epochs with subsequent averaging of spectral amplitude values separately for each of the experimental conditions. Conditions were compared directly, without subtracting a pre-stimulus baseline which in many paradigms already contained some anticipatory task-related activity. Differences between conditions were then evaluated with a statistical probability mapping based on randomization tests (Blair and Karniski, 1993) including corrections for multiple comparisons and for correlated data points. This procedure further included the requirement that two neighboring frequency bins differ significantly. This first analysis step served to identify the frequency ranges with the statistically most robust differences between conditions. Subsequently, signals were filtered in these frequency ranges with Gabor filters with a spectral width of 5Hz and a temporal width of 200ms at half amplitude, and demodulated with a Hilbert transform (Clochon et al., 1996). Filtered data were then again subjected to statistical probability mapping to identify the latency ranges and the sensors with the most pronounced effects.

In the studies reviewed in the following, only single areas of spectral amplitude enhancement were found at the sensor level. This suggested that the generators were not single dipoles which in MEG would produce dual areas of surface activation corresponding to the dipole's in- and outflowing flux wells, with the source dipole located in between the two surface areas. An alternative explanation for the observed surface patterns could be coupled or multiple dipoles eliciting a strong field over the area circumscribed by the dipoles but much weaker outer fields that would not be detectable with the present statistical probability mapping method. This would imply that the sources should be located close to the area below the sensor with the highest GBA (Kaiser et al., 2000, 2005b). Moreover, effects typically showed rather small amplitude differences around 1 fT, and effects were transient, lasting at most for several hundred milliseconds. The cortical generators of such weak and transient activations are difficult to determine. Techniques like beamformer (Hillebrand and Barnes, 2005; Hillebrand et al., 2005) may be applied successfully to the source localization of temporally more sustained activations with a better signal-to-noise ratio (Hoogenboom et al., 2006), whereas they may not yield reliable results for effects like the ones described in the following sections.

Whereas EEG studies have often described topographically broad GBA enhancements, intracranial recordings in presurgical patients have shown oscillatory activations to be highly local and to show task-specific differences between cortical areas (Lachaux et al., 2005; Tallon-Baudry et al., 2005). Considering the rather local activation patterns at the surface of the head that the MEG investigations described below have yielded, MEG may provide a link between invasive research on oscillatory activity on the one hand and the more global synchronizations observed in EEG. Possibly GBA in MEG represents a correlate of the task-specific synchronization of local network components rather than activations of entire object representations.

Auditory deviance detection

Our research on oscillatory activity during auditory processing was guided by the notion that similar to the visual system, there may be separate dorsal and ventral auditory streams specialized in the processing of spatial and non-spatial acoustic information, respectively. This hypothesis was originally formulated on the basis of single-cell recordings and anatomical work in monkeys (Rauschecker, 1998; Romanski et al., 1999; Tian et al., 2001), and has recently gained support from a growing number of human hemodynamic brain-imaging studies (Alain et al., 2001; Arnott et al., 2004; Warren and Griffiths, 2003). A first series of investigations assessed oscillatory responses during passive auditory change detection in "mismatch" paradigms (Näätänen et al., 1993), because rare deviant events are thought to activate brain structures that are involved in the processing of the stimulus feature underlying the deviance (Molholm et al., 2005).

To assess auditory spatial processing, we compared responses to infrequent right- and left-lateralized sounds with frequent midline presentations of the same sounds. Lateralized deviants gave rise to evoked mismatch fields peaking about 120 ms after stimulus onset with sources at the level of the supratemporal plane. Statistical probability mapping of spectral activity revealed subsequent GBA enhancements in sensors over posterior temporo-parietal cortex at latencies between 250 and 300 ms. Similar results were obtained in separate investigations both for syllables and animal vocalizations (Kaiser et al., 2000) and for complex distorted sounds (Kaiser et al., 2002a). Interestingly, right-lateralized stimuli gave rise to bilateral enhancements whereas sounds in the left hemifield were associated with GBA increases over the right hemisphere only, supporting a predominance of the right hemisphere for auditory spatial processing (Fig. 15.1a, b).

Deviations in acoustic patterns evoked mismatch fields at around 180 ms after stimulus onset, i.e. with longer latencies than spatial changes. For all three types of sounds – syllables, animal sounds, and meaningless distorted noises – pattern deviance was accompanied by increased gamma-band amplitudes in the range of 60–90 Hz over left anterior temporal and inferior frontal regions (Kaiser et al., 2002b) (Fig. 15.1c, d). As in the spatial mismatch studies, the latencies of GBA increases for pattern deviance detection followed the peak of the evoked mismatch responses by about 120 ms. The topography of GBA in these studies thus supported the notion of separate auditory dorsal "where" and ventral "what" processing streams. The results further suggest that synchronously oscillating networks in higher association areas along the putative dorsal and ventral pathways may perform a more detailed analysis of the sound feature underlying the deviance. The temporal delay between the superior temporal mismatch peak and the subsequent GBA increases over the auditory stream areas has provided first evidence for serial processing along these pathways in humans (Kaiser and Lutzenberger, 2003). Converging evidence for GBA increases to pattern deviants has recently been presented by Edwards et al. (2005)

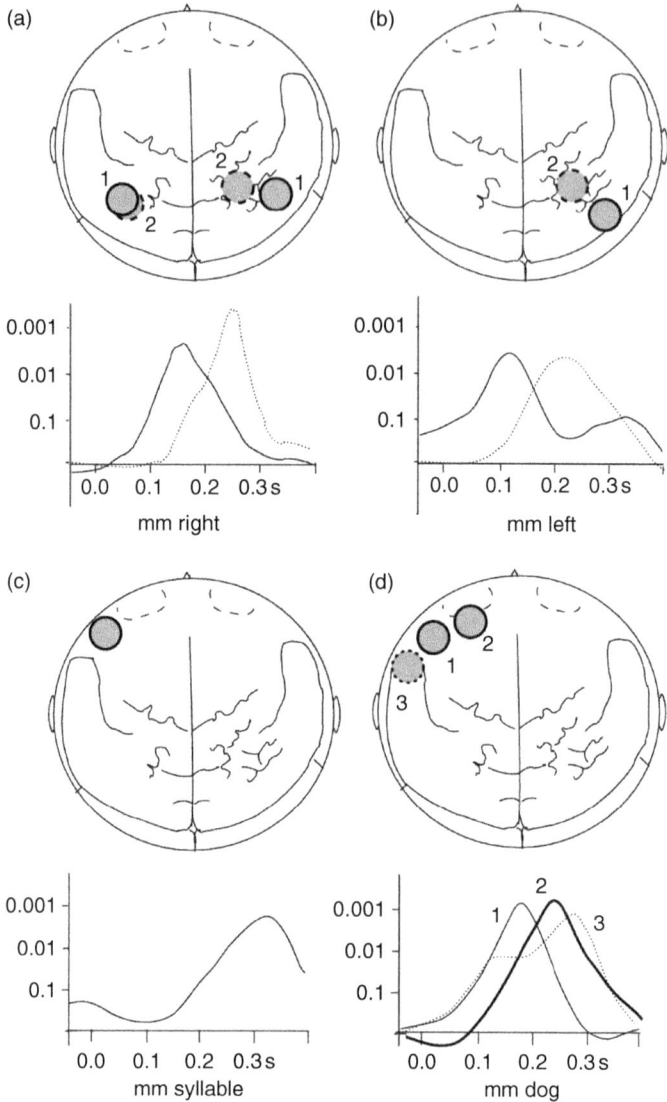

Figure 15.1 Topographies and time courses of magnetoencephalographic (MEG) GBA during different types of auditory processing. Maps show the locations of MEG sensors projected onto a two-dimensional cortical surface map to depict the topography of significant GBA increases. Graphs below each map show the time courses of statistical differences (p-values) between experimental and control conditions for each of the sensors; mm, mismatch. (a) Spatial mismatch: the comparison of deviant, right-lateralized syllables with standard midline syllables showed bilateral posterior temporo-parietal GBA increases at 53 ± 2.5 Hz. (b) Spatial mismatch: the comparison of deviant, left-lateralized syllables with standard midline syllables yielded right posterior temporo-parietal GBA enhancements at 53 ± 2.5 Hz. (c) Pattern mismatch: the comparison of the deviant syllable /ba/ with the standard

who found intracranial enhanced gamma oscillations at frequencies between 30 and 160 Hz at temporal and frontal recording sites during a pitch mismatch paradigm. This activity was interpreted in terms of assembly formation and neural communication by collective signaling that is elicited automatically by salient stimuli.

Gamma-band activity over left inferior frontal cortex was also found for the active detection of pattern deviants in an active oddball task with the same stimuli as in the animal vocalization mismatch study (Kaiser and Lutzenberger, 2004). Attention to deviant sounds and detection of pairs of two consecutive deviants elicited additional GBA components over dorsal prefrontal cortex at $\sim$200–300 ms after stimulus onset. These effects may have reflected the activation of executive networks involved in active memory processing and decision-making.

The processing of illusory acoustic changes was investigated in an audiovisual oddball study where standard stimuli were congruent combinations of a syllable and the corresponding mouth movement, and acoustic changes were induced by presenting either incongruent syllables (acoustic deviant) or incongruent mouth movements (visual deviant) (Kaiser et al., 2005a). The comparison of spectral responses to infrequent combinations of a visual /pa/ and an auditory /ta/ (visual deviants) with the standard stimulus yielded enhancements of GBA in frequencies of $\sim$73–80 Hz at sensors over posterior parietal areas (at $\sim$160 ms after the onset of the acoustic stimulus), over midline occipital cortex (at $\sim$270 ms), and over left inferior frontal cortex peaking at $\sim$320 ms. The posterior components may have reflected the activation of networks involved in the processing of visual speech-related movements and the top–down modulation of earlier visual networks by higher posterior parietal motion processing regions. The left inferior frontal component was localized over the same cortical area where purely acoustic phonetic mismatch had elicited GBA increases (Kaiser et al., 2002b), suggesting that local networks in this area signal acoustic pattern changes even in the absence of physical sound deviance.

Auditory short-term memory

The change processing studies described above have demonstrated the functional role of local networks in regions of the putative auditory dorsal and ventral streams for the representation of spatial and non-spatial sound features,

Caption for Figure 15.1 (cont.)
syllable /da/ elicited GBA at 83 ± 2.5 Hz over left inferior frontal cortex. (d) Pattern mismatch: the comparison of a deviant with a standard barking dog sound gave rise to GBA at 63 ± 2.5 Hz in three left anterior temporal and inferior frontal sensors. (Reprinted, with permission, from Kaiser and Lutzenberger, 2005.)

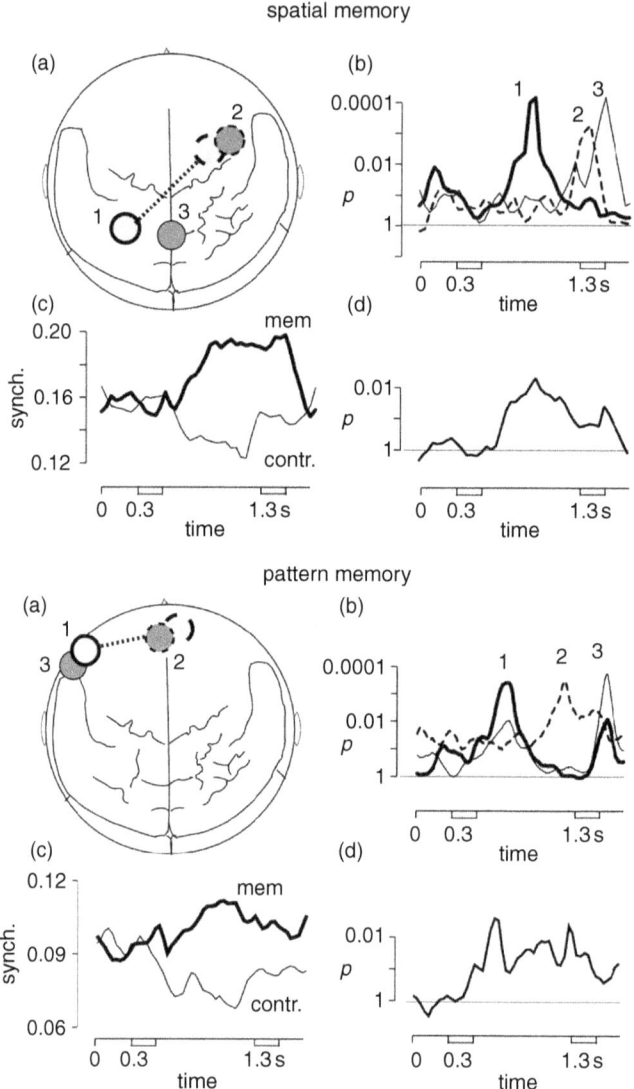

Figure 15.2 Comparison of gamma-band spectral amplitude and phase synchrony during an auditory spatial and an auditory pattern working memory task (upper and lower panels, respectively). (a) Magnetoencephalographic sensors showing spectral amplitude and coherence increases (broken lines) are projected onto schematic brain surface maps (seen from above, nose up). (b) Time courses of spectral amplitude differences for sensors showing significant effects (frequency ranges: 59 ± 2.5 Hz and 67 ± 2.5 Hz for spatial memory task, 67 ± 2.5 Hz for pattern memory task). The curves depict the p-values as results of t-test comparisons between memory and control conditions. The curves are numbered in the same way as the corresponding sensors on the map on the left. (c) Time course of gamma-band phase synchrony for memory and control conditions (bold and thin lines, respectively)

respectively. A recent set of investigations assessed top–down influences on oscillatory activity over these areas during active memory tasks. Two studies were conducted that employed delayed matching-to-sample tasks with an 800-ms delay phase requiring the memorization of either auditory spatial or auditory pattern information.

In the auditory spatial short-term memory study (Lutzenberger et al., 2002), S1 was a filtered noise convoluted with generic head-related transfer functions (Gardner and Martin, 1995) to create the impression of a sound source in external space at four lateralization angles; S2 was the same sound that could be presented either at the same lateralization angle as S1, or at more peripheral or more central angles in the same hemifield. Subjects had to respond to matching stimulus pairs which occurred with a probability of 20%. Only trials with non-matching pairs were included in the analyses to reduce the possible impact of motor processes. In the control task, subjects had to attend to the background noise and detect whether there was an intensity change at the end of the delay phase or not. Statistical probability mapping revealed GBA enhancements over posterior and over frontal regions (Fig. 15.2a, b, top panel). Left posterior temporo-parietal GBA was increased at ∼57–61 Hz during the delay phase, suggesting an involvement of the putative auditory dorsal stream in the encoding and maintenance of the spatial sounds. This was followed by a spectral amplitude enhancement at ∼65–69 Hz over right frontal cortex at the end of the delay phase, which may have reflected the activation of frontal comparison and decision-making networks. A third GBA component over midline parietal cortex peaked in response to S2. In addition to spectral amplitude increases, we also observed increased gamma coherence during the delay phase between the left temporo-parietal sensor and a sensor over right frontal cortex. Phase synchrony analysis (Lachaux et al., 1999) showed a sustained enhancement of phase-coupling between these sensors throughout the memorization phase, possibly reflecting the interaction between posterior sensory storage areas and frontal executive networks (Kaiser et al., 2005b) (Fig. 15.2c, d, top panel).

Caption for Figure 15.2 (cont.)
for the sensor pairs linked with the broken line in parts (a) of the figure. Phase synchrony increases were observed at 59 ± 2.5 Hz for the spatial memory task and at 67 ± 2.5 Hz for the pattern memory task. (d) Time courses of gamma-band phase synchrony differences (p-values) for the same sensor pair as in parts (c) of the figure. (Reprinted from Lutzenberger et al. (2002), with permission from Society for Neuroscience, from Kaiser et al. (2003), with permission from Elsevier, and Kaiser et al. (2005b), with permission from Elsevier.)

We used the same paradigm to investigate auditory pattern memory (Kaiser et al., 2003). Under the memory condition, same-different judgments had to be made about pairs of syllables that could differ either in their voice onset time or formant structure, while the control task involved the detection of possible spatial displacements in the background sound. Statistical probability mapping identified induced GBA increases during the memory task over left inferior frontal/anterior temporal cortex during the delay phase and in response to S2, and over prefrontal cortex at the end of the delay period (Fig. 15.2a, b, bottom panel). Furthermore, increased gamma coherence and phase synchronization between left inferior frontal and midline prefrontal sensors characterized the delay phase (Fig. 15.2c, d, bottom panel). The temporal dynamics of these effects mirrored the findings from the auditory spatial working memory study reported above. The encoding and memorization of information concerning sound identity was associated with oscillatory activity in networks belonging to the putative auditory ventral stream, and with interactions between this region and prefrontal networks.

The functional role of frontal GBA increases and coherence enhancements between higher sensory and frontal areas was tested in a spatial echoic memory study (Kaiser et al., 2005c) where we used the same stimuli as in the earlier spatial short-term memory paradigm (Lutzenberger et al., 2002) but a shorter delay phase of 200 instead of 800 ms. Here two temporally distinct spectral amplitude increases at ~80 Hz were found at sensors over posterior parietal cortex, whereas there was no frontal GBA enhancement. The first GBA component reached its maximum with the onset of S2, possibly enabling a direct comparison of the representation of S1 with S2 without requiring frontal executive networks to keep the representation of S1 active.

In the three memory studies summarized above, only non-matching S1–S2 pairs not requiring a motor response were included in the analyses. This raised the question whether the GBA increase in response to S2 reflected the activation of an additional network signaling that S2 differed from the memory representation of S1. This issue was investigated in a study again using the same stimuli as in the spatial memory study described above (Lutzenberger et al., 2002) but with equal numbers of matching and non-matching S1–S2 pairs and balanced response instructions across subjects (Leiberg et al., 2006). The main finding was that non-matching compared with matching pairs were accompanied by an additional GBA increase over posterior parietal cortex, supporting the hypothesis that stimulus discrimination relied on the activation of distinct neural representations. The non-matching stimulus required a specific neural representation which was associated with an additional GBA component.

In summary, this research has demonstrated that oscillatory activity in MEG distinguishes between different short-term memory functions. The comparison of auditory spatial and pattern memory showed that areas along the putative auditory dorsal and ventral streams, respectively, were involved in stimulus maintenance during the delay phases. At the same time, enhanced coherence between these regions and frontal areas may have reflected the integration of cortical subsystems necessary to keep the information active across a certain period of time.

Questions for future research

Behavioral relevance of oscillatory activity

While oscillatory synchronization has been put forward as a possible neural correlate of conscious awareness (Engel and Singer, 2001), in humans there has been little direct evidence for the behavioral relevance of GBA or for a relationship between GBA and conscious perception. During binocular rivalry, enhanced GBA preceding the response indicating a perceptual switch has been interpreted as reflecting the emergence of a new percept (Doesburg et al., 2005). Individuals reporting high rates of Necker cube reversals were distinguished from those with low reversal rates by increased gamma power at frontal EEG recording sites (Strüber et al., 2000). More direct evidence for the relevance of oscillatory synchrony for perceptual awareness was obtained from intracranial EEG recordings during a tactile detection task. Gamma coherence was increased in primary somatosensory areas ~150–300 ms after near-threshold contralateral hand stimuli that were perceived, but not for non-perceived stimuli with the same intensity (Meador et al., 2002). Further evidence that GBA reflects subjective perceptual experience has been yielded by a study assessing sound-induced illusory flash perception. Here, oscillatory activity at about 30–50 Hz was enhanced when a single light flash accompanied by two tone beeps was perceived as two flashes (Bhattacharya et al., 2002). Similarly, during a backward masking task, GBA was increased only when participants reported awareness of masked words (Summerfield et al., 2002). In a recent single-case study in a hemianopic patient, MEG gamma activity over left occipito-parietal cortex correlated with the subjective awareness of a visual stimulus in the blind hemifield (Schurger et al., 2006).

During the audiovisual oddball task reported above (Kaiser et al., 2005a), we observed a considerable variability in the subjects' ability to perceive acoustic deviance caused by changes in the visual or in the acoustic part of the audiovisual stimuli. Neural correlates of detection accuracy were assessed by applying a version of our statistical probability mapping to the search for the most pronounced correlations between stimulus-related changes in oscillatory activity and behavioral performance. The detection of illusory acoustic changes induced by visual deviants was associated with relative spectral amplitude increases at ~80 Hz over midline occipital cortex. In contrast, the detection of real acoustic deviants correlated positively with GBA enhancements at ~42 Hz over left superior temporal cortex, thus supporting the relevance of high-frequency oscillatory activity over early sensory areas for perceptual experience (Kaiser et al., 2006). While these findings are encouraging, future research may focus for example on the role of GBA for the detection of sensory events in noisy environments where successful feature binding should be particularly important. Furthermore, correlations between behavioral measures and oscillatory activity may help to reveal the processes underlying particular perceptual or cognitive functions.

Relationship between gamma and lower frequencies

The significance of different frequencies within the gamma range as well as relationships between gamma and lower-frequency bands have remained largely elusive. In the MEG studies described above, we have mainly found effects in the higher gamma band between ~50 and 90 Hz. This is above the core gamma frequencies typically found in EEG but corresponds well to findings of high-frequency oscillations in intracranial recordings (Crone et al., 1998, 2001; Lachaux et al., 2005). However, we have not found systematic relationships between the dominant frequency and task requirement or stimulus type. Tentatively, faster frequencies have been related to smaller networks (Pulvermüller et al., 1997; Herculano-Houzel et al., 1999), supporting the notion that MEG may be more sensitive to local network synchronization than EEG.

In the auditory memory studies reported above, lowering the significance criterion usually revealed theta as the next frequency band showing task-dependent modulations. For example, the delay periods of both the auditory spatial and pattern short-term memory tasks were characterized by prefrontal theta activity increases (Lutzenberger et al., 2002; Kaiser et al., 2003). The relevance of theta oscillations for working memory has been demonstrated by studies using EEG (Sauseng et al., 2004), MEG (Jensen and Tesche, 2002), and intracranial recordings (Raghavachari et al., 2001; Sederberg et al., 2003). Moreover, interrelationships between theta and gamma frequencies during working memory have been proposed by several authors (Jensen and Lisman, 1998; Schack et al., 2002; Demiralp et al., 2006). Our echoic memory task was also characterized by an increase of theta activity over posterior parietal regions (Kaiser et al., 2005c). Thus, theta spectral amplitude was enhanced over those areas that appeared particularly relevant for the respective memory process, i.e. over prefrontal cortex during short-term memory and over parietal cortex during spatial echoic memory.

Power increases in the alpha band have been proposed to indicate active processing during memory tasks, questioning the notion of alpha as a cortical "idling" rhythm. Memory-related alpha synchronization has been interpreted in terms of reflecting the active inhibition of potentially interfering, task-irrelevant, brain areas or processing systems (Klimesch *et al.*, 1999; Bastiaansen *et al.*, 2002; Herrmann *et al.*, 2004b). Systematic increases of alpha activity with memory load, i.e. the number of items held in working memory, have been reported during a visual Sternberg-type task over posterior and bilateral central regions in EEG (Jensen *et al.*, 2002) and during an auditory task over frontal areas in MEG (Leiberg *et al.*, 2006). Future research is warranted to further elucidate the exact roles of the different frequency bands and to clarify whether faster frequencies within the gamma range may be subdivided into functionally separate subbands.

Long-distance synchronization

Recent research has demonstrated increases in long-distance cortico-cortical synchronization in the gamma band in relation to cognitive processes and arousal states. This contradicts the predominant view that gamma synchronization is restricted to short distances, whereas long-range interactions rely on lower frequencies like theta, alpha, and beta (von Stein *et al.*, 1999, 2000). Long-distance phase synchrony increases in EEG have been reported during the perception of canonical "Mooney" faces (Rodriguez *et al.*, 1999), as well as for presentations of meaningful visual objects (Gruber and Müller, 2002, 2005; Gruber *et al.*, 2002). In MEG, we have found sustained enhancements of gamma phase synchronization between sensors over putative higher sensory storage areas and frontal sensors during the memorization phases of auditory delayed matching-to-sample tasks (Kaiser *et al.*, 2005b). These findings support the notion that cortico-cortical synchronization may represent a mechanism for transient coupling between neural assemblies representing an integral feature of memory processes (Axmacher *et al.*, 2006). The existence of long-distance gamma synchrony has also been supported by intracranial recording studies showing increased gamma coherence between cortical regions and between cortex and hippocampus during wakefulness but not during sleep (Cantero *et al.*, 2004). Moreover, gamma correlations across distant sites were observed in response to complex visual stimuli (Lachaux *et al.*, 2005). Obviously the current thinking concerning the relationship between frequency band and synchronization distance should be reconsidered in the light of these findings. Gamma synchronization may thus serve as a mechanism not only for local binding but also for communication across distant cortical areas.

Conclusions

Recent research on GBA using intracortical recordings, EEG, and MEG has demonstrated the important role of oscillatory synchronization for a variety of cognitive processes ranging from bottom-up driven perception of gestalt-like or meaningful information to top-down guided functions like selective attention and memory. Gamma-band activity thus appears to reflect the synchronization of cortical networks involved in task-dependent object representation. Oscillatory activity in MEG was found in response to auditory deviance detection and memory processing. Statistical probability mapping revealed the most robust experimental effects in the range of ~50–90 Hz. The topography of these effects was replicable across a series of independent studies and was consistent with the notion of parallel auditory "what" and "where" processing streams. While these findings reflected synchronization of local networks representing specific stimulus features, increased gamma coherence and phase synchrony between higher sensory storage systems and prefrontal regions during short-term memory indicated the task-specific coupling of cortical structures subserving different cognitive functions. Oscillatory activity may thus provide a mechanism for the integration of distributed brain processes.

References

Alain, C., Arnott, S. R., Hevenor, S., Graham, S., and Grady, C. L. (2001). "What" and "where" in the human auditory system. *Proc Natl Acad Sci USA* **98**:12301–12306.

Arnott, S. R., Binns, M. A., Grady, C. L., and Alain, C. (2004). Assessing the auditory dual-pathway model in humans. *Neuroimage* **22**:401–408.

Axmacher, N., Mormann, F., Fernandez, G., Elger, C. E., and Fell, J. (2006). Memory formation by neuronal synchronization. *Brain Res Rev* **52**:170–182.

Basar-Eroglu, C., Strüber, D., Schürmann, M., Stadler, M., and Basar, E. (1996). Gamma-band responses in the brain: a short review of psychophysiological correlates and functional significance. *Int J Psychophysiol* **24**:101–112.

Bastiaansen, M. C., Posthuma, D., Groot, P. F., and de Geus, E. J. (2002). Event-related alpha and theta responses in a visuo-spatial working memory task. *Clin Neurophysiol* **113**:1882–1893.

Bauer, M., Oostenveld, R., Peeters, M., and Fries, P. (2006). Tactile spatial attention enhances gamma-band activity in somatosensory cortex and reduces low-frequency activity in parieto-occipital areas. *J Neurosci* **26**:490–501.

Bhattacharya, J., Shams, L., and Shimojo, S. (2002). Sound-induced illusory flash perception: role of gamma band responses. *Neuroreport* **13**:1727–1730.

Blair, R. C. and Karniski, W. (1993). An alternative method for significance testing of waveform difference potentials. *Psychophysiology* **30**:518–524.

Busch, N. A., Debener, S., Kranczioch, C., Engel, A. K., and Herrmann, C. S. (2004). Size matters: effects of stimulus size, duration and eccentricity on the visual gamma-band response. *Clin Neurophysiol* **115**:1810–1820.

Cantero, J. L., Atienza, M., Madsen, J. R., and Stickgold, R. (2004). Gamma EEG dynamics in neocortex and hippocampus during human wakefulness and sleep. *Neuroimage* **22**:1271–1280.

Clochon, P., Fontbonne, J., Lebrun, N., and Etevenon, P. (1996). A new method for quantifying EEG event-related desynchronization:amplitude envelope analysis. *Electroencephalogr Clin Neurophysiol* **98**:126–129.

Crone, N. E., Miglioretti, D. L., Gordon, B., and Lesser, R. P. (1998). Functional mapping of human sensorimotor cortex with electrocorticographic spectral analysis. II. Event-related synchronization in the gamma band. *Brain* **121**:2301–2315.

Crone, N. E., Boatman, D., Gordon, B. and Hao, L. (2001). Induced electrocorticographic gamma activity during auditory perception. *Clin Neurophysiol*, **112**, 565–82.

Debener, S., Herrmann, C. S., Kranczioch, C., Gembris, D., and Engel, A. K. (2003). Top-down attentional processing enhances auditory evoked gamma band activity. *Neuroreport* **14**:683–686.

Demiralp, T., Bayraktaroglu, Z., Lenz, D., *et al.* (2006). Gamma amplitudes are coupled to theta phase in human EEG during visual perception. *Int J Psychophysiol* **64**:24–30.

Doesburg, S. M., Kitajo, K., and Ward, L. M. (2005). Increased gamma-band synchrony precedes switching of conscious perceptual objects in binocular rivalry. *Neuroreport* **16**:1139–1142.

Eckhorn, R., Bauer, R., Jordan, W., *et al.* (1988). Coherent oscillations: a mechanism of feature linking in the visual cortex? Multiple electrode and correlation analyses in the cat. *Biol Cybernet* **60**:121–130.

Edwards, E., Soltani, M., Deouell, L. Y., Berger, M. S., and Knight, R. T. (2005). High gamma activity in response to deviant auditory stimuli recorded directly from human cortex. *J Neurophysiol* **94**:4269–4280.

Engel, A. K. and Singer, W. (2001). Temporal binding and the neural correlates of sensory awareness. *Trends Cogn Sci* **5**:16–25.

Fell, J., Klaver, P., Lehnertz, K., *et al.* (2001). Human memory formation is accompanied by rhinal-hippocampal coupling and decoupling. *Nat Neurosci* **4**:1259–1264.

Fell, J., Fernandez, G., Klaver, P., Elger, C. E., and Fries, P. (2003). Is synchronized neuronal gamma activity relevant for selective attention? *Brain Res Rev* **42**:265–272.

Fiebach, C. J., Gruber, T., and Supp, G. G. (2005). Neuronal mechanisms of repetition priming in occipitotemporal cortex: spatiotemporal evidence from functional magnetic resonance imaging and electroencephalography. *J Neurosci* **25**:3414–3422.

Fries, P., Roelfsema, P. R., Engel, A. K., König, P., and Singer, W. (1997). Synchronization of oscillatory responses in visual cortex correlates with perception in interocular rivalry. *Proc Natl Acad Sci USA* **94**:12699–12704.

Fries, P., Reynolds, J. H., Rorie, A. E., and Desimone, R. (2001). Modulation of oscillatory neuronal synchronization by selective visual attention. *Science* **291**:1560–1563.

Gardner, W. G. and Martin, K. D. (1995). HRTF measurements of a KEMAR. *J Acoust Soc Am* **97**:3907–3908.

Gray, C. M. and Singer, W. (1989). Stimulus-specific neuronal oscillations in orientation columns of cat visual cortex. *Proc Natl Acad Sci USA* **86**:1698–1702.

Gray, C. M., König, P., Engel, A. K., and Singer, W. (1989). Oscillatory responses in cat visual cortex exhibit inter-columnar synchronization which reflects global stimulus properties. *Nature* **338**:334–337.

Gruber, T. and Müller, M. M. (2002). Effects of picture repetition on induced gamma band responses, evoked potentials, and phase synchrony in the human EEG. *Cogn Brain Res* **13**:377–392.

Gruber, T. and Müller, M. M. (2005). Oscillatory brain activity dissociates between associative stimulus content in a repetition priming task in the human EEG. *Cereb Cortex* **15**:109–116.

Gruber, T., Müller, M. M., Keil, A., and Elbert, T. (1999). Selective visual-spatial attention alters induced gamma band responses in the human EEG. *Clin Neurophysiol* **110**:2074–2085.

Gruber, T., Müller, M. M., and Keil, A. (2002). Modulation of induced gamma band responses in a perceptual learning task in the human EEG. *J Cogn Neurosci* **14**:732–744.

Gruber, T., Malinowski, P., and Müller, M. M. (2004a). Modulation of oscillatory brain activity and evoked potentials in a repetition priming task in the human EEG. *Eur J Neurosci* **19**:1073–1082.

Gruber, T., Tsivilis, D., Montaldi, D., and Müller, M. M. (2004b). Induced gamma band responses: an early marker of memory encoding and retrieval. *Neuroreport* **15**:1837–1841.

Herculano-Houzel, S., Munk, M. H., Neuenschwander, S., and Singer, W. (1999). Precisely synchronized oscillatory firing patterns require electroencephalographic activation. *J Neurosci* **19**:3992–4010.

Herrmann, C. S., Mecklinger, A., and Pfeifer, E. (1999). Gamma responses and ERPs in a visual classification task. *Clin Neurophysiol* **110**:636–642.

Herrmann, C. S., Munk, M. H., and Engel, A. K. (2004a). Cognitive functions of gamma-band activity: memory match and utilization. *Trends Cogn Sci* **8**:347–355.

Herrmann, C. S., Senkowski, D., and Rottger, S. (2004b). Phase-locking and amplitude modulations of EEG alpha: two measures reflect different cognitive processes in a working memory task. *Exp Psychol* **51**:311–318.

Hillebrand, A. and Barnes, G. R. (2005). Beamformer analysis of MEG data. *Int Rev Neurobiol* **68**:149–171.

Hillebrand, A., Singh, K. D., Holliday, I. E., Furlong, P. L., and Barnes, G. R. (2005). A new approach to neuroimaging with magnetoencephalography. *Hum Brain Map* **25**:199–211.

Hoogenboom, N., Schoffelen, J. M., Oostenveld, R., Parkes, L. M., and Fries, P. (2006). Localizing human visual gamma-band activity in frequency, time and space. *Neuroimage* **29**:764–773.

Jensen, O. and Lisman, J. E. (1998). An oscillatory short-term memory buffer model can account for data on the Sternberg task. *J Neurosci* **18**:10688–10699.

Jensen, O. and Tesche, C. D. (2002). Frontal theta activity in humans increases with memory load in a working memory task. *Eur J Neurosci* **15**:1395–1399.

Jensen, O., Gelfand, J., Kounios, J., and Lisman, J. E. (2002). Oscillations in the alpha band (9–12Hz) increase with memory load during retention in a short-term memory task. *Cereb Cortex* **12**:877–882.

Kaiser, J. and Lutzenberger, W. (2003). Induced gamma-band activity and human brain function. *Neuroscientist* **9**:475–484.

Kaiser, J. and Lutzenberger, W. (2004). Frontal gamma-band activity in magnetoencephalogram during auditory oddball processing. *Neuroreport* **15**:2185–2188.

Kaiser, J. and Lutzenberger, W. (2005). Human gamma-band activity: a window to cognitive processing. *Neuroreport* **16**:207–211.

Kaiser, J., Lutzenberger, W., Preissl, H., Ackermann, H., and Birbaumer, N. (2000). Right-hemisphere dominance for the processing of sound-source lateralization. *J Neurosci* **20**:6631–6639.

Kaiser, J., Birbaumer, N., and Lutzenberger, W. (2002a). Magnetic oscillatory responses to lateralization changes of natural and artificial sounds in humans. *Eur J Neurosci* **15**:345–354.

Kaiser, J., Lutzenberger, W., Ackermann, H., and Birbaumer, N. (2002b). Dynamics of gamma-band activity induced by auditory pattern changes in humans. *Cereb Cortex* **12**:212–221.

Kaiser, J., Ripper, B., Birbaumer, N., and Lutzenberger, W. (2003). Dynamics of gamma-band activity in human magnetoencephalogram during auditory pattern working memory. *Neuroimage* **20**:816–827.

Kaiser, J., Bühler, M., and Lutzenberger, W. (2004). Magnetoencephalographic gamma-band responses to illusory triangles in humans. *Neuroimage* **23**:551–560.

Kaiser, J., Hertrich, I., Ackermann, H., Mathiak, K., and Lutzenberger, W. (2005a). Hearing lips: gamma-band activity during audiovisual speech perception. *Cereb Cortex* **15**:646–653.

Kaiser, J., Leiberg, S., and Lutzenberger, W. (2005b). Let's talk together: memory traces revealed by cooperative activation in the cerebral cortex. *Int Rev Neurobiol* **68**:51–78.

Kaiser, J., Walker, F., Leiberg, S., and Lutzenberger, W. (2005c). Cortical oscillatory activity during spatial echoic memory. *Eur J Neurosci* **21**:587–590.

Kaiser, J., Hertrich, I., Ackermann, H., and Lutzenberger, W. (2006). Gamma-band activity over early sensory areas predicts detection of changes in audiovisual speech stimuli. *Neuroimage* **30**:1376–1382.

Karakas, S. and Basar, E. (1998). Early gamma response is sensory in origin: a conclusion based on cross-comparison of results from multiple experimental paradigms. *Int J Psychophysiol* **31**:13–31.

Keil, A., Müller, M. M., Ray, W. J., Gruber, T., and Elbert, T. (1999). Human gamma band activity and perception of a gestalt. *J Neurosci* **19**:7152–7161.

Klimesch, W., Doppelmayr, M., Schwaiger, J., Auinger, P., and Winkler, T. (1999). "Paradoxical" alpha synchronization in a memory task. *Cogn Brain Res* **7**:493–501.

Krause, C. M., Korpilahti, P., Porn, B., Jantti, J., and Lang, H. A. (1998). Automatic auditory word perception as measured by 40 Hz EEG responses. *Electroencephalogr Clin Neurophysiol* **107**:84–87.

Lachaux, J. P., Rodriguez, E., Martinerie, J., and Varela, F. J. (1999). Measuring phase synchrony in brain signals. *Hum Brain Map* **8**:194–208.

Lachaux, J. P., George, N., Tallon-Baudry, C., *et al.* (2005). The many faces of the gamma band response to complex visual stimuli. *Neuroimage* **25**:491–501.

Leiberg, S., Kaiser, J., and Lutzenberger, W. (2006). Gamma-band activity dissociates between matching and nonmatching stimulus pairs in an auditory delayed matching-to-sample task. *Neuroimage* **30**:1357–1364.

Lutzenberger, W., Pulvermüller, F., and Birbaumer, N. (1994). Words and pseudowords elicit distinct patterns of 30-Hz EEG responses in humans. *Neurosci Lett* **176**:115–118.

Lutzenberger, W., Pulvermüller, F., Elbert, T., and Birbaumer, N. (1995). Visual stimulation alters local 40-Hz responses in humans: an EEG study. *Neurosci Lett* **183**:39–42.

Lutzenberger, W., Ripper, B., Busse, L., Birbaumer, N., and Kaiser, J. (2002). Dynamics of gamma-band activity during an audiospatial working memory task in humans. *J Neurosci* **22**:5630–5638.

Meador, K. J., Ray, P. G., Echauz, J. R., Loring, D. W., and Vachtsevanos, G. J. (2002). Gamma coherence and conscious perception. *Neurology* **59**:847–854.

Molholm, S., Martinez, A., Ritter, W., Javitt, D. C., and Foxe, J. J. (2005). The neural circuitry of pre-attentive auditory change-detection: an fMRI study of pitch and duration mismatch negativity generators. *Cereb Cortex* **15**:545–551.

Müller, M. M. and Keil, A. (2004). Neuronal synchronization and selective color processing in the human brain. *J Cogn Neurosci* **16**:503–522.

Müller, M. M., Bosch, J., Elbert, T., *et al.* (1996). Visually induced gamma-band responses in human electroencephalographic activity: a link to animal studies. *Exp Brain Res* **112**:96–102.

Müller, M. M., Junghöfer, M., Elbert, T., and Rockstroh, B. (1997). Visually induced gamma-band responses to coherent and incoherent motion: a replication study. *Neuroreport* **8**:2575–2579.

Müller, M. M., Gruber, T., and Keil, A. (2000). Modulation of induced gamma band activity in the human EEG by attention and visual information processing. *Int J Psychophysiol* **38**:283–299.

Munk, M. H., Roelfsema, P. R., Konig, P., Engel, A. K., and Singer, W. (1996). Role of reticular activation in the modulation of intracortical synchronization. *Science* **272**:271–274.

Näätänen, R., Paavilainen, P., Tiitinen, H., Jiang, D., and Alho, K. (1993). Attention and mismatch negativity. *Psychophysiology* **30**:436–450.

Osipova, D., Takashima, A., Oostenveld, R., *et al.* (2006). Theta and gamma oscillations predict encoding and retrieval of declarative memory. *J Neurosci* **26**:7523–7531.

Pavlova, M., Lutzenberger, W., Sokolov, A., and Birbaumer, N. (2004). Dissociable cortical processing of recognizable and non-recognizable biological movement: analysing gamma MEG activity. *Cereb Cortex* **14**:181–188.

Pulvermüller, F., Preissl, H., Lutzenberger, W., and Birbaumer, N. (1996). Brain rhythms of language: nouns versus verbs. *Eur J Neurosci* **8**:937–941.

Pulvermüller, F., Birbaumer, N., Lutzenberger, W., and Mohr, B. (1997). High-frequency brain activity: its possible role in attention, perception and language processing. *Prog Neurobiol* **52**:427–445.

Raghavachari, S., Kahana, M. J., Rizzuto, D. S., *et al.* (2001). Gating of human theta oscillations by a working memory task. *J Neurosci* **21**:3175–3183.

Rauschecker, J. P. (1998). Cortical processing of complex sounds. *Curr Opin Neurobiol* **8**:516–521.

Rodriguez, E., George, N., Lachaux, J. P., *et al.* (1999). Perception's shadow: long-distance synchronization of human brain activity. *Nature* **397**:430–433.

Romanski, L. M., Tian, B., Fritz, J., *et al.* (1999). Dual streams of auditory afferents target multiple domains in the primate prefrontal cortex. *Nat Neurosci* **2**:1131–1136.

Rose, M. and Büchel, C. (2005). Neural coupling binds visual tokens to moving stimuli. *J Neurosci* **25**:10101–10104.

Rose, M., Sommer, T., and Buchel, C. (2006). Integration of local features to a global percept by neural coupling. *Cereb Cortex* **16**:1522–1528.

Sauseng, P., Klimesch, W., Doppelmayr, M., *et al.* (2004). Theta coupling in the human electroencephalogram during a working memory task. *Neurosci Lett* **354**:123–126.

Schack, B., Vath, N., Petsche, H., Geissler, H. G., and Moller, E. (2002). Phase-coupling of theta-gamma EEG rhythms during short-term memory processing. *Int J Psychophysiol* **44**:143–163.

Schurger, A., Cowey, A., and Tallon-Baudry, C. (2006). Induced gamma-band oscillations correlate with awareness in hemianopic patient GY. *Neuropsychologia* **44**:1796–1803.

Sederberg, P. B., Kahana, M. J., Howard, M. W., Donner, E. J., and Madsen, J. R. (2003). Theta and gamma oscillations during encoding predict subsequent recall. *J Neurosci* **23**:10809–10814.

Singer, W. (2001). Consciousness and the binding problem. *Ann N Y Acad Sci* **929**:123–146.

Singer, W., Engel, A. K., Kreiter, A., *et al.* (1997). Neuronal assemblies: necessity, signature and detectability. *Trends Cogn Sci* **1**:252–261.

Sokolov, A., Pavlova, M., Lutzenberger, W., and Birbaumer, N. (2004). Reciprocal modulation of neuromagnetic induced gamma activity by attention in the human visual and auditory cortex. *Neuroimage* **22**:521–529.

Strüber, D., Basar-Eroglu, C., Hoff, E., and Stadler, M. (2000). Reversal-rate dependent differences in the EEG gamma-band during multistable visual perception. *Int J Psychophysiol* **38**:243–252.

Summerfield, C., Jack, A. I., and Burgess, A. P. (2002). Induced gamma activity is associated with conscious awareness of pattern masked nouns. *Int J Psychophysiol* **44**:93–100.

Tallon-Baudry, C. and Bertrand, O. (1999). Oscillatory gamma activity in humans and its role in object representation. *Trends Cogn Sci* **3**:151–162.

Tallon-Baudry, C., Bertrand, O., Delpuech, C., and Pernier, J. (1996). Stimulus specificity of phase-locked and non-phase-locked 40 Hz visual responses in human. *J Neurosci* **16**:4240–4249.

Tallon-Baudry, C., Bertrand, O., Delpuech, C., and Pernier, J. (1997). Oscillatory gamma-band (30–70 Hz) activity induced by a visual search task in humans. *J Neurosci* **17**:722–734.

Tallon-Baudry, C., Bertrand, O., Peronnet, F., and Pernier, J. (1998). Induced gamma-band activity during the delay of a visual short-term memory task in humans. *J Neurosci* **18**:4244–4254.

Tallon-Baudry, C., Kreiter, A., and Bertrand, O. (1999). Sustained and transient oscillatory responses in the gamma and beta bands in a visual short-term memory task in humans. *Visual Neurosci* **16**:449–459.

Tallon-Baudry, C., Bertrand, O., Henaff, M. A., Isnard, J., and Fischer, C. (2005). Attention modulates gamma-band oscillations differently in the human lateral occipital cortex and fusiform gyrus. *Cereb Cortex* **15**:654–662.

Tian, B., Reser, D., Durham, A., Kustov, A., and Rauschecker, J. P. (2001). Functional specialization in rhesus monkey auditory cortex. *Science* **292**:290–293.

Tiitinen, H., Sinkkonen, J., Reinikainen, K., et al. (1993). Selective attention enhances the auditory 40-Hz transient response in humans. *Nature* **364**:59–60.

von Stein, A., Rappelsberger, P., Sarnthein, J., and Petsche, H. (1999). Synchronization between temporal and parietal cortex during multimodal object processing in man. *Cereb Cortex* **9**:137–150.

von Stein, A., Chiang, C., and König, P. (2000). Top–down processing mediated by interareal synchronization. *Proc Natl Acad Sci USA* **97**:14748–14753.

Warren, J. D. and Griffiths, T. D. (2003). Distinct mechanisms for processing spatial sequences and pitch sequences in the human auditory brain. *J Neurosci* **23**:5799–5804.

Part V DISTURBANCES OF POPULATION ACTIVITY AS THE BASIS OF SCHIZOPHRENIA

16

Neural coordination and psychotic disorganization

ANDRÉ A. FENTON

Introduction

I am trying to understand how ideas and concepts are generated and manipulated in networks of neurons; I want to understand how we think. You probably share my curiosity and believe that the human brain creates and processes mental objects like the ideas and concepts that make up thoughts. We probably also agree that a key to understanding these mental processes is to understand how neurons represent abstract information.

It is less certain we agree on what to do to discover how neurons represent this sort of information. While I suspect we will get quite far by studying mental processes in animals, I admit that I don't know whether or not animals have ideas, concepts, and thoughts. Such open questions do not invalidate the quest to understand thought because the pursuit is founded on the conviction that mental objects are properties of neural systems and that the neural systems in the human brain are fundamentally similar to the systems in the fascinating brains of lower mammals like the laboratory rat. If we restrict the discussion to the non-moral question of how neurons give rise to thought, then the question of animal mentalism need not be asked, because the answer is not important for directing a rigorous scientific effort to understand the neurophysiology of thought.

Some excellent scientists may not agree with the opinion that today, using laboratory animals, we can successfully study mental processes like thought (e.g. Vanderwolf and Cain, 1994; Vanderwolf, 2001) and while these people may

Information Processing by Neuronal Populations, ed. Christian Hölscher and Matthias Munk. Published by Cambridge University Press. © Cambridge University Press 2009.

one day be proven right, the question is still wide open. I am especially encouraged by the progress neuroscience has made with memory, a phenomenon that most of us would localize somewhere within the behaviorist's black box. The success with memory is owed to the deliberate application of the scientific method. Start with a neurophysiological theory of a mental function, develop a model for the appropriate neural phenomena, and study the phenomena deeply. I will begin the chapter with a very selective, biased, and brief review of the neurophysiological enquiry into memory. The point is to illustrate a neurophysiological success in understanding a mental process. I will then review how, using the same neurophysiological approach, my laboratory is trying to understand the forms of cognitive disorganization that are symptoms of psychosis (e.g. thought and behavioral disorganization, hallucination, and delusion).

On the value of theoretical and physiological models of mental processes

During a collaboration between Todd Sacktor's and my laboratories, we had the very good luck to discover that the molecule that maintains long-term potentiation of synaptic transmission (LTP) at hippocampal synapses (Ling et al., 2002) also maintains hippocampus-dependent memory (Pastalkova et al., 2006). Discovering that protein kinase M zeta (PKMζ) is the key molecular mechanism for maintaining both LTP and information storage was no accident. Like many others, we were searching for it by intensely studying LTP as a *model* of memory. To proceed, no one needed to believe that shocking a fiber bundle in the hippocampal slice produced memory, a mental object. We only needed to be confident the shock reliably produced a measurable physiological state, LTP, with key features of what a neurophysiological theory of memory predicted (review: Bliss and Collingridge, 1993). The theory that activity-dependent synaptic plasticity might underlie memory (Konorski, 1948; Hebb, 1949) related a physiological phenomenon with a mental phenomenon, providing the intellectual framework and a good deal of the motivation for the collaboration that led Bliss and Lomo to fully characterize LTP, a physiological phenomenon that in their words "was potentially useful for information storage" (Bliss and Lømo, 1973). It is very unlikely the properties of PKMζ, a key to the persistence of memory, would have been studied or recognized without a neurophysiological theory and model of memory.

We are trying to understand the neurophysiology of psychosis-related mental dysfunction and we are using the same approach that seems to have been so successful for understanding the neurobiology of memory. We started with a theory and sought a physiological model of the phenomenon to study deeply. We next describe the theoretical framework.

Theoretical framework: neural coordination

Our work is based on the notion that information is represented in the conjoint activity of many neurons and that the activity of a single neuron is at best an improvised reflection of what the system is representing. Admittedly, how neurons code information is still very much an open question. "Dedicated-coding" schemes (e.g. Barlow, 1972) are based on the concept of "cardinal units." These are cells dedicated to signaling high-order constellations of stimuli and concepts, like those defining one's grandmother or a giraffe. Averaging activity across a population of cardinal units has been used to decode neural activity successfully in some sensory and motor systems but with only limited success in most cognitive systems. The limited success is probably because many neural systems do not use a dedicated code. If they did, a fundamental limitation of capacity would arise from the fact that there are many more unitary concepts than neurons. This "combinatorial exhaustion" (von der Malsburg, 1981) is avoided by "ensemble-coding" (also called "distributed-coding") schemes, because each unit can participate in many representations.

Information in an ensemble code is represented by the coactivity of a functionally related group of cells, a "cell assembly." No single cell is essential to the code, and each cell participates in many different assembly representations (Fig. 16.1). After proposing that activity-dependent synaptic modification was the basis for the mental phenomena we call memory, Hebb (1949) developed the concept of a cell assembly as the basis of an "idea," a complex of associations, which in his words is "the unit of thought." He recognized that a crucial aspect of thought must be the flexible interplay of multiple ideas occurring simultaneously or in sequence. To advance understanding beyond the representation of simple individual experiences, we will need to consider the coexistence and coordination of the discharge of neurons forming one assembly with those of other assemblies.

Two cell assemblies that are co-active in the same neural system present a fundamental problem for ensemble-coding known as the "superposition catastrophe" (von der Malsburg, 1981). If two assemblies are simultaneously co-active and they have neurons in common, then they in effect become a single assembly (Fig. 16.1c). The result is confusion or inappropriate perception. The superposition catastrophe itself can be avoided by coordinating neural activity (review: Singer, 1999) so that discharge that defines one assembly is synchronous and desynchronized from the discharge of other assemblies with common cells. Stated another way, neural activity is normally coordinated so the group of co-active, grandmother-coding cells is constrained to discharge at different times than the group of co-active giraffe-coding cells. Accordingly, reliable patterns

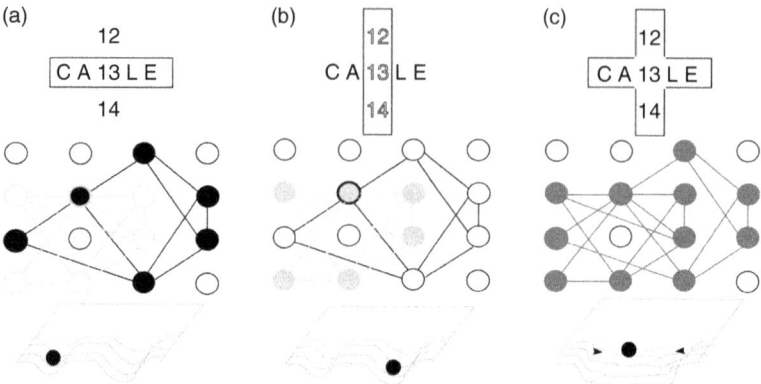

Figure 16.1 Cartoon of the central hypothesis illustrating neural network activity (middle) and the corresponding network state (bottom) in response to a stimulus (top). (a, b) Overlapping functionally coupled subpopulations in a network of neurons (circles represent network nodes) reliably co-activate to form a steady and stable network state (cell assembly) that represents an event, perception, or concept. A cell assembly is defined by co-active neurons (filled nodes). Reliably co-active neurons are assumed to be functionally coupled by active mono- and/or polysynaptic pathways (represented by the lines connecting the nodes). In neural network terms, a cell assembly corresponds to a "basin of attraction," a network state with a local minimum of energy. The state persists in response to small perturbations and changes in input. Only sufficiently large perturbations or changes in input can switch the network state to a different local minimum. The cell assembly for the word "cable" is distinct from other assemblies of different co-active subpopulations of mutually coupled cells that represent different events, perceptions, and concepts, such as the numerical sequence "12 13 14." The independent assemblies are not functionally coupled but they share nodes. The central symbol in the stimulus example can activate the cell assembly for "cable" or "12 13 14" appropriately depending on the context, set for example by the instruction to identify a word or a numerical sequence. Normally only one cell assembly will activate in response to an ambiguous or conflicting stimulus like the one shown. The ability to deal with conflicting information is called cognitive segregation. (c) Pathological coupling of independent assemblies by for instance disinhibition of the network, makes unrelated assemblies more likely to co-activate. This impairs cognitive segregation when independent assemblies interfere and/or merge resulting in lost fidelity of representing events, misperceptions, and disorganization of thought. In neural network terms, this corresponds to a "parasitic attractor" (Hoffman, 1987). A parasitic attractor is characterized by a broadening and deepening of the basin of attraction so that it (1) is more likely to find the network is the parasitic state, (2) is less precise which activity pattern defines the state, and (3) requires a larger perturbation to get out of the parasitic state (Olypher et al., 2006).

of positively, negatively, and independent pairwise discharge within a network of neurons is first-order evidence of neural coordination that is readily accessible to the experimentalist. In fact, within a neural network, the set of pairwise correlations may even account for the higher-order correlations in the network activity (Schneidman *et al.*, 2006).

Normal brains selectively activate representations that are currently appropriate and suppress ones that are currently inappropriate (Fig. 16.1). Hebb's concept of the cell assembly predicts that, for organized thought, multiple cell-assemblies must be activated together and in sequence. As evidence of this, we should be able to measure the organized correlated discharge of many cell pairs as "neural coordination." Neural coordination refers to the processes that coordinate the timing of responses by neurons participating in representing multiple distinct items, without changing the responses of such neurons to individual items. Others have called neural coordination "cognitive coordination" (review: Phillips and Singer, 1997) and more importantly, it is proposed that the core dysfunction in disorganized schizophrenia is an impairment of neural coordination (Tononi and Edelman, 2000; Phillips and Silverstein, 2003; Uhlhaas and Singer, 2006; review: Haenschel's chapter (Chapter 17) this volume).

Our central hypothesis is that the excessive co-activation of neurons in multiple cell assemblies causes the forms of cognitive disorganization seen in psychosis. Crucially, because neuronal co-activation can strengthen synapses, the inappropriate associations may be self-perpetuating. We call this the "hypersynchrony" hypothesis but emphasize that the hypothesis is not restricted to coordination phenomena that operate on the gamma (40 Hz) timescale, which is the focus of much work on field potentials and the binding problem (review: Engel and Singer, 2001; Haenschel's chapter (Chapter 17), Kaiser and Lutzenberger's chapter (Chapter 15)). To test the hypersynchrony hypothesis, we have characterized and manipulated the temporal coordination of neural discharge in the hippocampus.

We next describe neural coordination in the rodent hippocampus because this is the physiological framework within which our neurobiological investigations are based.

Physiological framework: neural coordination in hippocampal place cell firing

We focus on the place cell ensemble representation of space in the rodent hippocampus. A place cell increases firing when a rat moves through the cell's receptive ("place") field (O'Keefe, 1976); the conjoint activity of tens to hundreds of place cells is needed to accurately decode the rat's position (Wilson

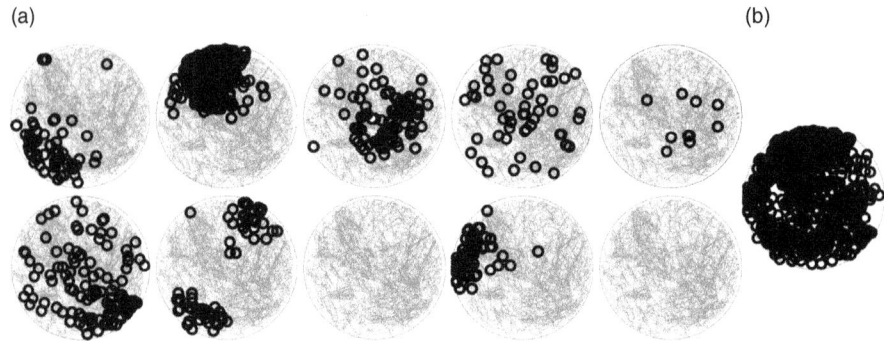

Figure 16.2 (a) Spike maps represent the positional discharge of ten simultaneously recorded hippocampal pyramidal cells. The track of the rat in the arena is represented by gray and the discharge of each cell is represented by the black circles in a single map. Each circle marks the location of the rat when one action potential was discharged. Six cells fired in the location-specific manner of place cells. One cell was active but not in a discrete region of the environment. Three cells fired only weakly (two not at all) in this space. (b) Superimposing the ten firing patterns illustrates that even such a small number covers most of the accessible space and that each location corresponds to a unique across-cell (ensemble) pattern of discharge.

and McNaughton, 1993; Fenton and Muller, 1998; Brown et al., 1998) (Fig. 16.2). Although place cell research from the beginning has concentrated on their receptive field properties, recently it has been described that place cell discharge is coordinated in a manner consistent with Hebb's cell assembly hypothesis. Two examples come from recording place cells as rats ran laps on a rectangular track. Decoding the rat's location on the track from place cell discharge was more accurate when the extent to which subsets of place cells did and did not fire together was considered (Harris et al., 2003). A second example is that the distance between firing fields on the track could be predicted by the probability of a pair of cells to co-activate on the timescale of the ~7-Hz theta oscillation in the hippocampal field potential. Cell pairs that fired within a few milliseconds of each other tended to have overlapping firing fields and cell pairs that fired one after the other with a delay of several tens of milliseconds tended to have firing fields that were sufficiently separate so it would take the rat seconds to move from the center of one field to the center of the other (Dragoi and Buzsáki, 2006). Though it is obvious that cells with overlapping firing fields should fire together, it was less obvious that cells that fire with a ~70-ms lag should have firing fields that were separated by 10–20 cm because a rat moving at 25 cm/s (i.e. running very fast) would cover less than 2 cm in 70 ms. While these data may suggest the existence of hippocampal cell assemblies, it may also be that the stimuli the rat uses from one moment to another can change (e.g. Olypher et al., 2002) due to processes like attention (Kentros et al., 2004). We took steps to control for such processes.

Behavioral place coordination: a behavioral model of thought

Spatial behavior has been traditionally studied in animals under static experimental conditions that favor the encoding and expression of only one representation of the environment. However, a place, like an idea, is an abstract and subjective complex of associations that in thought is coordinated with other places. For example, a person in an office may think about his or her place in the office (e.g. near the window), in the building (on the fifth floor), or in the city, which requires the coordinated activation of the neural ensembles that represent each concept of location. Below we describe that rats also coordinate their sense of place.

Our goal is to study impairments of neural coordination, but first we created a condition that allowed us to study the coordination of places at a behavioral level; this is our model of thought in the rat. We found circumstances that unambiguously required the rat to coordinate two representations of space. We now describe these behavioral conditions in detail.

A "two-frame" experimental system was developed to test if rats organize their behavior by selecting among multiple representations of simultaneously available stimuli. To provide the minimum of two subsets of simultaneously available stimuli, the well-accepted distinction between distal or "room frame" and proximal or "arena frame" cues was exploited. The arena (Fig. 16.3a) is an 82-cm diameter metal disk bounded by a 5-cm high rim, centered in a small room. Large visible objects off the disk provide room-frame stimuli. Arena-frame stimuli are created by the rat, which rapidly makes scent marks that can identify locations (Wallace et al., 2002). Such markings, including feces and urine and similar stimuli generated by the rat, are able to guide avoidance navigation (Bures et al., 1997; Wesierska et al., 2005). Any position the rat visits could be defined in reference to room cues, in reference to arena cues, or an arbitrary combination of stimuli from the two cue sets. In these conditions it is very difficult to determine which cues the rat uses to locate itself and events; moreover, there is no justification for assuming that even during constant conditions that the rat would always use the same cues for localization. The situation with the rat is hardly different from my own. As I sit writing this, an observer would only sometimes be correct to assume I am thinking about my position at the keyboard, because although my environment (a café) and my overt behavior in it (typing) is constant, for internal reasons I alternate between thinking about my position at the table, within the café, in the city, and even about what I have to eat on the kitchen counter at home. We assume

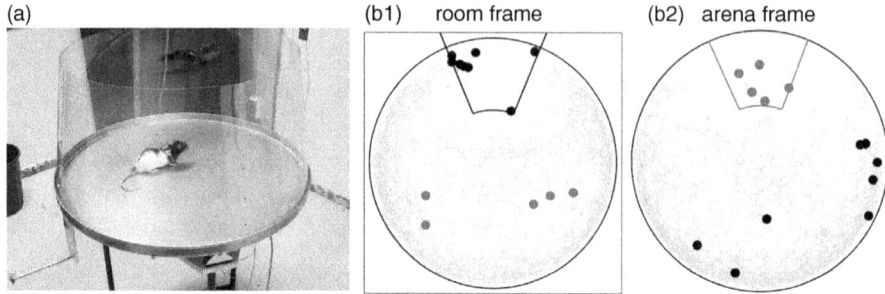

Figure 16.3 (a) Photo of the place avoidance apparatus. Room+Arena+ place avoidance on a continuously rotating (1 rpm) arena was used to assay cognitive segregation in the rat. The rat was reinforced to avoid shock in a part of the stationary room frame (black) as well as a part of the rotating arena frame (dark gray). (b) Plots of a rat's position (light gray) and the locations it was shocked during a single 10-min session. The same data are plotted in the stationary room frame (b1, square) and the rotating arena frame (b2, circle). Each shock for being in the room frame shock zone is indicated by a black dot and each shock for being in the arena shock zone is indicated by a dark gray dot. Notice that shocks for being in the room shock zone occurred in unpredictable locations on the arena. Similarly, shocks for being in the arena shock zone occurred in unpredictable places in the room. This Room+Arena+ task variant challenges the rat to segregate representations of the room space from representations of the arena space and selectively associate a shock with either room or arena locations.

that rats, like people, may have different points of view about where they are, and that it is adaptive to do this (Bures and Fenton, 2000).

We therefore used the subject's overt behavior to estimate whether or not the rat was localizing itself (and events) as separate positions in the room and arena frames. For this purpose, we continuously rotated the arena to explicitly dissociate the room and arena reference frames and we developed an avoidance task for the rat to perform during the rotation. The task has two 45° shock zones. One zone is fixed in the room frame and can be reliably located only by using room cues. The second shock zone is fixed in the arena frame and can be reliably located only by using rotating arena cues (Fenton *et al.*, 1998). Shocks that occur in the room shock zone will appear dispersed in the arena, and shocks that occur in the arena shock zone will appear dispersed in the room (Fig. 16.3b). Avoiding the separate room and arena shock zones during continuous rotation requires the rat to segregate representations of space based on room cues from those based on arena cues. Because the rat is in both a room location and simultaneously an arena location whenever it is shocked, the rat must only associate shock with the location in the corresponding shock zone otherwise it will experience and thus learn that shock is unpredictably dispersed

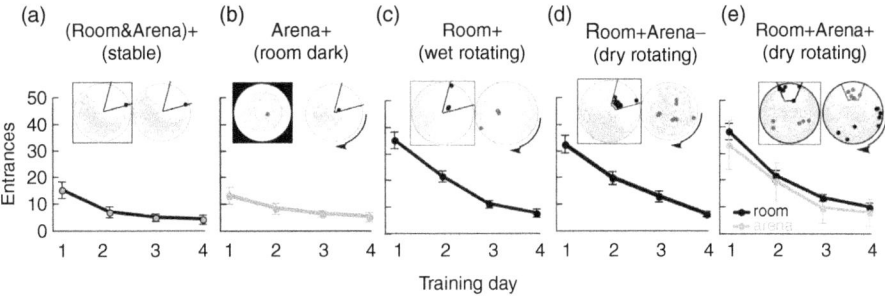

Figure 16.4 Initial learning of five variants of place avoidance during four 20-min sessions. The room frame (square) and arena frame (circle) tracks show performance on day 4; black dots are shocks delivered when the rat was in the room shock zone and gray dots show shocks delivered because the rat was in the arena shock zone. Note that during rotation, shocks that are clustered in one frame are dispersed in the other frame.

Table 16.1 *Experimental conditions of key place avoidance task variants*

Task variant	Arena rotating?	Lights	Water	Cues dissociated?
Room+Arena+	Yes	On	Dry	Yes
(Room&Arena)+	No	On	Dry	No
Arena+	Yes	Out	Dry	No
Room+	Yes	On	Wet	No
Room+Arena−	Yes	On	Dry	Yes

across the environment. To avoid the "trivial solution" of just standing still when rotation was shut off, hungry rats were trained to retrieve scattered food pellets. We call this the Room+Arena+ place avoidance task variant because avoiding a room location and a separate arena location are both reinforced.

Part of the value of the two-frame avoidance task is that it can be modified to test specific hypotheses (Bures *et al.*, 1997; Fenton *et al.*, 1998; Kubik and Fenton, 2005; Wesierska *et al.*, 2005). Performance improves (e.g. decrease in number of entrances per 5 min) within one 20-min session and behavior is asymptotic after four 20-min sessions for all task variants (Fig. 16.4). Figure 16.4 shows four other variants and Table 16.1 describes the physical conditions.

In the (Room&Arena)+ task (Fig. 16.4a) the arena does not rotate so the shock zone can be found using either room or arena cues or an arbitrary combination of both. A stable environment with both distal and local cues such as the one for the (Room&Arena)+ task is standard in behavioral and electrophysiological research. Unexpectedly, we found that despite no explicit reinforcement to form two avoidance memories, normal rats nonetheless learned two autonomous place avoidance memories in these standard conditions. The two memories were concurrently expressed (Fenton et al., 1998) and could be individually extinguished (Bures et al., 1997; Fenton and Bures, 2003). One memory was for the location of the shock zone in the room frame, the other for the location of the shock zone in the arena frame. Because the arena does not rotate in the (Room&Arena)+ variant, the two shock zones overlap and an experimenter cannot easily determine what spatial frame the rat is using to define the place it is avoiding. By dissociating the environment into the separate room and arena spatial frames we discovered that normal rats form separate, frame-specific spatial memories even if the memories are acquired without any explicit demand for separating room and arena cues into two spatial frames (Fenton et al., 1998).

Notice from Table 16.1 that the physical conditions of the Room+Arena+ and Room+Arena− task variants are identical. The tasks however, differ in a very important way. In the Room+Arena+ task there are two shock zones; one that is stationary and one that is rotating (Fig. 16.4e). In the Room+Arena− variant, there is only one shock zone (Fig. 16.4d). This requires that when the rat is shocked for being in the (room) shock zone it has to ignore where it is on the arena because, although each shock occurs at a place on the arena, these arena frame positions are irrelevant for predicting shock. It will appear as if the rat is being shocked unpredictably unless the rat is able to ignore arena positions, a process that first requires the rat to perceptually segregate room cues (and positions) from arena cues (and positions).

Continuously dissociating room and arena cues allows us to require that the rat segregates locations in the room and arena reference frames, and by arranging that only one of the frames can be perceived, we can create important control conditions that make the same overt behavioral demands with a minimal need to segregate and coordinate conflicting information. In the Arena+ task (Fig. 16.4b) the room cues are effectively removed by darkness so the rotating shock zone can be located only according to arena cues (Wesierska et al., 2005). In the Room+ task (Fig. 16.4c) arena cues are effectively removed by filling the disk with 1 cm of water so the shock zone can be located only by room cues (Wesierska et al., 2005).

Three place avoidance variants (Room&Arena)+, Room+, and Room+Arena− all require the rat to avoid a single shock zone that can be defined by stationary room cues, but the three tasks differ in their demand for the rat to perceptually segregate room and arena frame locations. In the (Room&Arena)+ task both room and arena locations can be perceived; because there is no requirement that one or the other is used selectively there is no demand to segregate room and arena cues for spatial perception or forming avoidance memory. The arena rotates in the Room+ variant, but arena cues are minimized by shallow water and so only self-motion cues conflict with the distal cues that are relevant for avoiding shock (Kubik and Fenton, 2005). With regards to cognitive demand, the Room+Arena− task is an especially interesting variant. This task has the same memory demand to avoid a single room-defined shock zone, and although the arena rotates just as in the Room+ variant, in the Room+Arena− task (Fig. 16.4d) the rat must also ignore the irrelevant arena cues. Of these three task variants, only the Room+Arena− condition requires the rat to segregate its perception of shock locations in the room and arena frames. An inability to perform the Room+Arena− task and a preserved ability to perform the Room+ task is taken to indicate a selective impairment of cognitive segregation. This impairment is selective because it manifests without impairments of the motivation or abilities to avoid shock, to navigate, or to store and recall spatial avoidance memory (Wesierska et al., 2005).

Having developed a means to measure place coordination, our behavioral model of thought, we recorded place cells during arena rotation in search of neural correlates of place coordination (Zinyuk et al., 2000; Fenton and Kelemen, 2006; Kelemen and Fenton, 2006). This chapter's main intent is to describe our physiological model of thought disorder, rather than place cell responses to dissociating the room and arena frames. For this purpose, it is only necessary to document that during conditions that require place coordination, we have evidence of neural coordination in the neural system that we will manipulate to model thought disorder. We recorded hippocampal neurons as rats performed the Room+Arena+ and (Room&Arena)+ task variants. Room and arena cues are relevant in both conditions but only during the rotating condition is there an explicit demand to segregate spatial perception and memory in the two frames. We have observed several neural coordination phenomena in these recordings (Fenton and Kelemen, 2006; Kelemen and Fenton, 2006). Perhaps the most straightforward is rotation-induced remapping.

Rotation-induced remapping: an example of neural coordination

Rotation-induced remapping demonstrates that hippocampal ensemble discharge contains two coordinated representations of the same environment. The subset of CA1 place cells that represent locations changes abruptly ("remaps": Muller and Kubie, 1987) when the arena is switched between stable and rotating, despite not physically disturbing the rat or the environment in any other way (Fig. 16.5). Initially, CA1 place cell representations were similar in the stable and rotating conditions, but as in Lever et al. (2002), after 1–2 weeks of training, the place cell representations of the stable and rotating conditions became distinct. Eventually, arena rotation induced place cell remapping of

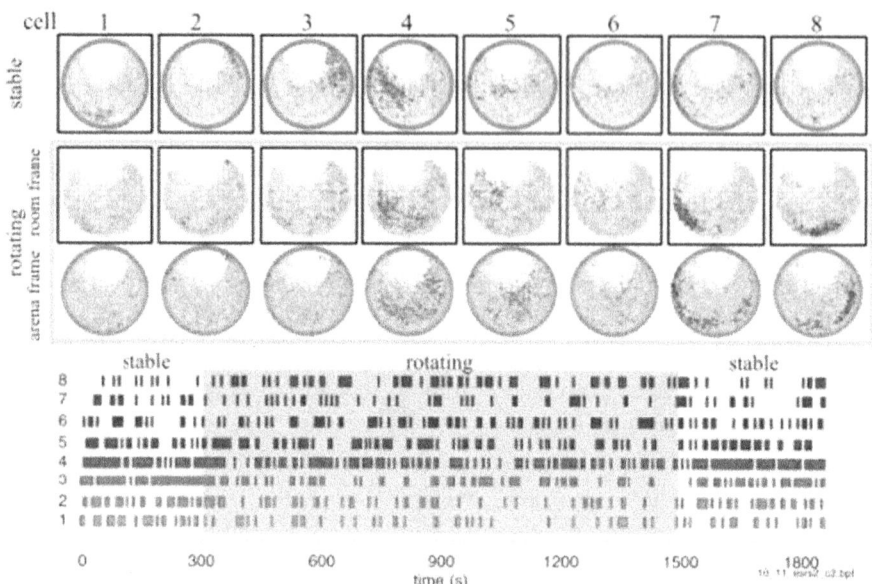

Figure 16.5 Rotation-induced remapping. (Top) Spike maps and raster plot (bottom) of discharge from eight simultaneously recorded CA1 cells illustrate rotation-induced remapping. Switching between the stable and rotating (shaded) conditions caused the ensemble pattern to change. Cells are ordered top-to-bottom by the ratio of firing rate in the two conditions. Cells 1–4 were more active when the arena was stable and cells 5–8 were more active when the arena was rotating. The rotation was switched on and off while the rat was on the arena doing (Room&Arena)+ and Room+Arena+ place avoidance.

the space although the stimuli themselves did not change (Fig. 16.5). This demonstrates unambiguously that the CA1 network contains two representations of the same physical stimuli, and that in normal rats when one representation is active the other can be suppressed. This is a form of neural coordination that can be observed even with small ensembles and from raster plots (Fig. 16.5).

Armed with the hypersynchrony hypothesis (that aberrant neural coordination underlies psychotic disorganization) and an experimental system that exhibits neural coordination (hippocampal pyramidal cell discharge), we were in a position to look for impaired neural coordination that could model the neural disorganization that underlies psychosis.

A model of the neural disorganization that underlies psychosis

The hypersynchrony hypothesis predicts that cognitive disorganization would be caused by manipulations that co-activate pyramidal cell pairs that were initially not co-active. It is well known that hippocampal pyramidal cell firing is strongly controlled by inhibition (only a minority, 30–40%, of pyramidal cells discharge while the rat is active: Wilson and McNaughton, 1993; Guzowski et al., 1999). We reasoned that disinhibiting the hippocampus might co-activate cell pairs that ordinarily would not fire together. Since we wanted to specifically impair neural coordination, while not changing basic discharge properties like firing rates, we sought a manipulation that might mildly disinhibit the hippocampus.

Two well-known facts were exploited. First, the intrahippocampal commissural projection is almost exclusively excitatory but it mediates feedforward inhibition (Buzsáki and Czech, 1981; Buzsáki and Eidelberg, 1981). This suggested that temporary inactivation of one hippocampus would block commissural excitation of the *uninjected* hippocampus and attenuate feedforward inhibition there (Fig. 16.6a). Second, neither permanent nor temporary unilateral hippocampal lesions have a major impact on hippocampus-dependent behavior, including water maze navigation, the gold standard for assessing hippocampal function (Fenton et al., 1993, 1995; Sutherland et al., 1983; Kubik and Fenton, 2005). This suggested that temporary inactivation of one hippocampus leaves basic hippocampal function intact; a prerequisite since we wanted to selectively impair neural coordination.

Our initial goal was to co-activate independently active pyramidal cell pairs by inactivating one hippocampus. We used the urethane-anesthetized rat preparation for two reasons. First, most of hippocampal physiology, including field potential oscillations, pyramidal cell and interneuron firing (Fox et al., 1986) and their temporal relations to the field oscillations are preserved under urethane anesthesia (Klausberger et al., 2003). Second, any changes in neural coordination caused by inactivating one hippocampus would not be secondary to altered behavior, perception, or sensation if the rat was anesthetized.

Under urethane anesthesia, injecting tetrodotoxin (TTX; 5 ng/µl) into one dorsal hippocampus silenced evoked and spontaneous neural discharge in both the dorsal hippocampus ~1 mm from the injection site (Klement et al., 2005) (Fig. 16.6b) as well as a site ~5 mm away in the ventral hippocampus (Olypher et al., 2006). The inactivation lasted over 8 h at the dorsal site, consistent with estimates from other brain regions (Zhuravin and Bures, 1991), and over 3 h in the distant ventral site. Since diffusion is isotropic in hippocampus (Hrabetova, 2005) and the 5-ng dose of TTX does not reach an effective concentration 2 mm from the injection site (Harlan et al., 1983; Zhuravin and Bures, 1991) the inactivation in the ventral site is probably due to blocking fimbria-fornix fibers near the injection site. This was enough to block action potential but not field potential oscillations in the ventral hippocampus (Fig. 16.6d). In fact, the TTX injection increased beta (12–25 Hz) field potential oscillations in the ventral

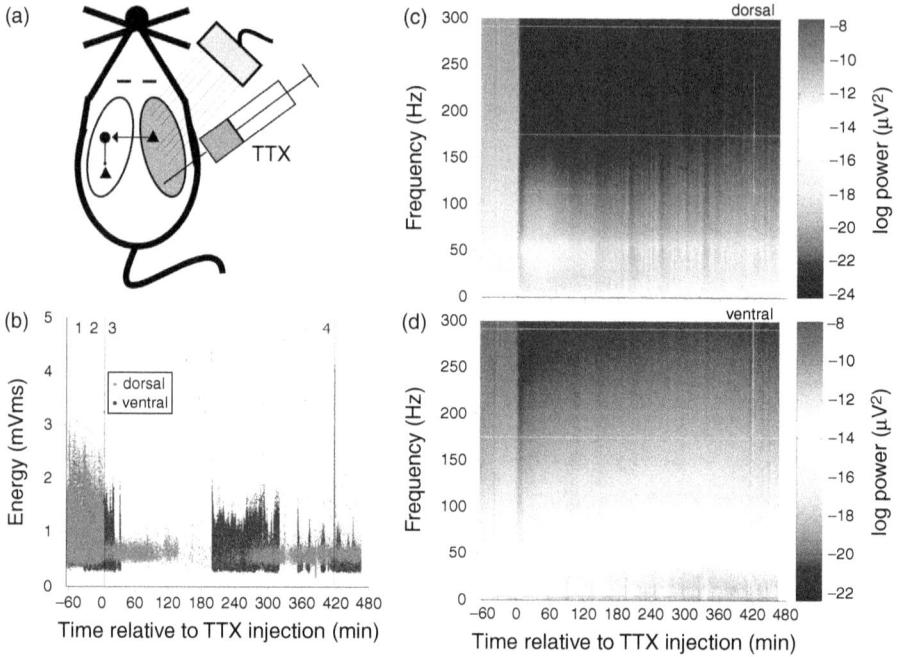

Figure 16.6 Recordings from the hippocampus injected with tetrodotoxin (TTX). (a) Cartoon of the recording set-up. The affected local circuit is indicated. Inactivating one hippocampus by TTX would block the commissural excitation projected by pyramidal cells (triangle) in the injected hippocampus. This commissural projection, although excitatory, activates inhibition via local interneurons (circles).
Two tetrodes recorded (b) action potential (c, d) and field potential activity from the urethane-anesthetized rat. The dorsal tetrode was ~1 mm from the TTX injection site in the dorsal hippocampus. The ventral tetrode was located in the ventral hippocampus, ~5 mm away from the injection site. (b) Action potential activity is depicted as the waveform energy (summed area under the curves on the four tetrode wires). Vertical line 1 indicates the time the injection cannula was inserted into the hippocampus; injection began and ended at the times marked by 2 and 3, respectively. The dorsal tetrode was moved 10 μm to mechanically stimulate discharge at time indicated by line 4. TTX blocked neural activity in the injected hippocampus, which the cartoon depicts would have blocked feedforward commissural inhibition of the uninjected hippocampus. The neural activity block was immediate at the dorsal site (gray) and delayed 30 min at the ventral site (black). Spontaneous ventral activity returned ~3 h after the injection. Although activity at the dorsal site could be evoked 7 h after the injection, spontaneous activity did not return by 8 h. Power spectra from field recordings at the dorsal (c) and ventral (d) sites indicate that field potentials were attenuated at the dorsal site for at least 8 h. At the ventral site field potentials in all frequency bands were maintained, except for beta band (12–25 Hz) oscillations, which were increased from 30 min after the TTX injection. This was also the time that spiking in the ventral hippocampus stopped. (Adapted from Olypher et al., 2006.)

hippocampus (Olypher et al., 2006). In the context of the hypersynchrony hypothesis, this is remarkable for two reasons. First, the presence of beta and maintained gamma oscillations in spiking activity reflects enhanced coupling between pyramidal cells in hippocampal slices (Faulkner et al., 1999). Second, increased beta is associated with cognitive disorganization, including frank thought disorder in schizophrenic patients (Spencer et al., 2004).

Recordings from the uninjected hippocampus indicated that injecting TTX into one hippocampus increased pyramidal cell firing in the uninjected side. The increase was specific to principal cells and transient, lasting about 15 min. Interneurons recorded simultaneously from stratum pyramidale did not change their rates. The transient firing rate increase was accompanied by decreased coupling between pyramidal cell and interneuron firing (Olypher et al., 2006), providing additional evidence that blocking commissural excitation to one hippocampus is disinhibiting. Although the TTX blockade of neural activity lasted several hours, the signs of disinhibition only lasted 15 min. This observation highlights the existence of feedforward and feedback neural circuits that maintain hippocampal discharge despite dramatic changes to the input (review: Buzsáki et al., 2002).

Once firing rates stabilized in the uninjected hippocampus, a remarkable change in neural coordination was observed. Pyramidal cell pairs that were unlikely to fire together before the TTX injection began to fire together after the injection. Those cell pairs that were likely to fire together before the injection did not change their coupling (Fig. 16.7).

These were very exciting observations for at least three reasons. First, it demonstrated in the urethane-anesthetized rat that injecting TTX into one hippocampus selectively impaired neural coordination in the uninjected hippocampus. Second, the impairment was specific; only cell pairs that previously did not fire together changed their coupling. This was precisely the change that the hypersynchrony hypothesis predicts would induce cognitive disorganization. We had already determined, using an artificial neural network in which two patterns of activity were stored, that increasing the coupling of initially weakly coupled cells caused failures to activate either pattern when there were conflicting inputs. Restoring the initially weak coupling restored the network's ability to switch between the two stored patterns according to which input was current and/or stronger (Olypher et al., 2006). Third, this model of neural disorganization that could underlie psychosis could be immediately used to test a major prediction of the hypothesis that increased co-activity underlies psychotic disorganization. The hypersynchrony hypothesis predicts that injecting TTX into one hippocampus will selectively impair place coordination, a behavioral model of thought.

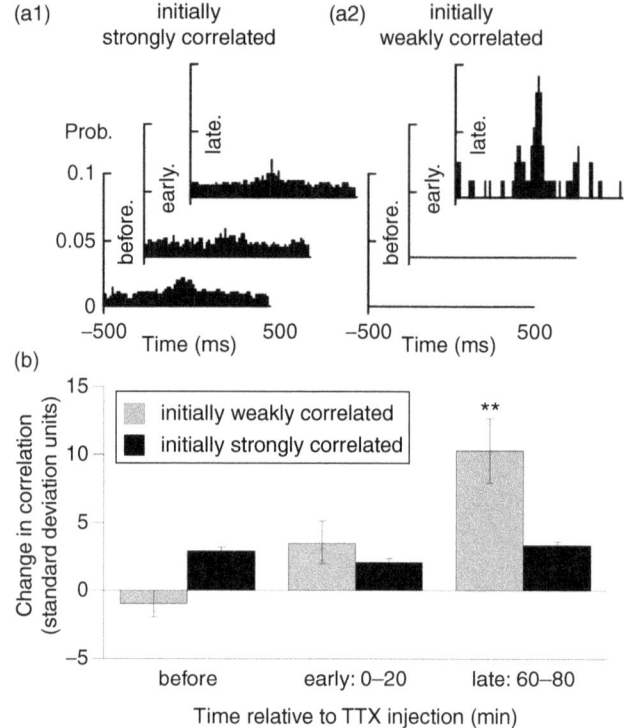

Figure 16.7 Injecting TTX into one hippocampus selectively increased co-activity of uncoupled neurons in the other hippocampus. (a) Example cross-correlation histograms of cell pairs with discharge that was initially strongly (a1) and weakly (a2) correlated. The histograms were computed during the 20 min before TTX injection, early (0–20 min) and late (60–80 min) after TTX injection. (b) The average change in Kendall's correlation for cell pairs that were initially weakly and strongly correlated. Only the initially weakly correlated cell pairs increased their functional coupling. This change was only evident after 40 min. Cell pairs were classified as initially weakly or strongly correlated by their Kendall correlation in the period before TTX injection. Cell pairs with correlations lower than the median were considered to be initially weakly correlated and the rest were considered to be initially strongly correlated. For the figure, correlations were calculated at 250 ms resolution ** $= p < 0.01$. (Adapted from Olypher et al., 2006.)

Testing the hypersynchrony hypothesis

We tested the hypothesis by examining how place avoidance behavior is affected when TTX is injected into one hippocampus. We knew that Room+ Arena− avoidance was impaired by the TTX injection (Cimadevilla et al., 2001) but this could have been the result of a non-specific TTX-induced memory or navigation deficit since solving spatial problems is sensitive to hippocampal dysfunction. The hypothesis and computer modeling (Olypher et al., 2006), predicted a more selective deficit. Specifically, the TTX injection should impair place avoidance in variants like the Room+ Arena− task that require cognitive segregation but spare place avoidance in other variants like the Room+ task that do not require cognitive segregation. Indeed, injecting TTX into one hippocampus did not impair a familiar Room+ avoidance (Kubik and Fenton, 2005; Wesierska et al., 2005) but the injection abolished the ability to learn, consolidate, or recall Room+Arena− avoidance (Cimadevilla et al., 2001; Wesierska et al., 2005). Importantly, the impairment was observed 60–80 min after TTX injection, the same time interval at which excessive co-activity was observed (Fig. 16.7).

While these data provide strong support for the hypothesis that excessive co-activity underlies psychotic disorganization, we point out two important caveats. First, TTX-induced co-activity was only demonstrated in urethane-anesthetized rats. It remains possible that the injection does not have the same effect in awake rats. Recordings from awake rats injected with TTX in one hippocampus would clarify this concern. The second, caveat is not so easily addressed. The pattern of impaired Room+Arena− avoidance and spared Room+ avoidance may be an inadequate model for psychotic cognitive disorganization. Unequivocal psychotic disorganization, like any mental state, can really only be demonstrated in people.

Analogous to how studying LTP as a model of memory led to identifying the key role of PKMζ in memory, we believe that important insights into mental dysfunction will be gained by studying hypersynchrony as a model for psychotic disorganization. We have therefore begun to examine whether psychotomimetics, drugs that induce or increase psychotic symptoms in people, induce excessive co-activity and impair cognitive segregation in rats. Our preliminary results, with phencyclidine (PCP), a potent psychotomimetic, are encouraging. A dose of PCP (5 mg/kg i.p.) that is sufficient to co-activate hippocampal pyramidal cells that did not fire together also impairs Room+Arena− but not Room+ place avoidance. A lower dose (3 mg/kg i.p.) was insufficient to alter hippocampal co-activity and it did not impair Room+Arena− avoidance (Fenton et al., 2006). This is precisely what the hypersynchrony hypothesis predicts.

We are also examining the hypersynchrony hypothesis in a well-studied neurodevelopmental model of schizophrenia (Lipska et al., 1992). The neurodevelopmental model of schizophrenia is based on the idea that a brain insult early in life can alter the maturation of prefrontal cortical circuits resulting in schizophrenia. Rats that receive a ventral hippocampal lesion as neonates exhibit several schizophrenia-like abnormalities when tested as adults. The abnormalities suggested a dysregulation of the dopamine system (Lipska et al., 1992). The rats were hyperactive in response to novelty and amphetamine

(Lipska et al., 1993). They were hypersensitive to the N-methyl-D-aspartate (NMDA) receptor antagonists MK-801 and PCP. They had deficits in pre-pulse inhibition of startle (Le Pen and Moreau, 2002), a social interaction assay (Sams-Dodd et al., 1997), and working memory (Lipska et al., 2002). Some of these abnormalities were alleviated by antipsychotics (review: Lipska and Weinberger, 2000). The hypersynchrony hypothesis predicts the neonatal lesion will induce both excessive co-activity and cognitive segregation deficits to the extent that the degree of co-activation will account for the cognitive deficit.

Ultimately the hypersynchrony hypothesis is best tested by recording neural activity from human subjects. The hypothesis predicts that excessive co-activity will be associated with psychotic symptoms. While there are serious ethical constraints to such a study, the hypothesis can be tested by analyzing action potential recordings from epileptic patients during presurgical screening for epileptic foci (Fried et al., 1999). These recordings are made from a subset of patients whose severe seizures cannot be localized to a brain region by other means. Some of these patients express post-ictal psychotic symptoms lasting days, which, according to the hypersynchrony hypothesis, should be accompanied by excessive post-ictal co-activity.

Summary

We have argued that the neurophysiological approach to understanding mental phenomena holds promise for understanding psychotic disorganization independent of whether or not experimental animals think. We have outlined our application of this approach to develop a physiological understanding of psychotic disorganization. As with LTP, the hypersynchrony hypothesis is based on a physiological theory of how neurons code information, and we have demonstrated means to selectively impair neural coordination, permitting experimental evaluation of the hypothesis. It may be premature to speculate on the impact of decisively confirming the hypothesis, but at the very least the hypersynchrony hypothesis provides the experimental basis for deliberate investigations into the events that underlie cognitive disorganization as well as the basis for rational novel approaches to developing antipsychotic therapies.

References

Barlow, H. B. (1972). Single units and sensation: a neuron doctrine for perceptual psychology? *Perception* **1**:371–394.

Bliss, T. V. P. and Collingridge, G. L. (1993). A synaptic model of memory: long-term potentiation in the hippocampus. *Nature* **361**:31–39.

Bliss, T. V. P. and Lømo, T. (1973). Long-lasting potentiation of synaptic transmission in the dentate area of the anaesthetized rabbit following stimulation of the perforant path. *J Physiol (Lond)* **232**:331–356.

Brown, E. N., Frank, L. N., Tang, D., Quirk, M. C., and Wilson, M. A. (1998). A statistical paradigm for neural spike train decoding applied to position prediction from ensemble firing patterns of rat hippocampal place cells. *J Neurosci* **18**:7411–7425.

Bures, J. and Fenton, A. A. (2000). Neurophysiology of spatial cognition. *News Physiol Sci* **15**:233–240.

Bures, J., Fenton, A. A., Kaminsky, Yu., et al. (1997). Dissociation of exteroceptive and idiothetic orientation cues: effect on hippocampal place cells and place navigation. *Phil Trans R Soc Lond B* **352**:1515–1524.

Buzsáki, G. and Czech, G. (1981). Commissural and perforant path interactions in the rat hippocampus: field potentials and unitary activity. *Exp Brain Res* **43**:429–438.

Buzsáki, G. and Eidelberg, E. (1981). Commissural projection to the dentate gyrus of the rat: evidence for feed-forward inhibition. *Brain Res* **230**:346–350.

Buzsáki, G., Csicsvari, J., Dragoi, G., et al. (2002). Homeostatic maintenance of neuronal excitability by burst discharges in vivo. *Cereb Cortex* **12**:893–899.

Cimadevilla, J. M., Wesierska, M., Fenton, A. A., and Bures, J. (2001). Inactivating one hippocampus impairs avoidance of a stable room-defined place during dissociation of arena cues from room cues by rotation of the arena. *Proc Natl Acad Sci USA* **98**:3531–3536.

Dragoi, G. and Buzsáki, G. (2006). Temporal encoding of place sequences by hippocampal cell assemblies. *Neuron* **50**:145–157.

Engel, A. K. and Singer, W. (2001). Temporal binding and the neural correlates of sensory awareness. *Trends Cogn Sci* **5**:16–25.

Faulkner, H. J., Traub, R. D., and Whittington, M. A. (1999). Anaesthetic/amnesic agents disrupt beta frequency oscillations associated with potentiation of excitatory synaptic potentials in the rat hippocampal slice. *Br J Pharmacol* **128**:1813–1825.

Fenton, A. A. and Bures, J. (1993). Place navigation in rats with unilateral tetrodotoxin inactivation of the dorsal hippocampus: place but not procedural learning can be lateralized to one hippocampus. *Behav Neurosci* **107**:552–564.

Fenton, A. A. and Bures, J. (2003). Navigation in the moving world. In: *The Neurobiology of Spatial Behaviour*, ed. Jeffery, K. J. Oxford, UK: Oxford University Press.

Fenton, A. A. and Kelemen, E. (2006). The discharge of hippocampal place cells with overlapping firing fields is coordinated on the timescale of seconds. *Soci Neurosci Abstr* **36**:211.09.

Fenton, A. A. and Muller, R. U. (1998). Place cell discharge is extremely variable during individual passes of the rat through the firing field. *Proc Natl Acad Sci USA* **95**:3182–3187.

Fenton, A. A., Arolfo, M. P., Nerad, L., and Bures, J. (1995). Interhippocampal synthesis of lateralized place navigation engrams. *Hippocampus* **5**:16–24.

Fenton, A. A., Wesierska, M., Kaminsky, Yu., and Bures, J. (1998). Both here and there: simultaneous expression of autonomous spatial memories. *Proc Natl Acad Sci USA* **95**:11493–11498.

Fenton, A. A., Kenney, J., and Kao, H.-Y. (2006). Phencyclidine impairs cognition if and only if it co-activates initially independently active neurons. *Soc Biol Psychiatry Meeting Abstract* 1477.

Fox, S. E., Wolfson, S., and Ranck, J. B., Jr. (1986). Hippocampal theta rhythm and the firing of neurons in walking and urethane anesthetized rats. *Exp Brain Res* 50:210–220.

Fried, I., Wilson, C. L., Maidment, N. T., et al. (1999). Cerebral microdialysis combined with single neuron and EEG recording in neurosurgical patients. *J Neurosurg* 91:697–705.

Guzowski, J. F., McNaughton, B. L., Barnes, C. A., and Worley, P. F. (1999). Environment-specific expression of the immediate-early gene *Arc* in hippocampal neuronal ensembles. *Nat Neurosci* 2:1120–1124.

Harlan, R. E., Shivers, B. D., Kow, L. M., and Pfaff, D. W. (1983). Estrogenic maintenance of lordotic responsiveness: requirement for hypothalamic action potentials. *Brain Res* 268:67–78.

Harris, K. D., Csicsvari, J., Hirase, H., Dragoi, G., and Buzsáki, G. (2003). Organization of cell assemblies in the hippocampus. *Nature* 424:552–556.

Hebb, D. O. (1949). *The Organization of Behavior*. New York: John Wiley.

Hoffman, R. E. (1987). Computer simulations of neural information processing and the schizophrenia–mania dichotomy. *Arch Gen Psychiat* 44:178–188.

Hrabetova, S. (2005). Extracellular diffusion is fast and isotropic in the stratum radiatum of hippocampal CA1 region in rat brain slices. *Hippocampus* 15:441–450.

Kelemen, E. and Fenton, A. A. (2006). Temporal coordination of hippocampal discharge during foraging in two continuously dissociated spaces. *Soc Neurosci Abstr* 36:211.12.

Kentros, C. G., Agnihotri, N. T., Streater, S., Hawkins, R. D., and Kandel, E. R. (2004). Increased attention to spatial context increases both place field stability and spatial memory. *Neuron* 42:283–295.

Klausberger, T., Magill, P. J., Marton, L. F., et al. (2003). Brain-state- and cell-type-specific firing of hippocampal interneurons in vivo. *Nature* 421:844–848.

Klement, D., Pastalkova, E., and Fenton, A. A. (2005). Tetrodotoxin infusions into the dorsal hippocampus block non-locomotor place recognition. *Hippocampus* 15:460–471.

Konorski, J. (1948). *Conditioned Reflexes and Neuron Organization*. Cambridge, UK: Cambridge University Press.

Kubik, S. and Fenton, A. A. (2005). Behavioral evidence that segregation and representation are dissociable hippocampal functions. *J Neurosci* 25:9205–9212.

Le Pen, G. and Moreau, J. L. (2002). Disruption of prepulse inhibition of startle reflex in a neurodevelopmental model of schizophrenia: reversal by clozapine, olanzapine and risperidone but not by haloperidol. *Neuropsychopharmacology* 27:1–11.

Lever, C., Wills, T., Caccuci, F., Burgess, N., and O'Keefe, J. (2002). Long-term plasticity in hippocampal place cells representation of environmental geometry. *Nature* 416:90–94.

Ling, D. S., Benardo, L. S., Serrano, P. A., *et al.* (2002). Protein kinase M zeta is necessary and sufficient for LTP maintenance. *Nat Neurosci* 5:295–296.

Lipska, B. K. and Weinberger, D. R., (2000). To model a psychiatric disorder in animals: schizophrenia as a reality test. *Neuropsychopharmacology* 23:223–239.

Lipska, B. K., Jaskiw, G. E., Chrapusta, S., Karoum, F., and Weinberger, D. R. (1992). Ibotenic acid lesion of the ventral hippocampus differentially affects dopamine and its metabolites in the nucleus accumbens and prefrontal cortex in the rat. *Brain Res* 585:1–6.

Lipska, B. K., Jaskiw, G. E., and Weinberger, D. R. (1993). Postpubertal emergence of hyperresponsiveness to stress and to amphetamine after neonatal hippocampal damage: a potential animal model of schizophrenia. *Neuropsychopharmacology* 9:67–75.

Lipska, B. K., Aultman, J. M., Verma, A., Weinberger, D. R., and Moghaddam, B. (2002). Neonatal damage of the ventral hippocampus impairs working memory in the rat. *Neuropsychopharmacology* 27:47–54.

Muller, R. U. and Kubie, J. L. (1987). The effects of changes in the environment on the spatial firing of hippocampal complex-spike cells. *J Neurosci* 7:1951–1968.

O'Keefe, J. (1976). Place units in the hippocampus of the freely moving rat. *Exp Neurol* 51:78–109.

Olypher, A. V., Lansky, P., and Fenton, A. A. (2002). Properties of the extra-positional signal in hippocampal place cell discharge derived from the overdispersion in location-specific firing. *Neuroscience* 111:553–656.

Olypher, A. V., Klement, D., and Fenton, A. A. (2006). Cognitive disorganization in hippocampus: a physiological model of the disorganization in psychosis. *J Neurosci* 26:158–168.

Pastalkova, E., Serrano, P., Pinkhasova, D., *et al.* (2006). Storage of spatial information by the maintenance mechanism of LTP. *Science* 313:1141–1144.

Phillips, W. A. and Silverstein, S. M. (2003). Convergence of biological and psychological perspectives on cognitive coordination in schizophrenia. *Behav Brain Sci* 26:65–82; discussion 82–137.

Phillips, W. A. and Singer, W. (1997). In search of common foundations for cortical computation. *Behav Brain Sci* 20:657–683; discussion 683–722.

Sams-Dodd, F., Lipska, B. K., and Weinberger, D. R. (1997). Neonatal lesions of the rat ventral hippocampus result in hyperlocomotion and deficits in social behaviour in adulthood. *Psychopharmacology* 132:303–310.

Schneidman, E., Berry, M. J. 2nd, Segev, R., and Bialek, W. (2006). Weak pairwise correlations imply strongly correlated network states in a neural population. *Nature* 440:1007–1012.

Singer, W. (1999). Time as coding space? *Curr Opin Neurobiol* 9:189–194.

Spencer, K. M., Nestor, P. G., Perlmutter, R., *et al.* (2004). Neural synchrony indexes disordered perception and cognition in schizophrenia. *Proc Natl Acad Sci USA* 101:17288–17293.

Sutherland, R. J., Whishaw, I. Q., and Kolb, B. (1983). A behavioural analysis of spatial localization following electrolytic, kainate- or colchicine-induced damage to the hippocampal formation in the rat. Behavioral. *Brain Res* **7**:133–153.

Tononi, G. and Edelman, G. M. (2000). Schizophrenia and the mechanisms of conscious integration. *Brain Res Brain Res Rev* **31**:391–400.

Uhlhaas, P. J. and Singer, W. (2006). Neural synchrony in brain disorders: relevance for cognitive dysfunctions and pathophysiology. *Neuron* **52**:155–168.

Vanderwolf, C. H. (2001). The hippocampus as an olfacto-motor mechanism: were the classical anatomists right after all? *Behav Brain Res* **127**:25–47.

Vanderwolf, C. H. and Cain, D. P. (1994). The behavioral neurobiology of learning and memory: a conceptual reorientation. *Brain Res Brain Res Rev* **19**:264–297.

von der Malsburg, C. (1981). *The Correlation Theory of Brain Function*, Technical Report 81-2. Frankfunt, Germany: Biophysical Chemistry, Max Planck Institute.

Wallace, D. G., Gorny, B., and Whishaw, I. Q. (2002). Rats can track odors, other rats, and themselves: implications for the study of spatial behavior. *Behav Brain Res* **131**:185–192.

Wesierska, M., Dockery, C., and Fenton, A. A. (2005). Beyond memory, navigation and inhibition: behavioural evidence for hippocampus-dependent cognitive coordination in the rat. *J Neurosci* **25**:2413–2419.

Wilson, M. A. and McNaughton, B. L. (1993). Dynamics of the hippocampal ensemble code for space. *Science* **261**:1055–1058.

Zhuravin, I. A. and Bures, J. (1991). Extent of the tetrodotoxin induced blockade examined by pupillary paralysis elicited by intracerebral injection of the drug. *Exp Brain Res* **83**:687–690.

Zinyuk, L., Kubik, S., Kaminsky, Yu., Fenton, A. A., and Bures, J. (2000). Understanding hippocampal activity using purposeful behavior: place navigation induces place cell discharge in both the task-relevant and task-irrelevant spatial reference frames. *Proc Natl Acad Sci USA* **97**:3771–3776.

17

The role of synchronous gamma-band activity in schizophrenia

CORINNA HAENSCHEL

Introduction

Despite almost 100 years of research into the pathophysiology of schizophrenia, the causes and mechanisms underlying the disease remain poorly understood. For a long time, biological research has focused on finding regionally specific pathophysiological processes in this disorder. In the last decade, however, theories of schizophrenia have laid emphasis upon pathophysiological mechanisms, which involve multiple cortical areas and their coordination. These theories suggest that the core impairment underlying both dysfunctional cognition and the overt symptoms of the disorder arise from a dysfunction in the integration and coordination of distributed neural activity (Andreasen, 1999; Friston, 1999; Phillips and Silverstein, 2003).

Interestingly, a disturbance of integrative processing had already been suggested by Bleuler (1950). He coined the term *schizophrenia* ("split mind") to highlight the fragmentation of mental functions. According to him, the fragmentation of mental functions constituted the primary disturbance in schizophrenia that represented a direct manifestation of the organic pathology whilst other symptoms, such as delusions and hallucinations, were accessory or secondary manifestations of the disease process. Contemporary models of schizophrenia for instance by Friston (1999) suggest that the core pathology is an impaired control of (experience-dependent) synaptic plasticity that manifests as abnormal functional integration of neural systems, i.e. dysconnectivity. Andreasen (1999) used "cognitive dysmetria" to refer to the fact that patients with diverse clinical

Information Processing by Neuronal Populations, ed. Christian Hölscher and Matthias Munk. Published by Cambridge University Press. © Cambridge University Press 2009.

and cognitive deficits share a common underlying deficit in the "timing or sequencing component of mental activity" across multiple brain regions. A possible candidate mechanism for the coordination of neural activity between and within functionally specialized brain regions into coherent ensembles that represent a specific cognitive content is synchronous oscillatory activity of neural responses (Singer, 1999; Varela et al., 2001). Such a dynamic integration into neuronal assemblies may range from local networks for early sensory processing to large-scale networks responsible for cognitive processes such as memory formation (Mesulam, 1994; Fuster, 1997; Kaiser and Lutzenberger Chapter 15, this volume). There is evidence that suggests that synchronous oscillatory activity is related to feature binding in perception (Rodriguez et al., 1999), working memory (Tallon-Baudry et al., 1998), and attention (Fries et al., 2001).

A vast number of studies show disturbances in these cognitive processes across the lifespan of individuals with schizophrenia, including childhood and adolescence (Davidson et al., 1999), and at the initial onset of psychosis (Saykin et al., 1994; Lewis et al., 2005). Importantly, the degree of cognitive dysfunction might be the best predictor of long-term functional outcome for patients with schizophrenia (Green, 1996).

This raises the question whether abnormalities in synchronous oscillatory activity (Lee et al., 2003) may provide a parsimonious account for many of the cognitive deficits observed in schizophrenia (Green, 1996; Phillips and Silverstein, 2003).

A substantial body of electroencephalogram (EEG) studies supports the hypothesis that the synchronization of oscillatory activity associated with perceptual and cognitive tasks is impaired in patients with schizophrenia (Kwon et al., 1999; Green et al., 2003; Gallinat et al., 2004; Uhlhaas and Singer, 2006; Uhlhaas et al., 2006). In addition, reductions in high-frequency oscillatory activity have been found in first-degree relatives of patients with schizophrenia (Hong et al., 2004b), in prodromal patients (D.H. Mathalon, unpublished observation), and in subjects with high schizotypy scores (Vernon et al., 2005). The underlying assumption of all these studies is that binding impairments in schizophrenia occur due to deficiencies in the ability to sustain precisely timed synchronized activity patterns or dysfunctional long-range coordination. However, the relevance of such research may go beyond the understanding of cognitive dysfunctions in schizophrenia. The study of dysfunctional binding mechanisms in schizophrenia could serve as a model for the functional role of neural synchronization in normal cognitive processes by demonstrating how a specific cognitive deficit is related to abnormal neural synchronization.

Currently, there are several unresolved issues with regards to the relationship between schizophrenia and neural synchrony:

(1) What are the best methods to investigate the temporal integration deficits in high-frequency spectral power and phase synchronization?
(2) Existing data suggest that patients with schizophrenia show abnormalities in high-frequency spectral power and phase synchronization in response to early perceptual and more complex cognitive stimuli. Do these deficits exist independently of each other or is there a relationship between them?
(3) Any account of a specific neural deficit in synchrony has to explain the heterogeneous symptoms associated with schizophrenia as well as the state and trait differences in patients, like symptom severity and illness duration (van der Stelt et al., 2004).
(4) Given that neuronal synchrony can be found in both humans and animals and can be studied at various levels of spatial analysis, from microscopic (e.g. single-unit recordings) to macroscopic (e.g. EEG) measurements (Singer and Gray, 1995; Varela et al., 2001) is it possible to link the results of clinical to pharmacological studies (van der Stelt et al., 2004)?

Methods for examining neural synchrony

The aim of this section is to describe the various approaches that have been used to investigate reductions in spectral power and synchrony. There are two main approaches: (1) the analysis of spectral power that only measures the spatial summation of cortical responses in the range of 1 cm, thus assessing more intra-area coordination (Varela et al., 2001); (2) the analysis of the stability of phase differences between electrodes (Lachaux et al., 1999) measuring long-range coordination between neural assemblies that are farther apart in the brain (>2cm) (Varela et al., 2001).

Local (intra-areal) synchronization in scalp EEG and magnetoencephalography (MEG) as well as in intracranial recordings is reflected in measures that examine the amplitude of oscillatory activity that arises from the synchronous and periodic discharges of thousands or millions of neurons and the associated synaptic events. The amplitude increases with the number of synchronously active neurons and the precision of synchrony. Each EEG electrode, or MEG sensor, measures the spatial sum of synchronous activity across a large number of oscillating neurons or neural circuits (Basar, 1980). This can be measured as spontaneous oscillatory activity, which is uncorrelated with the occurrence of any stimuli, or more importantly, as event-related oscillatory activity (Galambos, 1992). Two forms of event-related oscillatory activity need to be distinguished: (1) evoked and (2) induced. Evoked oscillations are tightly time- and phase-locked to the onset of the stimulus and can thus be seen in the average evoked response (ERP) and reflect stimulus-driven synchronized activity. Typically, evoked oscillations occur within the first 100 ms following stimulus onset (Pantev, 1995; Gruber et al., 1999; Herrmann et al., 1999) and may signal the precise temporal relationship of incoming stimuli (Engel et al., 1992). Evoked oscillations can be modulated by attention-dependent top–down processes (Tiitinen et al., 1993; Herrmann et al., 2004).

Intertrial phase-locking is the measurement of phase coupling to a transient event and represents an alternative, and probably more precise, way of defining the extent of temporal variation between incoming stimuli. Evoked activity theoretically exhibits a very tight phase relationship between the individual trials. Hence, intertrial phase-locking is useful for identifying both the decrease in the signal-to-noise ratio as seen in patients with schizophrenia (Light et al., 2006) and problems with entraining the stimuli (Kwon et al., 1999).

In contrast to evoked oscillations, induced oscillations are not phase-locked to the stimulus and their latency varies from trial to trial. Therefore, they are canceled by averaging and require single-trial analysis. Spectral induced activity reflects self-organized local synchronous activity. Induced oscillations have been related to (1) the construction of coherent object representations (Tallon-Baudry and Bertrand, 1999), (2) attention (Fries et al., 2001), (3) selective routing of activity (Fries et al., 2007), (4) cross-modal binding (Yuval-Greenberg and Deouell, 2007), and (5) maintenance of working memory contents (Tallon-Baudry et al., 1998, 2004).

The functional significance of evoked and induced gamma-band oscillations and their relationship to each other remains unclear. However, a reduction of these power measures suggest selective deficiencies in the ability of cortical networks or cortico-thalamo-cortical loops to engage in precisely synchronized high-frequency oscillations (Uhlhaas and Singer, 2006).

Phase synchrony may exist on a larger scale. Long-range synchrony is important for the dynamic coupling between anatomically distant but functionally closely related brain areas (Varela et al., 2001). Synchronous oscillatory activity supporting local integration within specialized areas tends to occur at higher frequencies (gamma band: >30 Hz) than synchronous activity supporting large-scale integration, which most often manifests itself in the beta (12–30 Hz) but also in the theta (4–8 Hz) and alpha (8–12 Hz) frequency ranges (von Stein and Sarnthein, 2000; Schnitzler and Gross, 2005). Oscillations in the theta, alpha, and beta frequency ranges may be particularly suited for long-range coordination because synchronization at lower frequencies tolerates longer conduction delays (Kopell et al., 2000). Such inter-electrode or inter-sensor synchrony is typically measured using coherence or phase-locking. Coherence does not separate the effects of amplitude and phase in the relations between two signals. In order to measure phase-locking between the temporal structures of EEG and MEG signals independent of signal amplitude new methods have been developed (Lachaux et al., 1999). These measure the stability of phase differences between electrodes across experimental trials. Indeed, it has been suggested that schizophrenia reflects an impairment in the long-range interactions of cortical areas that subserve the perceptual binding processes (Uhlhaas et al., 2006).

Learning more about the precise signature of event-related intertrial and inter-electrode phase synchrony and evoked and induced oscillatory power may help to unravel the nature of the abnormalities in neural synchrony associated with schizophrenia.

Neural synchrony in schizophrenia

Over the last ten years evidence has accumulated that patients with schizophrenia show aberrant cortical oscillatory activity.[1] There is evidence of impaired gamma-band oscillations from studies probing early sensory systems (e.g. steady state, sensory gating, and backward-masking experiments) and from more cognitive event-related paradigms (e.g. auditory oddball, visual binding, and working memory experiments). Some hypotheses relating abnormal synchrony and schizophrenia are that: (1) a lack of intertrial phase synchrony explains the decreased signal-to-noise ratio (also termed "cortical noise") (Gallinat et al., 2004; Winterer and Weinberger, 2004; Winterer et al., 2004; Light et al., 2006), (2) reduced evoked activity reflects either as before a decreased signal-to-noise ratio or a reduced entrainment in response to external stimuli (Clementz et al., 1997; Kwon et al., 1999; Krishnan et al., 2005), (3) reduced phase synchrony between electrodes can help to uncover deficiencies in the long-range synchronization of neural activity and thus in distributed cortical neural circuits (Spencer et al., 2003; Slewa-Younan et al., 2004; Symond et al., 2005; Uhlhaas et al., 2006), and (4) smaller induced activity is a sign of decreased self-organized local synchronization (Haig et al., 2000; Green et al., 2003).

Early sensory deficits: steady state, sensory gating, and backward masking

Studies on early sensory deficits have focused on investigating the relationship between the auditory and visual steady-state response, auditory sensory gating, and high-frequency oscillations in patients with schizophrenia. In the auditory domain, the steady-state response (SSR) is measured at the level of the middle latency auditory evoked potentials (Pantev et al., 1991). With higher

---

[1] Several studies have shown that abnormalities in synchrony and oscillatory activity occur in the lower frequency ranges (theta, alpha) as well.

rates of stimulus presentation the overlapping midlatency response (MLR) sums up to the auditory SSR. The entrainment of the EEG to externally driven rhythmic stimulation elicits the SSR that resonates at the stimulating frequency. The SSR is typically of largest amplitude when the stimulation is presented at 40 Hz. Lower or higher rates of stimulation produce a response of smaller amplitude (Galambos et al., 1981; Hari et al., 1989; Azzena et al., 1995; Brenner et al., 2003), suggesting that 40 Hz is a "preferred" working frequency of the auditory network, reflecting the increased phase-locking of individual trials (Artieda et al., 2004) and the synchronized activation of thalamocortical loops.

Patients with schizophrenia failed to show evoked EEG oscillations to steady-state auditory trains in the beta and gamma frequency range, but they were present at the lower frequencies (Kwon et al., 1999). Patients also exhibited delays in both the onset of phase synchronization and desynchronization in response to the auditory click trains. Light et al. (2006) extended these results by showing that the intertrial coherence was reduced in response to 30- and 40-Hz oscillations. In addition, they showed a modest correlation between reduced working memory performance (measured with the Letter–Number Sequencing Test) and the patients' 40-Hz intertrial phase synchronization. This suggests deficits in early sensory processing due to a failure in the entrainment of synchronized responses by higher-frequency stimuli. The reduction in intertrial synchrony was attributed to increased EEG response variability (Gallinat et al., 2004; Winterer and Weinberger, 2004; Winterer et al., 2004).

In addition to these deficits within auditory processing, there is also evidence of similar steady-state impairments in visual processing. Although the generators of the steady-state visual evoked potential (SSVEP) responses are still under investigation, several studies have shown the involvement of posterior cortical regions, primarily the occipital cortex. The SSVEP occurring at high temporal frequencies is thought to preferentially activate the magnocellular pathway, while responses at low frequencies are believed to activate the parvocellular pathway. Resonance of SSVEP is thought to reflect a combination of both resonances at local neural circuits and global resonance over the cortex (Silberstein et al., 1995). With the photic stimulation ranging from 4 to 40 Hz, patients again showed reduced signal power compared to healthy controls at frequencies above 17 Hz at occipital electrodes and higher levels of EEG noise during photic stimulation at frequencies below 20 Hz (Krishnan et al., 2005).

Further deficits in early sensory processing can be found in the auditory P50 event-related potential. The P50 click paradigm, where two auditory clicks follow each other rapidly, is used to investigate sensory gating. In healthy participants the P50 to the second click is suppressed compared to the P50 to first click, whereas patients with schizophrenia lack this P50 suppression.

Clementz et al. (1997) found a relationship between auditory early phase-locked gamma-band activity and P50 suppression and showed that the evoked gamma-band response may account for poor P50 suppression. Hong et al. (2004a) extended this finding by showing that reduced evoked beta-band activity to the first click was inversely correlated to the second click P50 response.

Finally, two studies examined the relationship between backward masking and gamma-band activity (Green et al., 2003; Wynn et al., 2005). Visual masking is a procedure that is used to assess the earliest components of visual processing. In backward masking, the identification of an initial stimulus (the target) is disrupted by a later stimulus (the mask). The masking function can be divided into an early component (e.g. up to about 60 ms) that reflects the involvement of sensory–perceptual processes, and a later component that reflects susceptibility to attentional disengagement as the mask diverts processing away from the representation of the target. Masking occurs when the transient channel activity elicited by the mask interrupts sustained channel processing elicited by the target. Patients with schizophrenia have consistently shown that they require longer intervals between the target and the mask to identify the target correctly and have shown a deficit for both the early and the late component. Green et al. (2003) showed that the control group, but not the patients, showed a burst of induced gamma-band activity 200–400 ms following target presentation. In contrast, patients exhibited an early gamma-band response (around 100 ms). In a follow-up study Wynn et al. (2005) replicated the finding of reduced gamma-band activity in patients compared to control participants.

Deficits in cognitive processing

There is also evidence for a reduction of gamma spectral power and phase synchronization during higher-level auditory and visual processing. Studies that have focused on cognitive deficits have used the auditory oddball paradigm (Haig et al., 2000; Lee et al., 2001; Gallinat et al., 2004; Slewa-Younan et al., 2004; Symond et al., 2005), tasks requiring visual binding (Spencer et al., 2003, 2004; Uhlhaas et al., 2006), and working memory (Basar-Eroglu et al., 2007). Studies using the auditory oddball paradigm reported a reduction of evoked gamma-band activity (1) in response to standards (Haig et al., 2000), (2) to novel targets associated with increased autonomic arousal (Lee et al., 2001), and (3) in a late latency range (220–350 ms) in response to targets in unmedicated schizophrenic patients (Gallinat et al., 2004).[2] Induced gamma-band activity was

---

[2] It is of interest to note that Gallinat et al. (2004) tested the effect of smoking in patients and found higher early gamma activity in response to targets in smokers compared to non-smokers.

reduced in response to targets over the left hemisphere and increased over the right hemisphere (Haig et al., 2000). Finally, reduced gamma-band phase synchronization was evident in response to the target in first-episode patients and chronic patients (Slewa-Younan et al., 2004; Symond et al., 2005).

In addition to deficits in the auditory oddball paradigm, there is also evidence for reduced visual evoked oscillatory activity in tasks requiring visual binding (Spencer et al., 2003, 2004; Uhlhaas et al., 2006). Spencer et al. (2003) used measures of phase-locking and phase coherence in response to gestalt stimuli (illusory Kanizsa triangles). Results showed reduced evoked gamma-band activity (around 80 ms) and abnormal and delayed onset of phase coherence in response to the illusory contours over visual cortices. In a recent study (Uhlhaas et al., 2006), we provided evidence for a close relationship between impaired neural synchrony in schizophrenia and specific cognitive deficits using "Mooney" face stimuli. Mooney faces consist of degraded pictures of human faces where all shades of gray are removed, leaving only black and white contours. Perception of Mooney faces requires the grouping of the fragmentary parts into coherent images. Schizophrenia patients exhibited a deficit in the perception of Mooney faces and reduced phase synchrony in the beta band (20–30 Hz) while the power of induced gamma-band oscillations was in the normal range. Finally, abnormal gamma-band activity has also been shown in response to working memory and executive function (Cho et al., 2006; Basar-Eroglu et al., 2007). Basar-Eroglu et al. (2007) used a working memory task in which three randomly selected numbers were presented in an N-back task. Controls showed a gradual increase of evoked gamma-band activity with increasing working memory load in the delay interval and following the current comparison stimulus. In contrast, patients showed high gamma-band activity independently of the working memory load. Using an executive control task Cho et al. (2006) showed that induced gamma-band activity was stronger in the high than in the low control condition over two frontal electrodes (AF8 and FC1) and that performance correlated with gamma-band activity in controls. By contrast, patients with schizophrenia did not show an increase in gamma-band activity during high control conditions at frontal electrodes and no correlation with performance.

The results across these studies suggest that schizophrenia patients may have abnormalities in establishing and maintaining both gamma spectral power and phase synchronization to simple auditory and visual stimuli. These abnormalities, in turn, may account for some of the perceptual and cognitive deficits in schizophrenia. It has recently been debated whether there is a relationship between abnormalities of early sensory processing and deficits in higher-order cognitive abilities like working memory (Haenschel et al., 2007). Since higher-order cognitive processes are largely dependent on the fidelity of information input from early sensory–perceptual stages of processing, it is likely that some of the cognitive dysfunctions are explained at least in part by sensory–perceptual deficits (Foxe *et al.*, 2005). Furthermore, some researchers have argued that early perceptual deficits should be seen as part of the reciprocal interactions between hierarchical levels of sensory systems (Friston, 2005) and that such a dysfunctional information integration (processing) system may affect higher-order processes.

The relationship between synchronous gamma-band activity and the symptoms of schizophrenia

A second approach to understanding the role of aberrant oscillatory gamma-band activity in schizophrenia is to examine their relationship with the diverse range of symptoms that occur in the disease. The symptoms can be classified as positive or negative. The former comprise hallucinations, delusions, thought disorder, and odd behavior and the latter include lack of content of speech, blunted affect, social withdrawal, lack of motivation, and diminished goal-directed behavior. There is some evidence that the abnormalities in the evoked and phase-synchronized high-frequency oscillations correlate with the Positive and Negative Syndrome Scale (PANSS) general symptom scores, with the subscales (positive, negative, general) and with specific items (delusions, conceptual disorganization, and poor attention) (Haig *et al.*, 2000; Spencer *et al.*, 2003; Uhlhaas *et al.*, 2006). For instance, psychomotor poverty has been associated with decreased synchrony, whereas reality distortion and disorganization coupled with increased gamma synchrony occurred in a sample of chronic patients with schizophrenia (Lee *et al.*, 2003).

Furthermore, some studies investigated the relationship between gamma-band oscillatory activity and hallucinations. Hallucinations are sensory perceptions without any external stimulation. Despite advances in neuroimaging, the physiological basis of abnormal human perceptions, such as hallucinations, has remained elusive. Recently, Behrendt and Young (2004) proposed that conscious perception is subserved by synchronization of gamma frequency oscillations in recurrent thalamocortical networks, and that hallucinations result from underconstrained activation in this circuitry. There are different ways to approach and measure hallucinations (Ford and Mathalon, 2005); to investigate the actual hallucinations, an underlying fundamental deficit, or disturbances in information processing resulting from hallucinations. Baldeweg *et al.* (1998) provided an example of symptom capture. In a single case study they associated increased somatic hallucinations with excessively high gamma power over somatosensory areas. A second example is the study by Lee *et al.* (2006). This study compared patients with a history of auditory hallucinations to patients without auditory hallucinations during rest using quantitative electroencephalography (qEEG) and low-resolution electromagnetic tomography (LORETA) source imaging. Results showed an increase in high-frequency oscillations frequency in the left inferior parietal lobule and the left medial frontal gyrus.

Ford and Mathalon (2005) used a fundamental deficit approach for the understanding of hallucinations. They hypothesized that a self-monitoring deficit reflects dysfunction of the efference copy/corollary discharge mechanism, which they related to auditory hallucinations. Results revealed that gamma-band coherence between frontal and temporal lobes was greater during talking than listening and was disrupted by distorted feedback during talking in healthy controls. In contrast, patients did not show this pattern for EEG gamma coherence. However, the relationship between neurobiological indicators of dysfunctional corollary discharge and the extent of current auditory hallucinations was not clear.

A third approach is the investigation of disturbed information processing secondary to hallucinations. Hallucinations tended for instance to reduce SSVEP responses (Krishnan *et al.*, 2005). Furthermore, Spencer *et al.* (2004) showed a correlation between the degree of gamma-band oscillations phase-locked to reaction time in response to gestalt stimuli and severity of visual hallucination in schizophrenia subjects. Hence, hallucinations may interrupt sensory input to the cortex. In summary, the evidence available points to the fact that depending on the symptoms high-frequency activity and synchrony may vary from reduced to augmented levels of activity. Pharmacological models of oscillatory activity may help to explain these differences in gamma-band oscillatory activity.

Neuropharmacological mechanisms underlying oscillatory activity

Several neurotransmitter systems have been linked to neuropharmacological changes in the brain in schizophrenia, the major ones being dopamine, N-methyl-D-aspartate (NMDA), gamma-aminobutyric acid (GABA), acetylcholine (ACh), and serotonine. For instance, both NMDA dysregulation and a defect of GABAergic interneuron disinhibition (Benes and Berretta, 2001) have been shown in schizophrenia. The question is how they link to current theories about the underlying mechanism of gamma-band activity and whether such dysfunctions might be informative for mechanisms that may be pharmacological targeted to treat cognitive dysfunctions in schizophrenia.

The precise mechanisms underlying the generation of oscillatory activity remain to be fully elucidated. There are however a numbers of potential theories that have been advanced, such as intrinsic oscillatory membrane properties of chattering cells (Gray and McCormick, 1996), thalamocortical resonant loops (Llinas and Ribary, 1993), and the emergence of oscillatory activity as a property of networks (Whittington et al., 1995). At a cellular and network level, cortical gamma-band activity can be generated by interaction of the glutamatergic pyramidal cells and GABAergic inhibitory interneurons (Traub and Whittington, 1999). There is, for instance, evidence that short-range gamma-band synchrony results from the mutual inhibition of GABAergic interneurons and long-range synchrony from the precise in-phase synchronization of interneuron networks and pyramidal cells (Traub et al., 2004). Furthermore, gamma-band oscillations can occur spontaneously during activation of metabotropic cholinergic receptors (Buhl et al., 1998), and can be induced transiently by activation of metabotropic glutamate receptors or by bursts of afferent stimulation (Whittington et al., 1995; Traub et al., 1999). Finally, there is some evidence that dopamine polymorphisms can modulate evoked oscillations as well (Demiralp et al., 2006).

Hence, the neuropharmacological mechanisms suggested for schizophrenia and underlying gamma-band oscillations bear some resemblance. Investigating the relationship between neuropharmacological mechanisms and high-frequency oscillatory activity can inform on both the underlying pathology of schizophrenia and the constraints on theories of high-frequency-band activity. In the following we will briefly link a specific system to schizophrenia, and then describe the effects on gamma-band oscillations and finally how these can be modulated by treatment.

Defects of GABAergic interneuron disinhibition (Benes and Berretta, 2001) have been shown in schizophrenia. Neuron counts revealed a deficit of interneurons in schizophrenia, which are mediated by $GABA_A$ receptor activity (Benes and Berretta, 2001). GABA antagonists are known to work as amnesic agents and may help to explain memory dysfunctions in schizophrenia. For instance, working-memory impairments have recently been linked to reduced inhibitory GABA neurons in the dorsolateral prefrontal cortex of individuals with schizophrenia (Lewis et al., 2005). The authors suggested that this may contribute to reduced gamma synchrony that is required for working memory function as shown by Cho et al. (2006). There is indeed evidence that $GABA_A$ is important for eliciting gamma-band oscillatory activity in inhibitory networks in vitro and that blocking $GABA_A$ receptor abolishes these oscillations (Whittington et al., 1995). Hence, the abnormalities of GABA neuronal networks in schizophrenic brains may contribute to a reduced gamma synchrony in these patients. Therefore, it has been proposed that an effective treatment of cognitive impairments in schizophrenia might result from targeted usage of GABA agonists (Lewis et al., 2005).

A role for NMDA receptor activity in schizophrenia is based on the finding that NMDA antagonists, such as ketamine or phencyclidine (PCP), mimic both the positive and negative symptoms of schizophrenia (Javitt and Zukin, 1991). Furthermore, altered gene expression of the NMDA receptor has been implicated by a prospective post-mortem study, which found the degree of neuropsychological impairment in everyday memory tested with the Rivermead Behavioral Memory Test (RBMT) (determined ante mortem) to be correlated with reduced mRNA for the NMDA receptor subunit NR1 in the temporal lobe (Humphries et al., 1996). Since NMDA-modulated GABAergic activity may play a critical role in gamma-band oscillatory activity, deficits in either of these receptors may result in aberrant cortical oscillatory activity. In vitro studies have demonstrated that the high-frequency oscillations were abolished by ketamine (Doheny et al., 2000). In combination with in vitro experiments, investigations into the nature of high-frequency oscillatory activity may clarify the role of NMDA/GABA activity in schizophrenia and may help to develop new treatments. For instance, it has been suggested that a viable therapeutic strategy may be the direct stimulation of the NMDA receptor via its glycine binding site (Heresco-Levy et al., 1996). Another strategy is the enhancement of NMDA-dependent plasticity using neuromodulators which act through pre- and postsynaptic mechanisms, such as acetylcholine, norepinephrine, and serotonin (Gu, 2002). The question is whether these strategies would normalize the abnormalities in high-frequency activity.

There are several lines of research indicating that the positive symptoms of schizophrenia are due to hyperactivity of the dopaminergic system in the mesolimbic pathway from midbrain ventral tegmental area (VTA) to limbic areas. In contrast, the negative and cognitive symptoms of schizophrenia are likely due to hypoactive dopamine transmission in the mesocortical pathway from midbrain to cerebral cortex. Dopamine agonists, such as amphetamine, tend to produce psychotic symptoms. Furthermore the efficacy of antipsychotic agents is thought to be related to their capacity to antagonize dopamine. Conversely, dopamine agonists do not produce marked formal thought disorder and negative symptoms (Javitt and Zukin, 1991; Krystal et al., 1994; Malhotra et al., 1996). Working memory impairments in schizophrenia have been related to a deficit in dopamine neurotransmission in the DLPFC (Weinberger, 1987). This is consistent with studies showing dopamine DA depletion in the monkey DLPFC markedly impairs working memory performance (Brozoski et al., 1979) and that working memory performance may be dependent on dopamine signaling through D1 receptors in the DLPFC D2 activity (Sawaguchi and Goldman-Rakic, 1991, 1994). It has further been suggested that dopamine/D1 signaling modulates the cortical signal-to-noise ratio by enhancing selective inputs to both pyramidal cells and inhibitory interneurons (Goldman-Rakic et al., 2000). However, it is not year clear how this relates to abnormalities in neuronal synchrony in schizophrenia. Winterer et al. (2006) examined the effect of a functional single nucleotide polymorphism (val(108/158)met) within the catechol-O-methyltransferase (COMT) gene which is involved in cortical synaptic dopamine metabolism on prefrontal "noise" measured with ERPs during an auditory oddball task. They found that this polymorphism was significantly associated with prefrontal "noise." In addition, there is some evidence that dopamine polymorphisms can modulate gamma-band oscillations in healthy human subjects (Demiralp et al., 2006).

Finally, alterations in cortical cholinergic systems have been found in patients with schizophrenia, such as reduced number of nicotinic and muscarinic acetylcholine receptors (reviewed in Friedman, 2004). Acetylcholine is a particularly important modulator of NMDA receptor function. Furthermore, the nicotinergic $\alpha 7$ receptor gene, a potential susceptibility gene for schizophrenia (Martin et al., 2004), is expressed strongly on glutamatergic synapses in human and rodent isocortex and hippocampus, both pre- and postsynaptically (Fabian-Fine et al., 2001). It has been suggested that schizophrenia patients use smoking as self-medication for symptom relief (Masterson and O'Shea, 1984; Lasser et al., 2000). As described above Gallinat et al. (2004) found higher early gamma-band activity in response to targets in smokers compared to non-smokers in patients and there is some evidence for an increase in amplitude and a decrease in latency in sensory gating in the gamma-band frequency range in healthy smokers (Crawford et al., 2002). Cholinergic modulation plays a role in the fast, state-dependent facilitation of gamma-band oscillations and in associated response synchronization as well as use-dependent long-term modification of cortical synchronization that favors synchronization of responses in the gamma frequency range (Rodriguez et al., 2004). Hence, impairments in "learning" and the development of cortical networks agree very well with the assumption that schizophrenia is a neurodevelopmental disorder. Indeed, cholinergic agonists such as nicotine have been proposed as a promising new treatment avenue for schizophrenia (Friedman, 2004; Martin et al., 2004). Acute administration of nicotine to patients with schizophrenia transiently improves their perceptual processing (sensory gating: Adler et al., 1993), attention (Harris et al., 2004), working memory (Jacobsen et al., 2004), and long-term memory performance (Myers et al., 2004).

Taken together, the balance between the different neurotransmitter systems within cortical networks is responsible for cognitive functioning.

Conclusion

The evidence reviewed in this chapter suggests that measures of neural high-frequency oscillatory spectral power and phase-locking may reflect a core deficit in the coordination of neural activity that underlies the specific cognitive dysfunctions associated with schizophrenia. Existing data point to the importance of evoked oscillatory activity associated with dysfunctions in the entrainment to sensory stimuli, intertrial phase-locking associated with an increase in "cortical noise," and inter-electrode phase synchronization measuring dysfunctions in long-range synchronization for the understanding of aberrant cortical oscillatory activity in schizophrenia.

The data reviewed here suggest that cognitive impairments in patients with schizophrenia correlate with deficits in high-frequency synchrony. These deficits in temporal integration in schizophrenia are not a unique phenomenon, but have been implicated in other neurological and psychiatric disorders as well (for review see Uhlhaas and Singer, 2006). Synchronization of neuronal activity within and across different brain regions is a fundamental property of cortical and subcortical networks that serves a variety of functions in cognitive processes. Future research needs to investigate the distinct and overlapping properties of altered synchronous oscillations and their relationship with cognitive abnormalities in specific disorders.

In this chapter we have concentrated on event-related high-frequency oscillatory activity. However, there is plenty of evidence of the contributions of lower frequencies (theta- and alpha-band activity) in response to cognitive tasks as well. For instance, theta activity has been related to working memory processes (Gevins et al., 1997; Jensen and Tesche, 2002) and there is evidence of impaired theta-band activity in patients with schizophrenia (Schmiedt et al., 2005). In healthy participants a relationship between both measures of power and phase in the gamma band and theta band has been established. Transient coupling between theta- and gamma-band activity seems to coordinate activity in distributed cortical areas and may provide a mechanism for communication during cognitive processing (Schack et al., 2002; Sederberg et al., 2003; Canolty et al., 2006; Demiralp et al., 2007). This relationship between different frequency bands and their functional significance in patients with schizophrenia needs to be further investigated.

The data reviewed in this chapter indicate that measures of neural synchrony may be very useful for the diagnosis of cognitive dysfunction. There is also some evidence for a symptom specific change of high-frequency synchrony: whereas the negative symptoms correlate with reduced gamma-band activity, the positive symptoms correlate with increased gamma-band activity. Future studies will show whether synchronous oscillatory activity can provide a suitable endophenotypic marker for the study of candidates risk genes, in particular those that affect neuromodulation. Neurotransmitter alterations may directly interfere with the mechanisms that support synchronization of neuronal responses. Hence, the relationship between schizophrenia and aberrant synchronization may in the future aid the diagnosis of schizophrenia and may advance the development of neuropharmacological drugs that target specific symptoms of the disorder. For instance, one of the major aims in future research will be to develop treatments that target the cognitive impairments in schizophrenia. It will be important to establish whether novel treatments with improved pharmacological profiles and new atypical antipsychotic drugs can ameliorate those cognitive deficits as well as the abnormal synchrony found in this disorder. An additional question is the degree to which high-frequency measures can be used as biomarkers for assessing medication response (Braff and Light, 2004). The (long-lasting) effects of medication and their impact on oscillatory activity are not well understood, but the current evidence indicates that the new generation antipsychotic medications (for instance clozapine, olanzapine, risperidone) may elevate gamma-band synchronization (Hong *et al.*, 2004b).

The data reviewed here confirm that high-frequency oscillatory activity may be a useful tool in elucidating the neural effects of such treatments and in relating them to the profile of cognitive and aberrant high-frequency oscillatory activity observed in patients.

References

Adler, L. E., Hoffer, L. D., Wiser, A., and Freedman, R. (1993). Normalization of auditory physiology by cigarette smoking in schizophrenic patients. *Am J Psychiat* **150**:1856–1861.

Andreasen, N. C. (1999). A unitary model of schizophrenia: Bleuler's "fragmented phrene" as schizencephaly. *Arch Gen Psychiat* **56**:781–787.

Artieda, J., Valencia, M., Alegre, M., *et al.* (2004). Potentials evoked by chirp-modulated tones: a new technique to evaluate oscillatory activity in the auditory pathway. *Clin Neurophysiol* **115**:699–709.

Awh, E., Barton, B., and Vogel, E. K. (2007). Visual working memory represents a fixed number of items regardless of complexity. *Psychol Sci* **18**:622–628.

Azzena, G. B., Conti, G., Santarelli, R., *et al.* (1995). Generation of human auditory steady-state responses (SSRs). I. Stimulus rate effects. *Hear Res* **83**:1–8.

Baldeweg, T., Spence, S., Hirsch, S. R., and Gruzelier, J. (1998). Gamma-band electroencephalographic oscillations in a patient with somatic hallucinations. *Lancet* **352**:620–621.

Basar, E. (1980). *EEG-Brain Dynamics: Relation between EEG and Brain Evoked Potentials.* Amsterdam, the Netherlands: Elsevier.

Basar-Eroglu, C., Brand, A., Hildebrandt, H., et al. (2007). Working memory related gamma oscillations in schizophrenia patients. *Int J Psychophysiol* **64**:39–45.

Behrendt, R. P. and Young, C. (2004). Hallucinations in schizophrenia, sensory impairment, and brain disease: a unifying model. *Behav Brain Sci* **27**:771–787; discussion 787–830.

Benes, F. M. and Berretta, S. (2001). GABAergic interneurons: implications for understanding schizophrenia and bipolar disorder. *Neuropsychopharmacology* **25**:1–27.

Bleuler, E. (1950). *Dementia Praecox or the Group of Schizophrenias*. New York: International University Press.

Braff, D. L. and Light, G. A. (2004). Preattentional and attentional cognitive deficits as targets for treating schizophrenia. *Psychopharmacology* **174**:75–85.

Brenner, C. A., Sporns, O., Lysaker, P. H., and O'Donnell, B. F. (2003). EEG synchronization to modulated auditory tones in schizophrenia, schizoaffective disorder, and schizotypal personality disorder. *Am J Psychiat* **160**:2238–2240.

Brozoski, T. J., Brown, R. M., Rosvold, H. E., and Goldman, P. S. (1979). Cognitive deficit caused by regional depletion of dopamine in prefrontal cortex of rhesus monkey. *Science* **205**:929–932.

Buhl, E. H., Tamas, G., and Fisahn, A. (1998). Cholinergic activation and tonic excitation induce persistent gamma oscillations in mouse somatosensory cortex in vitro. *J Physiol* **513**:117–126.

Canolty, R. T., Edwards, E., Dalal, S. S., et al. (2006). High gamma power is phase-locked to theta oscillations in human neocortex. *Science* **313**:1626–1628.

Cho, R. Y., Konecky, R. O., and Carter, C. S. (2006). Impairments in frontal cortical gamma synchrony and cognitive control in schizophrenia. *Proc Natl Acad Sci USA* **103**:19878–19883.

Clementz, B. A., Blumenfeld, L. D., and Cobb, S. (1997). The gamma band response may account for poor P50 suppression in schizophrenia. *Neuroreport* **8**:3889–3893.

Crawford, H. J., McClain-Furmanski, D., Castagnoli, N., Jr., and Castagnoli, K. (2002). Enhancement of auditory sensory gating and stimulus-bound gamma band (40 Hz) oscillations in heavy tobacco smokers. *Neurosci Lett* **317**:151–155.

Davidson, M., Reichenberg, A., Rabinowitz, J., et al. (1999). Behavioral and intellectual markers for schizophrenia in apparently healthy male adolescents. *Am J Psychiat* **156**:1328–1335.

Demiralp, T., Herrmann, C. S., Erdal, M. E., et al. (2006). DRD4 and DAT1 polymorphisms modulate human gamma band responses. *Cereb Cortex* **17**:1007–1019.

Demiralp, T., Bayraktaroglu, Z., Lenz, D., et al. (2007). Gamma amplitudes are coupled to theta phase in human EEG during visual perception. *Int J Psychophysiol* **64**:24–30.

Doheny, H. C., Faulkner, H. J., Gruzelier, J. H., Baldeweg, T., and Whittington, M. A. (2000). Pathway-specific habituation of induced gamma oscillations in the hippocampal slice. *Neuroreport* **11**:2629–2633.

Engel, A. K., Konig, P., Kreiter, A. K., Schillen, T. B., and Singer, W. (1992). Temporal coding in the visual cortex: new vistas on integration in the nervous system. *Trends Neurosci* **15**:218–226.

Fabian-Fine, R., Skehel, P., Errington, M. L., et al. (2001). Ultrastructural distribution of the alpha7 nicotinic acetylcholine receptor subunit in rat hippocampus. *J Neurosci* **21**:7993–8003.

Ford, J. M. and Mathalon, D. H. (2005). Corollary discharge dysfunction in schizophrenia: can it explain auditory hallucinations? *Int J Psychophysiol* **58**:179–189.

Foxe, J. J., Murray, M. M., and Javitt, D. C. (2005). Filling-in in schizophrenia: a high-density electrical mapping and source-analysis investigation of illusory contour processing. *Cereb Cortex* **15**:1914–1927.

Friedman, J. I. (2004). Cholinergic targets for cognitive enhancement in schizophrenia: focus on cholinesterase inhibitors and muscarinic agonists. *Psychopharmacology* **174**:45–53.

Fries, P., Reynolds, J. H., Rorie, A. E., and Desimone, R. (2001). Modulation of oscillatory neuronal synchronization by selective visual attention. *Science* **291**:1560–1563.

Fries, P., Nikolic, D., and Singer, W. (2007). The gamma cycle. *Trends Neurosci* **30**:309–316.

Friston, K. J. (1999). Schizophrenia and the disconnection hypothesis. *Acta Psychiatr Scand (Suppl)* **395**:68–79.

Friston, K. (2005). Disconnection and cognitive dysmetria in schizophrenia. *Am J Psychiat* **162**:429–432.

Fuster, J. M. (1997). Network memory. *Trends Neurosci* **20**:451–459.

Galambos, R. (1992). A comparison of certain gamma band (40-Hz) brain rhythms in cat and man. In: *Induced Rhythms in the Brain*, ed. Basar, E. and Bullock, T. H., pp. 201–216. Boston, MA: Birkhauser.

Galambos, R., Makeig, S., and Talmachoff, P. J. (1981). A 40-Hz auditory potential recorded from the human scalp. *Proc Natl Acad Sci USA* **78**:2643–2647.

Gallinat, J., Winterer, G., Herrmann, C. S., and Senkowski, D. (2004). Reduced oscillatory gamma-band responses in unmedicated schizophrenic patients indicate impaired frontal network processing. *Clin Neurophysiol* **115**:1863–1874.

Gevins, A., Smith, M. E., McEvoy, L., and Yu, D. (1997). High-resolution EEG mapping of cortical activation related to working memory: effects of task difficulty, type of processing, and practice. *Cereb Cortex* **7**:374–385.

Goldman-Rakic, P. S., Muly, E. C., 3rd, and Williams, G. V. (2000). D(1) receptors in prefrontal cells and circuits. *Brain Res Brain Res Rev* **31**:295–301.

Gray, C. M. and McCormick, D. A. (1996). Chattering cells: superficial pyramidal neurons contributing to the generation of synchronous oscillations in the visual cortex. *Science* **274**:109–113.

Green, M. F. (1996). What are the functional consequences of neurocognitive deficits in schizophrenia? *Am J Psychiat* **153**:321–330.

Green, M. F., Mintz, J., Salveson, D., et al. (2003). Visual masking as a probe for abnormal gamma range activity in schizophrenia. *Biol Psychiat* **53**:1113–1119.

Gruber, T., Muller, M. M., Keil, A., and Elbert, T. (1999). Selective visual–spatial attention alters induced gamma band responses in the human EEG. *Clin Neurophysiol* **110**:2074–2085.

Gu, Q. (2002). Neuromodulatory transmitter systems in the cortex and their role in cortical plasticity. *Neuroscience* **111**:815–835.

Haenschel, C., Bittner, R. A., Haertling, F., *et al.* (2007). Contribution of impaired early-stage visual processing to working memory dysfunction in adolescents with schizophrenia: a study with event-related potentials and functional magnetic resonance imaging. *Arch Gen Psychiat* **64**:1229–1240.

Haig, A. R., Gordon, E., De Pascalis, V., *et al.* (2000). Gamma activity in schizophrenia: evidence of impaired network binding? *Clin Neurophysiol* **111**:1461–1468.

Hari, R., Hamalainen, M., and Joutsiniemi, S. L. (1989). Neuromagnetic steady-state responses to auditory stimuli. *J Acoust Soc Am* **86**:1033–1039.

Harris, J. G., Kongs, S., Allensworth, D., *et al.* (2004). Effects of nicotine on cognitive deficits in schizophrenia. *Neuropsychopharmacology* **29**:1378–1385.

Heresco-Levy, U., Silipo, G., and Javitt, D. C. (1996). Glycinergic augmentation of NMDA receptor-mediated neurotransmission in the treatment of schizophrenia. *Psychopharmacol Bull* **32**:731–740.

Herrmann, C. S., Mecklinger, A., and Pfeifer, E. (1999). Gamma responses and ERPs in a visual classification task. *Clin Neurophysiol* **110**:636–642.

Herrmann, C. S., Munk, M. H., and Engel, A. K. (2004). Cognitive functions of gamma-band activity: memory match and utilization. *Trends Cogn Sci* **8**:347–355.

Hong, L. E., Summerfelt, A., McMahon, R. P., Thaker, G. K., and Buchanan, R. W. (2004a). Gamma/beta oscillation and sensory gating deficit in schizophrenia. *Neuroreport* **15**:155–159.

Hong, L. E., Summerfelt, A., McMahon, R., *et al.* (2004b). Evoked gamma band synchronization and the liability for schizophrenia. *Schizophr Res* **70**:293–302.

Humphries, C., Mortimer, A., Hirsch, S., and de Belleroche, J. (1996). NMDA receptor mRNA correlation with antemortem cognitive impairment in schizophrenia. *Neuroreport* **7**:2051–2055.

Jacobsen, L. K., D'Souza, D. C., Mencl, W. E., *et al.* (2004). Nicotine effects on brain function and functional connectivity in schizophrenia. *Biol Psychiat* **55**:850–858.

Javitt, D. C. and Zukin, S. R. (1991). Recent advances in the phencyclidine model of schizophrenia. *Am J Psychiatry* **148**:1301–1308.

Jensen, O. and Tesche, C. D. (2002). Frontal theta activity in humans increases with memory load in a working memory task. *Eur J Neurosci* **15**:1395–1399.

Kopell, N., Ermentrout, G. B., Whittington, M. A., and Traub, R. D. (2000). Gamma rhythms and beta rhythms have different synchronization properties. *Proc Natl Acad Sci USA* **97**:1867–1872.

Krishnan, G. P., Vohs, J. L., Hetrick, W. P., *et al.* (2005). Steady state visual evoked potential abnormalities in schizophrenia. *Clin Neurophysiol* **116**:614–624.

Krystal, J. H., Karper, L. P., Seibyl, J. P., *et al.* (1994). Subanesthetic effects of the noncompetitive NMDA antagonist, ketamine, in humans: psychotomimetic, perceptual, cognitive, and neuroendocrine responses. *Arch Gen Psychiat* **51**:199–214.

Kwon, J. S., O'Donnell, B. F., Wallenstein, G. V., et al. (1999). Gamma frequency-range abnormalities to auditory stimulation in schizophrenia. *Arch Gen Psychiat* **56**:1001–1005.

Lachaux, J. P., Rodriguez, E., Martinerie, J., and Varela, F. J. (1999). Measuring phase synchrony in brain signals. *Hum Brain Map* **8**:194–208.

Lasser, K., Boyd, J. W., Woolhandler, S., et al. (2000). Smoking and mental illness: A population-based prevalence study. *J Am Med Assoc* **284**:2606–2610.

Lee, K. H., Williams, L. M., Haig, A., Goldberg, E., and Gordon, E. (2001). An integration of 40 Hz Gamma and phasic arousal: novelty and routinization processing in schizophrenia. *Clin Neurophysiol* **112**:1499–1507.

Lee, K. H., Williams, L. M., Breakspear, M., and Gordon, E. (2003). Synchronous gamma activity: a review and contribution to an integrative neuroscience model of schizophrenia. *Brain Res Brain Res Rev* **41**:57–78.

Lee, S. H., Wynn, J. K., Green, M. F., et al. (2006). Quantitative EEG and low resolution electromagnetic tomography (LORETA) imaging of patients with persistent auditory hallucinations. *Schizophr Res* **83**:111–119.

Lewis, D. A., Hashimoto, T., and Volk, D. W. (2005). Cortical inhibitory neurons and schizophrenia. *Nat Rev Neurosci* **6**:312–324.

Light, G. A., Hsu, J. L., Hsieh, M. H., et al. (2006). Gamma band oscillations reveal neural network cortical coherence dysfunction in schizophrenia patients. *Biol Psychiat* **60**:1231–1240.

Llinas, R. and Ribary, U. (1993). Coherent 40-Hz oscillation characterizes dream state in humans. *Proc Natl Acad Sci USA* **90**:2078–2081.

Malhotra, A. K., Pinals, D. A., Weingartner, H., et al. (1996). NMDA receptor function and human cognition: the effects of ketamine in healthy volunteers. *Neuropsychopharmacology* **14**:301–307.

Martin, L. F., Kem, W. R., and Freedman, R. (2004). Alpha-7 nicotinic receptor agonists: potential new candidates for the treatment of schizophrenia. *Psychopharmacology* **174**:54–64.

Masterson, E. and O'Shea, B. (1984). Smoking and malignancy in schizophrenia. *Br J Psychiat* **145**:429–432.

Mesulam, M. (1994). Neurocognitive networks and selectively distributed processing. *Rev Neurol* **150**:564–569.

Myers, C. S., Robles, O., Kakoyannis, A. N., et al. (2004). Nicotine improves delayed recognition in schizophrenic patients. *Psychopharmacology* **174**:334–340.

Pantev, C. (1995). Evoked and induced gamma-band activity of the human cortex. *Brain Topogr* **7**:321–330.

Pantev, C., Makeig, S., Hoke, M., et al. (1991). Human auditory evoked gamma-band magnetic fields. *Proc Natl Acad Sci USA* **88**:8996–9000.

Phillips, W. A. and Silverstein, S. M. (2003). Convergence of biological and psychological perspectives on cognitive coordination in schizophrenia. *Behav Brain Sci* **26**:65–82; discussion 82–137.

Rodriguez, E., George, N., Lachaux, J. P., et al. (1999). Perception's shadow: long-distance synchronization of human brain activity. *Nature* **397**:430–433.

Rodriguez, R., Kallenbach, U., Singer, W., and Munk, M. H. (2004). Short- and long-term effects of cholinergic modulation on gamma oscillations and response synchronization in the visual cortex. *J Neurosci* **24**:10369–10378.

Sawaguchi, T. and Goldman-Rakic, P. S. (1991). D1 dopamine receptors in prefrontal cortex: involvement in working memory. *Science* **251**:947–950.

Sawaguchi, T. and Goldman-Rakic, P. S. (1994). The role of D1-dopamine receptor in working memory: local injections of dopamine antagonists into the prefrontal cortex of rhesus monkeys performing an oculomotor delayed-response task. *J Neurophysiol* **71**:515–528.

Saykin, A. J., Shtasel, D. L., Gur, R. E., et al. (1994). Neuropsychological deficits in neuroleptic naive patients with first-episode schizophrenia. *Arch Gen Psychiat* **51**:124–131.

Schack, B., Vath, N., Petsche, H., Geissler, H. G., and Moller, E. (2002). Phase-coupling of theta-gamma EEG rhythms during short-term memory processing. *Int J Psychophysiol* **44**:143–163.

Schmiedt, C., Brand, A., Hildebrandt, H., and Basar-Eroglu, C. (2005). Event-related theta oscillations during working memory tasks in patients with schizophrenia and healthy controls. *Brain Res Cogn Brain Res* **25**:936–947.

Schnitzler, A. and Gross, J. (2005). Normal and pathological oscillatory communication in the brain. *Nat Rev Neurosci* **6**:285–296.

Sederberg, P. B., Kahana, M. J., Howard, M. W., Donner, E. J., and Madsen, J. R. (2003). Theta and gamma oscillations during encoding predict subsequent recall. *J Neurosci* **23**:10809–10814.

Silberstein, R. B., Ciorciari, J., and Pipingas, A. (1995). Steady-state visually evoked potential topography during the Wisconsin card sorting test. *Electroencephalogr Clin Neurophysiol* **96**:24–35.

Singer, W. (1999). Neuronal synchrony: a versatile code of the definition of relations? *Neuron* **24**:49–65.

Singer, W. and Gray, C. M. (1995). Visual feature integration and the temporal correlation hypothesis. *Annu Rev Neurosci* **18**:555–586.

Slewa-Younan, S., Gordon, E., Harris, A. W., et al. (2004). Sex differences in functional connectivity in first-episode and chronic schizophrenia patients. *Am J Psychiat* **161**:1595–1602.

Spencer, K. M., Nestor, P. G., Niznikiewicz, M. A., et al. (2003). Abnormal neural synchrony in schizophrenia. *J Neurosci* **23**:7407–7411.

Spencer, K. M., Nestor, P. G., Perlmutter, R., et al. (2004). Neural synchrony indexes disordered perception and cognition in schizophrenia. *Proc Natl Acad Sci USA* **101**:17288–17293.

Symond, M. P., Harris, A. W., Gordon, E., and Williams, L. M. (2005). "Gamma synchrony" in first-episode schizophrenia: a disorder of temporal connectivity? *Am J Psychiat* **162**:459–465.

Tallon-Baudry, C. and Bertrand, O. (1999). Oscillatory gamma activity in humans and its role in object representation. *Trends Cogn Sci* **3**:151–162.

Tallon-Baudry, C., Bertrand, O., Peronnet, F., and Pernier, J. (1998). Induced gamma-band activity during the delay of a visual short-term memory task in humans. *J Neurosci* **18**:4244–4254.

Tallon-Baudry, C., Mandon, S., Freiwald, W.A., and Kreiter, A.K. (2004). Oscillatory synchrony in the monkey temporal lobe correlates with performance in a visual short-term memory task. *Cereb Cortex* **14**:713–720.

Tiitinen, H., Sinkkonen, J., Reinikainen, K., *et al.* (1993). Selective attention enhances the auditory 40-Hz transient response in humans. *Nature* **364**:59–60.

Traub, R.D. and Whittington, M.A. (1999). *Fast Oscillations in Cortical Circuits*. Cambridge, MA: MIT Press.

Traub, R.D., Whittington, M.A., Buhl, E.H., Jefferys, J.G. and Faulkner, H.J. (1999). On the mechanism of the gamma → beta frequency shift in neuronal oscillations induced in rat hippocampal slices by tetanic stimulation. *J Neurosci* **19**:1088–1105.

Traub, R.D., Bibbig, A., LeBeau, F.E., Buhl, E.H., and Whittington, M.A. (2004). Cellular mechanisms of neuronal population oscillations in the hippocampus in vitro. *Annu Rev Neurosci* **27**:247–278.

Uhlhaas, P.J. and Singer, W. (2006). Neural synchrony in brain disorders: relevance for cognitive dysfunctions and pathophysiology. *Neuron* **52**:155–168.

Uhlhaas, P.J., Linden, D.E., Singer, W., *et al.* (2006). Dysfunctional long-range coordination of neural activity during gestalt perception in schizophrenia. *J Neurosci* **26**:8168–8175.

van der Stelt, O., Belger, A., and Lieberman, J.A. (2004). Macroscopic fast neuronal oscillations and synchrony in schizophrenia. *Proc Natl Acad Sci USA* **101**:17567–17568.

Varela, F., Lachaux, J.P., Rodriguez, E., and Martinerie, J. (2001). The brain web: phase synchronization and large-scale integration. *Nat Rev Neurosci* **2**:229–239.

Vernon, D., Haenschel, C., Dwivedi, P., and Gruzelier, J. (2005). Slow habituation of induced gamma and beta oscillations in association with unreality experiences in schizotypy. *Int J Psychophysiol* **56**:15–24.

von Stein, A. and Sarnthein, J. (2000). Different frequencies for different scales of cortical integration: from local gamma to long range alpha/theta synchronization. *Int J Psychophysiol* **38**:301–313.

Weinberger, D.R. (1987). Implications of normal brain development for the pathogenesis of schizophrenia. *Arch Gen Psychiat* **44**:660–669.

Whittington, M.A., Traub, R.D., and Jefferys, J.G. (1995). Synchronized oscillations in interneuron networks driven by metabotropic glutamate receptor activation. *Nature* **373**:612–615.

Winterer, G. and Weinberger, D.R. (2004). Genes, dopamine and cortical signal-to-noise ratio in schizophrenia. *Trends Neurosci* **27**:683–690.

Winterer, G., Coppola, R., Goldberg, T.E., *et al.* (2004). Prefrontal broadband noise, working memory, and genetic risk for schizophrenia. *Am J Psychiat* **161**:490–500.

Winterer, G., Egan, M. F., Kolachana, B. S., *et al.* (2006). Prefrontal electrophysiologic "noise" and catechol-*O*-methyltransferase genotype in schizophrenia. *Biol Psychiat* **60**:578–584.

Wynn, J. K., Light, G. A., Breitmeyer, B., Nuechterlein, K. H., and Green, M. F. (2005). Event-related gamma activity in schizophrenia patients during a visual backward-masking task. *Am J Psychiat* **162**:2330–2336.

Yuval-Greenberg, S. and Deouell, L. Y. (2007). What you see is not (always) what you hear: induced gamma band responses reflect cross-modal interactions in familiar object recognition. *J Neurosci* **27**:1090–1096.

Part VI SUMMARY, CONCLUSION, AND FUTURE TARGETS

18

Summary of chapters, conclusion, and future targets

CHRISTIAN HÖLSCHER AND MATTHIAS MUNK

Summary of chapters

The different chapters discuss a wide range of concepts, techniques, and strategies of how to investigate the issue of information encoding in neuronal populations. The diversity clearly shows that the question is of central interest, that there is a lively competition of ideas and concepts, and that it is now possible to address the issue adequately by using new technology, the lack of which hampered discoveries in the past. The multileveled concepts described here show that there will not be a simple model or coding mechanism that can adequately explain how the brain works.

Part II: Organization of neuronal activity in neuronal populations

In this section, discussion focused on what general rules and physiological processes are in place that govern information encoding, processing, network formation, and laying down of memory traces.

In Chapter 2, Edward Mann and Ole Paulsen gave a detailed overview of the cellular mechanisms that underlie the establishment of oscillating networks. The fact that a multitude of specialized ion channels, interneuron subtypes, and neuronal projections are in place to establish defined oscillations in a controlled way clearly supports their concept that this mechanism is important for organizing brain processes. This point was also well illustrated by Brian Bland in Chapter 12, where he showed which "purpose-built" basal brain nuclei and

Information Processing by Neuronal Populations, ed. Christian Hölscher and Matthias Munk.
Published by Cambridge University Press. © Cambridge University Press 2009.

projections are involved in the induction and control of theta oscillations in areas throughout the brain.

Mann and Paulsen showed that neurons can have intrinsic biophysical properties that are determined by the type and distribution of ion channels, and which give them defined oscillatory properties. These neurons are tuned to specific frequencies which not only support the induction of defined frequencies but also buffer or filter out others. In addition, the connections between neurons can be tuned to support specific frequencies. Electrical synaptic coupling is one way to enable the transmission of high-frequency oscillations. To control spike timing, specific interneurons appear to control excitatory spiking by inhibitory synaptic activity. Inhibitory and excitatory neurons can be coupled to form feedback loops which can establish oscillations in a fast and well-controlled fashion. Such networks in the hippocampus and entorhinal cortex have the capacity to intrinsically generate network oscillations at theta and gamma frequencies. Preferred frequencies can easily be built into the network by choosing the right parameters of axonal length and synaptic transmission speed. Mann and Paulsen then described specific basal brain projections to the hippocampus that act as pacemakers and control points for theta induction. They showed that one of the functions of this elaborate system of inducing oscillations is to control spike timing, which is crucial for inducing synaptic plasticity. This Hebbian long-term change of synaptic weights then can store sequences of neuronal activities. This point was also made and illustrated further in Chapters 4, 7, 15, and 16, and others.

Kenneth Harris in Chapter 3 discussed the issues of Hebbian synaptic plasticity and how it could support the formation of cell assemblies which encode and compute information. Assemblies could be connected to form cascades and these cascades of assemblies could represent steps of serial computational sequences. He proposed that the function of such assemblies of neurons is not solely to represent sensory input, but also to compute the information, and to take part in internal cognitive processes. He then listed a set of four characteristics or "signatures," that assemblies would have to show if they are the building blocks of cognitive processes. For example, he suggested that the reason why spike trains vary widely in response to identical stimuli is not simply noise and inherent unreliability of the system, but a reflection of the fact that these neurons are engaged in other computations in addition to the representation of the available information.

His main line of argument is that the observed spike timing precision (e.g. of place cells in the hippocampus) has been interpreted as evidence for temporal coding, but that there can be other explanations. He suggests that there is no need for phase coding, and that the sequences of cell assembly activity can encode the information by itself, without the need for precise spike-timing control. He goes on to point out that this system would not need to decode the phase codes at target neurons. He also describes several observations that suggest that phase coding is at least not always present in stimulus-evoked neuronal responses. One example is the fact that neurons can fire even after presentation of stimuli, which to him indicates that these neuronal assemblies are involved in internal processing of information, and not solely in the encoding of sensory input. He also points out that temporal order is not always detectable in neurons that receive sensory input. In the olfactory system, downstream neurons do not seem to be "receptive" to temporal information contained in the activity of primary sensory neurons that project onto them, and that these secondary neurons simply fire when the required amount of excitatory input is reached (however, for a detailed account of the olfactory system, see Chapter 11).

Harris further listed a string of arguments and experimental observations to support the concept that phase coding is not inherently required for information encoding, and that his proposed system of cell assemblies could be capable of information processing without the need for precise control of spike timing.

Robert E. Hampson and Sam A. Deadwyler described in Chapter 4 the properties of single neurons in the hippocampus during the performance of a complex task. They pointed out that the information conveyed by these neurons not only correlates with temporal sequences of a delayed matching-to-sample memory task, and not only correlates with the spatial position of the rat in space, but also appears to reflect complex computations that suggest that they are involved in decision-making. They also showed that representation of information is task selective and does not simply mirror all available sensory information of the situation. Furthermore, neuronal activity not merely mirrors task critical information, but predicts future behavioral choices made by the animal. The authors interpret these findings in such a way that hippocampal networks are directly involved in decision-making processes that control the behavior of the animal. To support this point they stated that neuronal activity can predict a correct choice or a wrong choice made by the animal. This direct relationship between neuronal activity and behavior makes it very likely that these neurons are part of a planning and decision-making system.

In a series of elegant experiments using arrays of electrodes to record from multiple neurons in the hippocampus, they showed that in a delayed matching-to-sample task, neurons can be identified that encode different parts of the overall task. Some neurons encode simple aspects such as the type of stimulus shown (match vs. nonmatch), or the left lever position versus the right lever position. Other cells combine several task-specific aspects to form more complex representations, e.g. neurons that only fire during the left nonmatch stimulus presentation, but not during the left match or right nonmatch situation. They identified ten classes of neuronal responses in this task that describe 90% of all cells found. The complete information of the task is represented in the overall network of all of these cells. When analyzing where these cells are located in the hippocampus, they found clusters of neurons that perform similar tasks. They assumed that information encoding in the hippocampus therefore is not following a sparse code, but a dense (population) code which distributes the information over a group of neurons. However, by pointing out what information is lost when ensembles were assembled by arbitrary selection spike trains, they demonstrate the importance of spatiotemporal patterns (i.e. across neurons and time) and the capacity of simultaneously recorded ensembles to encode additional information, which is more consistent with a sparse code. The authors conclude that hippocampal neurons encode task-specific information via a code that shares the features of both dense and sparsely distributed models. When analyzing neuronal ensemble activity further, they showed that in single trials, a correlation can be drawn between ensemble patterns recorded during correctly performed trials and during trials where an error was made by the animal. Patterns observed during the sample phase will predict the animal's behavior during the nonmatch presentation phase. This finding demonstrates that hippocampal neurons not only encode task-specific information, but also information reflecting cognitive processes that are part of the planning and decision-making networks. The information encoded in the networks is processed to compute behavioral decisions and target-oriented behavioral programs.

Their finding also demonstrates the importance of multi-cell recording techniques of population activity, since the sum of individual neuronal spike trains did not contain the same information as the ensemble activity of simultaneously recorded neurons.

Adam Johnson, Jadin C. Jackson, and A. David Redish discussed this point in Chapter 5 on an even more detailed level. Using computational models to analyze the information content of neuronal populations, they demonstrated that neural activity can represent many types of information – from sensory descriptions of the world to motor planning for behavior and even to the cognitive processes in between. They describe mathematical methods for examining highly dynamic cognitive processes through observation of neural representations with multiple dynamics. Reconstruction alone cannot be used to infer internal states of an animal's sensory and cognitive networks such as the difference between random firing and well-represented variables. This is particularly important when considering issues of memory and recall. Similar to Harris (Chapter 3) and Hampson and Deadwyler (Chapter 4), the authors of Chapter 6 (Laubach, Narayanan, and Kimchi) emphasized that in information processing subserving cognitive processes, neuronal activity needs to be detached from responses to sensory stimuli, at least for some of the time.

They also underscored the importance to record from many cells simultaneously, as neuronal representations are distributed, and in order to obtain a more complete representation of the encoded information, single-cell recording would not be sufficient. Furthermore, they raised the important point that the representation of information is not a self-serving process but needs to serve a purpose. An important aspect is that information must be "read out" and translated into sensible behavioral patterns. This property imposes restrictions and requirements on the code that can be used to represent and compute information.

The authors presented a series of mathematical descriptions and models to evaluate what limitations and properties a system must have to fulfill the criteria outlined above. Furthermore they tested some of their proposals using simulation in order to test the robustness and dynamics of their models.

Part III: Neuronal population information coding and plasticity within brain areas

In Part III, the interaction between neurons within brain areas is discussed. How are neurons functionally connected to form a network? What mechanisms can be identified to control this process?

As an introduction to this topic, Christian Hölscher described in Chapter 7 which general properties can be found in different brain regions when recording from behaving animals during the performance of tasks. He first cited Hebb's postulates, to remind readers that ideas about network organization have been around for a long time. Hebb not only proposed a mechanism of how synaptic weights could change in order to store information in previously active cell assemblies, but also proposed the concept of synchrony and simultaneous neuronal activity in populations that encode the same information. Hölscher then went on to describe neurons that are active in different segments of a delayed match-to-sample task. These neurons appear to encode one part of the task, e.g. the start, the presentation of the stimulus, or the presentation of the match. These data are very similar to the data found in the rodent hippocampus, as described by Hampson and Deadwyler (Chapter 4). When adding the information encoded in these different neurons (neuronal networks, really) together, it becomes clear that all aspects of the task are encoded. How are these cells linked to form an information-encoding network that would be capable of controlling behavior in this task? According to Hebb, they should be linked somehow, e.g. by oscillations as suggested by others. Analyzing firing probabilities of neurons that respond to objects in the task, Hölscher could show by using cross-correlation and autocorrelation analysis that they are synchronized in the gamma-frequency range. Cells that encode similar information are synchronized by gamma, while cells that do not encode similar information are not. That suggests that cells or networks are brought together by gamma rhythms and fire either simultaneously or in a series of active assemblies that is controlled in time by gamma field potential oscillations. Further research using large-scale electrode arrays would enable us to directly test these theories in real time.

Looking at hippocampal "place cells," Hölscher showed that these cells are under tight control of theta and gamma activity. The firing probability is increased when the local field potential is in the disinhibiting (depolarizing) phase. This also has the advantage of increasing the likelihood that several projection neurons activate a target neuron simultaneously. This in turn increases the likelihood of inducing synaptic plasticity, as has been shown in numerous experiments, and therefore satisfies Hebb's first postulate on the induction of long-term potentiation. Furthermore, recordings of several cells simultaneously over time periods show that these cells (or networks that they belong to) are part of a sequence that can be entrained and repeated. This observation directly supports Hebb's second postulate of cell assemblies working in serial order, which can store complex dynamic processes and can guide target-oriented behavior. Hölscher concluded that Hebb had it right and that field potential oscillations in the theta and gamma range most likely play a multitude of roles in coordinating neuronal activity, associating them into networks, and also coordinating series of networks into sequences, and additionally in controlling synaptic plasticity to store these activity patterns in memory.

James Hyman and Michael Hasselmo described in detail in Chapter 8 what roles hippocampal theta rhythm could play in coordinating neuronal activity and synaptic plasticity. They showed that spiking activity of hippocampal neurons is strongly controlled by theta activity, and that not only long-term potentiation (LTP) (an upregulation of synaptic activity) but also long-term depression (LTD) (the downregulation of synaptic activity) induction is under the control of theta rhythm. Long-term depression could also serve as a form of memory by selectively inhibiting or disconnecting neurons within a network that do not cooperate nor encode the same information. The authors showed that the process of LTP and LTD induction is driven by calcium levels in the dendrites of CA neurons which are oscillating between moderate and high levels from peak to trough phase of theta. Since theta is mainly a field potential oscillation, it will slightly inhibit (or hyperpolarize) the neurons on the peak phase, while slightly depolarizing them on the valley phase. This will have important effects on calcium channel opening probabilities, and as had been shown by others, medium levels of calcium will induce LTD, while high levels will induce LTP. Hence, the influence of theta on synaptic plasticity is large and could be utilized to connect neurons that work together on a more permanent scale, while keeping neurons that are not cooperating in neuronal networks separate to avoid interference or "contamination" of information. Indeed, the dangers lie in the area of bringing together active networks that encode separate content. In the worst-case scenario, the networks would merge into one large network, losing all information that had been encoded in each network.

Matthias Munk described in Chapter 9 what general principles may govern information encoding and maintenance in the neocortex, based on what we know about structural and functional features of cortical networks. He first outlined the difficulties to find meaningful codes for the most complex biological information processing structure, the neocortex. He argued that such difficulties may be related to the highly self-referencing connectivity of the cortex leading to a very high proportion of self-maintaining activity. Another unresolved problem of structure–function complexity is the fact that representations of content and context or the embedding of specific information processing into the lifelong ongoing self-organizing brain process make it difficult to identify the rules with which information is represented in neuronal signals. He then discussed what a neuronal code might be and what needs to be done in order to qualify a neuronal signal as an expression of a certain code. The crucial point of deciphering a code, no matter in what system, is that the transformation rules through which the carrier signals convey information are not only identified, but must be verified. This can either be done by knowing the readout mechanism and reconstructing the signal transformation in a model or, experimentally, by reversible manipulation of an essential, for the code under study, activity feature.

After reviewing a range of established and putative codes, Munk analysed what options exist for more advanced codes that might be able to handle higher-order information. As the author is convinced that activity patterns at the level of mesoscopic networks are still the least-understood and potentially very useful carrier signals, he reviewed three classes of population activity. Neuronal *avalanches* represent a relatively new observation of synchronous population activity made at first in field potentials of brain slices and cell cultures, which can express dynamical behavior that can be classified as self-organizing critical. They bear the potential to provide a measure of the functional state of networks which are directly relevant to neuronal computations driving specific behavior. Originating from a structural concept of neocortical function, *synfire chains* represent synchronous population activity based on precise spike firing and therefore operate at a much more restricted temporal scale than avalanches. Their great potential lies in the multiplexing of circuits, because in principle every spike in a single neuron could result from a different synfire process, opening up the possibility that huge numbers of these processes operate in parallel scanning rapidly through extended circuits in order to monitor and or read out the results of local computations. Last, but not least, *oscillations*, the best-known brain activity pattern reflecting functional states, are discussed as an interesting mechanism for coding and signal routing. The examples presented and discussed for distributed sensory and memory-related representations show how limited the current understanding of high-dimensional coding really is. The bottom line for the moment is that distributed representations are highly sparse and that the brain must dispose of an elaborate readout mechanism for integrating the complete information about, for example, visual objects.

In Chapter 10, Frank W. Ohl and Henning Scheich discussed the information-processing principles in the auditory cortex. Until the 1980s the auditory cortex was seen as the top-hierarchy level of bottom–up processing of auditory stimuli. Furthermore, plastic changes on anatomical and functional levels were only considered relevant for developmental processes, and the adult cortex was considered to be "finished" and inflexible in its anatomical connections and processing mechanisms. However, this view has been replaced by a dynamic view of the auditory cortex as a structure that is not the "top level" of all information processing, but that it holds a strategic position in the interaction between bottom–up and top–down processing. The interactions with other processing regions are highly flexible and dynamic. The authors presented extensive experimental evidence from the analysis of gerbil and macaque auditory cortex that has led to this change of view.

The authors concluded that the abstraction of features of auditory stimuli as it occurs in *category learning* (*concept formation*) is a complex but fundamental cortical mechanism of information processing and memory formation for which the auditory cortex is the ideal structure that is equipped with the required functional and anatomical systems.

The authors described several experiments in which animals have to perform a auditory tasks. Recordings in the auditory cortex showed that not only auditory stimuli drive cells, but also non-auditory, task-related information. This shows the dynamic plastic relationship that the auditory cortex plays, and that it is not simply an area for sound analysis. The results compare to those presented by Hampson and Deadwyler (Chapter 4) or Hölscher (Chapter 7), where neurons in the hippocampus or in the perirhinal cortex are found that encode several qualities of a task. More importantly, the studies showed that the neuronal activity in the auditory cortex changed over time after learning the auditory tasks, and started to develop movement-related properties. The sensory input of the tones became associated with target-directed movements in the task that the monkey had learned. This prediction (and assumption), that neuronal representation also must incorporate instructions for target-oriented movements in order to form a useful and effective system, had been described earlier by David Redish and colleagues in Chapter 5.

In Chapter 11, Thomas A. Cleland described information representation and processing in the olfactory system. He presented a general theory of early olfactory sensory processing in the primary olfactory epithelium and olfactory bulb. The primary olfactory representation is mediated by the activation pattern of primary olfactory sensory neurons in the sensory epithelium. The rate-coded activity of these primary sensory neurons is selective for different aspects of odor quality, but the quality of this signal is unavoidably limited by additive and antagonistic interference among ambient odorants as well as other confounding influences. The secondary olfactory representation is similarly mediated by the activation pattern across the population of principal neurons immediately postsynaptic to the olfactory sensory neurons (the mitral cells). The transformation between the primary and secondary representation is a robust, intricate, two-stage process that corrects for artefacts that can hinder the recognition of odor qualities, regulates stimulus selectivity, and transduces the underlying mechanics from a robust but costly rate-coding scheme on a slow timescale to a sparse dynamical representation operating on the beta- and gamma-band timescales which is suitable for integration with other central high-level processes. Therefore, a conversion appears to be made from the rate-coded activity of the primary sensory neurons down to the bulbar output activity which is sparse and temporally structured.

This model follows the previously described general trend which states that sensory input and motor output tends to be rate-coded, while central processing follows phase-coding principles. The model proposed by Cleland therefore nicely describes the dynamics and principles that govern information representation in neuronal networks.

Another important anatomical aspect of neuronal network architecture can be learned from the olfactory bulb. There is very little collateral connectivity between the chemoselective glomeruli on the bulbar surface, and they show a more columnar architecture. Consequently, the physical distribution of chemoselective glomeruli on the bulbar surface is irrelevant to the representation or processing of stimulus quality. This independence from topology is an important feature, as topologies of odor similarity depend on the statistics and features of the current odor environment and hence are unpredictable. Unlike other cortical structures that process information using a two-dimensional array of population information encoding (e.g. the somatosensory cortex), the precise locations of these glomeruli are irrelevant and will not contain any information. Two-dimensional activity patterns of glomeruli are therefore not part of olfactory representation. This principle can also be found in other brain areas, in particular in most nuclei (e.g. the basal ganglia) which appear to have a random distribution of information processing neurons in the tissue. A similar sparse architecture has been proposed for the hippocampus, which would explain why "place cells" are actually distributed randomly in the hippocampus, with no relation to the dimensions and topography of the real-world spatial information encoded in these cells.

Part IV: Functional integration of different brain areas in information processing and plasticity

In Part IV, the interaction between brain areas was discussed. Additional to the problems of organizing neuronal populations to encode information, brain areas have to be functionally linked with other areas in order to make information processing possible. This requires systems that somehow associate active neuronal populations across distances in a controlled manner.

Brian Bland starts off this section in Chapter 12 by describing in fine detail how theta rhythm is induced, controlled, and modulated. Specific basal brain nuclei that project far to target brain areas, and which mostly use acetylcholine as their "private" transmitter, enable the controlled activation of brain areas required to process the information. The main example which is described in detail is the septohippocampal projection from the nucleus basalis to the hippocampus. Additional projections such as the theta inputs that ascend from the pontine region to the midline diencephalic region, travel through to the medial septum, and then input to the hippocampus are described. These projections are the main driving force for the induction of theta rhythm in the hippocampus. This "master switch" not only controls the general activation of cortical areas, but is also capable of functionally connecting and synchronizing different brain areas when required. Similar to thalamic nuclei, the basal brain nuclei are capable of inducing theta in most cortical areas, and can synchronize or desynchronize these areas when required.

The functional importance of this process was illustrated by Bland in the description of his sensorimotor model of integration of information. It is apparent that sensory information which is received by primary sensory brain areas and is processed in association areas such as the hippocampus has to be able to drive and coordinate motor programs. Furthermore, the activated motor programs have to be continuously updated according to sensory input (e.g. during the performance of a running task in a maze). Motor activity has to be controlled in a target-oriented way. Bland described the need to integrate many brain areas that have to cooperate during such tasks, but also have to be disconnected if these areas are involved in separate processes.

Bland describes data recorded from the different nuclei that are involved in inducing, modulating, and controlling theta rhythm. The detailed description of this complex system underscores the fact that these oscillations are a vital part of the system, and clearly play important roles in the overall control of information processing.

In Chapter 13, Kari Hoffmann described the process of face recognition in the brain, and which areas are involved in this. Face recognition is an essential skill in non-human primates as well as in humans. In humans, areas that are involved in face recognition include the lateral fusiform gyrus (also called the fusiform "face" area), the inferior occipital gyrus and sulcus, sometimes referred to as the lateral occipital complex, the superior temporal sulcus, and even the amygdala (probably for the perception of fear-inducing facial expressions). In the macaque, there are homologous cortical areas that perform these functions. These different brain areas will have to be coordinated and integrated in order to successfully perform the face recognition operation. Electrophysiological investigations as well as optical recording have shown that neurons with similar responses are clustered together within a brain region. Neurons with one response pattern tend to be surrounded by similar neurons, creating dozens of patches of neurons responding exclusively to faces but not objects, to objects but not faces, or to both stimulus classes. This suggests a columnar organization of information representation. However, the author goes on to describe that functional magnetic resonance imaging (fMRI) methodology will bias the analysis of brain activity towards clustering results and will underreport distributed network activity, simply because the activity of such networks might be too low to activate a voxel. Hoffmann argued eloquently that this notion of "face modules" is not convincing, and that these neurons are not exclusively tuned for faces. In fact, on closer inspection, there are more data to support the notion that encoding of faces is sparse and distributed. This means that a given neuron will fire strongly to only a few stimuli, and a given stimulus will elicit activity in many neurons. Hoffmann also noted that sparsity leads to greater capacity – neuronal population activity will not "saturate" as quickly. Distributed responses enable generalization and pattern completion from incomplete or degraded stimuli, or from biological noise (see Chapter 1 for details). Moreover, distributed representations in principle can be decoded more quickly than local, "grandmother cell" representations.

The next questions that need to be addressed are, when moving from groups of neurons within focal areas to the presence of multiple modules, how do they communicate with each other, and are all such groups needed for face processing? How are associations achieved? Several principals can potentially be recruited to perform these tasks. Avalanches of neuronal assemblies that are linked in sequence (see e.g. Chapters 4 and 9) and operate with a population code, or other assembly architectures could be drawn in to explain these processes. Hoffmann discussed the merits of cell assemblies and why they would be needed to identify faces under difficult viewing conditions. Simple stimulus-driven responses of tuned neurons would be insufficient to explain the view-invariant properties of cell groups. Here, cell assemblies would be required to solve this problem. In addition, to solve the binding problem of which face belongs to which monkey (or human), temporal synchrony as suggested by Hebb and others would help to disambiguate associations. Hoffmann then argued that in order to stabilize recognition and association of disparate information to a coherent complex, cell assembly principles would not be sufficient. Considering that there is a large body of evidence for oscillation and synchronization of neuronal responses on every level of the visual system, it is salient to conclude that oscillations perform the important task of bringing together separate cell assemblies that are required to communicate with each other.

Hoffmann then went on to describe future explorations in this area to test specific concepts of image recognition, view invariance, and binding of objects that are represented separately in the brain, but connected in real space.

In Chapter 14, Nielsen and Rainer described in more detail what type of information local field potentials, single-unit responses, and multi-unit activity encode. They described examples of recording experiments in different visual cortical areas in the primate. Local field potentials resemble the overall population activity of excitatory and inhibitory neurons, including membrane potential fluctuations that occur within a short distance of the electrode tip. Single-unit activity will resemble the activity of one neuron only, and the authors pointed out that there is a large bias towards recording from excitatory neurons which produce strong action potentials. Multi-unit activity is composed of the neuronal activity of many single cells and provides an "overview" of the population action potentials within that area. Nielsen and Rainer went on to point out important differences of what type of information is encoded by these signals. According to current estimates, over 70% of excitatory synapses in the cortex are connected to other local neurons, and only about 30% target distant brain regions. Since both multi-unit and single-unit activity capture spiking activity, they show mainly local processing *within* a cortical column and to a smaller part the long-range *output* that targets distant brain regions. Local field potentials, on the other hand, measure extracellular field potentials generated by membrane currents originating from axons, somata, and dendrites. Synchronized dendritic activity is thought to have the largest contribution to the local field potential, making it a measure of local information processing within the specific cortical region, as well as of the *inputs* that the region receives. These differences in encoding can be used to make assessments about information representations in the cortex: which part of the information is already present in the input, which part is locally processed, and what is the actual output of that region.

The authors went on to describe a list of experiments in visual areas of the cortex that analyze information encoding found in these different types of recording signals. For example, both single-neuron activity and field potentials recorded in area V1 display sensitivity to the orientation of grating patterns and grating contrasts. However, field potential frequencies in the gamma band are more sensitive to the stimulus parameters. Similar results have been obtained for area MT in the visual cortex, which is involved in motion perception. Both field potentials (in particular above 40 Hz) and single MT neuron activity carries information about the direction and speed of a moving stimulus. In the inferotemporal (IT) cortex, which represents the final stage of the ventral visual processing stream, a similar picture emerges. Both single-neuron activity and field potentials can show strong preferences for complex objects in several tasks. However, this is not always the case, and the authors presented their own work which shows results where the correlation between both types of signals was not found. When comparing the best single-unit response with the field potential response, there was not always a direct correlation between these signals. Therefore, the stimulus preference of the field potential can in general not be inferred from the stimulus preference of a locally recorded single neuron.

Similar observations were made when recording in a variety of motor cortex areas. In the motor cortex, the field potential can be used to successfully decode a movement direction. Furthermore, combining the local field potential signal with either single-unit or multi-unit activity results in higher decoding accuracy than possible based on any signal alone, suggesting that independent information is coded in these signals. In a different area, only the field potential can be used to decode the transition from planning an eye movement to executing it. The authors went on to describe similar results in areas V1, V2, and V4.

All these findings highlight the fact that local field potentials, multi-unit recording, and single-neuron recordings reflect different information-processing mechanisms. Since the field potential signal reflects input and local processing within a region, and single- and multi-unit activity represent the output of that region, the differences in the information found in these signals can be used to identify how different brain regions participate in the processing of information. The authors presented data for which they have mapped the IT region to compare the information contents of single-unit activity and field potentials during the presentation of visual stimuli. Interestingly, the information content of field potentials increases from posterior to anterior recording locations. The information content of single-unit activity did not change across the whole area recorded.

The authors concluded that information is encoded in the output of single neurons throughout IT. However, the encoding at the input level increases from posterior to anterior IT. This technique of analysis permits the researcher to identify what contribution the cortical area has in the processing of information, in addition to the analysis of what type of information is encoded. The insights given by these joint analyses go a step beyond traditional brain–behavior correlations based on each of the signals considered alone. It furthermore highlights the importance in analyzing the network activity of neuronal population, which encodes more than just the sum of individual neuronal encodings.

In Chapter 15, Jochen Kaiser and Werner Lutzenberger gave an overview of how different cortical regions in the brain cooperate in behavioral tasks, using electroencephalograph (EEG) and magnetoencephalograph (MEG) techniques. Brain areas have to be coordinated in their efforts in order to process high-level visual and auditory information. As described in previous chapters, the synchronization of neuronal networks that cooperate in this process is a possibility how to quickly and reversible associate networks. The authors reviewed a wide range of results that support this notion, where the processing of complex stimuli correlated directly with the onset of gamma rhythm activity. The techniques of EEG and MEG recordings can supplement the information gained by recordings made inside the brain.

In addition to the processing of complex stimuli, the authors described results that show that increased attention also increases gamma activity. The authors conclude that induced gamma activity is not only related to perceptual processes but may also reflect the modulation of cortical network synchronization by top–down processes such as selective attention. Gamma activity also increases in memory retrieval tasks. Since the perception of an object activates the corresponding memory trace, gamma activity could be the tool to activate the complete network that encodes the memory trace. For example, initial presentations of line drawings of common objects elicited increases of induced gamma phase synchronization between distant electrode sites. These effects were reduced to repeated presentations of the same stimuli, perhaps reflecting network "tuning" to represent only the memorized item, and not similar items that are co-activated initially.

The authors went on to describe their own work using the MEG technique, which has several technical advantages over the EEG recording technique. In experiments that analyzed auditory spatial processing, they compared responses to infrequent right- and left-lateralized sounds with frequent midline presentations of the same sounds. Lateralized deviant stimuli gave rise to evoked mismatch fields that peaked about 120 ms after stimulus onset. Gamma activity was increased over the posterior temporo-parietal cortex at latencies between 250 and 300 ms. Similar results were obtained in separate investigations for syllables and animal vocalizations. Interestingly, right-lateralized stimuli gave rise to bilateral enhancements, whereas sounds in the left hemifield were associated with gamma increases over the right hemisphere only, supporting a predominance of the right hemisphere for auditory spatial processing. Deviations in acoustic patterns evoked mismatch fields at around 180 ms after stimulus onset, i.e. with longer latencies than spatial analysis. For all three types of sounds – syllables, animal sounds and meaningless distorted noises – pattern deviance was accompanied by increased gamma-band amplitudes in the range of 60–90 Hz over left anterior temporal and inferior frontal regions. The temporal delay between the superior temporal mismatch peak and the subsequent gamma increases over the "auditory stream" areas has provided first evidence for serial processing along these pathways in humans.

In an active oddball task, gamma activity in the left inferior frontal cortex was seen during the detection of acoustic pattern mismatch. Attention to deviant sounds and to the detection of pairs of two consecutive deviants elicited additional gamma rhythm activity over the dorsal prefrontal cortex at ~200–300 ms after stimulus onset. The authors interpreted this increase of a gamma rhythm as a reflection of the activation of networks which are involved in working memory, information processing, and decision-making.

In other experiments, subjects were presented with combinations of a syllable and the corresponding mouth movements shown on a monitor. Mismatch situations were included where either incongruent syllables (acoustic deviant) or incongruent mouth movements (visual deviant) were shown. The visual deviant stimulus produced enhancements of gamma in frequencies between ~73 and 80 Hz over the posterior parietal areas (at ~160 ms after the onset of the acoustic stimulus), over midline occipital cortex (at ~270 ms), and over left inferior frontal cortex peaking at ~320 ms.

The involvement of short-term auditory memory was investigated in delayed matching-to-sample tasks. It was found that left posterior temporal–parietal gamma activity was increased during the delay phase. This was followed by an increase in gamma over the right frontal cortex at the end of the delay phase, which could reflect the activation of frontal comparison and decision-making networks. A third gamma component over the midline parietal cortex was also observed.

More importantly, in addition to spectral amplitude increases, the authors also observed increased gamma coherence during the delay phase between the left temporal–parietal and the right frontal cortex. Phase synchrony analysis showed a sustained enhancement of phase coupling between these regions throughout the memory delay, possibly reflecting the interaction between posterior sensory storage areas and frontal executive networks. Very similar results were observed in auditory pattern memory tasks. These results show that not only is gamma activity enhanced during the processing of these complex stimuli, but cortical areas that cooperate in these tasks are phase-locked during the processing events.

The authors then asked central questions that will be a focus for future research. For example, what behavioral relevance does gamma activity have? The authors gave a list of experimental results that show a correlation between subjective states of awareness and gamma activity. While this is initially encouraging, the issue will have to be addressed in further detail. In particular, investigations to what degree gamma activity is required for the proper functioning of the network associations and processing of complex information would be most welcome here.

Finally, the authors briefly discussed the issue of long-distance synchronization of gamma oscillations. They have found sustained enhancements of gamma phase synchronization between higher sensory cortical areas and frontal areas during the memory delay of auditory delayed matching-to-sample tasks. These findings support the notion that cortico-cortical synchronization may represent a mechanism for transient coupling between neural assemblies representing an integral feature of memory processes. Long-distance gamma synchrony has also been reported by others, e.g. in intracranial recording studies that showed increased gamma coherence between cortical regions and between cortex and hippocampus. These findings demonstrate that synchronization of neurons in the gamma frequency is not only found over short distances, but that this mechanism is potentially capable of functionally connecting large areas over large distances in the brain, and to act as a universal carrier signal for cooperation between areas in the brain.

Part VI: Disturbances of population activity as the basis of schizophrenia

This Part addressed the central issue whether field potential oscillations as described in the previous chapters actually have a functional role, and whether the disturbance of gamma rhythm has functional implications and could explain dysfunctions of memory formation, attention, and processing of complex stimuli.

André A. Fenton described in Chapter 16 what role gamma and theta oscillations could play, and how any disturbances of these oscillations could affect neuronal network coordination, information processing, and also cognitive states. In the introduction, he reminded the readers that the concept of long-term potentiation (LTP) of synaptic transmission as a cellular model for memory formation in the brain has been very helpful to guide researchers towards designing novel experiments that shed light on these mechanisms. Building on this, he proposed a concept that could be useful to guide research into the direction of recording neuronal populations in behaving animals to identify what mechanisms underlie cognitive processes. As described in earlier chapters, Donald Hebb already had suggested that neurons that work together in encoding information might be associated by synchronous activity. This synchronous discharge of excitatory output then could induce a shift in synaptic weights on the target neurons and thereby store the information of the active network in the form of LTP. Fenton then briefly discussed previously proposed methods of information encoding, e.g. by the mechanism of information conversion onto highly specialized "cardinal cells." He pointed out the major problems that such systems would have to face, such as the issue of "combinational exhaustion," the problem that there would be not enough neurons to encode every possible object and category that is imaginable. He also described Donald Hebb's cell assemblies (as discussed in Chapters 4, 7, and others). He points out that networks that encode information in a distributed fashion also have problems to deal with, e.g. the danger of merging two active networks into one, which would create a completely new network and with the price of the loss of all previous information (the "superposition catastrophy"). The result is a set of wrong associations, which can be interpreted as confusion or inappropriate perception on the cognitive level. Fenton proposed that the superposition catastrophe itself can be avoided by coordinating neural activity in a synchronous way, connecting neurons that cooperate in information processing, while keeping neurons separate that encode different content. On a cognitive level, ordinarily, brains selectively activate representations of objects and scenes that are appropriate and suppress representations that are inappropriate. Hence, Fenton suggested that a lack of coordination of neuronal associations can lead to false cognitive representations and associations. He proposed that the core dysfunction in disorganized schizophrenia is an impairment of neural coordination which makes inappropriate associations possible.

Fenton's central hypothesis is that the excessive co-activation of neurons in multiple cell assemblies causes the forms of cognitive disorganization seen in psychosis. Crucially, because neuronal co-activation can strengthen synapses, the inappropriate associations may be self-perpetuating and become fixed in memory. He called this the "hypersynchrony" hypothesis.

Fenton then went on to describe several experiments that underpin his theory. He used a tetrode system for identifying single cells in the hippocampus of behaving rats. As described in Chapters 4 and 7, neurons in the hippocampus encode spatial information. These cells are driven by external cues at the walls, and a turn of such cues by 45° will also turn the firing field of the cell by the same value. He analysed the associations of hippocampal cell activity to external frames of references. The animals were situated on a turntable, which acts as a moving frame of reference, while the external laboratory cues act as a stationary frame of reference. In this set-up, Fenton could identify cells that are driven by either the stationary frame of reference or by the rotating frame of reference. These cells can be seen as a cognitive representation of the two different frames of reference. In order to navigate successfully on the rotating platform to avoid a shock, the animals must develop a representation of these two frames of reference. It now can be tested what happens if the separation shown in the hippocampal cell activity is disturbed.

According to the hypersynchrony hypothesis, a disturbance in the synchronization should result in a disturbance in association of cell assemblies, and a breakdown of the clearly separated representation of the two frames of reference by hippocampal cells. On the behavioral (cognitive) level, this should result in an impairment of successfully navigating the rotating platform task. At first, Fenton could show in initial experiments in anesthetized rats that unilateral injection of tetrodoxin (TTX) can disinhibit neurons recorded in the contralateral hippocampus. This increase in activity had several consequences. First, the TTX injection increased beta (12–25 Hz) field potential oscillations in the ventral hippocampus. This reflects an increase of coupling between pyramidal cells. Increased beta rhythm is also found in schizophrenic patients. Second, previously only weakly coupled neurons (as assessed in cross-correlation of firing activity) now have become strongly coupled. This suggests that networks that previously had been dissociated now become associated.

In a behavioral task, place avoidance, injecting TTX into one hippocampus did not impair a familiar avoidance task on the rotating platform, but abolished the ability to learn, consolidate, or recall an avoidance task. This suggests that the hypersynchronization of previously dissociated cells does not impair basic navigation, but does impair the formation of novel memory or of the recall of the required spatial information of the two frames of reference. Future experiments will include recording of neuronal activity in the hippocampus of behaving rats after the injection of drugs such as with phencyclidine, a potent psychotomimetic which can induce schizophrenic symptoms in humans.

In conclusion, the initial experimental results have validated the experimental set-up and added vital support to the concept that synchronization of neurons is an essential element in the functional association into networks, and the dissociation of separate networks.

In Chapter 17, Corinna Haenschel summarized current thinking of what processes might underlie schizophrenic pathophysiology. The concept that a main factor in schizophrenia is the fragmentation or dissociation of thought processes has already been proposed a hundred years ago. Current models of schizophrenia suggest that the core impairment underlying both dysfunctional cognition and the overt symptoms of the disorder are to be found in deficient coordination of distributed brain activity. In addition, the impaired control of synaptic plasticity induced by experience has been implicated in this process. On the cellular level, it has been proposed that a mechanism for the coordination of neural activity between and within functionally specialized brain regions is oscillatory activity which synchronizes neural responses. This mechanism can be in control of a dynamic integration into neuronal assemblies and be scaled up all the way to large-scale networks responsible for cognitive processes and memory formation.

Cognitive abnormalities are found throughout the life of schizophrenics, starting in childhood and continuing throughout adolescence to adulthood. Importantly, the degree of cognitive dysfunction might be the best predictor of long-term prognosis for patients with schizophrenia. They show disturbances in a range of cognitive processes, such as attention, memory, and goal-directed planning, and there is a considerable debate as to which of these deficits represent the core features of schizophrenia. Thus, impaired neuronal synchrony may provide a parsimonious account of many of the cognitive deficits observed in schizophrenia.

Numerous investigations in humans provided a growing body of evidence for abnormalities in high-frequency oscillations and synchronization in the EEGs of patients with schizophrenia. Haenschel and colleagues provided evidence for a close relationship between impaired neural synchrony and specific cognitive deficits in patients with schizophrenia. They showed that schizophrenic patients showed a deficit in a gestalt perception task and also showed a strong reduction in the long-range synchronization of neural responses in the beta band. This could reflect an impairment in the association of neural activity and underlie the specific cognitive dysfunctions associated with the disorder.

Haenschel then listed five open questions that focus on the relationship between schizophrenia and neural synchrony. These include technical questions of how to measure oscillations, whether deregulated synchronization would explain all of the symptoms found in schizophrenia, if measurements of oscillations could be used as a marker for diagnosing subjects, and the question whether medication that normalizes oscillations would help to treat the condition. Initial studies reported evidence to show that a new generation of antipsychotic medications increased gamma-band synchronization.

Haenschel continued by describing technical aspects and parameters of oscillation measurements in humans, which could explain observations like a reduced signal-to-noise ratio found in some studies. She continued to describe the finding in the last ten years in the field of aberrant oscillatory activity in schizophrenia. There is evidence of impaired gamma oscillations from a wide range of experiments that have analyzed early sensory systems (like steady state, sensory gating, and backward masking experiments), as well as from more cognitive paradigms (like the auditory oddball and visual binding experiments). In a series of experiments it has been demonstrated that patients with schizophrenia show deficits in early sensory processing due to a failure in the entrainment of intrinsic gamma-frequency oscillators. When stimulated with auditory or visual stimuli in the 40-Hz range, schizophrenics showed much reduced stimulus-induced EEG oscillations in this frequency range. Sensory gating is also affected as shown in the classic double-click auditory task. The evoked cortical response (P50) is measured. In healthy participants the P50 to the second click is suppressed compared to the P50 to first click, whereas patients with schizophrenia lack this P50 suppression. During this task, a relationship between reduced early phase-locked gamma and P50 suppression was found in schizophrenics. Beta activity also was affected (see previous chapter on this effect). In visual masking tasks, the identification of an initial stimulus (the target) is disrupted by a later stimulus (the mask). Patients with schizophrenia have consistently shown that they require longer intervals between the target and the mask to correctly identify the target and have shown a deficit for both the early and the late component. Control subjects showed a burst of induced gamma activity after target presentation, while schizophrenics exhibited a much earlier and reduced gamma response. In more complex auditory and visual tasks that require more cognitive high-level processing, several changes in EEG activity were observed. In auditory oddball tasks, induced gamma activity was reduced in response to targets over the left hemisphere and increased over right hemisphere. In visual gestalt type "binding" tasks, measures of phase-locking and phase coherence in the EEG were affected. Results showed reduced evoked gamma activity and abnormal and delayed onset of phase coherence in response to the stimuli. The author concluded that the results across these studies suggest that schizophrenic patients are impaired in establishing or maintaining both gamma activity and phase synchronization, which could account for some of the perceptual and cognitive deficits observed.

When analyzing the relationship between gamma activity in schizophrenia and the diverse range of symptoms that are observed in the disease, it had been concluded that the lower the frequency of the early evoked gamma-band response, the greater the degree of the classic "off" (inactivity) symptoms. For the "on" (hyperactivity) symptoms, the overactivation of gamma appears to be the main problem. This is illustrated by the results of studies that investigated the relationship between gamma oscillatory activity and hallucinations. The hypothesis is that conscious perception is subserved by synchronization of gamma-frequency oscillations in recurrent thalamocortical networks, and that hallucinations result from uncontrolled activation in this circuitry. Results of investigations of auditory hallucinations showed an increase in high-frequency oscillations in the left inferior parietal lobule and the left medial frontal gyrus. In addition, the fine-tuning and control over these enhanced oscillations appear to be affected. Results of other studies revealed that gamma-band phase-locking between frontal and temporal lobes was greater during talking than listening in healthy controls. In contrast, patients did not show such gamma coherence.

Haenschel then reviewed the literature to summarize which pharmacological and transmitter systems are involved in inducing and supporting EEG oscillations. As described in other chapters, the systems involve GABAergic interneurons and glutamatergic excitatory neurons that are connected via negative feedback loops. These networks are capable of supporting oscillatory activity, while drugs that modulate GABA or glutamate receptor subtypes can influence gamma activity. Altered GABAergic interneuron disinhibition has been shown in schizophrenic patients. Neuron counts also revealed a reduced number of interneurons in schizophrenia, which might explain the abnormal oscillatory dynamics. Glutamate receptors play a role, since metabotropic glutamate receptors can induce oscillations in hippocampal slices. A role for N-methyl-D-aspartate (NMDA) receptor activity in schizophrenia is based on the finding that NMDA antagonists, such as ketamine or phencyclidine, mimic both the positive and negative symptoms of schizophrenia. There are numerous reports indicating that the positive symptoms of schizophrenia are due to hyperactivity of the dopaminergic system. Dopamine agonists such as amphetamine can produce psychotic symptoms. Furthermore the efficacy of antipsychotic agents is thought to be related to their capacity to antagonize dopamine. However, dopamine agonists do not produce marked formal thought disorder and negative symptoms as observed in schizophrenia. Finally, changes in cortical cholinergic transmission have been found in patients with schizophrenia, such as reduced number of nicotinic and muscarinic acetylcholine receptors, and changes in the nicotinic α7 receptor gene, a risk factor gene for schizophrenia. Cholinergic projections control and induce theta rhythm in target areas (as described in previous chapters), and plays a role in the fast, state-dependent facilitation of gamma oscillations, as well as use-dependent long-term modification of cortical synchronization in the gamma range. Acute administration of nicotine to patients with schizophrenia transiently improves some of their symptoms, e.g. in sensory gating, working memory, and long-term memory.

The author concludes that there is a direct correlation between disturbances of high-frequency EEG oscillations and cognitive states in schizophrenia. The hope is that measures of neuronal synchrony may be very useful for the diagnosis of cognitive dysfunction, and that pharmacological intervention to target this dysregulation could open the door to novel treatments of schizophrenia.

Conclusion

So how do higher structures of the brain encode information?

The overview of all chapters shows a wide range of results from research at almost all levels of neuroscience. The conclusions drawn from these results show some degree of overlap, but also emphasize different properties and concepts that govern the processing and encoding of information at different levels of neuronal organisation in the brain. We will attempt to synthesize the findings and concepts into a model that will incorporate most aspects. The fact that such a wide range of results and concepts exists already documents that no single coding mechanism will be able to explain in general how information is represented in the brain and how the brain computes behavior. As described in the introductory chapter (Chapter 1), we will go through a list of possible encoding principles and try to identify which principles are supported by empirical evidence.

Rate-coding

Most chapters in this book incorporate rate-coding into their models at least to some degree, emphasizing that this coding mechanism is a fundamental part of information processing in the brain. In particular, as Cleland and others describe, the peripheral input and motor output appear to be mainly rate-coded. It appears that the input and output of information to the brain very often follow a rate-coding principle, while later stages of processing seem to translate their input into signals using various other types of codes which probably allows for more degrees of freedom in the complexity of operations and for detaching processes dealing with external information from processes creating and updating internal information.

Labeled line coding

In general, labeled line coding is also very much present at the input stage but then most likely loses its character of exclusiveness in the CNS. Kari Hoffmann described this issue in Chapter 13, discussing to what degree "face cells" really are face cells (or face recognition areas in the cortex). The same points were raised by other authors (in Chapters 4, 7, and 16) when discussing the properties of "place cells" in the hippocampus, which are not exclusively place encoding cells at all. Instead, they integrate several types of sensory and other information to establish representations of space, movements, directions, and goal-directed behavioral programs. The need to synthesize multisensory information to establish such behavioral programs makes it unlikely that neurons in the higher processing areas retain their labeled line exclusiveness.

Cell assemblies

Several authors suggested that coding of information is achieved by neuronal processes that are more or less compatible to the cell assembly concept which had originally been proposed by Donald Hebb (see Chapters 3, 4, 9, and 16). Chapter 9 for example described the principles that govern neuronal assemblies some of which are expressed as *avalanches*. In avalanches, distributed groups of neurons activate a second layer of distributed target neurons, which then activate further neurons. There is no evidence that neurons in different layers are synchronized or phase-coupled oscillators which coordinate or restrain neuronal activity, as would be required for a phase-coding system. Harris (Chapter 3) described what information processing such assemblies could perform. The amount of information and the speed and precision of processing that such a system can offer is sufficient to encode large amounts of information as required in the brain. Additionally, Hams described experimental observations to support the concept that phase-coding is not inherently required for information encoding, and that his proposed system of cell assemblies could be capable of information processing without the need for precise control of spike timing. Hoffmann (Chapter 13), however, pointed out that the avalanche mechanism will not perform well under noisy conditions, and it would be difficult to associate with or integrate into other avalanches when required in more complex tasks, e.g. when associating visual with auditory information.

Information encoding in neuronal populations has the advantage of increasing the signal-to-noise ratio, making the rather noisy neuronal communication much more reliable. The death of a sensory neuron or a motor neuron can have direct consequences, while the death of a member of a neuronal assembly has no functional consequences. More importantly, the encoding of information in more holistic style network systems will add novel properties of pattern completion and robustness against damage and information loss. Neurons can be used for several networks, cutting down the numbers required for information processing. Novel associations between networks can be created within milliseconds, which is impossible in "grandmother cell"-style hierarchical systems. There, every association will have to be direct (grandmother – chair, grandmother – white cup, grandmother – ringing telephone). Such directly wired associations sometimes would have to be across large areas of cortex, creating the need for many fiber bundles that can be only used for one job. Also, completely novel associations would be impossible or extremely time-consuming in such a hierarchical system. An artist with such a brain architecture could never dream up the concept of a blue duck, as no previous association had been formed between those neurons. In cell assemblies, associations can have a certain amount of fuzziness, enabling a low level of novel associations that usually does not affect information processing, but offers the potential of producing completely novel associations.

Phase-coding

A number of chapters focus on this principle, and there is good empirical evidence to support the hypothesis that this encoding principle is used in the brain. One extreme example was described in Chapter 9 where among other mechanisms of neuronal assemblies the concept and the most probable correlate of *synfire chains* were depicted. Synfire chains are reflected in millisecond-precise spatiotemporal spike patterns which repeat with surprising precision and therefore operate at a much more restricted temporal scale than avalanches. The latter have so far mostly been described as scale-free, which means their duration and size has to cover several orders of magnitude and therefore qualifies them as a self-organizing critical process.

The great potential of synfire chains lies in the possibility that they can multiplex cortical circuits, because in principle every spike in a single neuron could result from a different synfire process, opening up the possibility that huge numbers of these processes might operate in parallel scanning rapidly through extended circuits in order to monitor and or read out the results of local computations. In addition, due to the temporal constriction, association with other synfire chains would be possible if the phase code is utilized to associate appropriate (e.g. temporally congruent) synfire chains into larger complexes.

The role of oscillations

As described in several chapters, fast EEG field potential oscillations have profound effects on neuronal activity, firing probability, and synaptic plasticity (see e.g. Chapters 7, 8, 9, and 16). As described in Chapter 2, the fact that a multitude of specialized ion channels, interneuron subtypes, and neuronal projections are in place to establish defined oscillations in a controlled way clearly supports the concept that this mechanism plays important roles in the brain (see also Chapters 7, 12, and 16). One important function of this elaborate system of inducing oscillations is to control spike timing, which is crucial for inducing synaptic plasticity. This Hebbian long-term change of synaptic weights then can store sequences of neuronal activities. This point has also been made and illustrated further in Chapters 4, 7, 8, 15, and 16, and others. Data presented in Chapter 7 show that neurons that are collaborating in a task are synchronized in the gamma frequency range, while cells that do not encode similar information are not. In Chapter 11, it was described in detail that the primary olfactory representation is mediated by the activation pattern of primary sensory neurons in the sensory epithelium. The transformation between the primary and secondary representation is a two-stage process which translates information codes from a rate-coding scheme on a slow timescale to a sparse dynamical representation operating at the beta and gamma frequencies. These findings support the previously described general trend which states that sensory input and motor output tends to be rate-coded, while central processing follows phase-coding principles. Chapter 14 described results that show when combining the local field potential signal with either single-unit or multi-unit activity results in higher decoding accuracy than possible based on any signal alone, suggesting that independent information is coded in these signals. All these findings highlight the fact that local field potentials, multi-unit recording, and single-neuron recordings reflect different mechanisms of information processing.

Chapter 16 also discussed what role gamma and theta oscillations could play, and how any disturbances of these oscillations could affect neuronal network coordination, information processing, and even cognitive states. Chapter 17 presented data to demonstrate that disturbances in synchronized oscillatory activity could result in a loss of information processing, cognitive processes, and memory formation.

Phase-locking of oscillations across brain areas

Synchrony of EEG field potentials will have direct effects on neuronal firing properties (see Chapter 7), and is capable of coordinating networks across the brain. However, if the synchronization is to happen in real time (as would be required for the induction of LTP), phase-locking should occur even over larger distances.

Chapters 7 and 12 describe how gamma and theta activity can control the phase-locking of neuronal activity, and how this mechanism can join neurons in their activity to form a network. In addition, gamma rhythm can separate networks that are working on separate information items and must be kept apart. Chapter 16 described examples of how the separation can be disturbed, and how this could lead to cognitive impairments that might underlie mental illnesses. Chapter 17 provided evidence for a close relationship between impaired neural synchrony and specific cognitive deficits in patients with schizophrenia. Schizophrenic patients are impaired in establishing or maintaining both gamma activity and phase synchronization, which could account for the perceptual and cognitive deficits observed. Chapter 15 also reviewed a wide range of results that support and add to these findings. In MEG and EEG studies, the authors observed increased gamma phase-locking during cognitive tasks between involved cortical areas.

Neuronal network type of encoding

Interestingly enough, relatively little has been said in the various chapters about this type of information encoding, probably because concepts and experimental approach are considerably more complicated than for most other forms of encoding. As described in Chapter 1, neuronal networks have unique properties that are also found in brain activities. The error correction and pattern completion properties, and the sometimes surprising lack of performance loss after cortical damage in the brain, point towards a distributed "holistic" type of information of the neuronal network type. The great technical disadvantage that we have to face at present is that a large number of neurons would need to be recorded in order to analyze the exact nature of information-encoding principles in large neuronal networks that are sparsely distributed across cortical areas. Nevertheless, several indications have been described in some chapters that show the distributed character of information in particular areas. Chapter 4 noted the fact that simultaneously recorded ensembles encoded synergistic information in the network additional to that which would be predicted from the sum of individual neurons. Chapter 5 underscored the importance to record from many cells simultaneously, as neuronal representation is distributed, and in order to obtain a more complete representation of the encoded information, single-cell recording would not be sufficient. Chapter 9 described distributed sensory and memory-related representations which show how limited our current understanding of high-dimensional coding really is. Different network operations are described which have the potential for supporting or implementing assembly codes. Although neuronal assemblies cannot be decoded by subsequent processing stages in the same straight forward fashion as e.g. information arriving on sensory afferents, we need to unravel the mechanisms for integrating the complete information about their content by deriving elaborate readout schemes and causally proving their effectiveness. In Chapter 14 it was shown that when combining the local field potential signal with either single-unit or multi-unit activity, a higher decoding accuracy than is possible based on any signal alone is achieved, suggesting that to some extent independent information is coded in these signals.

It furthermore highlights the importance in analyzing the network activity of neuronal populations, which encodes more than just the sum of individual neuronal encodings.

Encoding of sequences

Several chapters discussed how temporal sequences could be built up in the brain. In order to achieve this, neuronal chains of events have to be established, which goes far beyond the simple stimulus–response type of encoding that many researchers investigate. Avalanches and synfire chains would be capable of developing dynamic temporal sequences to control e.g. motor programs. Chapters 4, 7, and 16 and others show data of individual neurons that encode different aspects of a specific task. When adding the information encoded in these different neurons (neuronal networks, really) together, it becomes apparent that all aspects of the task are encoded. How are these cells linked to form an information-encoding network that would be capable of controlling behavior in such a task? Chapter 7 presented data that show by using cross-correlation and autocorrelation analysis that they are synchronized in the gamma frequency range. Cells that encode similar information are synchronized by gamma, while cells that do not encode similar information are not. Furthermore, when recording from several hippocampal neurons over time, reproducible sequences of neuronal activities can be found. Such sequences could encode behavioral sequences in goal-directed behavior. Similar to Hebb's prediction, field potential oscillations in the theta and gamma range most likely play a multitude of roles in coordinating neuronal activity, associating them into networks, and also coordinating series of networks into sequences, and additionally to control synaptic plasticity to store these activity patterns in memory.

Sparse versus rich code

As described in detail in Chapters 11 and 13 and others, phase-coded networks often use a sparse code, which provides them with unique properties such as a high storage capacity. Chapter 9 described mechanisms for coding and signal routing distributed representations that are highly sparse. Chapter 11 described the conversion from the rate-coded activity of the primary sensory neurons down to the bulbar output activity that is sparse and temporally structured. Hence, most authors did find results in their studies that are consistent with sparse coding. However, in a detailed study of the hippocampus described in Chapter 4, the authors found evidence not only for sparse coding but also for a dense (population) code that distributes the information over a group of neurons. However, they also point out the fact that simultaneously recorded ensembles encoded additional information that was lost when ensembles were simply assembled by combining non-simultaneous spike trains. This demonstrates the importance of the spatiotemporal pattern (i.e. across neurons and time) of neuronal activity, which is more consistent with a sparse code. The authors conclude that the hippocampus uses a code that shares the features of both dense and sparsely distributed models.

In Chapter 13, it was reported that electrophysiological investigations as well as optical recording have shown that neurons with similar responses cluster together within a brain region. Neurons with one response pattern tend to be surrounded by similar neurons, creating dozens of patches of neurons that respond almost exclusively to faces but hardly to objects, or vice versa. These observations point towards a population code that is not sparse. Yet, on closer inspection, there are more data to support the notion that encoding of faces is sparse and distributed. This means that a given neuron will fire strongly to only a few stimuli and a given stimulus will elicit activity in many neurons.

In conclusion, there is evidence for several types of sparse and rich coding principles, with the possibility that some of these results have been biased by inadequate resolution of neuronal activity, suggesting a rich code type of encoding when there really is a sparse or a mixed type of encoding.

Where do cognitive processes take place?

One aspect that has been mentioned by several authors is the fact that neuronal assemblies in higher-order information-processing areas have to have some type of endogenous activity that is independent from sensory input. Harris proposed in Chapter 3 that the function of such assemblies of neurons is not solely the representation of sensory input, but also to compute the information, and to take part in internal cognitive processes. This point was also made in Chapter 4, where neuronal classes in the hippocampus were described that not only represent sensory input but correlate with activities that are part of the planning processes. The authors of Chapter 5 also emphasized that in the information-processing cognitive processes, neuronal activity needs to be detached from responses to sensory stimuli, at least for some of the time.

Information processing has evolved to be goal-directed

No matter which higher brain function is the object of detailed analyses, whether it is sensory encoding, orientation in space, or abstract reasoning, an important aspect one should always keep in mind is the fact that during evolution, the development of all brain functions was exposed to constant selection pressure and therefore was constrained by the need to achieve goals. Brains do not simply represent or process sensory information. Neuronal representations very often incorporate instructions for target-oriented movements in order to form a useful and effective system. The need to react quickly to sensory input changes in the environment and to satisfy the basic needs of organisms (food, shelter, social interaction) sets clear constraints on how brains have to function. A representation must be embedded in the individual's current needs, interests, and goals, and must drive goal-directed behavior. Any hypothesis of how information is processed in the brain has to incorporate this property. Several authors mentioned this important aspect.

Chapter 12 described in detail how somatosensory information can be integrated with motor activity. The authors of Chapter 5 also raised the important point that representation of information is not a self-serving process but needs to serve a purpose. An important aspect is that information must be "read out" and translated into sensible behavioral patterns. This property imposes restrictions and requirements on the code types that can be used to represent and compute information. Chapter 10 confirmed this point for a different modality by showing that neuronal activity in the auditory cortex is not solely occupied with the presentation of stimuli. Recordings in the auditory cortex showed that not only auditory stimuli drive cells, but also non-auditory, task-related information. This points towards a holistic distribution of information, and towards a direct association of sensory encoding networks with motor output networks. This association would make motor responses very fast and eliminated the need to "translate" sensory information code into motor codes. As described in Chapter 7, the initial concept of a "cognitive map" in the hippocampus as represented by the sum of all place neurons opens up the question how this map could be read and translated into motor activity. As Chapters 4, 7, 10, and 12 described, several task-related and motor-related aspects are already encoded in hippocampal neurons, making a long-winded and time-consuming translation of codes or a transfer of information to other brain areas unnecessary.

Reassembling the image: how are the individual elements of distributed representations brought together again?

Several authors offered answers to the question of how separate information encoded in neurons is associated to form representations of objects, activities of concepts. Chapter 3 showed that cell assemblies already have the capacity to associate different aspects to form neuronal populations that represent complex objects. The disadvantage of such assemblies is that they would be inflexible and would respond in a more direct stimulus–response type of neuronal activity. The avalanche will then set off further neuronal responses downstream, where further computation and association will take place. This would work well for the processing of one item, but would start to cause problems when several aspects need to be integrated. A system that coordinates independent avalanches and associates them when required (and keeps them apart when not) will be needed for more complex associations. Chapter 13 discussed this issue and outlined the need for more elaborate mechanisms to allow for more flexibility in associating different representations, and to make the system more robust to noise interference. Many authors suggest a flexible association of neurons and networks by the use of phase codes. This would add a time label to the avalanche, which could serve as an index for associating parallel or consecutive items or actions in time and space. Chapters 4 and 7 showed neuronal responses in spatial tasks that encoded several aspects of the task. When adding the information encoded in these different neurons, most aspects of the task are encoded. How are these cells linked to form an information-encoding network that would be capable of controlling behavior in this task? By using cross-correlation and autocorrelation analysis it was found that the neurons are synchronized in the gamma frequency range. Also, data are shown to demonstrate that neurons are grouped in time sequences that can be stored and repeated. These sequences have the ability to store behavioral sequences, which could be utilized to encode goal-directed behavioral programs. By adding the phase code time label, cell assemblies can be associated, dissociated, and ordered in a temporal sequence. Chapter 15 described several experiments that suggest that the synchronization of neurons in different cortical areas can bring together networks in a task-specific and reversible manner. Chapters 16 and 17 described observations that show that hypersynchronization of neuronal activity can interfere with such spatial and temporal ordering of neuronal activity, which affects cognitive processes and might be responsible for hyperassociations seen in schizophrenic patients.

Can we start to figure out how the brain works as a whole?

If we want to synthesize these findings in a more global model, we will have to accept first that there is no single mechanism or principle that can account for all of these observations. The brain is the result of millions of years of evolution, and it is very likely that all available and useful principles will be put to use wherever they offer more robustness, higher storage capacity, lower energy costs, or lower numbers of required neurons. Many of the heated discussions in the past were based on the assumption that there is a single "correct" principle that governs all brain activities. The wealth of information presented in this book clearly demonstrates that this is not the case.

Rate-coding is found throughout the brain, as mentioned earlier. However, when associating many individual aspects to concepts and schemes, rate-coding would become too time-consuming (having to wait for the trains of firing activity to finish), too noise sensitive, and too inflexible. Cell assemblies such as avalanches will offer several advantages, as described above. Moreover, rate-coding that serves the purpose for peripheral sensory input (e.g. the touch receptors) is translated into phase code. This code offers very fast processing speed, makes the representation more robust, and adds a time label to neuronal activity. This can be used to associate neuronal populations that need to cooperate when building up a more complex representation, e.g. associating the color yellow to a car. It is feasible (as suggested by some authors) that for simple representations, simple cell assemblies are completely sufficient for doing the job. Chapter 3 emphasized that oscillations may not really be required for simple representations, and Chapter 12 showed that the amplitudes of oscillations and the level of phase synchronization is directly correlated with the complexity of the task. This suggests that the more complex the task is, the more the networks need to be synchronized by oscillation. Therefore, cell assemblies such as avalanches might well be utilized to convey simple representations if that is all that is required, and additional mechanism come into play when more computation power is required. The advantage of avalanches is that they are fast as they have a strong feedforward component and constitute and decompose very quickly.

As described above, the time constraints superimposed on neuronal firing probabilities in a phase-coding design could add several new properties to the system. A time label for example offers the ability to associate different networks when required, even across larger areas in the brain. Also, as Hebb had already proposed, the synchronous firing of neurons can easily induce synaptic changes, thereby fixing the spatial pattern of activity in memory. Synaptic weights of asynchronous input are lowered, separating networks that are not related to each other even further, and increasing the signal-to-noise ratio.

This phase-coding system offers a number of important properties that we find in the brain. It speeds up information processing, offers neuronal network type of information distribution which can perform error correction and pattern completion in a fast and unsupervised way, is robust to physical damage and noise of individual neurons, and can create fast and novel (creative) associations that are difficult to obtain in other designs.

The phase code finally will have to be translated back into a rate code, as required for the activation of muscles in time and space. This rate code also has phase code properties overlaid, which is used for arranging and storing muscle activation patterns of motor activity programs (not discussed in this book).

Other aspects of the brain mechanisms that we need to incorporate into our universal model involve the top–down and bottom–up principle of information processing. This can also be described as part of neuronal network properties, where several seemingly anatomically distinct brain areas function as a network, communicating back and forth, and may even be unified for some of the time into one network. Examples for this were given in Chapter 12, where structures as diverse as basal nuclei, thalamic nuclei, and cortical structures form a coherent functional network that has no "instructor" and no final stage of representation. Instead, all levels play a role in balancing the overall activity and controlling the end result. The old model of a hierarchical processing and representation of information has no place here any more. Another example is the well-developed anatomical and functional thalamus–cortex association, with the most prominent example being the lateral geniculate–area V1 connection. Here, the feedback projections are as powerful as the feedforward connections, a finding that has puzzled neuroscientists and anatomists for decades. Other examples are direct connections from V1 to V5, or from V1 to premotor regions in the dorsal stream, which have also caused much confusion to some researchers. The direct associations between primary sensory information and motor areas seemed counter-intuitive but make complete sense in the light of network dynamics and the need to quickly initiate behavioral output with very tight feedback in both directions, as described previously.

Another important aspect of the way the brain operates is that it needs to be equipped with endogenous "drive" that can activate and process information independently of sensory input. Several authors emphasized the point that the brain is not a simple stimulus–response unit. Brain areas must be able to retrieve stored information independently of current input, and process information according to need and requirements. The brain is an active machine that independently plans ahead and anticipates events before they happen, and while they still can be influenced.

Furthermore, an important aspect that modelers sometimes overlook is the need for a set of instructions, rules, "needs," and "values." Evolution has endowed brains with a robust set of "dos" and "don'ts," of "urges" and "fears" that make us do what we need (want) to do. No model of the brain would be complete without a clear set of principles and a mechanism of how sensory input and motor output is judged and guided. Several brain structures have been identified that are involved in controlling brain activity, and focusing cortical activity (attention) on relevant aspects of the overall picture. Similar structures will have to be implemented that emphasize the survival relevant aspects of information processing and goal directed behavioral planning.

The model sketched out here is clearly incomplete, but is not meant to be a final textbook-type model, but more a guide for future research. The multileveled approach reflects the multileveled findings from research reported in this book, and indicates the radical changes that have been made from concepts of the 1950s to today. What it also demonstrates is that the seemingly contradictory findings and ideas presented in research do not have to be contradictory at all, but might be only the result of limited data availability, limited analysis of the information content of these data, or interpretations of the available data that are too restrictive, or the result of analyzing the neuronal activity of only one brain region, where only some information processing principles are in use.

What questions remain?

The findings presented by the various authors show the diversity and dynamics of information processing mechanisms in the brain, but also underline the need for further data to identify how exactly these mechanisms operate. The question of how sparse information is encoded (e.g. in the hippocampus) still needs further clarification. As Chapters 4 and 13 demonstrate, there is most likely a mix of encoding principles, with sparse coding dominating overall. Does this apply to other brain areas? Another important question raised in Chapters 4, 7, and 13 is how dedicated neuronal networks or units are in processing information (as described in the "labeled line code" section in Chapter 1). Previous research results interpreted the correlation between neuronal activity and stimulus quality as being very rigid. Chapters 4, 7, and 11 provide evidence for the possibility that "labeled" networks are much more flexible in their information processing ability than previously thought, and that a limited range of stimuli biased the results towards rigid responses to stimuli (as in the face recognition module concept, or in the "place cell" concept). The use of more varied sets of stimuli or more complex tasks then showed the true dynamic nature of these neurons that then suddenly displayed a much wider range of responses to these new stimuli sets (as demonstrated convincingly in Chapters 4 and 11). New experimental designs will answer the question of how dedicated such cortical modules really are in processing specific information.

The prize question clearly is the all-governing mechanism of how information is associated to form coherent representations and new goal-directed behavioral programs. This is the most complex of all performances achieved by the brain. Few people argue over how individual sensory qualities are encoded in the input to the CNS, but how the information is finally associated in the CNS still evokes heated and controversial discussions.

Future experiments will require the recording of large numbers of neurons in several brain areas to investigate how networks are functionally joined up during performance of complex tasks. The data presented in this book already describe several aspects and principles of how this can be achieved, but at present there are not enough data available to make clear-cut statements. Other properties of the brain that have been mentioned but poorly researched will then be more open to detailed analysis. How is information distributed in networks, is it a population code where neurons with similar (but not identical) information content are bundled, or is it a more holistic distribution as described as neuronal network encoding in Chapter 1? Perhaps both principles are in use in different brain areas, maybe even at the same time. Do we see avalanche-type of stimulus-induced population activity in the brain? At what time do neurons become synchronized, and populations of neurons in different brain areas phase-locked? Do we find disturbances of such synchronization in people that are not able to perform in specific tasks? When networks are synchronized that should operate in an uncorrelated way, can we see cognitive impairments, false associations, even hallucinations? How can we show that one is related to the other?

The multitude of novel and exciting findings obtained from recording larger populations of neurons show how fertile this approach is, how much we have learned from it in the past, and how much we will learn by using these techniques. Clearly, we are still only at the beginning of finding out how the brain encodes information. However, the speed with which new concepts and findings have revolutionized the thinking in this research field makes it hopeful that the next decade will provide us with a wealth of information that will shed more light on information encoding systems and mechanisms. Further large-scale recordings will be required, and the different chapters in this book show the advantages of single-cell (array) recording, field potential recording, as well as fMRI and MEG recording. More complex goal-directed tasks will be used to analyze what information is processed in real time, and how dynamic and flexible the system really is. The properties of "modules" or sparse networks will be further illuminated, and hopefully the connection mechanisms of active networks in the brain will be described in finer detail, but at the same time with a wider scope with respect to their functional relevance.

Why do this?

Finding out what machine language the brain speaks is clearly more than a philosophical task. With increasing numbers of brain diseases and degenerations that occur impair cognitive functions not only in aging populations (e.g. Alzheimer's diseases), it would be very desirable for the physicians in charge to have an objective basis for their diagnoses and all the knowledge for causal therapeutic regimes. Ultimately, it will be possible to assist impaired functions and maybe even replace lost functions by implanting chips or other hybrid systems even within the CNS. However, it will be essential to understand how information is processed in the brain.

Bionic man

Currently, great advances have been made to replace lost auditory or visual sensory input by engineered electronic systems. Degeneration of the retina can be treated by implanting chips that are light sensitive and transmit signals to the brain. A lost cochlea can be replaced by a microphone and signal transducer/amplifier which can be implanted to relay auditory input to the brain. This has produced spectacular results in the past. However, the success of the current systems is not due to the intelligent engineering that can translate sound into signals that the brain understands, but instead the extremely versatile and dynamic properties of the brain is exploited by simply offering signals to the cortex and leave the hard work of translating the input (making sense of it) to the cortex itself. There are clear limitations to this strategy, and any improvement to it will require detailed knowledge of how information is really encoded in the brain.

Loss of motor output is compensated for by using a similar strategy, e.g. by implanting sensors in the motor cortex and connecting it to muscles of a limb. Again, the brain has to learn how to activate the muscles in a sensible fashion. No engineer of such a system would claim to know how the brain does this amazing feat, yet would give a lot to find out how the brain actually does it. If signals are sent to the cortex that resemble much closer actual information encoding patterns of the brain, it is hoped that this artificial information input will be understood and utilized much better.

Of course, there are and there will be ethical limits to applying neurotechnology to human beings. This does not only apply to therapeutic measures as described above for the first simple sensory prosthesis, but will most likely be very relevant when hybrid systems will allow for interaction of artificial systems with higher brain functions. Then it will not only be a challenge to define rational rules for avoiding brain doping and other forms of fraud, but to prevent misuse. In this respect it is comforting that progress is slow and there is some time left for ethicians and the legal professions to develop useful and practicable rules.

The ultimate goal?

Analyzing the mechanisms and principles of how the brain processes information could also be of great use for building computer systems that have entirely novel properties and capabilities. In theory, we should be able to build computers that are able to process information on a similar level to the average human being. It is unlikely that computers will be built as exact copies of human brains. However, new information-processing principles could be used to construct computer systems that have novel properties that current systems lack. The downfall of current systems is that their architecture is completely deterministic and dependent on the software that runs the computer, and on the data that are fed into it. This architecture does not permit any novel associations and genuine novel and "creative" computation. The average computer that we use is – in principle – no different from a fancy pocket calculator. It obeys the "cardinal cell" logic and architecture of information processing and projection to defined and deterministic and often hard-wired information stores and units of information representation. We have described in this book what the shortfalls of such systems are. There is no freedom of association, the system is too rigid and thereby error intolerant. It is perfect for processing large amounts of data using a predetermined algorithm (a pocket calculator can work out the square root of any number in microseconds). However, novel and non-deterministic associations are not possible; the computational product will never be more than the sum of its parts. This puts severe restrictions on the capabilities of these systems. Teaching a computer how to recognize a face is already an almost insurmountable obstacle for current-generation computers, a job that we can perform effortlessly. Once we understand more about how information is represented and associated in the human brain, we could make use of this knowledge to create new computing systems that allow distributed information representation and associations between similar (yet not identical) parts of this information. Such a system would not be completely deterministic, and could form novel associations, concepts, and "ideas" that are *not* already included in the data that had been fed to the computer. We have heard promises of the creation of such systems before, using neuronal networks, genetic algorithms, expert systems based on statistical analysis, etc. Up to now, the results have been rather sobering. Computing programs based on pure neuronal network algorithms turned out to have very limited practical use. While they show non-deterministic properties, it became clear that the actual developments of information representation in these networks become impossible to control and to guide. A computer that cannot be controlled is of little practical use. Such findings only underline the fact that neither the brain nor any such intelligent system can be simple, nor can it be based on a single information analysis mechanism. The authors in this book went to great lengths to outline the complexity of information analysis systems in the brain, and often emphasized the fact that not just one but several systems are implemented. Any truly creative system will require self-instructing (non-hard-wired) association mechanisms that can draw "conclusions" from the input data that are genuinely novel. Perhaps the information gained from analyzing the brain can give us important pointers of how the linear deterministic input/output computational principle can be married with a non-deterministic (and therefore hard to control) system, to produce a new computing system that has all the properties that we require.

These new computing principles and architectures could be able to provide us with intelligent systems that actually *can* recognize faces, drive trains, analyze complex problems, create new concepts, model future cities, and have genuinely creative ideas that humans could not create (e.g. because such a system would involve keeping amounts of information in memory that no human ever could). The advantages of computers are that they can work day and night, do not have to be motivated (e.g. by high salaries or fancy cars), can be loaded with large amounts of information that can be stored and kept in working memory practically error-free, can be connected to form ever larger units across the globe, and can be tailored according to requirements and financial limits.

Evolution has provided us with an exceptional computation unit within our skulls, but the limits are plain to see to everyone. We take years to learn a language and decades to realize that carbon emission could endanger us by changing global climates. Perhaps the next man-made catastrophe will be forecast by an intelligent system that diagnoses the early signs within days or hours, and not in decades as required by humans.

This is for us to find out . . .

Index

analogue states, 4
auditory system, 13
avalanches, 205

Bayesian framework, 109
binding problem, 16, 338, 354, 363, 410
bottom-up, 224, 242

chemical synaptic coupling, 28
coherency, 100, 339, 373

Descartes 3, 4
decoding, 97, 120, 134
delayed nonmatch to sample task, 77, 82, 87, 153
desynchronizing, 304, 389, 414
digital states, 4

electrical synaptic coupling, 26
encoding of sequences, 13, 167
ensembles *see* neuronal assembly

frequency coding, 7, 236

gamma-frequency oscillations, 31, 36, 155, 166, 199, 207, 269, 337, 339, 363, 409, 410

synchronization, 35, 157, 374, 416

Hebb, 21, 151, 162, 164, 336, 389
hippocampus, 21, 30, 39, 67, 74–75, 82–83, 89, 95, 110, 120, 174, 215, 287, 328, 388, 391
homunculus, 3

labeled line, 8
long-term potentiation (LTP) *see* synaptic plasticity

multineuron recording (multi-unit recording, multi-electrode recording), 6, 74, 183

network
 oscillations, 23, 25, 30, 205
 properties, 5
 synchrony, 21, 155, 337, 376, 389, 417
neural circuits, 49, 198, 401, 413
neural representation, 96, 194
neuronal assembly (cell assembly, ensembles), 6, 35, 49, 50, 56, 63, 65, 76,

77, 82, 87, 110, 120, 139, 197, 235, 337, 389
neuronal populations, 5, 36, 196
neuronal synchrony, 64, 157, 268, 283, 335, 413

optical recording, 13

parallel processing, 6, 192, 196, 199
phase coding, 9
population
 activity, 5, 95, 105, 211, 256, 338, 354
 coding, 11, 74, 77, 120, 192, 210, 247
principal component analysis, 126

rate coding, 4, 152, 197, 247, 248, 258, 334

sequence learning, 22
serial computation, 49
sharp wave–ripples, 32, 35
simulations, 105
slow oscillations, 31, 271, 339, 363, 414
sparse, distributed coding, 74–75, 82–83, 89, 95, 192, 198, 335, 389

470

spike train temporal
    structure, 50, 56,
    268, 334
synaptic plasticity (LTP), 162,
    225, 268, 388
synfire chains, 204, 206, 336

temporal correlations, 35
temporal patterns, 6, 273,
    364
theta frequency, 33, 39,
    162, 164, 174, 209,
    283, 376, 392

top-down, 64, 338
topological, topographical
    coding, 8, 76, 192,
    236, 264, 364
tuning curves, 97, 225,
    256, 336

For EU product safety concerns, contact us at Calle de José Abascal, 56–1°,
28003 Madrid, Spain or eugpsr@cambridge.org.

www.ingramcontent.com/pod-product-compliance
Lightning Source LLC
LaVergne TN
LVHW081523060526
838200LV00044B/1979